Springer Water

The book series Springer Water comprises a broad portfolio of multi- and interdisciplinary scientific books, aiming at researchers, students, and everyone interested in water-related science. The series includes peer-reviewed monographs, edited volumes, textbooks, and conference proceedings. Its volumes combine all kinds of water-related research areas, such as: the movement, distribution and quality of freshwater; water resources; the quality and pollution of water and its influence on health; the water industry including drinking water, wastewater, and desalination services and technologies; water history; as well as water management and the governmental, political, developmental, and ethical aspects of water.

Gao Dawen · Mohammad Nabi

Novel Approaches Towards Wastewater Treatment

Effective Strategies and Techniques

Springer

Gao Dawen
School of Environment and Energy Engineering
Beijing University of Civil Engineering and Architecture
Beijing, China

Mohammad Nabi
School of Environment and Energy Engineering
Beijing University of Civil Engineering and Architecture
Beijing, China

ISSN 2364-6934 ISSN 2364-8198 (electronic)
Springer Water
ISBN 978-3-031-55188-8 ISBN 978-3-031-55189-5 (eBook)
https://doi.org/10.1007/978-3-031-55189-5

This Springer imprint is published by the registered company Springer Nature Switzerland AG
The registered company address is: Gewerbestrasse 11, 6330 Cham, Switzerland

Paper in this product is recyclable.

Contents

Chapter 1
Anaerobic Ammonium Oxidation Process

1.1 Introduction

The eutrophication of waterbody caused by the excess of nitrogen and phosphorus has attracted worldwide attention. The traditional biological nitrogen removal technology, known as nitrification and denitrification, have found extensive application in the denitrification of various water bodies (such as domestic sewage, and industrial wastewater). With increasingly strict nitrogen discharge standards, biological nitrogen removal technology has been promoted significantly. Anaerobic ammonia oxidation (anammox), a pioneering autotrophic biological process in nitrogen removal, has garnered significant interest among scholars since its discovery in 1995 [216]. The typical reaction is shown below (Eq. 1.1). In this process, anaerobic ammonium-oxidizing bacteria (AnAOB) have the ability to efficiently convert ammonium (NH_4^+–N) into nitrogen gas (N_2) by utilizing nitrite (NO_2^-–N) under anaerobic conditions (Fig. 1.1) [68, 150, 151, 155, 157]. Until now, three main pathways have been identified for converting NH_4^+–N and NO_2^-–N to N_2 in the process (Fig. 1.1) [5, 6]. The first pathway involves the reduction of NO_2^-–N into nitric oxide (NO) through nitrite reductase *(Nir)*, followed by the conversion of NO into hydroxylamine (NH_2OH). *Nir* and the ammonium enzyme (*Nrf*) can also reduce NO_2^-–N to NH_4^+–N, while NH_2OH is formed at the same enzyme complex's cytoplasmic site [116]. Hydroxylamine oxidoreductase (HAO) then facilitates the synthesis of N_2H_4 by combining NH_2OH and NH_4^+–N. This N_2H_4 compound is subsequently oxidized to N_2 with the assistance of a hydrazine oxidizing enzyme (HZO) [6]. In the second pathway, NO_2^-–N is reduced to NH_2OH through the involvement of the hydroxylamine ubiquinone redox module (HURM) [81]. Subsequently, a membrane-bound enzyme complex facilitates the conversion of NH_4^+–N and NH_2OH into N_2H_4. The oxidation of N_2H_4 to N_2 takes place in the periplasmic region. Supporting evidence for this anammox metabolism pathway suggests that electrons generated during the oxidation of N_2H_4 are transferred via an electron transport chain by a Nir in the cytoplasm. This electron transport chain enables the transfer of electrons generated during

G. Dawen and M. Nabi, *Novel Approaches Towards Wastewater Treatment*,
Springer Water, https://doi.org/10.1007/978-3-031-55189-5_1

the oxidation of N_2H_4. Within the cytoplasm, NO_2^-–N is reduced to NH_2OH. Additionally, the membrane-bound enzyme complex enables the conversion of NH_4^+–N and NH_2OH into N_2H_4 [186]. In the third pathway, NO and NH_4^+–N are combined through the activity of a HZO, resulting in the formation of N_2H_4. Subsequently, the same enzyme oxidizes N_2H_4 to N_2 [116]. Hydrazine, a vital product in the anammox process, is an energetically potent compound that holds the potential to serve as a valuable fuel source [127]. In fact, hydrazine has been widely used to enhance the Anammox-mediated process. However, its considerable potency is counteracted by its potential toxicity, presenting inherent difficulties in harnessing it as a viable energy source [272]. Due to its distinctive characteristics compared to conventional bacteria, the specialized branch of bacteria known as AnAOB has garnered growing attention for the past few years. In comparison to conventional nitrification and denitrification processes, the anammox process offers notable advantages. It can reduce the demands for dissolved oxygen (DO) by 60%, without the need for additional organic matter. Additionally, it significantly diminishes sludge production by 70% and reduces alkalinity by 45% [6]. Moreover, by inhibiting nitrous oxide (N_2O) generation and achieving a 90% decrease in carbon dioxide (CO_2) emissions during AnAOB metabolism, the anammox process also makes a substantial contribution to the reduction of greenhouse gas (GHG) production [6, 87, 208]. Meanwhile, with their high nitrogen removal efficiency and the utilization of energy-conserving, anammox-based systems have received enough attention for achieving sustainable development goals [6]. In total, anammox exhibits several advantages, including the absence of the need for additional organic carbon sources, high nitrogen removal capacity, low operational costs, and a small physical footprint [243]. In a word, anammox has emerged as a globally significant topic of interest.

$$\begin{aligned} 1NH_4^+ + 1.32NO_2^- + 0.066HCO_3^- + 0.13H^+ &\rightarrow 1.02N_2 \uparrow \\ + 0.26NO_3^- + 0.066CH_2O_{0.5}N_{0.15} + 2.03H_2O & \end{aligned} \quad (1.1)$$

Although the anammox process boasts numerous advantages, it also presents significant disadvantages and challenges. On one hand, the growth rate of AnAOB is slow, with a doubling time of 10–12 days. Additionally, the process exhibits a low cell yield and requires a long start-up cycle [270]. Du et al. [46] has found that the anammox sludge granulation is feasible for facilitating the implementation of the anammox process. This granulation method proves effective in achieving biomass retention, shortening start-up time, and enhancing resilience to unfavorable conditions. So, promoting anammox granulation may be an effective means to increase the growth of AnAOB. On the other hand, the electron acceptors (usually NO_2^-–N) of the anammox process are often insufficient with the composition of actual wastewater. Ammonia-oxidizing bacteria/archaea (AOB/AOA) conduct the process of partial nitritation (PN), while denitrifying and dissimilatory nitrate (NO_3^-–N) reduction to ammonium (DNRA) occurs as part of the partial denitrification (PD). In addition, archaea/bacteria involved in denitrifying anaerobic methane

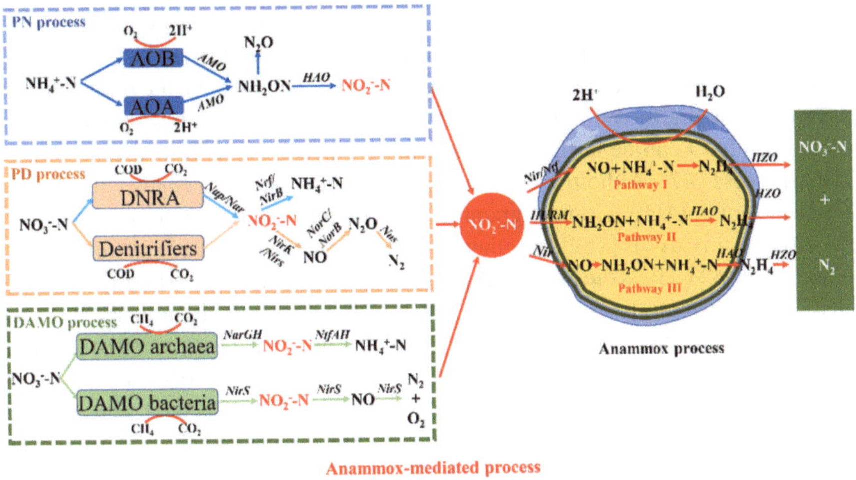

Fig. 1.1 Comprehensive schematic of the anammox process including various microbial functions (Partial nitration process by AOB and AOA, PN: NH_4^+–N → NO_2^-–N; Partial denitrification process by DNRA and denitrifiers, PD: NO_3^-–N → NO_2^-–N; Denitrifying anaerobic methane oxidation process by DAMO archaea and DAMO bacteria, DAMO: NO_3^-–N → NO_2^-–N)

oxidation (DAMO) contribute to the production of NO_2^-–N for AnAOB [6, 67, 274]. These microbial communities play a significant role in various systems, such as one-stage or two-stage PN/Anammox (PN/A), PD/Anammox (PD/A), and DAMO/anammox (DAMO/A), where they can successfully integrate with AnAOB (Fig. 1.1). The PN/A process has been successfully implemented in sidestream systems of wastewater treatment plants (WWTPs) [129]. However, the PN-A process may not be practical in mainstream systems with low concentrations of free ammonia (FA), low temperatures, and high concentrations of chemical oxygen demand (COD). Nevertheless, recent research has shown that the PN/A process can be effectively applied for treating mainstream wastewater, employing a complex control strategy that includes optimizing aeration modes, setting appropriate dissolved oxygen (DO) levels, and maintaining optimal temperature conditions. This strategy has proven successful in achieving high removal efficiencies [247, 249, 260]. Despite this success, the PN-A process may not be a profitable option for many WWTPs. An alternative to the PN-A process is the simultaneous partial nitrification, Anammox, and denitrification (SNAD) process, which has been shown to be effective in treating high concentrations of NH_4^+–N and COD wastewater under low DO conditions, achieving high removal efficiencies of total nitrogen (> 80%) and COD (about 90%) by the combined activity of AOB, AnAOB, and denitrifying bacteria [218, 219]. In recent times, the PD/A system has garnered attention as a promising alternative for achieving highly efficient nitrogen removal in wastewater treatment. According to reports, the PD process can realize optimal sustainability in mainstream wastewater treatment, providing NO_2^-–N for the anammox process efficiently, with friendly control and without the need for aeration and DO [171]. Additionally, the PD/A process demonstrates efficient reduction in organic matter demand and sludge production [67, 274].

Moreover, PD-A process has the advantage of reducing GHG production in contrast to the traditional nitrification–denitrification process. It has shown that the production N_2O can be reduced though utilizing heterotrophs to consume N_2O in PD/A process [47]. Although the PD-A process may face challenges with the low abundance of AnAOB, biomass granulation techniques have been found to enhance the activity of AnAOB [169]. Another potential implementation of anammox-based processes in WWTPs is the DAMO-A process. Even though CH_4 is a highly impactful GHG (about twenty-five times more potent than CO_2), DAMO-archaea have the ability to reduce NO_3^-–N to NO_2^-–N. This process provides a means to significantly mitigate greenhouse gas emissions [60]. The DAMO-A process encompasses three primary categories of microorganisms: DAMO-archaea/bacteria, and AnAOB. While DAMO-archaea can specifically convert NO_3^-–N to NO_2^-–N, AnAOB play a crucial role in reducing NH_4^+–N and NO_2^-–N to N_2 [82]. However, DAMO-bacteria also contribute to the removal of nitrogen, although they are not involved in the anammox metabolism pathways [82]. Moreover, AnAOB are known to be sensitive to various environmental conditions, including temperature, dissolved oxygen (DO), pH, light, antibiotics [200, 201], salinity [280, 281, 284], and heavy metals [170]. These sensitivities have posed challenges to the widespread adoption of the anammox process in the applications of practical engineering. As a result, researchers are actively working to overcome these obstacles and make the anammox process more applicable in wastewater treatment. In recent years, there has been extensive laboratory and practical research conducted on the anammox process, leading to significant advancements in our understanding. However, there is still a lack of comprehensive monographs or articles that compile and systematically elucidate the characteristics of anammox and its application. In conclusion, there are numerous aspects of anammox that require further investigation and clarification regarding its comprehensive characteristics and its suitability for various types of wastewaters. Therefore, it is essential to perform a thorough examination of the traits of anammox to enable its successful application in mainstream wastewater treatment.

Previously, documented literature has predominantly concentrated on studying the application of anammox and its individual coupling processes for treating specific types of wastewaters. A comprehensive chapter of anammox (from characteristics to its the practical application of wastewater treatment) is lacking. Firstly, the chapter offers an overview of pure anammox processes and highlights the diversity of AnAOB. It explores how different types of influencing factors, such as organic or inorganic pollutants, can affect the choice between integrated and separated process types. The discussion also encompasses the crucial theoretical factors that are indispensable for ensuring the stable operation of the anammox process. Furthermore, a comprehensive review is conducted on the implementation of anammox technology in treating different wastewater compositions. The advantages and limitations of anammox technology are thoroughly evaluated, drawing upon both laboratory experiments and real-world applications. This evaluation offers a comprehensive comprehend of the strengths and weaknesses of anammox technology in the context of denitrification processes. In conclusion, the acquired knowledge offers a promising prospect for the rapid and widespread adoption of anammox technology. This advancement presents an opportunity to enhance the energy efficiency and sustainability of current WWTPs.

1.2 Microbiology of Anammox Bacteria

1.2.1 The Diversity of Anammox Bacteria

The exploration of anammox process is a captivating journey. Early indications of the anaerobic oxidation of NH_4^+–N were observed in oxygen-depleted aquatic environments many years ago. However, during that period, the exact biological pathway responsible for these observations remained elusive. The initial demonstration of the anammox process took place in Delft, the Netherlands. During the operation of a denitrifying fluidized bed reactor (FBR), it was observed that the production of N_2 increased, coinciding with a reduction in ammonia levels in the reactor effluent [96]. This process was executed by AnAOB, which belong to the Planctomycetes phylum. Subsequent to the initial discovery, additional instances of AnAOB existence were reported worldwide. The majority of these anammox bacteria were detected in WWTPs that handle wastewater containing high levels of nitrogen compounds [205]. Remarkably, Broda made predictions regarding two plausible chemo lithotrophic nitrogen pathways by analyzing the Gibbs free energy changes of reactions. Among these pathways, one involved the intriguing process of anaerobic ammonia oxidation. The existence of this particular bacterium remained uncertain until fortuitous observations were made by Arnold Mulder in a pilot plant dedicated to denitrification. The occurrence of anammox was methodically replicated in a denitrifying fluidized bed reactor, establishing its biological mediation. Consequently, this process was officially patented as anoxic ammonium oxidation [68].

Water is an important factor in the enrichment of AnAOB because the existence of water could create an anaerobic environment or carry substrates of the AnAOB meeting the basic needs of AnAOB and improving the anammox activity [255, 262]. AnAOB were first found in wastewater, with the development of research and sequencing technology, the AnAOB are widely distributed in diverse natural ecosystems, including marine sediments [68, 200, 201], wetland [44, 68], oxygen minimum zones [255, 262], etc., AnAOB play a significant role in the nitrogen cycle within the global ecosystem, serving as a primary participant in nitrogen transformation processes [3]. AnAOB are classified within the phylum Planctomycetales and are positioned in a distinct branch. Within this phylum, AnAOB form a unique group distinguished by the presence of anammoxosomes, specialized structures associated with the anammox process. This structural characteristic sets AnAOB apart from other bacteria within the Planctomycetales phylum [200, 201]. Up to now, scientists have identified six primary genera and over 20 distinct species of anammox bacteria (Table 1.1). The six main genera includes *Candidatus Kuenenia, Candidatus Brocadia Candidatus Jettenia, Candidatus Anammoximicrobium, Candidatus Anammoxoglobus* and *Candidatus Scalindua* [160]. Among them, *Candidatus Brocadia, Candidatus Jettenia,* and *Candidatus Kuenenia* are tare regarded as the prominent genera of anammox bacteria [43].

AnAOB belonging to different genera exhibit physiological and ecological characteristics. It comprehensively shows the optimal conditions for different anammox genera (Table 1.1). *Candidatus Brocadia* is the most common anammox genus. It has been reported that *Candidatus Brocadia* demonstrates suitability for environments characterized by high concentrations of NH_4^+–N and NO_2^-–N substrates [43]. Oppositely, *Candidatus Jettenia* shows resistance to high substrate concentration environment, especially high NO_2^-–N [92, 93]. Moreover, *Candidatus Kuenenia* and *Candidatus Anammoxoglobus* plays an important role in degrading organic matter (formic acid and propionic acid), which is particularly well-adapted to environments that harbor organic-rich substances like landfill leachates, as well as effluents from livestock and poultry breeding [92, 93, 200, 201]. Compared to other genera, the enrichment of *Candidatus Scalindua* usually appears in environments with elevated salinity levels and colder temperatures [35, 124]. The optimal temperature range for *Candidatus Scalindua* is considerably lower compared to other types of AnAOB (Table 1.1). Interestingly, the existence of *Candidatus Anammoximicrobium* is often found in anammox processes, but their typical biological characteristics need further investigation. Even so, the application of nitrogen removal with anammox in different type wastewater has certain advantages. Remarkably, in a study, a novel genus of AnAOB was discovered in a full-scale biofilm system employed for treating landfill leachate in wastewater treatment. Through the utilization of metagenome-assembled genome (MAG) and metatranscriptome analyses, the researchers successfully developed a metabolic framework for this novel anammox bacterium, the provisional name assigned to this bacterium is "*Ca. Loosdrechtia aerotolerans*" [291]. It notably covered a novel ferredoxin-dependent nitrate reductase, which is absent in other anammox genome. This characteristic may bestow a selective advantage in terms of nitrate assimilation. It is a milestone in the development of anammox. In conclusion, due to the metabolic diversity of AnAOB influences their microbial composition, anammox is of vital importance for the global nitrogen cycle.

1.2.2 Cell Structure and Physiological Characteristics of Anammox Bacteria

Prokaryotic cells, in general, exhibit a comparable cellular organization, a cell envelope, cytoplasm, and plasma membrane. Bacterial cells are commonly classified into two distinct groups, Gram-negative and Gram-positive group, based on variations in their cell envelope structure [96, 102]. Gram-positive bacteria possess a cell envelope consisting of a thick layer of peptidoglycan intertwined with the cytoplasmic

Table 1.1 Six anammox bacterial genera and their favorable growth condition

Genus	Species	Temperature (°C)	pH	Sewage type	References
Candidatus Brocadia	*Candidatus Brocadia anammoxidans* *Candidatus Brocadia fulgida* *Candidatus Brocadia sinica* *Candidatus Brocadia brasiliensis* *Candidatus Brocadia caroliniensis*	34–35	7.5	Synthetic wastewater	[73]
Candidatus Kuenenia	*Candidatus Kuenenia stuttgartiensis*	25–37	6.5–9.0	Wastewater	[212]
Candidatus Jettenia	*Candidatus Jettenia asiatica* *Candidatus Jettenia moscovienalis* *Candidatus Jettenia caeni*	30–35	8.0–8.5	Wastewater	[92, 93]
Candidatus Anammoximicrobium	*Candidatus Anammoximicrobium moscowii*	19–22	1.8–8.3	Wastewater	[245, 250, 252, 261]
Candidatus Anammoxoglobus	*Candidatus Anammoxoglobus propionicus* *Candidatus Anammoxoglobus sulfate*	35	8.0–8.2	Synthetic wastewater	[68]
Candidatus Scalindua	*Candidatus Scalindua brodae* *Candidatus Scalindua wagneri* *Candidatus Scalindua rubra* *Candidatus Scalindua marina* *Candidatus Scalindua sorokinii* *Candidatus Scalindua arabica* *Candidatus Scalindua sinooifield* *Candidatus Scalindua zhenghei* *Candidatus Scalindua Profunda* *Candidatus Scalindua richardsii*	15–45	2.0–8.5	Marine water	[206]

membrane. In contrast, Gram-negative bacteria have a more complex cell envelope, which includes the cytoplasmic membrane, a periplasmic space housing a loosely cross-linked peptidoglycan layer, and an outer membrane. The outer membrane is a bilayered structure consisting of phospholipids facing the periplasmic side and lipopolysaccharides on the outer surface. While the presence of intracellular membrane-bound organelles is a characteristic feature of eukaryotic cells, it is worth noting that certain prokaryotes also possess intracellular membrane systems [102]. Regarding AnAOB, it has been demonstrated that their cell wall is primarily proteinaceous, distinguishing them from Gram-negative bacteria by the absence of peptidoglycan and the characteristic outer membrane [188]. Consequently, Anammox bacteria are generally classified as Gram-negative bacteria.

AnAOB exhibits a wide ecological distribution, resulting in the occurrence of irregular cellular morphologies, such as spherical or ovoid shapes [6]. One distinctive feature of AnAOB is their carmine color, which indicates their high activity. While Anammox granules typically exhibit a carmine coloration, their hues can range from carmine to shades of brown or black, primarily due to differences in heme c concentration within the bacteria. AnAOB with elevated levels of heme c exhibit a gray coloration, whereas AnAOB lacking heme c appear black in appearance [2]. AnAOB exhibit distinctive cellular structures, and the cell can be divided into three primary components: the anammoxosome, ribosomal cytoplasm, and outer chamber cytoplasm (Fig. 1.2) [236, 237]. The anammoxosome, which occupies a significant portion of the cell volume, is often referred to as a "prokaryotic organelle" that bears resemblance to the eukaryotic mitochondrion. The anammox metabolism takes place specifically within the anammoxosome, facilitated by the curved membrane structure [6]. Recent observations using scanning electron microscopy have revealed that Anammox cells have a highly compact spherical shape (Fig. 1.2). When subjected to examination using a transmission electron microscope (TEM), these spheres typically have a diameter ranging from 0.6 to 1.0 μm. Furthermore, the compartmentalization of AnAOB has been extensively studied, with the enclosed structure delimited by a bilayer membrane being referred to as the "anammoxosome."

In AnAOB, the cytoplasm primarily consists of three distinct compartments, each separated by single bilayer membranes [179, 236, 237]. These compartments are as follows: (1) The peripheral region, known as the paryphoplasm, comprises the outer periphery and is characterized by the presence of the cytoplasmic membrane and cell wall on its external surface, while the intracytoplasmic membrane defines its internal boundary; (2) The riboplasm cytoplasm, or riboplasm, harbors DNA, ribosomes, and reserve substances such as glycogen granules; and (3) The central ribosome-excluded compartment, called the anammoxosome, is the most significant portion, accounting for approximately 50–70% of the overall cell volume. It is bounded by the anammoxosome membrane and plays a crucial role in anammox metabolism. The membrane enclosing the anammoxosome is frequently characterized by pronounced curvature, potentially serving to enhance the ratio of surface area to volume [179]. Unlike typical Gram-negative bacteria, anammox bacteria do not possess an outer membrane surrounding the cell wall along with an inner membrane. Instead, their cell wall is characterized by the presence of two inner membranes, with no outer

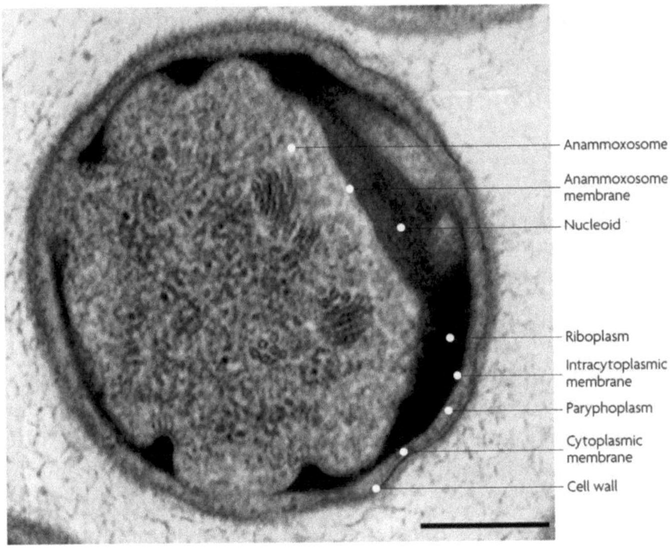

Fig. 1.2 Typical cellular structure of anammox bacteria (*Candidatus Kuenenia*) [127]

membrane [102]. The composition of the anammox cell wall primarily comprises proteins, with the outermost of the two inner membranes closely associated with the cell wall structure. This unique cell wall structure distinguishes anammox bacteria from other Gram-negative bacterial groups and underscores their specialized adaptation to their ecological niche. It is worth noting that electron microscopy studies have revealed a distinct feature of AnAOB, the absence of peptidoglycan in their cell wall. This distinctive arrangement sets AnAOB apart from other Gram-negative bacterial counterparts and emphasizes their specialized adaptation to their specific ecological niche. Through the use of immunogold labeling, the outermost membrane closely positioned to the cell wall in AnAOB has been identified as the cytoplasmic membrane. This determination was made based on the direct detection of RNA on the inner side of this membrane [179]. The innermost membrane, located inside the cytoplasmic membrane, is commonly known as the intracytoplasmic membrane [179]. The region between these two membranes is known as the "paryphoplasm", which corresponds to the Gram-negative periplasm but is enclosed within the essential cell boundary [102]. These structural characteristics, including the presence of both cytoplasmic and intracytoplasmic membranes, are shared among Planctomycetes, such as Pirellula and Isosphaera, distinguishing them apart from other types of archaeal cell walls [96]. Therefore, concerning the structural constitution of their cell walls, AnAOB cannot be classified as either Gram-positive or Gram-negative bacteria.

1.3 Enrichment Strategies of Anammox Bacteria

1.3.1 Enrichment of Anammox Bacteria in Bioreactor Systems

The successfully establishment of the anammox process can be achieved through the enrichment technique employing a bioreactor [96]. It is feasible to cultivate a predominant anammox consortium in which AnAOB dominate 95% of the total microbial population in the biomass [49, 233]. Nonetheless, the sluggish proliferation and metabolism of AnAOB present challenges in their enrichment process. To achieve a dominant AnAOB culture, bioreactors with effective biomass retaining ability at low substrate concentrations are required for long-term operations [119]. The reactor configuration is a crucial factor in the enrichment of AnAOB and is considered a significant factor influencing the initiation of the anammox process. An efficient anammox reactor will offer a substantial specific surface area, which is critical for facilitating mass transfer of NO_2^-–N in the biofilm, thus increasing the nitrogen removal capacity in the anammox system [108, 254, 259]. Previous research has demonstrated successful cultivation of AnAOB using both suspended and attached growth-type bioreactors, which are specifically designed to accommodate the growth characteristics of these slow-growing bacteria. Suspended-type bioreactors, for instance, the sequencing batch reactor (SBR) [30, 130], UASB [109–111, 242], membrane bioreactor (MBR) [109–111, 242], and the gas-lift reactor [37, 99]. The attached growth-type bioreactors including the fixed-bed reactor (FBR) [258], rotating biological contactor (RBC) [261] and up-flow biofilter (UBF) [195, 196], employ biofilm carriers to maintain the anammox microbial community. Following are the typical examples of AnAOB enrichment processes using various bioreactor configurations:

1.3.1.1 Sequencing Batch Reactor (SBR)

Successful enrichment of AnAOB can be successfully achieved through the implementation of a SBR, which is a modified variant of the activated sludge method. The enrichment of AnAOB can be successfully achieved through the implementation of a SBR, which is a modified version of the activated sludge method (Fig. 1.3). Operating in batch mode with mixing capabilities, the SBR conducts all operational steps within a single tank. SBR has demonstrated its effectiveness in promoting a uniform dispersion of biomass throughout the system, thereby preventing the accumulation of NO_2^-–N [130, 215]. It is crucial for AnAOB enrichment to the uniform distribution of substrate and biomass in the reactor, preventing substrate inhibition of AnAOB due to the bacteria being affected by high NO_2^-–N concentrations [96]. Using the SBR system, the cultivation of AnAOB has been effectively achieved, demonstrating a rapid initiation of anammox activity within 60 days and an impressive NRE of 82% [38]. In an experimental setup of the SBR system, incorporating

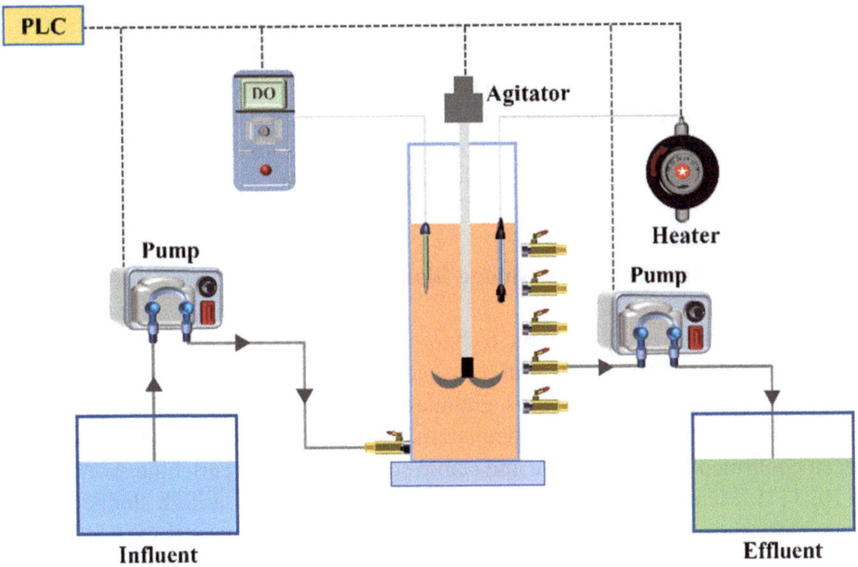

Fig. 1.3 Schematic diagram of sequential batch reactor (SBR)

biomass immobilization and a biofilm-based approach, a consistent substrate-limited enrichment condition was established, allowing the sustained cultivation exceeding one year [215]. The utilization of zeolite as carrier offers significant improvements to the SBR system, resulting in a decrease in biomass washout in the effluent to 3 mg/L, while maintaining a specific anammox activity (SAA) of 0.5 g/g/d [58]. Additionally, the simultaneous implementation of denitrification and anammox processes in an SBR system has demonstrated remarkable efficacy in achieving an impressive NRE of 97.47% [48]. Maintaining optimal operational conditions is crucial, with the recommended pH range typically set between 7.0 and 8.0, and the temperature maintained within the range of 30–35 °C. When operating an SBR system with anammox, applying a nitrogen loading rate (NLR) between 0.1 and 0.7 kg/m^3/d ensures complete removal of NO_2^-–N. However, caution should be exercised as an NLR of 0.75 kg/m^3/d may lead to the onset of biomass flotation [37]. However, it is crucial to note that an examination of shear effects in anammox processes has shown that excessive agitation in SBR (about $\geq$ 250 rpm) can adversely affect AnAOB activity by reducing the size of AnAOB granules and compromising system efficiency [9]. Hence, to achieve optimal conditions during the enrichment process and minimize the detrimental effects of shear stress, it is advisable to keep the stirring speed in the SBR below the maximum mechanical stress threshold tolerated by AnAOB. This approach ensures that the system operates within a range that promotes favorable conditions for the enrichment process while minimizing the potential negative impacts caused by excessive shear stress. Additionally, the use of biofilm systems and carrier materials in SBR system can also improve the AnAOB enrichment process, creating a stable substrate-limiting condition that is conducive to long-term cultivation. Overall, SBR is an effective system for AnAOB enrichment due to capability to ensure a uniform dispersion of biomass within the reactor preventing NO_2^-–N accumulation. The use of carrier materials, biofilm systems, and lower stirring speeds can also improve the AnAOB enrichment process, creating a stable and substrate-limiting environment, ideal for long-term cultivation of AnAOB.

1.3.1.2 Up-Flow Anaerobic Sludge Blanket (UASB)

UASB is a cutting-edge anaerobic technology that effectively processes wastewater using anaerobic microorganism in a suspended growth configuration (Fig. 1.4). To ensure the retention of anammox granular, the reactor is engineered with proper separation of gas, liquid and particulate matter. Importantly, the UASB reactor has been confirmed a most effective and stable configuration for enriching AnAOB [96, 110, 111]. Meanwhile, the UASB is a reliable system with a high sludge settling capability and a high NRE achieved through TN removal of up to 91.82% [198]. The anammox process implemented in a UASB reactor has demonstrated a remarkable performance, achieving a NLR of 74.3–76.7 kg/m^3/d acquired through high-loaded anammox treatment [223]. A study conducted on the anammox process in granular UASB systems revealed the optimal conditions for achieving high efficiency. These conditions include the influent concentration of substrate ranging from 644 to 728 mg/L, an optimum HRT ranging from 0.90 to 1.25 h, and an up-flow velocity of 0.60 to 1.79 m/h [277]. A study conducted on the anammox process in granular UASB systems revealed the optimal conditions for achieving high efficiency. Under these optimal conditions, a remarkable NRE of 91.82% was achieved within two months, facilitated by maintaining a pH range of 7.5–7.8 and a temperature range of 32–34 °C [198]. It is noteworthy that temperature and HRT play significant roles in the effectiveness of the anammox process within a UASB. While the most common temperature range for UASB systems is 32–36 °C, successful cultivation of AnAOB has been achieved within a period of 4.5 months at a lower temperature of 24 °C [49]. Furthermore, the effectiveness of the anammox process is influenced through HRT and reactor volume. Studies have shown that a longer HRT can enhance anammox performance, based on the dimensions of the reactor and NLR. For instance, in a 1.1 L UASB reactor, the best performance was achieved with the HRT exceeded 1.58 h and the NLR reached 10.5 kg/m^3/d, resulting in a remarkable 90% reduction in NH_4^+–N concentration. Nevertheless, when the HRT was decreased to 0.21 h, the performance declined [223]. To prevent the accumulation of NO_2^-–N during the initial operation, a strategy was implemented in a UASB reactor with a volume of 6.25 L. The reactor was run with an HRT of 1–5 days, which enabled the establishment of stable anammox activity. Subsequently, the HRT was kept at 3.5 days to sustain the continued stability of the process [230]. It has been reported that granulation could significantly increase the biomass of AnAOB resulting in a higher rate of volumetric removal in the anammox process within UASB reactors [278]. Notably, the stable UASB configuration can also prevent substrate inhibition caused

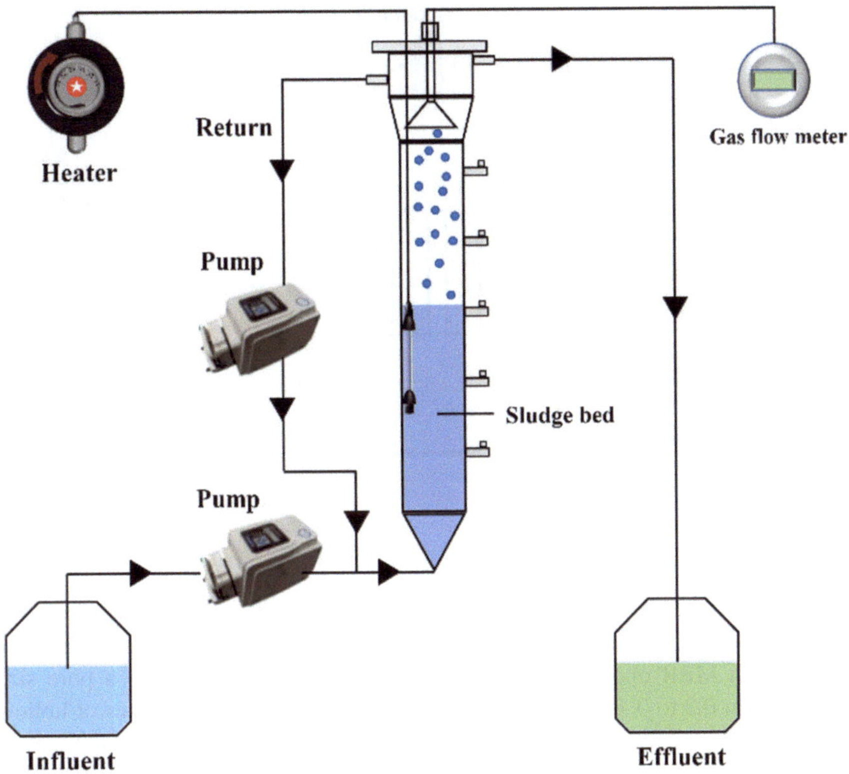

Fig. 1.4 Schematic diagram of up-flow anaerobic sludge blanket (UASB)

by NO_2^-–N concentration shock [96]. Therefore, with its ability to generate high biomass concentrations, the UASB reactor is an efficient and effective technology for enriching AnAOB in the wastewater treatment process. It can be concluded that UASB reactor is widely recognized as a highly effective technology for treating wastewater due to its reliable system and high sludge settling capability. With its ability to generate high biomass concentrations, the UASB reactor demonstrates efficiency and effectiveness in cultivating AnAOB for the treatment of wastewater in future processes.

1.3.1.3 Membrane Bioreactor (MBR)

The MBR is a widely utilized technology for wastewater treatment that combines suspended bioreactor technology with membrane filtration processes (Fig. 1.5) [224]. There are two types of MBR: submerged and external based on the location of the membrane component [143]. During the membrane process in an MBR, an impermeable membrane material is employed for microfiltration, effectively retaining

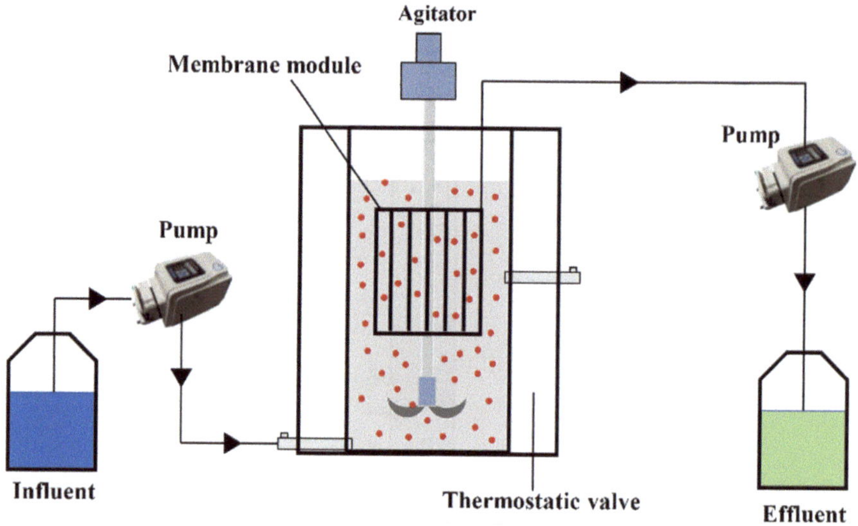

Fig. 1.5 Schematic diagram of membrane bioreactor (MBR)

biomass. When considering the reactor configuration, the fiber membrane is widely employed in MBR operating systems. This type of membrane, with a pore size ranging from 0.2 to 0.4 μm, can be in the form of flat sheet membranes or hollow membrane fibers [108, 144, 239]. An MBR can also operate via the application of an anaerobic technique, called anaerobic membrane bioreactor (AnMBR) [96]. Another alternative to the MBR system is the rotating flat sheet membrane bioreactor (RFMBR) [108]. The RFMBR exhibits a higher impact from shear stress when operated at a stirring speed of 20 rpm, due to its rotatable multi-flat sheet structure, making it more effective than a conventional MBR operating at a stirring speed of 60 rpm. In a study, a rapid initiation of anammox activity was accomplished within a week with a 97.6% enrichment purity in a 15 L MBR system. The process conditions involved a sludge retention time (SRT) of 12 days, an agitation rate of 160 rpm, and the regulation of the influent NO_2^-–N concentration within the range of 100–120 mM [213]. Using MBR for full biomass retention in AnAOB enrichment process has been identified as an alternative technique [96]. While the MBR can produce elevated levels of mixed liquor suspended solids (MLSS), and reduce surplus sludge production, its application is limited with membrane fouling and the higher expense of submerged MBR in terms of membrane materials [108, 180]. Meanwhile, fouling of the membrane in submerged MBRs is primarily caused by the accumulation of sludge flocs on the membrane surface and solute-membrane adsorption, leading to reduced membrane permeability [45]. Moreover, the buildup and settling of microorganisms, colloids, solutes, and cell debris on the membrane material can lead to elevated transmembrane pressure or reduce the permeate flux [180]. The fouling problem in MBRs is dependent on the size and distribution of biomass particles, with smaller particles leading to more pronounced fouling [143]. However, an effective strategy of regularly flushing the membrane during backwashing may resolve the membrane fouling problem. In conclusion, while membrane contamination can be a challenge, the significance of MBR in facilitating the efficient enrichment of AnAOB cannot be overlooked.

1.3.1.4 Fluidized Bed Reactor (FBR)

The FBR is a bioreactor system that utilizes an attached growth method, employing a fluidization process where biomass flows through a porous plate or distributor containing a granular solid material, typically a catalyst (Fig. 1.6) [96, 195, 196]. By maintaining an adequate velocity, the solid material is suspended, behaving like a fluid, ensuring the biomass particles remain in a fluidized state. This fluidization state enhances the characteristics of mass transfer compared to other systems utilizing attached growth [96]. A 2.5 L FBR was utilized to assess the performance of an anammox process with optimal conditions of 36 °C and pH 7, as determined by batch tests [238]. To sustain the fluidization state of the bed, anoxic fluid recirculation from the top of the reactor was employed at a flow rate of 47 L/h, resulting in a superficial liquid velocity of 24 m/h. The microorganisms, which form a biofilm on small support particles, remained in a state of fluidization in the reactor. Typical materials employed as carriers for biomass attachment in anammox FBR include sand particles ranging from 0.3 to 0.6 mm and non-woven fabric material (Fig. 1.6) [214, 238]. To suspend the catalyst particles, the FBR is usually designed with ample reactor volumes and a sufficient flow rate. In research, FBR was reported to accelerate anammox granules formation and mitigate the loss of suspended biomass during operation. It can improve the substrate transport process for the activity of AnAOB though keeping anammox granule a complete mixed-bulking biomass. Therefore, the utilization of a FBR system is well-suited for the growth of bacteria with slow growth rates and can accelerate the initial phase of reactor operation [66]. This is particularly advantageous when enriching AnAOB populations. For example, in the treatment of sludge digestion effluent using an anammox process in a 2.5 L FBR, a nitrogen removal rate (NRR) of 0.18 kg/kg/d was accomplished [214]. This demonstrates the efficacy of the FBR in fostering the proliferation and performance of AnAOB. Nonetheless, it is important to note that an unstable fluidization process can potentially result in significant microorganism loss, leading to a decrease in the NRE of the system [66, 214]. Although FBR can greatly expedite accelerate the initiation of an anammox process and improve the enrichment of AnAOB, the substantial expenses associated with constructing and maintaining the system is significant challenge for its widespread application.

1.3.1.5 Up-Flow Biofilter (UBF)

The UBF bioreactor system is widely employed in wastewater treatment. It operates by directing the upward flow of wastewater by a sediment basin containing packing material or media (Fig. 1.7) [96, 195, 196]. This type of bioreactor allows for the

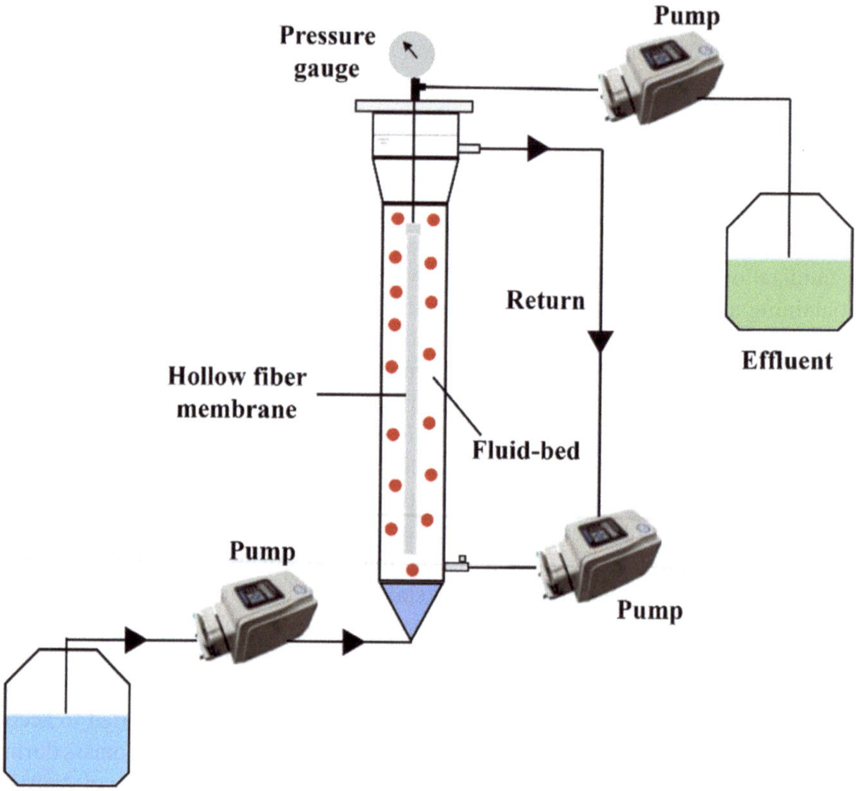

Fig. 1.6 Schematic diagram of fluidized bed reactor (FBR)

attachment of biofilm on the surface area of the packing material, facilitating the removal process of nitrogen and promoting retention of slow-growing bacteria [195, 196]. The UBF system has been effectively utilized for the enrichment of AnAOB and the operation of anammox process, which involves the nitrogen removal from wastewater, using conventional WWTPs sludge as the inoculum operated at pH between 7.0 and 8.0 and temperatures ranging from 30 to 36 °C [24, 110, 111]. The anammox process in the UBF system has demonstrated a significant NH_4^+–N removal effectiveness of 89% at HRT greater than 4.6 h. Nevertheless, reducing the HRT further resulted in a decline in the NRE [113]. The implementation of the UBF system, incorporating hollow bamboo balls as carriers for the biomass, has proven its efficacy in facilitating the development of anammox granules and biofilms. This system operates at a relatively high temperature of 35 °C and incorporates appropriate effluent recirculation. These factors have contributed to the successful enrichment of AnAOB and have allowed for a high SAA even at a demanding NLR of 34.5 kg/m^3/d [24]. In a study, assessing stability of anammox systems and the start-up time under substrate loading shocks, the SBR and UBF systems were considered

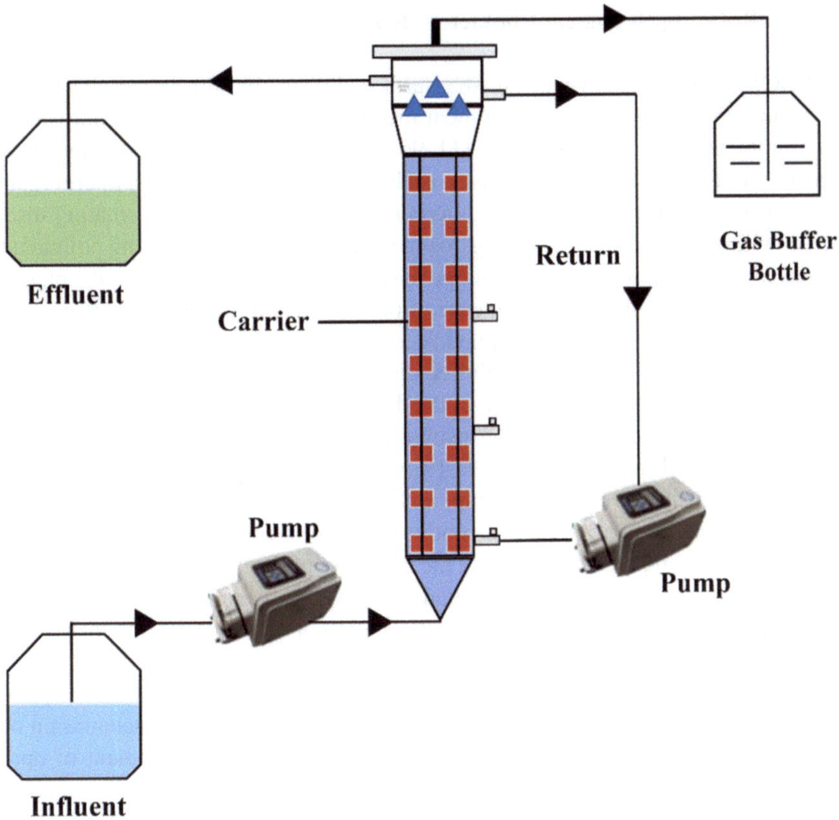

Fig. 1.7 Schematic diagram of up-flow biofilter (UBF)

to be better reactor configuration for the AnAOB enrichment process, with the UBF being the superior option [113]. However, based on a similar study, the biomass production in the UBF was lower than that in the SBR system [195, 196]. Despite the UBF's success in promoting AnAOB enrichment and nitrogen removal efficiency, several challenges still hinder its wide-scale application. One major challenge is the high cost of maintenance, as regular cleaning and replacement of the packing material are necessary to prevent clogging and maintain high performance. Additionally, the UBF's complexity and relatively large footprint may also limit its use in smaller treatment facilities. Therefore, further research is necessary to address these challenges and optimize the UBF system's performance in wastewater treatment applications.

1.3.1.6 Rotating Biological Contactor (RBC)

The RBC system, also known as a media, surface, disc and biofilm reactor, is an innovative and effective wastewater treatment method that has gained increasing attention over the past few years (Fig. 1.8) [11, 96, 261]. RBCs are typically composed of multiple parallel packs of rotating discs, each of which is covered with microorganisms that form biofilms [11]. These microorganisms attached to the rotating discs interact with the wastewater, breaking down organic matter and removing pollutants. Compared to traditional activated sludge systems, RBCs offer several advantages, including low energy consumption, ease of operation, and minimal sludge production. The RBC system has been applied successfully for AnAOB enrichment using a 6.2 L reactor with 13 polyvinyl chloride discs, providing a combined surface area of 0.32 m^2, leading to 87% coverage of the disc surface [147]. Under the maximum influent surface load of 12 $g/m^3/d$, the results of the reactor start-up in the RBC system are promising, with a high NRE of 92.1%. The microbiology of the RBC biofilm system is primarily governed by compound mass transfer, which is influenced by several factors including operational conditions, biofilm architecture, biofilm development and shedding mechanisms, and the thickness of the biofilm's boundary layer [271]. Although the RBC system has found extensive application in wastewater treatment, but there is limited research on its application in AnAOB enrichment process at the laboratory scale [271]. To further improve the efficiency of the RBC system for AnAOB enrichment, it is important to explore the effects of different operational factors such as temperature, pH, HRT, and influent nitrogen concentration on the performance of the anammox process. Furthermore, the development of optimized biofilm structure and biofilm attachment or detachment mechanisms can help enhance the mass transfer of nitrogen compounds and improve the overall efficiency of the RBC system. Further research should understand the potential of the RBC system for AnAOB enrichment and to optimize its design and operation parameters for maximum efficiency. The RBC system holds great promise as a valuable tool for AnAOB enrichment in the future, contributing to more sustainable and efficient wastewater treatment practices.

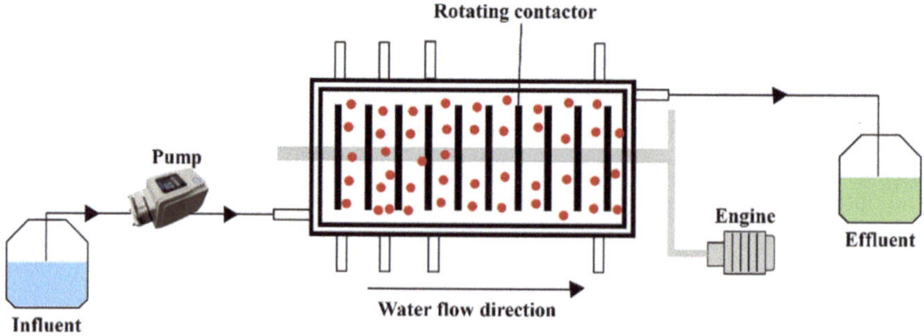

Fig. 1.8 Schematic diagram of rotating biological contactor (RBC)

1.3.1.7 Gas-Lift Reactor

The gas-lift reactor is a type of suspended growth bioreactor used to enhance the anammox process (Fig. 1.9) [96]. In this system, the circulation and mixing of biomass, as well as the maintenance of anaerobic conditions, are achieved through the utilization of compressed air, gas bubbles, water vapor, or other vapor bubbles (Fig. 1.9). To create the necessary turbulence for the biomass, a tube (referred to as a riser) is employed within the reactor, facilitating the bubbling process. This turbulence results in a circular flow of both the biomass and the liquid [99]. Additionally, to sustain fluidization of the biomass and ensure anaerobic conditions, argon flushing is utilized. This process effectively preserves the appropriate operational parameters required for cultivating in the reactor [37]. The gas-lift reactor system has demonstrated a favorable gas–liquid exchange capacity, which is necessary for the growth of AnAOB. Studies have shown that this system can achieve a high 8.9 kg/m^3/d NRR with a NLR of 10.7 kg/m^3/d and HRT of 1 day [80]. However, one of the challenges of using this system is the accumulation of N_2, which can cause biomass flotation problems as a result of the buoyancy of large aggregates of flocs and granules. This can lead to reduced biomass settling and eventual biomass washout, which can affect the efficiency of the reactor [37]. The presence of settling granules is crucial for the proper functioning of the up-flow and gas-lift reactor system. Biomass retention plays a vital role in this regard, as it ensures the continuous presence of settling granules throughout the operation of the reactor. Therefore, biomass retention is critical during the operation of the gas-lift reactor system, as it helps to maintain the presence of settling granules. Overall, the gas-lift reactor system has shown promise as a viable alternative for AnAOB enrichment. However, further investigation needed to enhance its performance and address the challenges associated with its operation, including biomass flotation and washout.

In summary, SBR and UASB are two commonly used anaerobic treatment technologies for enriching AnAOB to treat nitrogenous wastewater. SBR is typically a batch processing system that allows wastewater to be exposed to both aerobic and anaerobic conditions at different stages, supporting the growth and metabolism of different types of microbial communities. UASB, on the other hand, utilizes the upflow to maintain anaerobic conditions, with anaerobic microorganisms attached to high-density sludge particles for reaction. Both of these reactors are able to efficiently remove nitrogen compounds. In addition, MBR incorporates a membrane filtration setup to segregate processed water from the microbial biomass in the reactor, resulting in high-quality effluent and a compact system footprint. FBR operates by creating a fluidized bed of biomass, which provides a large surface area for microbial growth and allows for efficient mass transfer. UBF utilizes a filter bed filled with media that provides a surface for biofilm formation and enables the elimination of both organic and inorganic contaminants. RBC consists of rotating discs with biofilm attached, which provides aeration and a surface area for microbial growth. Gas-lift reactor uses gas to provide circulation, promoting microbial the growth of AnAOB and facilitating the treatment of wastewater. Overall, these reactors provide various options

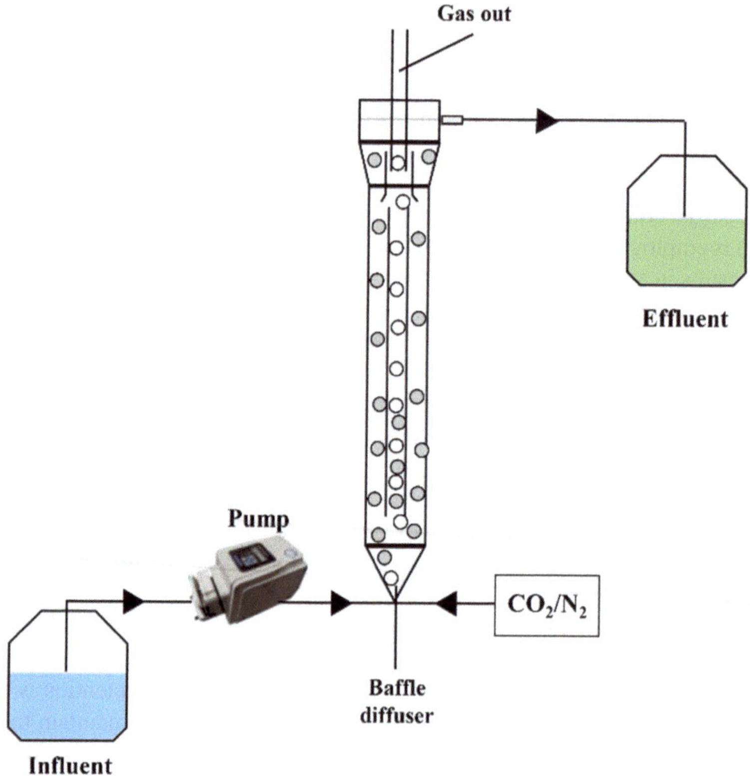

Fig. 1.9 Schematic diagram of gas-lift reactor

for enriching AnAOB and treating wastewater. Each technology has its unique advantages and limitations, and the selection of a specific technology should be based on the specific wastewater characteristics, treatment goals, and site-specific considerations.

1.3.2 Enrichment of Anammox Bacteria by Carriers Addition

To effectively enrich AnAOB in sludge and increase their abundance and activity, it is essential to maintain the anammox process within favorable environmental parameters. Nevertheless, except for the inherent limitation of AnAOB, operating the reactor can result in biomass loss, due to various factors such as bubble adhesion, bubble breakage, coalescence of flocculated sludge, and inadequate HRT, which can also limit the proliferation of AnAOB [313–325, 317–319]. Hence, ensuring an adequate amount of sludge biomass within the reactor is the crucial for stable and efficient enrichment of AnAOB [182].

The use of carriers has been reported to have a crucial impact on reducing biomass loss and improving the efficiency of AnAOB enrichment [252, 253]. The incorporation of carriers expands the overall surface area available for biomass retention, consequently boosting the activity of AnAOB [96]. Carriers provide attachment sites for AnAOB to attach and prevent the displacement of slow-growing bacteria [17]. The extracellular polymeric substance produced by AnAOB fulfills multiple functions. It functions as a pathway for ion transport, facilitating communication between bacteria and the surrounding environment. Additionally, EPS possesses adhesive characteristics that facilitate the binding of cells to the surface of carriers, enhancing their overall adherence. Up to now, the methods of anammox biomass retention have been proposed, including granule formation, biofilm development on carrier materials, gel encapsulation, and membrane-based techniques [313–325, 317–319] (Fig. 1.10a). Utilizing carrier-based biofilm immobilization has emerged as a viable strategy to support the proliferation of AnAOB and enhance the property of anammox-based systems [247, 249, 260]. The introduction of carriers can facilitate the optimization of microorganism population distribution (Fig. 1.10b). Specifically, the use of microporous carriers promotes the aggregation of AnAOB and denitrifiers within biofilms, enabling the synergistic integration of denitrification and anammox processes. This coupling not only enhances resistance to disturbances but also has the potential to enhance the anammox process itself, leading to efficient removal of NO_3^-–N from the effluent [292].

Various types of carrier-based biofilm have been extensively studied for enriching bacteria in wastewater treatment processes [65]. In general, biofilm carriers can be categorized into two types: inorganic and organic. Inorganic carriers such as volcanic and furnace slag have been commonly used in biological filters and constructed wetlands [104, 142] (Table 1.2). Meanwhile, polymer inert organic materials have become commercially available and popular due to large specific surface area, excellent mechanical strength, and much lower density in full-scale WWTPs [65, 329]. In

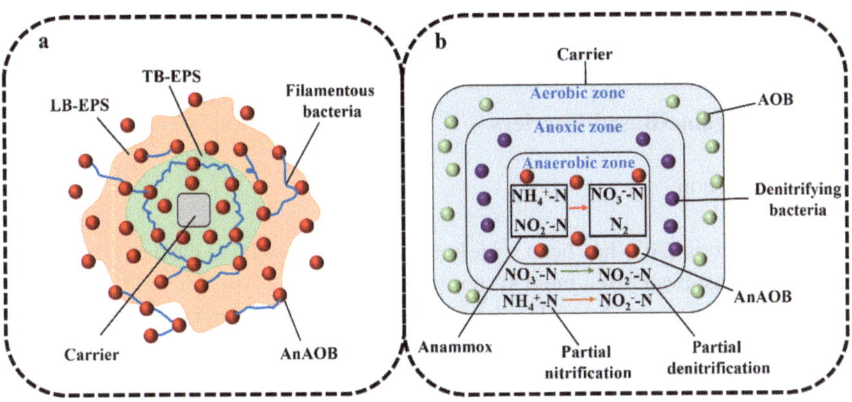

Fig. 1.10 **a** Mechanism of AnAOB immobilization via carriers; **b** Schematic diagram of the anammox-based process for nitrogen removal via carriers

Table 1.2 Advantages and disadvantages of biofilm carriers using in the anammox-based systems

Material category		Biofilm type	Characteristic	Advantages	Disadvantages
Inorganic	Typical carrier Specialized functional carrier	Volcanic, ceramics, etc., Zeolite Tourmaline Limestone, marble chips	High density	Microporous High ammonium microenvironment Promoting activity, increasing alkalinity Increase alkalinity	Filter clogging
Organic	Polymer inert carrier Natural organic carrier	Polyurethane, K-carrier, etc., Biochar, corncob, etc., Gel carrier	Low density, inert Low density, biodegradable Various forms	Providing spatial structure, enhancing the biomass retention Porous, low cost Resilience to an adverse environment, enhancing the biomass retention	Smooth surface, the internal clogging problem Promoting denitrification Low mechanical strength, poreclogging

contrast, biochar and corncob tend to have a larger surface area and porous surface structure compared to polymer organic carriers [18, 289, 292]. Studies have compared the effectiveness of AnAOB enrichment using different carriers [77, 292]. A previous study found that polyurethane sponges were able to enrich higher biomass and achieve better system performance compared to plastic ring carriers [292]. Another study reveals that anammox reactor established with the gel carriers exhibited more resistance to increasing organic matter stress than granular activated carbon carriers. These findings suggest that the choice of carrier greatly affects the operation performance of anammox process and the effectiveness of AnAOB enrichment. Therefore, it is crucial to select the appropriate carrier promote for enrichment of AnAOB and ensure the stable operation of anammox-based systems.

1.3.2.1 Organic-Based Carriers

Activated Carbon

Activated carbon (AC) has been widely utilized as an effective carrier to enrich the AnAOB in wastewater treatment systems [313–325, 317–319]. The porous and inert nature AC can promote AnAOB adhesion and growth. In particular, AC can promote EPS secretion, which selectively attach AnAOB to the carrier material [167]. The incorporation of hydroxyapatite (HAP) in the anammox system offers a significant advantage by promoting the stability of sludge durability and nitrogen removal. This feature effectively prevents sludge flotation and enhances the applicability of HAP as an inoculum, ensuring consistent performance in the system [148]. In a recent study, it was found that the richness of *Ca. Brocadia, a* typical anammox bacteria, increased significantly from 2.1 to 20% with a corresponding nitrogen removal rate of 88%. The reactor with the 10 g/L of AC acquired a NRE of 95.34%, surpassing that of the control reactor (94.26%). Moreover, EPS in the presence of AC rose to 109.83 mg/g VSS with the addition of AC, which was 17.9% higher than the control. Addtionally, *Ca. Kuenenia,* which was not present in the seed sludge increased sharply to 11.71% [161, 162]. In addition, AC can also act as an electron shuttle to promote nitrogen removal process through composing Feammox and anammox. Several studies have shown that the addition of AC to anammox reactors increases the richness of AnAOB and the nitrogen removal effect. According to some reports, the system achieved a nitrogen removal rate of 0.46 kg/(m^3 d), resulting in an approximate 82.9% TN removal efficiency. Importantly, the most abundant AnAOB is *Ca. Brocadia* (1.18%) and *Ca. Jettenia* (2.10%) [89]. Overall, while the use of AC as a carrier is a promising approach to promote the activity of AnAOB and promote their growth, the complexity and high cost of the activated carbon production process are the primary factors that limit its widespread use in the enrichment of AnAOB and the implementation of anammox-based processes.

Biochar

Biochar, a carbon-rich material produced by the pyrolysis of organic matter in the absence of oxygen (Fig. 1.11) [279–281]. Biochar is a novel scientific term. Compared to activated carbon, the process of producing biochar is simpler and more cost-effective, which has made it a popular choice for use in wastewater treatment processes. When biomass, such as wood, manure, or leaves, are heated in a closed container with little or no air, such result is described in early research as "a carbon rich product" [189]. The management of the waste stream coming from plants or animals offers a lot of potential for reducing the pollution burden to the environment. Utilizing waste biomass for the creation of biochar is not only practical but also advantageous [263]. Due to its substantial surface area, high charge density, and low ecotoxicity risk, biochar offers numerous benefits including climate change mitigation and energy production and has been suggested as the most potential carrier material for AnAOB enrichment [279, 283]. In a research study, it was found that adding biochar amendment contributed to a significant increase in the richness of AnAOB. The observed increase ranged from 4 to 12.3 times compared to the control condition [167]. This result demonstrates the enrichment effect of biochar on AnAOB.

As a carrier, biochar has been shown to enhance the biological retention of AnAOB, making tin a more stable and environmental disturbance-resistant nitrogen removal mechanism (Fig. 1.12). It was reported that the addition of 5% biochar (made by coconut) in Anammox system reduced the start-up time from 69 to 46 d with an average ammonia removal efficiency (ARE) of 96% [1]. In a study, adding various types of biochar, including peach, bamboo, and jujube biochar, was found to enhance the system's performance. The maximum ARE was 84.28%, while the average TN removal efficiency was 88.39%, surpassing the performance of the control group

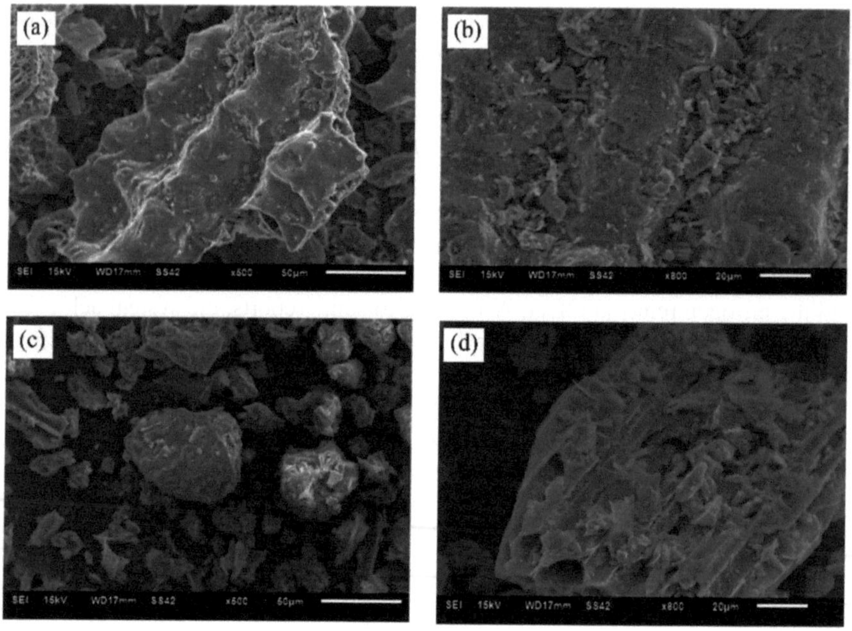

Fig. 1.11 Structure of biochar (made of rice husk) (a) and (b); structure of biochar (made of rice husk) (c) and (d)

(ARE: 77.00%; TN: 83.81%). And, the relative abundance of Planctomycetes rose from 4.6 to 7.1% in bioreactors added with jujube biochar, increasing by 54.50% [275]. In addition to improving ammonia removal efficiency, biochar can decrease the inhibitory effect of COD and NO_2^-–N for AnAOB. Guo et al. [76] has found that adding bamboo biochar increased the average ARE by 5.3% with the 150 mg/L of COD [76]. In another study, it was observed that a high concentration of NO_2^-–N (600 mg/L) had a pronounced inhibitory effect on the growth of AnAOB. However, the addition of three different types of pyrolytic biochar led to significant improvements in NRE. The NRE increased by 231%, 149%, and 51.0% when incorporating each respective type of biochar. As an effective carrier material for the enrichment of AnAOB, biochar enhances the anammox process stability, improves nitrogen removal effect, and reduces inhibition by COD and NO_2^-–N. The use of biochar in enriching AnAOB is a promising area for future research and application.

Polymer Inert Organic Carrier

Over the past few decades, various polymer inert organic materials have been utilized for AnAOB enrichment, including polyurethane sponge [190], non-woven fabric [251] (Fig. 1.13a), Kaldnes-type carriers [330, 331] (Fig. 1.13b), semi-soft fiber

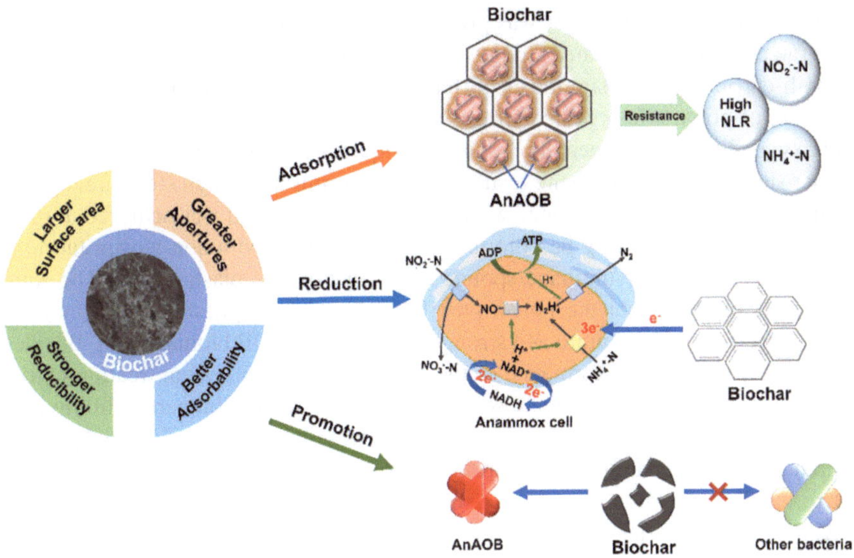

Fig. 1.12 Mechanism of biochar in facilitating anammox performance and anammox bacteria growth

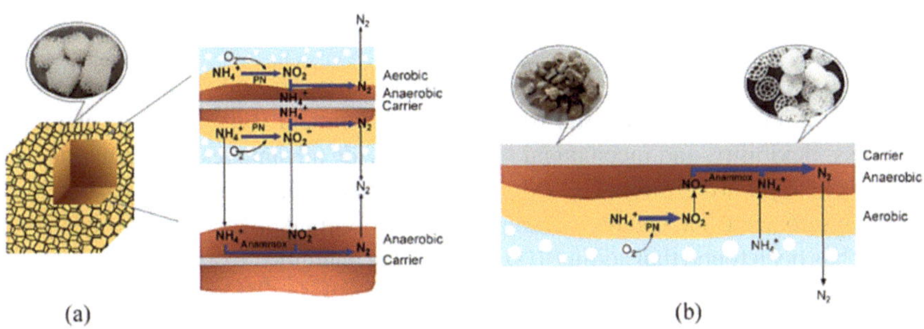

Fig. 1.13 Differences in biofilm due to different structural types of carriers: **a** Carriers with three-dimensional reticular spatial structure (polyurethane sponge, non-woven fabric); **b** Carriers without three-dimensional reticular spatial structure (inorganic and Kaldnes type carriers)

[264], ring-laced string [226], net-type acryl resin carriers [221], polypropylene hollow cylinder [77], and polyacrylonitrile-based activated carbon fiber [18].

Compared with material characteristics, structure, sizes, and packing ratio of polymer organic biomass carriers may have a more significant impacts on the composition and abundance of microorganisms in the systems due to their resistance to biodegradation and high hydrophobicity. For example, the three-dimensional porous structure is more preferential to maintain more biomass, resulting a thick enough biofilm to provide a more anaerobic environment for the growth of AnAOB (Fig. 1.13a). A previous study showed that the thicker of the anoxic/anaerobic biofilm layer in polyurethane sponges was greater than in plastic ring carriers at full-scale integrated fixed-film activated sludge (IFAS) system, while the aerobic biofilm layer in both carriers has no significant difference [292]. Thus, it is perceived that the structure of polymeric organic carriers significantly affected the efficiency of mass and oxygen transfer between the biofilm and liquid. The non-woven fabric is another carrier showed that a promising result due to its porous and rough structure, leading to more satisfactory performance with a NRR of 0.4 kg/m^3/d was achieved at room temperature in the moving bed biofilm reactor (MBBR) with non-woven fabric as carriers instead of polyurethane sponge and polyethylene plastics [77]. The same result was reported that the non-woven fabric with the highest specific area and dense structure retained more and stronger attached AnAOB biomass [31]. Besides, the nitrogen removal performance of the deammonification system treating lagoon supernatant was remarkably increased by increasing the biofilm carrier filling ratio from 35 to 55% [138, 145, 146]. However, the literature on carrier sizes and packing ratio in the enrichment of AnAOB are still infrequently reported, indicating the need for research in this area Overall, the structure, sizes, and packing ratio of polymer organic biomass carriers play a crucial role in AnAOB enrichment.

1.3.2.2 Inorganic Carriers

Graphene

Graphene oxide (GO) is a widely studied carbon support with numerous benefits that make it a promising material for AnAOB enrichment (Fig. 1.14). Its chemical inertness, flexibility, and mechanical strength provide advantages over other carbon supports [167]. Furthermore, GO exhibits exceptional electron transfer capabilities, enabling it to act as an electron shuttle between important anammox enzymes [167, 301]. It has been found that adding 100 mg/L GO in anammox reactor increased NRE by 10.26% [254, 259]. Moreover, AnAOB can transform GO to reduced graphene oxide (RGO), and a 100 mg/L GO increase anammox property by 17.3% [301]. Additionally, GO can also impact the key enzyme activity and the secretion of EPS in AnAOB. Compared to without GO, the activity of hydrazine dehydrogenase (HDH) was increased by 19.9%, while nitrite reductase (NIR) increased by 27.9%, respectively [313–325, 317–319]. The optimal dosage of GO at 10 mg/L demonstrated a nearly 46% increase in NRE and a 1.1-fold increase in HDH enzyme activity. In addition, the presence of GO promoted the aggregation of AnAOB through facilitating the hydrophobic interaction of EPS, and the relative abundance of *Ca. Jettenia* increased from 11.9 to 12.0% [52]. In batch testing, it was confirmed that the optimum dose of GO at 100 mg/L increased TN removal rate by 17.2%. Additionally, the concentration of EPS was approximately 48.6%, which is better than that of the control when 0.1 g/L of GO was added. Under the variational doses of RGO, the activities of HDH, NIR, and NO_3^-–N reductase was increased by 0.70–2.75 times compared to the control [301]. The use of TEM revealed that GO with a large surface area facilitated to the attachment of cell [167]. These results suggest that the adding GO

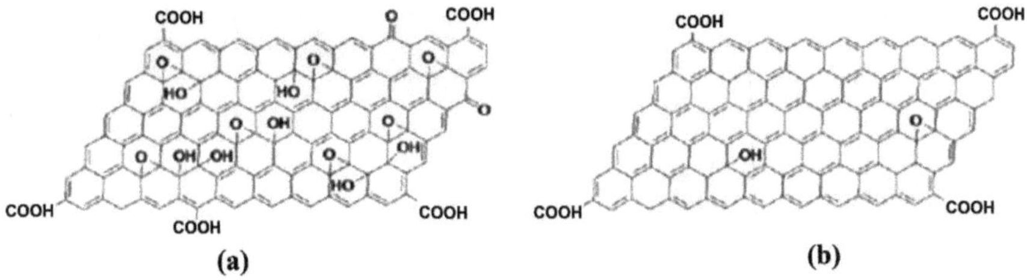

Fig. 1.14 **a** Molecular structure of graphene oxide (GO); **b** Molecular structure of reduced graphene oxide (RGO)

is a highly efficient method for enriching AnAOB. However, the price of GO has forced researchers to consider cost issues.

Zeolite

Zeolite, with its large surface area and extensive internal porosity, has been found to be an ideal carrier material for AnAOB enrichment (Fig. 1.15) [14]. Zeolites are hydrated aluminosilicate compounds, composed of three-dimensional tetrahedral networks of SiO_4 and AlO_4 connected by shared oxygen atoms [14, 313–325, 317–319]. Using zeolite particles as carriers in reactors with suspended biomass effectually enhances the granulation of anammox, resulting in a substantial increase in biomass concentration. Employing zeolite as carrier for the formation of anammox biofilm reduces biodegradable effluent decrease from 12 to 3 mg VSS/L, while significantly boosting biomass concentration. SAA was also observed to increase from 0.35 to 0.5 g N/ (g VSS d) [58]. Furthermore, the use of zeolite-carrier has been found to increase the nitrogen removal efficiency by 25%, with a 2.48% abundance of Ca. *Jettenia* acquired, increasing PN/PS by 21.4%. These results demonstrate that the sludge adhered to the surface of the zeolite exhibits enhanced aggregation and biofilm formation capabilities [166]. In addition, natural zeolites have gained significant popularity as effective selective ion exchangers for NH_4^+–N removal from wastewater. Compared to the control group with a 62 mg/L TN in effluent with the zeolite dosage increased from 0.05 to 1.0 g/L resulted in a significant reduction in TN concentration, from 51.8 to 7.78 mg/L. Moreover, the gene copy of anammox, which serves as a marker for the community characteristics of AnAOB, exhibited a 37.5% increase compared to uncoated membrane reactors [95]. This indicates that zeolite-coated membranes create a favorable condition for AnAOB, indicating that they provide a niche for their growth. These findings highlight that the ability of zeolite-containing membranes or surfaces to enhance the retention of AnAOB and effectively remove TN from wastewater. In short, using zeolite as a carrier for AnAOB enrichment has shown great potential in improving the ability of the sludge adhered to the surface of the carrier.

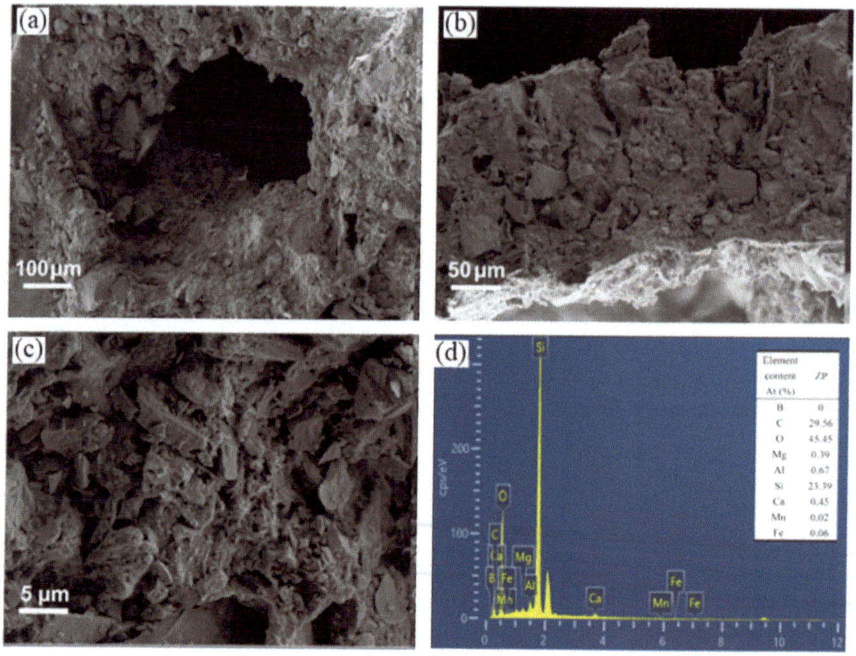

Fig. 1.15 Scanning electron microscope (SEM) image of zeolite (**a**)–(**c**); XPS spectra and element content of zeolite (**d**)

1.3.2.3 Gel Carriers

Gel entrapment has gained significant attention as a promising immobilized technique for wastewater treatment [65, 329]. Many previous studies demonstrated that gel immobilization exhibits several unique strengths, including enhanced biomass retention and protection against the adverse environment [99, 103]. Given these advantages, the entrapped and immobilization technologies are particularly suitable for the rapid startup anammox-based process and enrichment of AnAOB.

Compared to other carriers, the gel carriers have proven to be effective in reducing initial seeding biomass concentration to establish anammox-based systems and facilitating the initial biofilm formation, thereby enriching the AnAOB effectively (Table 1.3). The start-up time of different immobilization anammox-based systems were summarized (Table 1.3). It can be demonstrated that the time required to establish the immobilization anammox-based systems was significantly shorter than the previously reported values in other biofilm systems. Meanwhile, the start-up time of the gel anammox-based systems was remarkably affected by the type and amount of inoculation sludge. One study has shown that the anammox system with activated sludge as inoculum needs more time to start up than the system with anammox sludge [27, 28]. Four different kinds of gel carriers, including carboxymethylcellulose (CMC), sodium alginate (SA), polyvinyl alcohol (PVA) and polyvinyl alcohol-sodium alginate (PVA-SA), were evaluated in anammox systems [65, 336]. PVA-SA was considered to the best gel material with the best systems performance and spatial structure. As the most common immobilized material, PVA-SA has significant advantages in reducing the seeding biomass concentration and start-up time. In the gel immobilized reactor, the seeding biomass concentration could be significantly lowered to 0.33 g/L within 35 days [7]. The start-up time of the deammonification was shortened to 10 days with only 0.4 g/L seeding biomass [290, 293, 294]. A similar result was reported that the partial nitrification-anammox system was established with the NRR of 0.3 kg/m^3/d within 13 days using PVA-SA gel carriers [105–107].

Furthermore, in addition to facilitating rapid start-up, gel entrapment provides powerful protection for AnAOB from inhibitory effects induced by organics, substrates, and other environmental factors [150, 157, 257, 339]. High-strength wastewater treatment can negatively affect the anammox activity is due to high concentration of nitrite and ammonium. The protective effect of polyvinyl alcohol and waterborne polyurethane (WPU) gel against high nitrite concentrations has been experimentally and theoretically verified [175]. The gel entrapment of anammox-based systems also presents strong and consistent adaptability to unfavorable conditions, such as high salinity, organic matters and low temperature [105–107, 150, 157, 257]. Recent studies have shown that the anammox system with PVA gel carrier possesses more excellent organics resistance than granular activated carbon (GAC) systems [103]. Additionally, the negative effect of pH and temperature was significantly improved using PVA-SA gel carrier [339]. The unique biofilm structure of gel carriers plays a crucial role in their protective effect, which the biomass mainly grows inside the immobilized gel. The protective effect of gel carriers was very likely because this biofilm/carriers structure could effectively separate the AnAOB cell from the inhibitory factors (including high concentration substrate, organics, heavy metal, oxygen) with the surrounding environment [175]. This statement could be evidenced by several studies, where the diffusion of dissolved oxygen (DO) and FA inside the biofilm were limited by outer gel material [120]. Although a full-scale two-stage gel anammox system was established in two months with an NRR of 3.0 kg/m^3/d, most studies concerning the gel anammox systems were still carried out at bottle-scale continuous stirred tank reactor (CSTR) (Table 1.3). Despite this limitation, the successful application of gel entrapment technology in various wastewater treatment systems for enriching AnAOB.

1.3.2.4 Modified/Composite Carriers

Different types of carriers exhibit comparative advantages and disadvantages in promoting the operation of the anammox process and the enrichment of AnAOB. Due to the AnAOB possessing an extremely low cellular yield and slow growth rate, the effect of different carriers on the anammox system performance depended on its effect on biofilm formation [131]. Therefore, the primary purpose of carrier

Table 1.3 Comparison of the start-up time and inoculum sludge amount in different immobilization anammox-based systems

Immobilization material type	Reactor format	Influent (mg/L)		NRR (kg/m^3/d)	System	Inoculum sludge type	Inoculum VSS (g/L)	Start-up time (d)	References
		Inf. NH_4^+–N	Inf. NO_2^-–N/ Inf. NO_3^-–N						
PVA-SA	Up-flow column	105	105	10.8	Anammox	Anammox	0.33	35	[7]
PVA-SA	Column	93.3 ± 8.4	103.8 ± 8.5	0.43 ± 0.03	Anammox	Anammox	Not reported	35	[164]
PVA-SA	CSTR	99.1 ± 6.8	157.4 ± 12.3	0.21	Anammox	Anammox	0.85 ± 0.02	9	[27, 29]
cryoPVAG	CSTR	70–150	90–180	0.58 ± 0.01	Anammox	Anammox	11.3 ± 0.7	30	[175]
WPU	UASB	120–200	120–200	0.505	Anammox	Anammox	3.76	49	[19]
PEG	CSTR	690	–	3	Two stage PN/A	Anammox	Not reported	60	[97]
PVA	Up-flow reactor	60–80	–	0.09	PN/A	PN/A	Not reported	27	[115]
SA	Up-flow reactor	50	–	0.53	PN/A	PN/A	0.4	10	[290, 293, 294]
PVA/SA	UASB	51.7 ± 0.85	–	0.24–0.27	PN/A	PN/A	Not reported	22	[298]
PVA/SA	SBR	92.1 ± 1.0	124.9 ± 2.8	0.3	PD/A	AS/A	Not reported	13	[105–107]

PVAG poly (vinyl alcohol) cryogel, *PVA-PP* Polyvinyl alcohol-polypropylene, *PEG* Polyethylene glycol, *A* Anammox, *AS* Activated sludge, *PN/A* Partial nitrification and anammox, *PD/A* Partial denitrification and anammox modification is related to rapid and sufficient biofilm formation, which is connected with the biofilm formation process and some specific carrier problems. Detailed descriptive information of each study has been summarized and available in this work (Table 1.4). It was shown that the characteristic of carrier surface played a crucial role in the initial attachment of AnAOB. Carriers surface modification is the most common way to facilitate the immobilization of AnAOB. Synthetic polymer organic material has been extensively studied in surface modification due to its property, such as high hydrophobic, low bio-affinity, and resistance to biodegradation [185]. A novel modified carrier applied in integrated floating-film activated sludge was proposed. Surface modification using zero-valent iron significantly improved the surface characteristic, promoting the AnAOB attachment and biofilm development [152, 154]. In a study, nitrogen removal performance was achieved with an NRR of 2.5 $kg/m^3/d$ within 56 days through pre-seeding a layer of partial nitrification biofilm on the carrier surface [125, 126]. A similar result was acquired in another study, proceeding denitrifying biofilm on carriers surface has also been successful in enhancing nitrogen removal performance, including the specific anammox growth rate and NH_4^+–N removal rate [227]. Furthermore, a previous study compared the enhancing effects of two methods, including seeding denitrifying biofilm and coating granular active carbon on the carriers [125, 126]. These results indicated that the bio-primer method could modify carrier surface with a higher increase of AnAOB activity. The biofilm carrier with hydrophilic modified surface has been extensively studied in enriching AnAOB [329]. However, the hydrophilic modification of carrier surface with 10% $KMnO_4$ and gelatin has failed to benefit the AnAOB immobilization [27, 28]. This might be owing to the fact that the EPS produced by AnAOB possess high hydrophobicity and a slight decrease of surface water angle could not significantly improve the AnAOB attachment [27, 28]. But no other reports have provided strong evidence for the evaluation. The effect and mechanism of hydrophilic surface modification requires further investigation. Zeolite has been applied to enhance the anammox-based system in many studies due to its exceptional ability to selectively absorb ammonium, while the iron nanopowder could facilitate the AnAOB activity and system performance in both gel and polymer inert organic modified carriers and improve carrier properties (Table 1.4). Chemicals addition in gel entrapment technology, glutaraldehyde and iron nanopowder have also been shown to improve carrier properties, mechanical strength, and reduce AnAOB loss while maintaining unimpeded mass transfer during the long-term operation of anammox-based processes [248, 257]. Furthermore, instead of immobilizing AnAOB, a new hollow gel capsule was proposed to enclose biomass, effectively protecting AnAOB from the high nitrite inhibition of 600 mg/L [217]. Designed sandwich-structured carriers biofilm carrier was also presented to effectively immobilize AnAOB biomass and fast start up of the system without a lag phase [313–325, 317–319]. For waste utilization, several previous studies also attempted a novel biomass carrier to establish the simultaneous partial nitrification, anammox, and denitrification process by utilizing waste activated sludge, red soil, and chemical additive vesicant as raw material [41]. In short, modifying the surface of biofilm carriers is an effective method to enrich the AnAOB.

Table 1.4 Modified method and purpose of modified and novel composite biofilm carriers in anammox-based systems

Carrier type	Modified method	Purpose	Modification purpose classification	System	References
K5	Surface dextranfunctionalized Surface silica-functionalized Seeded denitrifying biofilm	Increase bacteria affinity Increase surface roughness As a bio-primer	Initial attachment enhancement	Anammox	[228]
K3	Seeded heterotrophic biofilm Coated with GAC	As a bio-primer Increase surface roughness and CEC		Anammox	[125, 126]
HDPE	The addition of INP	Improve the AnAOB activity	Initial attachment and biofilm Development enhancement	Anammox	[152, 154]
PS	The addition of Fe and AC	Provide anaerobic microenvironment and Improve the surface environment	Initial attachment and biofilm development enhancement	Anammox	[25, 26]
PVA/SA	The addition of foaming Agent	Improve the gas permeability	Carrier properties enhancement	PN/A	[32]
PVA	A new capsule	Avoid damaging bacteria		Anammox	[33]
Novel composite Biofilm carrier	Sandwich-structured carriers WAS, RS and CAV Zeolite particles with floating Material Coated with zeolite particles Melt blend and etch of PE, PEO, and zeolite	Improve biomass retention – Provide high ammonium microenvironment Provide high ammonium microenvironment	Carrier properties enhancement Waste utilization Biofilm development enhancement Biofilm development enhancement	Anammox SNAD PN/A A PN/A	[40, 55, 65, 95, 106]

A Anammox, *PN* Partial nitrification, *SNAD* Simultaneous partial nitrification, Anammox, and denitrification, *GAC* Granular active carbon, *AC* activated carbon, *PE* Polyethylene, *HDPE* Polyethylene, *CEC* Cation exchange capacity, *PVA* Polyvinyl alcohol; *SA* Sodium alginate; *CS* Chitosan; *INP* iron nano powder, *NW* non-woven fabric, *WAS, RS and CAV* waste activated sludge, red soil and chemical additive vesicant, *SS* suspended sludge, *AB* Anammox biofilm, *AAS* aerobic activated sludge, *PE* polyethylene, *PEO* poly(ethylene oxide), *PS* polyurethane sponge carrier

In summary, achieving the enrichment of AnAOB (Anaerobic Ammonium-Oxidizing Bacteria) requires an efficient and persistent carrier. Consequently, it is crucial to carefully select suitable biofilm carriers that facilitate the enrichment of AnAOB. However, it is important to acknowledge that there might not be an absolute "perfect" carrier that suits all fields of application for AnAOB enrichment. Therefore, to effectively enrich AnAOB and enable widespread anammox applications, future research should focus on investigating various approaches for carrier modification, novel composite carrier preparation, and understanding their mechanisms in facilitating the AnAOB enrichment [65]. The novel desirable carriers are recommended to possess more of four main characteristics as much as possible, including spatial configuration, improving growth environment, and reinforcing the functional bacteria activity for the AnAOB enrichment (Fig. 1.16). In addition, it should improve the nitrogen removal contribution of the anammox process as high as possible with the help of the novel carriers, instead of pursuing the absolutely dominant anammox process [69]. Therefore, the novel carriers possessing those four desirable characteristics would have great potential at the full-scale WWTPs in improving the activity of AnAOB in the anammox process.

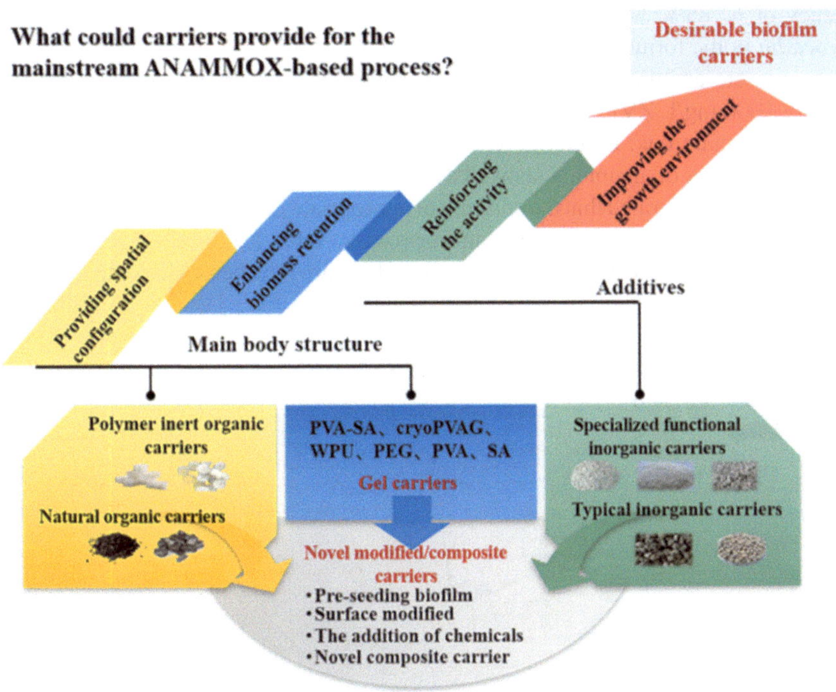

Fig. 1.16 Theoretical structure of the desirable biofilm carrier for mainstream anammox-based process

1.3.3 *Enrichment of Anammox Bacteria by Metals and Intermediates*

1.3.3.1 Iron Material

Zero-valent iron (ZVI) is an economical and environment-friendly element that effectively promotes the growth of AnAOB without secondary contamination or toxicity without causing secondary environmental contamination or toxicity [141, 313–325, 317–319]. Due to its high reducibility (− 700 to 350 mV), ZVI is commonly utilized in wastewater treatment to enhance microbial communities, physiological functions, and overall activity, thereby facilitating the growth of AnAOB [306, 316, 320]. The corrosion of ZVI often coincides with the depletion of DO in water, creating a favorable condition for the proliferation of AnAOB [141]. ZVI also releases ferrous iron (Fe^{2+}) into the water, facilitating the compression and neutralization of the electric double layer. This process facilitates particles aggregation, ultimately resulting in the granulation of sludge (Fig. 1.17). Wang et al. [245, 246, 250, 252] found that the addition of 50 mg/L nano zero-valent iron (nZVI) significantly improved the NH_4^+–N removal efficiency, With the inclusion of nZVI resulted in a removal efficiency of 67.85%, whereas without nZVI, it was only 58.38%. Notably, the Ca. *Brocadia* in the former system increased from 42.1 to 52.5% significantly. ZVI can also promote anammox process quick establish. In a study, it was observed that adding 1000 mg/L ZVI in the system generated a decrease of 11 days in the startup process of anammox. Furthermore, the activity of AnAOB was increased by 20%, and the biomass of anammox showed a significant increase of 54% [75].

It has been illustrated that the addition of appropriate iron ion can promote AnAOB growth, increasing the NRE of anammox process (Table 1.5). Iron ion also enhance the formation of heme c and ferritin in cells, which are indispensable components for

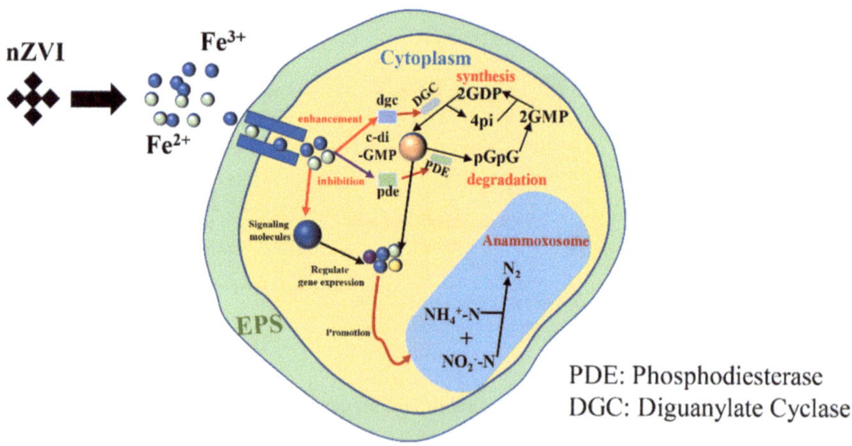

Fig. 1.17 Enhancement mechanisms of iron element addition in anammox process

the metabolic pathway of AnAOB [167]. AnAOB harbors a considerable abundance of compact iron particles, which are likely linked to iron respiration. Therefore, it is hypothesized that the participation of Fe is an appropriate measure for facilitating the enrichment of AnAOB. The existence of Fe^{2+} acquired the reinforcement of marine anammox bacteria (MAB), NH_4^+–N and NO_2^-–N were observed at 4 mg/L in effluent [56]. Furthermore, the removal efficiency of TN increased by 7.77% with Fe^{2+} concentration maintained in 4.5 mg/L in anammox system, in which *Ca. Jettenia* was increased from 2.18 to 3.17% [286]. Suitable quantities of Fe^{3+} improves the activity of anammox significantly. The existence of Fe^{3+} facilitated the growth of MAB under a fluctuating nitrogen loading. Ca. *Scalindua* became the predominant genus in the reactor contained Fe^{3+}, proportionating 27.70%, which was observably better than the control (17.07%) [300].

Ferrihydrite is an insufficiently crystalline Fe^{3+} mineral connected with numerous crucial nutrients (Fig. 1.17). It has the potential to stimulate a unique microbial process involving Fe^{3+} reduction coupled with anammox, known as feammox, which presents a novel pathway for nitrogen removal [56, 300, 313–325, 317–319]. Based on network-based analyses, ferrihydrite was reported to facilitate the development of densely interconnected and complex microbial networks, thereby enhancing the resilience of anammox bacteria to environmental disturbances and facilitating the involvement of cellular pathways. Notably, ferrihydrite also enhanced the relative abundance of Planctomycetes in the reactor, leading to an 11% increase [167]. The addition of nano-scale magnetite was observed to significantly enhance nitrogen removal, with a maximum increase of 33.3% in maximum SAA. This suggests that the presence of nano-scale magnetite may expedite the release of Fe^{2+} or Fe^{3+}, thereby improving the nitrogen removal efficiency in anammox processes. However, the presence of iron ion can also have a negative effect. The study found that $FeCl_3$ or ferric ammonium citrate, as iron sources, exhibited virtually no reaction [313–325, 317–319]. These results showed that Fe^{2+} and Fe^{3+} can impede substance mobilization and metabolic processes [296]. Iron-carbon micro-electrolysis (ICME) is an increasingly popular wastewater treatment technology known for its affordability, operational convenience, and high effectiveness [308, 309, 313–325, 317–319, 321, 322]. The addition of Fe–C particles to anammox systems has been shown to increase AnAOB growth, NRE and ARE was increased to 88.84% and 83.90%, respectively. And, TN removal efficiency increased by 8.4% [141]. The integration system showed an increase of 39.63% in TN removal efficiency. This was accompanied by a shift in the predominant anammox genera, with Ca. *Kuenenia* and Ca. *Brocadia* becoming dominant [276]. In a word, the addition of appropriate concentration iron ions and minerals can promote the enrichment of AnAOB.

1.3.3.2 Other Metals Material

In the anammox system, it is essential that specific trace metal ions participated in coenzymes, synthesizing enzymes, and cofactors components in AnAOB metabolism and proliferation [313–325, 317–319]. Mn^{2+} is an essential trace element required

Table 1.5 Effect of iron ion additives on the anammox performance and anammox bacterial activity

Iron ion	System	Reactor	Substrate concentration (mg/L)	EPS content	Effect	References
ZVI (100 mg/L)	Anammox	UASB	NH_4^+–N:80; NO_2^-–N:80	84.8 mg/g VSS	ARE and NRE attained 88.19% and 88.73%, the control was 58.71% and 50.38%	[141]
ZVI (4 g/L)	Anammox	UASB	NH_4^+–N:208; NO_2^-–N:213	–	NRE was increased from 82 to 86% Ca. *Brocadia* was 18.13%	[313–325, 317–319]
ZVI (1000 mg/L)	Anammox	SBR	NH_4^+–N:100; NO_2^-–N:120	–	The start-up time was decrease by 11 days; the growth rate and quantity of anammox was increased by 20% and 54%, respectively	[75]
nZVI (50 mg/L)	Anammox	UASB	NH_4^+–N:70.23; NO_2^-–N:84.45	143.96 mg/g sludge (PN); 134.00 mg/g sludge (PS)	ARE and NRE was reached by 10% and 15%, respectively	[245, 246, 250, 252, 253]
Fe^{2+} (5 mg/L)	Anammox	UASB	NH_4^+–N:28.54; NO_2^-–N:37.68	189.4 ± 22.3 mg/(g VSS)	SAA was increased to 10.49 mg TN/(g VSS·h), great than control (7.82 mg TN/g VSS/h)	[15, 16, 23]
Fe^{2+} (5 mg/L)	Anammaox	SBR	NH_4^+–N:50; NO_2^-–N:60	46.8 mg/g protein; 4.3 mg/g polysaccharide	NRE increased by approximately 10%	[313–325, 317–319]
Fe^{2+} (1 mM/L)	Anammox	SBR	NH_4^+–N:40; NO_2^-–N:50	43.39 mg/g VSS (PS); 110.4 mg/g VSS (PN)	NRR enhanced from 76.6 to 81.5 g N/(m^3 d) Ca. Brocadia was 29.3%	[210]

(continued)

Table 1.5 (continued)

Iron ion	System	Reactor	Substrate concentration (mg/L)	EPS content	Effect	References
Fe^{3+} (6 mg/L)	Anammox	SBR	NH_4^+–N:85; NO_2^-–N:110	–	SAA was increased by 33.41%	[134, 139]
Fe^{3+} (0.14 mmol/L)	Anammox	EGSB	NH_4^+–N:35.6; NO_2^-–N:60.3	–	NRR was improved by 1.94 times of the control Ca. Brocadia was 3.85%	[245, 246, 250, 252, 253]
Fe^{3+} (3.68 mg/L)	Anammox	UASB	NH_4^+–N:100; NO_2^-–N:100	–	The maximum SAA experienced a increase of 533.2%	[21]

for the metabolism of AnAOB. Aklujkar et al. [4] reported that Mn^{2+} is a significant component of a specific type of superoxide dismutase. Appropriate Mn^{2+} will suppress the proliferation of autotrophic and heterotrophic bacteria, indirectly promote the growth of AnAOB [4]. In a study, it was found that adding 0.05 mM Mn^{2+} resulted in an 8% increase in ARE and NRE [92, 93]. Nickel (Ni^{2+}) shares properties and functions to iron in cellular growth processes. It plays a crucial role in protein synthesis, promoting cell growth, and stimulating metabolic activities within the cells [26, 25]. It was illustrated that in the short-term, Ni^{2+} could significantly restrain the activity of AnAOB. However, over the long-term, AnAOB developed increased tolerance as the dosage of Ni^{2+} decreased. The anammox system achieved an outstanding nitrogen removal efficiency (NRE) of 93.00% with 10 mg/L Ni^{2+} in the long-term operation, still keeping good performance. But, a significant decrease of 38.07% in the abundance of Ca. *Kuenenia* [268, 269]. These findings showed that the substantial impact of Ni^{2+} on the dynamics and composition of the AnAOB population.

In addition to $Mn^{2+}$and Ni^{2+}, appropriate concentrations of K^+, Ca^{2+}, Cu^{2+}, Zn^{2+}, Co^{2+}, and Mo^{2+} can enhance SAA and NRE. However, excessive concentrations have an inhibition effect on microbial. Due to elevated levels of free Ca^{2+}in the bacterial population, when an anammox system is inhibited, the recovery rate and viability of AnAOB was increased with increased the concentrations of Ca^{2+}, indicating a mechanism for the recovery of impaired AnAOB [159]. Hydroxyapatite (HAP) serves as an inorganic carrier for the formation of anammox biofilms, is affected by the concentration of Ca^{2+}, can increase the settling performance and shear strength of anammox granulation, enhancing the stability of the reactor indirectly [308, 309, 321, 322]. Additionally, Ca^{2+} can stimulate the secretion of EPS, thereby enhancing anammox granulation [304]. Adding Ca^{2+} also exhibits an antagonistic effect on Na^+, mitigating the inhibitory impact of salinity. It was reported that the Ca^{2+} with the concentration of 200 mg/L increased the SAA by 78% of AnAOB activity. The addition of Ca^{2+} increases the sequestration of Na^+ within the soluble EPS, thereby decreasing the detrimental effects of salinity. The AnAOB is particularly sensitive to inorganic salts in wastewater. Prolonged exposure to high salinity not only reduces removal property but also compromises the sedimentation property of AnAOB [193]. While to treatment of wastewater containing high NH_4^+–N and salinity using anammox process, the prolonged presence of K^+ hindered the transformation of Na^+, keeping proper cell osmotic pressure. Therefore, the exogenous addition of K^+ may be a batter choice for the enrichment of AnAOB in treating wastewater containing a high salinity.

In general, the addition time, type, and concentration of metal ions will affect the anammox process. The appropriate concentration acclimation time of metal ions can enhance the SAA and NRE of anammox process. Nevertheless, to fully understand the correlations between the activity AnAOB and metal ions, additional research is needed to investigate the additive effects of metal ions and the metabolic characteristics of AnAOB following their introduction. These studies will provide important understanding of the complex relationships between metal ions and the performance of the anammox process.

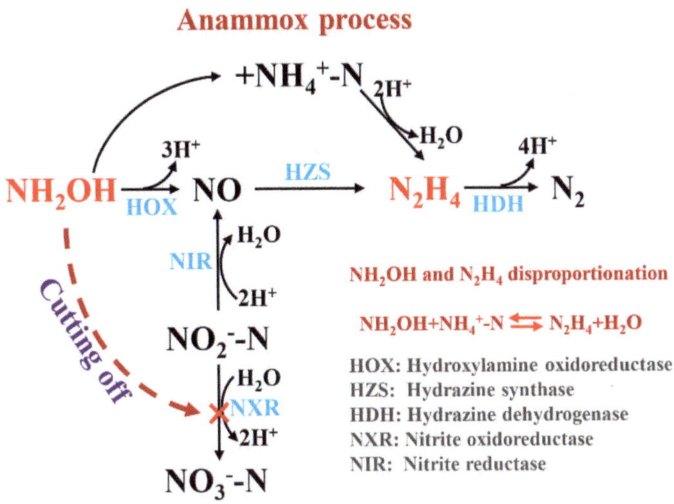

Fig. 1.18 Mechanisms of enhancing anammox metabolic activity through NH_2OH and N_2H_4 addition

1.3.3.3 Intermediates (Hydroxylamine, Hydrazine)

NH_2OH and N_2H_4 is considered a vital intermediate compound in the anammox process. These intermediates play a crucial role in the utilization of the substrate in anammox process (Fig. 1.18). NH_2OH is important for the synthesis of N_2H_4 in the process of NH_4^+–N removal, HOX-catalyzed transformation to NO, disproportionation to N_2H_4, and enhancement of NRE [167, 313–325, 317–319]. In research, adding 8 mg/L NH_2OH contributed to a significant recovery of SAA compared to the control. The SAA values increased from 0.11 to 0.24 kg/ (kg VSS·d). Additionally, the content of EPS decreased from 129.5 to 111.2 mg/g VSS [57]. These findings indicate that the introduction of NH_2OH had a superior recovery effect on the anammox system. Furthermore, NH_2OH at specific concentrations has been found to restrain the activity of nitrite-oxidizing bacteria (NOB), thereby promoting the richness of AnAOB. NH_2OH can facilitate NO_3^-–N removal in the PD system and NO_2^-–N oxidation in the PN system. Consequently, the addition of NH_2OH in both systems leads to the accumulation of NO_2^-–N, offering essential substrates for the anammox process [220, 313–325, 317–319]. It has been found that the PN-A process can be completely restored after 5 days with the addition of 10 mg/L NH_2OH. In the PD system, adding 20 mg/L NH_2OH contribute to a 2.56-fold enhancement in the rate of NO_2^-–N accumulation. However, it was found that adding more than 35 mg/L NH_2OH did not produce a remarkable effect on the reduction of NO_3^-–N [306, 316, 320]. The results indicated that a low concentration of NH_2OH, acts as an electron shuttle, could penetrate the cell membrane of denitrifying bacteria, greatly enhancing NAR activity, while weakening NAR activity in the high concentrations of NH_2OH.

Supplemented N_2H_4 can promote the enrichment of AnAOB and the TN removal, thereby enriching AnAOB effectively [273]. Addition of N_2H_4 in the presence of HDH facilitates the removal of electrons, replacing the electrons participated in in the oxidation of NO_2^-–N to NO_3^-–N. This electron transfer process replenishes the anabolism of AnAOB while reducing the production of NO_3^-–N (Fig. 1.18). In research, adding 7.5 mg/L N_2H_4 can increase the SAA of anammox from 0.42 to 0.72 mg/g/h, which demonstrated that adding N_2H_4 did not demonstrate any noticeable impact on growth of AnAOB. Notably, the initiation period of anammox system reduced by 4 weeks [305]. Similarly, N_2H_4 also have a significant impact on enhancing AnAOB growth and inhibiting NOB. In anammox systems inhibited by NO_2^-–N, adding NH_2OH and N_2H_4 has been found to be more advantageous [10, 313–325, 317–319]. The impact of NH_2OH and N_2H_4 additions on the functional gene *hzo* in the anammox process showed no significant difference. However, N_2H_4 exhibited a more direct and rapid promotion effect on AnAOB compared to NH_2OH. N_2H_4 exhibited a more direct and rapid promotion effect on AnAOB compared to NH_2OH [220].

Appropriate addition of intermediates can inhibit NOB activity, promote the SAA of anammox, stabilize anammox, and achieve effective aggregation of anammox bacteria. Although NH_2OH and N_2H_4 have been demonstrated to reinforce the enrichment of AnAOB, further comprehensive investigations are required to thoroughly examine the potential toxicity of these substances to the metabolic pathways of AnAOB. This research will provide valuable theoretical guidance for enhancing the applications of the anammox process in practice.

1.4 Factors Affecting Anammox Performance

1.4.1 Substrate-NH_4^+–N and/or NO_2^-–N

NH_4^+–N and NO_2^-–N, as essential substrates in anammox process, are especially important in influent [91, 241, 270]. Nevertheless, the high concentrations of NH_4^+–N and NO_2^-–N inhibits activity of AnAOB. The size of sludge particles and the composition of EPS are influenced by the high substrate concentration. As the influent NH_4^+–N concentration was reduced from 300 to 50 mg/L, the NLR of the anammox system exhibited a decrease from 7.42 ± 0.74 to 1.24 ± 0.03 kg/m^3/d. Meanwhile, both the granulation degree of AnAOB and the mass transfer experienced significant reductions [255, 262]. NH_4^+–N, NO_2^-–N, free nitrous acid (FNA) and free ammonia (FA) was reported to inhibit the activity of AnAOB under high concentrations [153, 156, 290, 293, 294]. 100 mg/L NO_2^-–N has been reported to inhibit the growth of AnAOB [297]. And, the anammox process exhibited an inhibitory concentration at half-maximal level (IC50) of 50 mg/L for NO_2^-–N [12]. The anammox process displayed a lower inhibition threshold for NO_2^-–N [340]. NH_4^+–N is found abundantly in various sources such as domestic sewage, printing industries, and landfill leachate [270]. While NH_4^+–N serves as a primary nutrient element in most WWTPs and plays a crucial role in nitrogen removal processes, elevated levels of NH_4^+–N can negatively impact the property of the anammox process. The activity of AnAOB can be inhibited with above 1 g/L NH_4^+–N concentrations. Moreover, when pH exceeding 8.7 in the NH_4^+–N-rich wastewater, and will produce FA, which generates a more inhibitory effect than the NH_4^+–N for AnAOB [153, 156]. Generally, the optimal NO_2^-–N and NH_4^+–N ratio for the anammox process is theoretically estimated to be 1.32. However, in practical scenarios, NO_2^-–N is often prone to consumption through denitrification or oxidation pathway, and the NO_2^-–N and NH_4^+–N ratio is typically exceeding theoretical threshold [20, 22]. Recent studies showed that zeolite presented in the water would adsorb excess NH_4^+–N, buffering the unbalance of the NO_2^-–N and NH_4^+–N ratio in anammox process [95]. The existence of NO_2^-–N can guarantee the anammox process. But, AnAOB can be inhibited with high NO_2^-–N concentrations while without presenting NH_4^+–N. It has been reported that NO_2^-–N had a biotoxic effect, and FNA, the by-product of generated via NO_2^-–N acidification, could change the intracellular pH, disrupt transmembrane potential, lead to cell lysis, as well as inhibit the metabolic activity of AnAOB, influencing the anammox performance [128, 167, 270]. However, Cho et al. [29] reported that the influence of NO_2^-–N on AnAOB is primarily observed as an inhibitory effect, without causing damage to the physical form of AnAOB, such as biofilm or granules. The critical concentrations of NO_2^-–N that can inhibit the anammox process were reported to be in the range from 100 to 400 mg/L, with noticeable inhibitions observed at NO_2^-–N concentrations exceeding 100 mg/L [112]. Yao et al. [297] reported that the anammox process experiences inhibition when exposed to a 100 mg/L NO_2^-–N. In another research, the activity of AnAOB was inhibited when NO_2^-–N concentrations exceed 280 mg/L [270]. The impact of NO_2^-–N on the growth of AnAOB is influenced by the concentration of NO_2^-–N and the physiological conditions of AnAOB, determining the inhibitory effect. Adding intermediates, for instance N_2H_4 at 0.1 mM and NH_2OH, successfully mitigated the detrimental effects of high-concentration substrates on AnAOB [13]. Moreover, alternative approaches offered enough inorganic carbon sources to minimize harm, including the supplementation of exogenous inorganic carbon, which enhanced the NH_4^+–N removal rate and the N_2H_4 accumulation.

1.4.2 Temperature, DO and pH

Temperature is a crucial factor that influences anammox process, with different AnAOB thriving at varying temperatures [128]. The optimal working temperatures range for AnAOB is between 35 and 45 °C. Interestingly, AnAOB has been discovered to thrive under unusual temperature conditions, such as lower temperatures (− 5 to 4 °C) and higher temperatures (60–80 °C) [29]. The secreted EPS can provide a shielding potential to enable AnAOB grow normally under colder temperatures (18 and 13 °C). It was widely believed that Ca. *Kuenenia* was most prevalent bacteria at low temperature (13 °C) [265]. Meanwhile, Ca.*Brocadia* and Ca. *Jettenia*, as the dominant species of AnAOB, were also reported that they can work at 13°C with over 90% TNRR [265]. Due to the cold shock proteins and core enzymes, the Ca. *Brocadiafulgida* were acclimatized at a lower temperature [72, 270]. Certainly, when AnAOB is enriched at a relatively colder temperatures (6 and 25 °C), it operates in a suboptimal range and exhibits lower NRE. However, the decreased nitrogen removal efficiency in the anammox process in lower temperature conditions is improved by closely monitoring pH [270]. The findings illustrated that the anammox process is adaptable to diverse nitrogen-rich wastewater streams, regardless of variations in temperature profiles. Notably, when exposed to the higher temperatures (at 45 °C), the AnAOB can experience irreversible damage due to cytochrome removal. This can significantly impact their ability to carry out anammox activity even if further substrate feeding.

The precise control of DO levels in the reactor system is vital for the proliferation of AnAOB [112]. AnAOB grow and carry out metabolic activities more favorably under strictly anoxic conditions. Studies have indicated that anammox activity can be inhibited by oxygen levels ranging from 0.5 to 2% of air saturation, and recovery is possible when oxygen levels are below 1% of air saturation. Conversely, when the oxygen levels in the air-saturated environment surpass 18%, the suppressive effect on the AnAOB's activity becomes irreversible [50]. Erdim et al. [53] demonstrated that when anammox activity is suppressed by DO, a brief application of nanoscale zero-valent iron can recover the activity of AnAOB rapidly during stable and resilient anammox process operations. Additionally, exposure to DO can own regulate the transcription levels of central metabolism-driven genes in *K. stuttgartiensis*, diverting the energy towards metal transportation and the assembly of flagella as a survival mechanism during anammox activity inhibitions [270].

AnAOB flourishes within a pH range of 6.0–9.0, with the highest NRE achieved at pH 8.0. In both alkaline and acidic conditions, anammox activity reduces but doesn't fully stop. Optimal activity occurs around neutral to slightly acidic pH (6.5–7.5). Extreme pH levels hinder their function, but some tolerance allows limited activity [112, 270]. Notably, the change of pH value can also affect the production of FA and FNA. At pH values below 7, FA concentration tend to decrease while FNA concentration increases; whereas at pH values above 7, FA production increases and FNA decreases [328]. Therefore, it is essential to adjust the pH to neutral levels to preventing FA and FNA from inhibiting the activity of AnAOB.

Overall, understanding the effect of temperature, DO, and pH on the anammox process is crucial to guarantee optimal NRE in wastewater treatment process.

1.4.3 Organic Carbon Source

AnAOB refers to autotrophic microorganism that do not mandatorily require the organic carbon for the nitrogen removal [112, 199, 270]. And, AnAOB are chemoautotrophic microorganisms, they can utilize CO_2 as their sole carbon source [64, 225].

It is widely accepted that organic matter as a negative effect on the anammox process [167, 275]. Meanwhile, the abundant existence of organic matter in the influent stream, such as landfill leachate and monosodium glutamate wastewater can interfere with the AnAOB metabolism, thus affecting the anammox activity [43, 54, 123]. Organic carbon can also facilitate the growth of heterotrophic microorganisms like denitrifying bacteria, which can outcompete possess AnAOB for substrates and limit the relative abundance of AnAOB [167]. However, an appropriate concentration organic matter can facilitate AnAOB enrichment (Table 1.6). Moreover, specific organic compounds like formic acid (FA) and humic acid (HA) have been found to effectively enhance the metabolic of AnAOB [313–325, 317–319]. For instance, it was found that the SAA of AnAOB was increased by 15.5% in the short term by the addition of 5 mM FA [150, 157]. Moreover, compared to the control (42.1%), the TN removal efficiency in FA-amended system reached 58.6%, and the SAA was increased by 2.69 times, with Ca. *Kuenenia* increasing from 9.74 to 19.8% continuously [150, 157]. Adding FA was found to make Ca. *Kuenenia* more competitive in another research as well [84]. Similarly, compared to without HA, the NRE in the anammox reactor increased from 84.1 to 95.3% with COD increased to 684.39 mg/L, while NRR increased from 1.21 to 1.61 kg/(m^3 d) [161, 162]. Furthermore, the richness of Planctomycetes increased with fed 500 mg/L of COD. These results illuminated that HA could increase the relative abundance of the functional bacteria and promote the growth of AnAOB [161, 162]. Specifically, FA and HA have the ability of transfer electrons, which can involve in the bacteria biochemical reaction [211]. Further research has shown that quinones serve as the primary redox molecules in humic substances, playing a critical role in facilitating electron exchange in respiration of microbial. The introduction of quinoid redox mediators at concentrations ranging from 0.05 to 0.8 mM has been found to significantly enhance the activity of essential enzymes in AnAOB, resulting in a 0.6 to threefold increase in their functionality [194]. Therefore, to ensure the long-term stability of anammox process, the content of organic carbon source should be strictly controlled.

1.4.4 Inorganic Carbon Source

AnAOB, as autotrophic bacteria, means they can obtain energy for their growth through inorganic carbon (IC) [14, 167, 200, 201]. In addition, inorganic carbon is also crucial in maintaining pH during the nitrogen removal of anammox process [98, 174]. Previous studies have demonstrated that insufficient levels of inorganic carbon in the influent can decrease anammox activity as a result of the suboptimal utilization of inorganic carbon by AnAOB below specific critical levels [167]. Therefore, it is crucial to supply adequate amounts of inorganic carbon to ensure the development of AnAOB and the long-term stability of the anammox process. Several studies have shown that adding adequate IC sources, such as $NaHCO_3$ and $KHCO_3$, can enhance the activity of AnAOB in anammox-based wastewater treatment systems. For instance, it was reported that adding adequate $NaHCO_3$ in the reactor, the nitrogen

Table 1.6 Effect of organic matter addition on the performance of anammox process and the activity of anammox bacteria

Organic matter type	Concentration (mg/L)	System	Reactor type	Substrate concentration (mg/L)	Effect on nitrogen removal	References
Sodium acetate	20 mg/L	Anammox	UASB	NH_4^+–N:50; NO_2^-–N:50	Anammox reaction was encouraged within 6 h	[20, 22]
Sodium acetate	97.5 mg/L	Anammox	SBR	NH_4^+–N:90; NO_2^-–N:110	The TN removal efficiency experienced a 15% increase	[85]
Glucose	70 mg/L	Anammox	ABR	NH_4^+–N:55.6; NO_2^-–N:59.2	TN removal efficiency improved by 7%	[192]
Glucose	56.4 mg/L	Anammox-denitrification	UBF	NH_4^+–N:82.7; NO_2^-–N:90.8	TN removal efficiency enhanced from 95.3 to 96.7%	[195, 196]
Acetate	0.05–0.2 mM/L	Anammox	–	–	The activity of AnAOB was increased to 116%	[78]
Humic acid	0.25 mM/L	Anammox	Biofilm reactor	NH_4^+–N:100; NO_2^-–N:132	TN removal efficiency improved by 40%	[151, 155]
Humic acid	576 mg/L	Anammox-partial nitritation	–	NH_4^+–N:200	NRR boosted from 1.27 to 1.61 kg/(m^3 d)	[161, 162]
Fulvic acid	1 mM	Anammox	Biofilm reactor	NH_4^+–N:20; NO_2^-–N:20	TN removal efficiency improve by 7%	[150, 157]

removal rate could increase dramatically from 5.2 to 11.8 g/m^3 in only 32 days [295]. Similarly, in another study, it was demonstrated that a considerable decrease in nitrogen removal efficiency (NRE) occurred when the initial concentration of $NaHCO_3$ was reduced from 60 to 2 mg/L. However, the nitrogen removal property gradually recovered after the resumption of IC supplementation [121]. These findings suggest that sufficient IC is essential for the optimal performance of AnAOB. Furthermore, extended supplementation of inorganic carbon can substantially enhance the efficiency of anammox systems. $KHCO_3$ has also been found to be an effective IC source for the enrichment of AnAOB. The introduction of inorganic carbon can effectively minimize pH fluctuations and enhance the buffer capacity of the reactor, thereby enhancing the performance of anammox system [167]. Therefore, adding sufficient IC can not only supports the growth of AnAOB but also protects the system against adverse pH fluctuations, which can improve the effectiveness of anammox process. However, excessive inorganic carbon addition can lead to the production of $CaCO_3$ on the surface of AnAOB, which can affect the diffusion conditions of substrates [167]. High concentrations of inorganic carbon have also been found to decrease the abundance of particular bacteria based on microbial community analysis (Yu et al. 2019). In a word, adding an appropriate amount of IC is a crucial factor for the AnAOB's enrichment of in wastewater treatment systems.

1.4.5 Salinity

Saline wastewaters, resulting from the utilization of seawater as an input, are significant byproducts generated by various industries and municipal water sources. This is primarily attributed to the scarcity of freshwater resources [61, 299]. Industries such as petroleum refineries, shipboard operations, the leather industry, and the fish canning industry are major contributors to saline wastewater, characterized by high salt [94]. High salinity will lead to high osmotic pressure, which can adversely affect microorganism or cause them to be plasmolyzed and dormant [334]. AnAOB are no exception, and high osmotic pressure in concentrated wastewater can significantly hinder the activity of AnAOB [112]. While biological treatment processes are commonly affected by high salinity levels, the anammox process remains a highly promising method for the treatment of wastewater containing elevated salts. Researchers have successful acclimated freshwater AnAOB to elevated salinity levels or enriched AnAOB from marine sediments due to AnAOB existed in anoxic marine environments [218, 219, 280, 281, 284]. It has been reported that the presence of certain salinity levels (NaCl 3–15 g/L) could increase the retention of bacteria in the reactor, promoting the formation of anammox granular, but high salinity inhibited anammox activity [86, 200, 201, 280, 281, 284, 311]. The metabolic activity of the AnAOB can be affected by the higher salinity conditions of wastewater, as the differential osmotic pressure can lead to the loss of cellular water, eventually disrupting structure of the metabolic enzyme structure of the AnAOB [91]. Nevertheless, the anammox process demonstrated improved performance under high salinity conditions through acclimation [334]. Regulating the inhibitory effects of salinity is achieved by providing inorganic ions to the AnAOB or by the acclimatization of AnAOB to high salt conditions. However, both processes are energy-consuming [311]. Some studies have reported successful approaches to enriching marine AnAOB and acclimating freshwater AnAOB [134, 139]. Kartal et al. [117] reported that documented that AnAOB originating from a freshwater environment effectively acclimated to high salt levels of up to 30 g/L. The concentration of salts (90% NaCl and 10% KCl) was raised using concentration increments of 5, 10, 30, 45, 60, 75 and 90 g/L. Compared to the freshwater sludge, the adapted sludge showed comparable anammox activity and efficiency in nitrogen removal. Although the AnAOB can be deactivated at a salt concentration of approximately 45 g/L, the inhibition from high salt was reversible [117]. Notably, the performance of the anammox system may decline over time when operating in a high salinity environment, but acclimatization can improve the tolerance of AnAOB to salt stress [39]. Jin et al. [109] found that discovered that acclimatization could decrease the salinity inhibition of anammox from 67.5% to 43.1%. Ca. *Kuenenia*, which has been extensively studied, has the ability to withstand low levels of salinity in wastewater during the nitrogen removal process. However, the salt adaption ability of other anammox species is unclear and requires further investigation in future.

Two commonly used strategies to mitigate the suppressive impact of salinity on microorganisms are (1) introducing inorganic ions into cells to equilibrate the osmotic pressure of the extracellular environment and (2) producing or accumulating organic solutes [96, 134, 139]. Compatible solutes are small organic molecules that microorganisms can absorb to regulate osmotic pressure and maintain cellular equilibrium [91, 337]. These solutes can be synthesized by the cells themselves or acquired from the surrounding extracellular environment. By acting as protein protectors, they play a vital role in preserving the structure and functionality of intracellular proteins when cells are subjected to salinity-induced inhibition [91]. Glycine betaine stands out as a particularly valuable and efficient compatible solute. It can be readily absorbed by bacteria in saline environments, where it helps improve bacterial activity by maintaining osmotic balance. Notably, a study demonstrated that the prolonged addition of 0.4 mM GB resulted in a significant 30% enhancement in anammox activity [337]. In short, additional research is required to investigate the salt tolerance capacity of other Anammox bacteria species in the future studies.

1.4.6 Antibiotic-Resistant Bacteria (ARB)/ Antibiotic-Resistant Gene and Heavy Metals

Environmental pollutants are also important factors that can affect the functions of AnAOB [270]. Among the various environmental pollutants, ARB (antibiotic-resistant bacteria)/ARG (antibiotic-resistant gene), heavy metals are the most influential. Lower concentration of environmental pollutants has been reported to improve the anammox performance, while high concentrations can inhibit the activity of AnAOB, which can cause harmful impacts on the anammox process [91].

The antibiotic is a typical environmental pollutant that can significantly affect the activity of the AnAOB [63]. Antibiotics are extensively used in the healthcare sector, livestock farming, and aquaculture, leading to their release into the environment [91]. Low concentrations of antibiotic can noticeably enhance the population of anammox and denitrifying bacteria and improve the property of the anammox process, while high concentrations have the opposite effect [247, 249, 260]. Previous studies have shown that administering antibiotics at low doses can significantly boost the abundance of AnAOB and denitrifying bacteria, resulting in enhanced nitrogen removal performance [181]. Whereas, at high antibiotic concentrations, the relative abundances of AnAOB and denitrifying bacteria decrease and the abundance of nitrifying bacteria increase, leading to a decline in the nitrogen removal performance [181]. Furthermore, different antibiotics exhibit varying inhibitory concentrations. For example, tetracycline (1000 μg/L), chlortetracycline (20 mg/L), acetaminophen (400 mg/L), and doxycycline (400 mg/L) have been reported [62]. Bacterial growth is enhanced at lower antibiotic concentrations due to the secretion of EPS, which protect cell structures. However, at higher concentrations, such as oxytetracycline (OTC) at 1000 mg/L, EPS content decreases, impacting the performance of anammox [308, 309, 321, 322]. With adding 10 mg/L tetracycline hydrochloride, SAA shows a 60% reduction in activity, and the NLR decreases significantly [181, 207]. Furthermore, an OTC concentration of 2 mg/L leads to a 50.1% decrease in cytochrome c concentration and complete inhibition of anammox activity at 23–100 mg/L OTC [307]. Antibiotic stress can result in the buildup of ARGs/ARB, which can be transferred to other bacteria through vertical or horizontal gene transfers. This represents a significant risk to human and environmental health [270, 310, 313]. Anammox bioaugmentation can regulate antibiotic stress in AnAOB [62]. Additionally, UV radiation-based disinfection and electrochemical techniques have shown promise in removing ARGs/ARB from anammox systems, providing new insights into antibiotic removal in the anammox process [266, 267].

Metals are commonly found in wastewater and have beneficial or detrimental effects on the activity of AnAOB Low concentration of metal ions, such as Cd^{2+}, Cu^{2+}, Co^{2+}, Zn^{2+}, can promote the activity of AnAOB, while high concentrations will inhibit it, thus reducing the anammox efficiency [153, 156, 183, 310, 313]. Metal ions enter the cell of microorganisms, destroying the structures of enzyme and reducing the cytochrome c levels, thereby reducing the efficiency of AnAOB. However, the inhibition caused by heavy metals can be reversed [280, 281, 284]. Cd^{2+} is known to cause both immediate and prolonged suppression of the anammox process. The IC50 of Cd^{2+} was found to be 5.43 mg/L for acute exposure and the long-term inhibition threshold was 20 mg/L [308, 309, 321, 322]. Co^{2+} improved the performance of the anammox under low concentrations. But it can inhibit the activity of the AnAOB under high concentrations [133, 137]. Cr^{6+} can also reduce the activity of the microorganisms [183]. Zn^{2+} and Cu^{2+} have garnered significant attention due to their potent inhibitory effects and bioavailability to AnAOB. Zn^{2+} has been reported to possess antibacterial properties and the ability to absorb UV radiation

[203]. Additionally, Zn^{2+} serves as a crucial micronutrient for key coenzymes of metal proteinases and specific enzymes [308, 309, 321, 322]. However, high levels of Zn^{2+} can amass within cells and attach to proteins that participate in electron transport systems, thereby reducing the activity of anammox [172, 173]. It has been found that the short-term IC50 of Zn^{2+} was 6.9 mg/L and the long-term inhibition level was 3.39 mg/L [41]. Recent investigations have expanded beyond studying the individual effects of metals on AnAOB to examining the collective impacts of pollutants, such as metals and antibiotics [312]. Notably, one study revealed that the availability and inhibitory effect of Cu^{2+} decreased upon binding with antibiotics [312]. Furthermore, in recent years, research on metals has transitioned from solely focusing on the anammox inhibition to encompass genomic analyses [167]. Restoring AnAOB activity impaired by metal suppression can be accomplished through various approaches. One method involves the recovery of heavy metals from the anammox system. Alternatively, low-intensity ultrasound biostimulation, EDTA washing, and brine addition have also shown promise. EDTA, known for its high attraction to metal ions, which allows it to create metal complexes that could potentially be absorbed onto other materials. This process enables the separation of metals from these substances. Furthermore, EDTA has shown to effectively remove EPS generated by the anammox community [133, 137]. These methods have been employed for the remediation of wastewater containing non-bioavailable metals and have shown promising results in treating bioavailable metals. The periodic introduction of betaine has also been proposed as a superior substitute for EDTA, which is not readily biodegradable [133, 137].

In summary, the enrichment of AnAOB is affected by several factors. The availability of substrates, such as NH_4^+–N and/or NO_2^-–N, is essential for their growth and metabolism. Optimal temperature ranges support their activity, while deviations can hinder their performance. DO levels and pH conditions impact their enzymatic activities and overall metabolic processes. The presence of suitable organic and inorganic carbon sources is crucial for their energy and carbon requirements. Salinity levels play a role in their tolerance and adaptation. The presence of ARB/ARG and heavy metals can have detrimental effects on their growth and metabolic processes. Understanding these factors is important for optimizing the enrichment of AnAOB, particularly in wastewater treatment applications. By controlling substrate availability, maintaining appropriate temperature, DO, and pH conditions, providing suitable carbon sources, and mitigating the impact of salinity, ARBs/ARGs, and heavy metals, it is possible to improve the growth and activity of AnAOB, facilitating efficient and sustainable nitrogen removal.

1.4.7 Microplastics

As urbanization continues to accelerate, emerging pollutants, such as NPs and microplastics, have become a growing concern [222]. They are ubiquitously found in the surroundings, ranging from freshwater bodies to facilities that treat wastewater.

It has been illustrated that these emerging pollutants have a significant impact also on the normal growth and reproduction of AnAOB. Regrettably, the corresponding control strategies are unclear.

Microplastics, which refer to plastic particles that are smaller than 5 mm and are created through the fragmentation of larger plastic items, are a global concern, as they have been found in every corner of the planet, including the deepest parts of the ocean and even in the air we breathe [91, 178]. Microplastics can have a notable impact on the NRE of the anammox process and the structure of microbial community at elevated concentrations [222]. Up to now, microplastics exist in the environment have been divided into two categories: biodegradable and non-biodegradable. Biodegradable microplastics, including fibers and proteins like polylactic acid and polybutenic acid, have been identified [91]. The presence of microplastics can disrupt the structural of EPS and hinder the enrichment of AnAOB. Additionally, nanoscale microplastics have been observed to bind to cell membranes, leading to acute inhibition of key enzymes, disruption of metabolic intermediates and end products, and damage to cellular structures and DNA [306, 316, 320]. Microplastics possess a high specific surface area, which enables them to assimilate diverse contaminants, including polycyclic aromatic hydrocarbons (PAHs), polychlorinated biphenyls (PCBs), metals, and antibiotics. When desorbed from microplastic surfaces, these pollutants can negatively impact AnAOB [91]. Microplastics also act as carriers for contaminants such as antibiotics, heavy metals, and persistent organic pollutants, which can affect the activity of AnAOB when desorbed from the surfaces of microplastic [306, 316, 320].

1.4.8 Nanoparticles

Nanoparticles (NPs) are commonly released into the environment through various industrial applications, including cosmetics, coatings, electronic products, energy materials, and catalytic materials [279, 283, 288]. The NPs mainly cover nZVI, GO, TiO_2 NPs, CuO NPs, ZnO NPs, Fe_3O_4 NPs, carbon nanotubes (CNTs), and Ag NPs. Nevertheless, the impact of NPs on the anammox process varies lied on concentration. NPs can have a positive effect by enhancing the performance of nitrogen removal and increasing the diversity microbial community under low concentrations [152, 154]. For instance, certain NPs like nZVI have been observed to aid in the restore of anammox activity following DO inhibition [287]. Similarly, graphene nanocrystals (GNs) and γ-Fe_2O_3 NPs have been shown to improve nitrogen removal performance by facilitating electron transfer and enhancing the metabolism of hydrazine dehydrogenase (HDH). Due to GNs can shorten the distance of electron transport, and as an electron mediator, γ-Fe_2O_3 nanoparticles can facilitate electron transfer by shuttling between various enzyme [51]. The process of electron transfer plays a critical role, and GNs and NPs can greatly enhance this process, thereby promoting the performance of anammox process. However, NPs restrain the activity of AnAOB under high concentrations. For example, ZnO NPs, commonly applied in pigments, coatings, and cosmetics. This could result in cell membrane damage and a decline in the survival rate of AnAOB [308, 309, 321, 322, 335]. The IC_{50} of ZnO NPs, which represents the concentration causing a 50% reduction in activity, was reported as 84.7 mg/L, with a reversible inhibition concentration of 100 mg/L [203]. Similarly, Ag NPs can adsorb metals like Cu^{2+} and Zn^{2+}, which may subsequently desorb into the environment, leading to increased metal concentrations in the water. This can be harmful to the efficiency of anammox process [158]. Similarly, TiO_2 nanoparticles exhibit a notable Cd absorption capability, whereas Cu nanoparticles interact with EPS by selectively adhering to tyrosine or tryptophan-containing groups via C–O–C and –OH moieties [231]. While NPs can affect the enrichment of AnAOB, their negative impact is relatively easier to control compared to that of microplastics.

In summary, microplastics and nanoparticles are two types of pollutants that can affect the enrichment of AnAOB. Microplastics, whether biodegradable or non-biodegradable, possess a significant surface area and can absorb various pollutants. Consequently, this leads to severe inhibitory impacts on key enzymes, metabolic byproducts, and intermediates of AnAOB metabolism. The anammox process can be hindered by nanoparticles, with the extent of inhibition being dependent on their concentration. While some NPs can facilitate electron transfer and promote the enrichment of the AnAOB, others can lead to increase the production of reactive oxygen species within cells, cause leakage, and ultimately decrease the resilience of AnAOB. Overall, both microplastics and nanoparticles are global concerns that require strict control in the enrichment process of AnAOB to prevent irreversible impacts on their performance.

1.5 Biological Morphology of Anammox Bacteria

AnAOB can exist in various morphologies in reactor systems. It should be noted that different morphologies of AnAOB have been utilized for nitrogen removal. Anammox biomass can be categorized into free cell or flocculent sludge, biofilm, and granular sludge [323, 326]. Flocculent sludge is a common form of AnAOB morphology, and in the majority of full-scale anammox applications, suspended sludge is utilized. The use of granular sludge is advantageous as it facilitates a longer HRT for the biomass with its superior settling velocity and density. This reduces the washout of the AnAOB from the system. Additionally, the biofilms can increase the remain of anammox microorganism and are capable of withstanding adverse conditions, making them particularly suitable for AnAOB [270]. The details are described as below.

1.5.1 Floc-Based Sludge System

During the past few years, there has been a growing interest among researchers in using floc-based sludge as a method for removing nitrogen from wastewater with high levels NH_4^+–N. The full-scale implementation of the PN/A approach in SBR systems that use individual flocs has been successfully demonstrated at various WWTPs. Studies have illustrated that large floc (D > 0.2 mm) exhibits high AnAOB (around 51%) compared to the smaller flocs [290, 293, 294]. In flocs, the AnAOB can access an anchorage-based anaerobic microenvironment, facilitating a higher abundance of AnAOB and hassle-free growth rate [270]. The NRE of anammox was reported to be highest in suspended sludge-based systems, compared to other bacterial forms. Biofilm reactors integrated with flocs have emerged as a promising approach to enhance the performance of nitrogen removal in the anammox process. Malovanyy et al. [177] and Veuillet et al. [240] demonstrated that effectiveness of integrated fixed-biofilm activated sludge (IFAS) reactors in achieving significantly higher nitrogen removal rates compared to MBBRs. In fact, the NRR in IFAS reactors was reported to be three to fourfold increase in efficiency compared to that of MBBRs. This suggests that IFAS reactors are a feasible configuration for implementing the PN/A process. The success of IFAS reactors relies on the cooperative interactions among microorganisms in both biofilms and flocs. Biofilm structures, characterized by their intricate organization and extended detention time, promote the proliferation of AnAOB, whereas flocs tend to promote AOB growth primarily due to their reduced resistance to oxygen transfer [332]. Various operational strategies have been devised to effectively control the separation of microbial growth. For example, selectively discharging flocs can restrain NOB's growth and prevent AnAOB's washout, thereby optimizing the performance of the PN/A process [132]. The diversity of AnAOB in flocs is significantly distinct form that in biofilm, the major genera of AnAOB were *with Candidatus Brocadia* and *Candidatus Kuenenia being the major genera of AnAOB in large flocs and biofilm*, and *Candidatus Brocadia* being the predominant genus of AnAOB in small flocs [290, 293, 294]. Finally, effective enrichment of AnAOB often requires a combination of flocculent and biofilm due to the unique growth characteristics of AnAOB.

1.5.2 Granular-Based Sludge System

Granular-based anammox systems are the most common form of AnAOB, except for biofilm, which is another form that can support dense bacterial communities for uninterrupted denitrification processes (Fig. 1.19) [79, 285]. Granular sludges detected in reactors are generally illustrated by high biomass densities and a uniform external structure compared to flocs (Fig. 1.20) [66]. Granules are more compact and exhibit a robust microbial structure denser, and their high settling velocities, allow for the employment of high hydraulic loads without significant biomass loss. The granular sludge comprises microbial cells, EPS, and porous inorganic substances, among others components [2, 88, 163]. The process of granular sludge in anammox process formation development is depicted (Fig. 1.19). Anammox granular sludge (AnGS) is a type of anaerobic granular sludge that offers several advantages over other forms

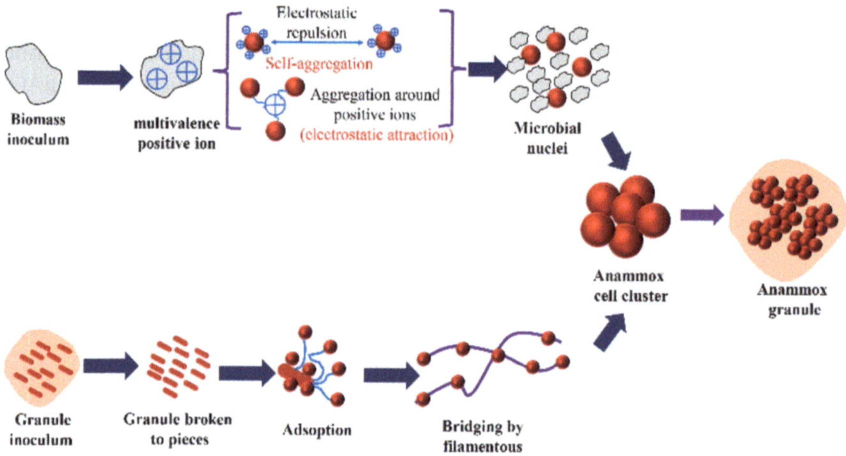

Fig. 1.19 Illustration of the detailed process of anammox granule formation

of AnAOB, such as complete biomass immobilization and minimal biomass production. Using a granular sludge system provides numerous benefits, such as higher volumetric loading rate [324], enhanced specific surface area, improved mixing capabilities and reduced overall investment cost [2]. As mentioned earlier, AnAOB can exist in various forms, including biofilm, microbial clusters, granules, or microbial clusters such as gel carriers [112]. Whereas, in contrast to the requirement for carriers in biofilm or gel entrapment systems, AnGS ensures complete biomass retention without the use of such carriers, unlike biofilm or gel entrapment systems [163]. In recent years, WWTPs employing AnGS with different granule diameter (0.5–25 mm) have shown consistent treatment efficiency for up to 11 years [338]. One effective method of retaining the biomass of anammox within the reactor is by granulating it. This approach has proven to be successful in treating ammonium-rich wastewater at the scale of laboratory, pilot and full [2]. In fact, the initial implementation of a large-scale AnGS took place in Rotterdam, The Netherlands in 2002, and achieved a final NRR of 9.5 kg/m^3/d after 3.5 years, as opposed to the traditional anammox process required two years [232]. Similarly, a successful startup was achieved by seeding anaerobic sludge in EGSB a brewery wastewater treatment facility, resulting in continuous NH_4^+–N removal after three months of operation [244]. A pilot-scale anammox reactor was inoculated with foreign manure granules, resulting in a high nitrogen removal rate of 27.8 kg/m^3/d after only two weeks. Real-time polymerase chain reaction (PCR) analysis showed that 91.4–92.4% microbial community was composed of AnAOB (Ni et al. 2013). Researchers have also investigated the efficiency of anammox in UASB reactors under low ambient temperatures [110, 111]. For example, introducing 2 ml of exceptionally potent anammox sludge every 2 days at a temperature of 9.1 °C, a NRR of 6.6 kg/m^3/d was recorded [169]. In a lab-scale UASB, a super high-rate performance was achieved, with a NRR ranging from 74.3 to 76.7 kg/m^3/d [223]. Over the years, to help several models have been developed

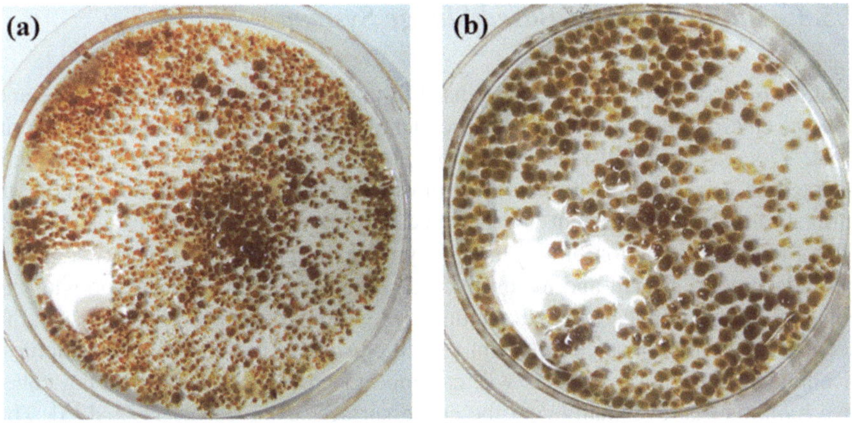

Fig. 1.20 The morphology of granular-based anammox

to explain the granulation process of anammox, including the selection pressure model, extracellular polymers (ECP) bonding model, inert nuclei model, synthetic and natural polymer-bonding model, attrition model, secondary adsorption minimum model, multivalent cation-binding model, and interfacial tension model [2].

Compared with biofilms, granulated-based anammox has higher stability due to the high resistance of the aggregated particles. For instance, the granular-based sludge systems could acquire around 90% NRE within 45 days operation at under temperatures of about 13 °C [265]. The shapes of anammox granule are irregular, ranging from three-dimensional spheres to uneven surface structure. The carmine-colored granules usually have high anammox activity. Furthermore, the diameter of AnAOB granules influences settling performance of bioreactor [338]. Chen et al. [15, 16, 23] reported that increasing particle size resulted in the microbial community's richness and diversity within the reactor, followed by an increase. However, if the particle size is too small, it can be challenging for the anammox sludge to increase biomass. It has been confirmed that the anammox granular ranging in size from 0.5 to 0.9 mm is suitable for treating sewage with low ammonia nitrogen content [338]. In another investigation, the highest biomass and proportion of AnAOB were observed in granules with particle sizes ranging from 0.5 to 1.0 mm [15, 16, 23]. Notably, the presence of larger-sized granules can produce the formation of voids or empty spaces within the sludge, potentially causing buoyancy issues [245, 246, 250, 252, 253]. However, these granules possess distinctive benefits in terms of their activity and stability. In a word, granulation is a crucial method for enhancing the enrichment of AnAOB.

1.5.3 Biofilm Carrier-Based Sludge System

Biofilm is a common form of microbial life that helps microorganisms with sand adverse environments and facilitates product and signal exchange between bacteria [172, 173, 327]. In full-scale anammox processes, biofilm formation is widely adopted as it is considered one of the most practical and widespread methods to mitigate biomass detachment and enhance the microbial retention capacity [138, 145, 146]. The formation of biofilm is a dynamic progression that involves microbial attachment, growth and mature [65]. In summary, individual microorganisms adhere to the carrier, leading to the initial aggregation of biomass. Subsequently, the accumulated biomass can develop into small clusters of biofilms or microcolonies. Eventually, these microcolonies will continue to grow and merge, forming cohesive biofilms that eventually coat the entire surface of the carrier [172, 173]. The quality of biofilm significantly influences the stability of the biofilm system, although limited studies have focused on the formation of superior. Key measurable parameters to assess biofilm quality include formation duration, ecological and mechanical resilience, and the predominance of key microbial species. These parameters can be influenced by various factors, including microbial attachment, the compatibility between microorganisms and the carrier material, surface texture, hydraulic conditions, and even the use of physical aids [202, 268, 269]. Enhancements in the selection and optimization of materials or microorganisms can address challenges associated with intrinsic factors. Nonetheless, the design of biological systems can still be tailored systematically according to the traits exhibited by the predominant bacteria. A multitude of research has demonstrated that AnAOB can thrive within a biofilm structure by aggregating with other bacteria [199]. In different types of reaction systems, the utilization of biofilm carriers has emerged as the most practical approach to enhance the anammox performance. This includes promoting the growth of anammox populations and reducing the time required for system start-up. Among them, carriers play a vital role as essential substrates and support materials for enriching AnAOB [292]. In previous studies, the effectiveness of different substrates (polyurethane sponges, volcanic rocks, zeolites, ceramics, and plastic rings) has been explored to effectively enrich AnAOB [74]. With the high surface area of carrier could facilitate better anti-shock loading, its intricate structure can also pose limitations to mass transfer, hindering the growth of AnAOB [36]. Therefore, researchers are developing novel carriers to expedite the establishment of anammox biofilms and enhance the performance of AnAOB, aiming to achieve enhanced nitrogen removal efficiency. Adding redox-active biochar could prevent biomass loss and maintain stable immobilization of AnAOB, while promoting the secretion of EPS, thereby enhancing the proliferation of AnAOB [282]. Similar results were confirmed in another research, where the addition of iron-based modified carriers was reported to promote higher biomass growth with enhanced EPS production, enriching effectively AnAOB, strengthening the anammox performance. Recently, Li et al. [136] proposed a novel modified carrier made of polyurethane sponge adhered mixing zeolite and tourmaline 3:1, which could achieve a NRR of 0.150 kg/(m^3 d) and

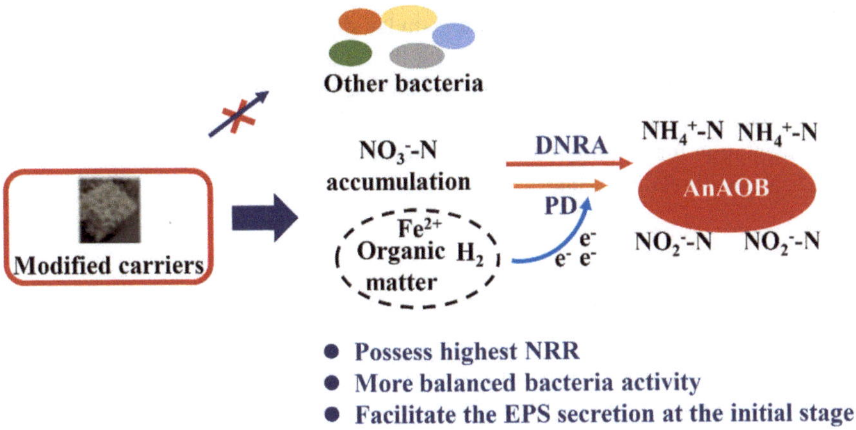

Fig. 1.21 Detail mechanism of modified carrier (polyurethane sponge adhered mixing zeolite and tourmaline with ratio of 3:1) promoting anammox [136]

establish anammox process can establish anammox process within 28 days. Importantly, this carrier can effectively enrich AnAOB in activated sludge containing a variety of microorganisms (Fig. 1.21). This study may develop a new method to enrich AnAOB, namely targeted screening of AnAOB. However, the mechanism underlying this process remains unclear and requires further studied in the future. In conclusion, biofilm serves as an effective means to rapidly enrich AnAOB, and the development of new environmentally carriers is a crucial step for their future application.

In summary, each form of AnAOB has its unique characteristics. Floccule, for example, has a larger surface area to volume ratio, making it more efficient in removing nitrogen. Granular is known for its high biomass and high stability. Biofilm, on the other hand, is more resistant to environmental changes and more stable, making it ideal for long-term nitrogen removal. Understanding the biological morphology of AnAOB is essential for enriching AnAOB and optimizing the performance of anammox-based process. Future research can focus on exploring the interaction between the biological morphology of AnAOB and their metabolism, as well as developing new bioreactor designs that can effectively enhance the growth and activity of AnAOB in different morphological forms.

1.6 Application of Anammox in Wastewater Treatment

With the popularization of anaerobic ammonium oxidation technology, this new type of self-sustaining denitrification approach has been effectively implemented for the treatment of diverse wastewater streams, such as sludge digestion liquid, landfill leachate, saline wastewater, pharmaceutical wastewater, urban sewage, etc.

Among them, the most widely used are sludge digestion liquid and urban sewage. The utilization of anammox technology for the remediation of industrial wastewater, including but not limited to leather processing, semiconductor and food processing, coking, pharmaceutical, and swine wastewater, has gained traction. However, its implementation for treating industrial wastewater containing high concentrations of NH_4^+–N is still relatively rare.

1.6.1 Sludge Digested Liquor

Sludge anaerobic digestion liquid and sludge dewatering filtrate are complex wastewater streams characterized by low C/N ratios [200, 201]. Sludge digestion liquid contains 500–2000 mg/L NH_4^+–N and organic matter at approximately 2000–3000 mg/L, which is poorly biodegradable [256]. The pH range of sludge digestion liquid generally falls between 7.0 and 8.5, and the temperature ranges from 30 to 37 °C, providing favorable conditions for the proliferation of AnAOB and AOB [256]. In fact, the applicability of PN coupled anammox processes was first explored for treating clarified sludge liquid from the Dokhaven wastewater treatment plant in the Netherlands [234] (Fig. 1.22). The effective denitrification results obtained from treating clarified sludge liquid from the Dokhaven wastewater treatment plant prompted the construction of the world's first operational PN/A integrated reactor. This reactor had a capacity of 70 m^3 and could process 750 kg/d [232] (Fig. 1.22). Since then, the implementation of the anammox process for sludge treatment has gained popularity in Europe. In a study, stable PN process was achieved by carefully controlling the concentrations of FN and FNA, resulting in a NRR of 1.23 kg/m/d in the pilot-scale reactor. The treatment efficiency reached 98.82% for wastewater with ranging from 2100 to 2200 mg/L NH_4^+–N [256]. Sludge liquid treatment has become increasingly popular for applying the anammox process due to its advantageous features, including elevated liquid temperatures, reduced liquid volumes, concentrated NH_4^+–N levels, and low C/N ratio. Currently, around 75% of the global anammox engineering systems are specifically designed for treating sludge liquid, indicating the maturity and the widespread adoption of anammox in this domain [200, 201]. Nevertheless, some technical challenges remain in the widely application of the anammox process for sludge liquid treatment. These challenges include investigating the impact of sulfide anaerobic digestion on anammox process, understanding the links between nitrogen oxide generation, and developing effective measures to reduce emissions. Future research and development in this field could further optimize the anammox process for treating complex wastewater.

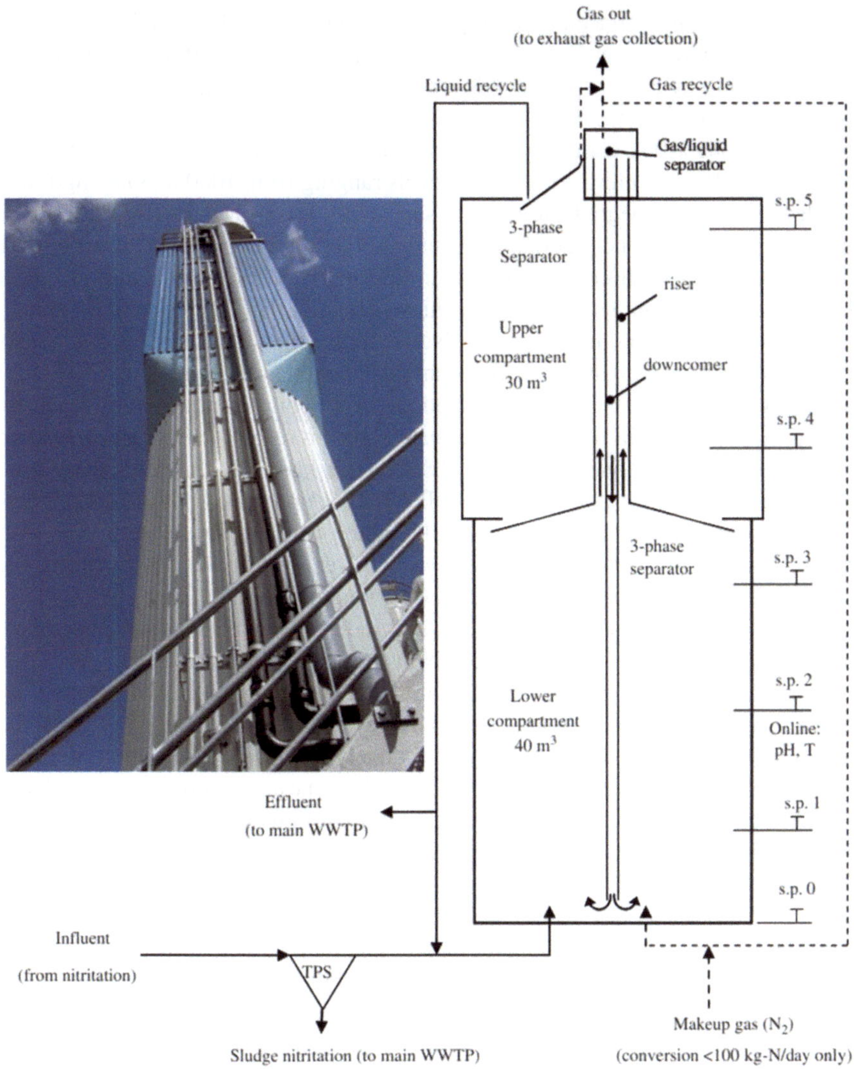

Fig. 1.22 The full scale (70 m^3) anammox reactor in Rotterdam for treating 500 kg/d sludge liquid [232]

1.6.2 Landfill Leachate

Landfill leachate is a complex wastewater with high concentrations of organic matter, toxic substances including heavy metals, variable water quality, high NH_4^+–N, salinity, and low biodegradability [43, 101, 167, 200, 201]. The landfill leachate is usually categorized into young (≥ 0.6), middle (0.1–0.6), and mature (≤ 0.1) based on the BOD_5/COD [165]. The anammox process was found to be suppressed by the elevated by the high concentration organic content in landfill leachate (Fig. 1.23). Even so, anammox can well adapt for treating landfill leachate. When the anammox process used to treat landfill leachates, Ca. *Kuenenia* existed stably [184, 191]. This may be Ca. *Kuenenia* has the high affinity for substrate, which is more adaptable to landfill leachate [100]. Currently, significant quantities of mature waste leachate are generated daily, with elevated NH_4^+–N levels ranging from 1000 to 3000 mg/L and low levels of easily degradable organic substances [165]. Due to landfill leachate include a high TN concentration, universally 1000–3000 mg/L, the effluent of pure anammox process will contain more than 200 mg/L of NO_3^-–N, failing to meet the discharge regulations. Hence, to achieve effective landfill leachate treatment, anammox requires a combination of diverse approaches, for instance, anammox-based PN or PD and anammox-based constructed wetlands. Jiang et al. [105–107] and Ren et al. [200, 201] both investigated the treatment of mature landfill leachate. The former study utilized a sequencing batch biofilm reactor (SBBR), achieved an NRE of 98.6 ± 0.2% and an NRR of 2.83 ± 0.06 kg/m^3/d. The latter study used an integrated fixed film activated sludge (IFAS) PN/A process, obtained a total inorganic nitrogen removal efficiency of 98.1% and a COD removal rate 52.9% with the operation of 301–405 days. In recent years, several researches have explored the use of various techniques for treating mature landfill leachate. Zhang et al. [313–325, 317–319] used a combination of three independent SBR to develop a new process called PNFA, which integrates PN in situ fermentation, and anammox. This integrated system demonstrated remarkable NRE of 99.2 ± 0.1%. Furthermore, by implementing in-situ liquid sludge partial denitrification and anammox process, the build-up of NH_4^+–N and NO_3^-–N was effectively mitigated, resulting in excellent NRE (99.6%) and NRR (822 gN/(m^3 d)) [105–107]. The diverse composition of landfill leachate, coupled with the existence of microbial films, can mitigate the concentration of harmful substances, thereby creating favorable conditions for the coexistence of denitrification bacteria and AnAOB. This enhanced interaction enhances the performance of the denitrification coupled anammox. Other studies have observed significant progress in the treatment of leachate wastewater, with reported mean removal rates of NH_4^+–N, TN, and COD at 94%, 88%, and 26% respectively. Additionally, an impressive NRE of 99% has been achieved [105–107]. Furthermore, with continued research, was by the Hubei Shiyan landfill leachate treatment plant implemented a two-stage UASB based anammox process, which utilizes anammox, a membrane bioreactor, and reverse osmosis. This integrated process effectively regulated the NH_4^+–N concentration in the effluent to 25 mg/L [270]. The implementation of the two-stage UASB based anammox process has proven to be a groundbreaking approach in effectively treating landfill leachate with a low C/N ratio. This innovative method showcases the energy efficiency and positive attributes of anammox technology within the scientific community.

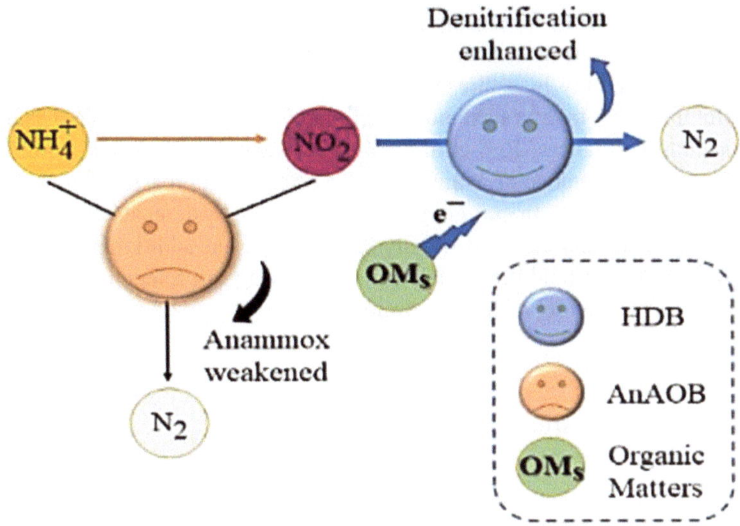

Fig. 1.23 Inhibition of the anammox process by high concentration organic matter in landfill leachate. It can be concluded that the integration process based on anammox process is commonly used to treat landfill leachate. And researchers are continuously exploring various combination of technologies. However, the inhibition substances of landfill leachates produced on the impact of the AnAOB and the effective mitigation strategies need to be further studied

1.6.3 Saline Wastewater

As industrialization continues to develop, the release of industrial effluents into waterbodies has become a growing concern. These effluents, originating from industries such as seafood processing, textile printing, pharmaceutical industries, tanning, farming and landfill leachate, often exhibit elevated concentrations of NH_4^+–N and salinity [149]. Research has demonstrated that augmenting the K^+ concentration in saline industrial wastewater can effectively enhance the treatment efficiency of anammox systems for such effluents [114]. The presence of high salt concentrations can greatly influence the activity and stability of AnAOB. In research, it was reported that on an integrated process that couples nitridation with anammox to employed to assess the nitrogen removal efficiency of freshwater nitrifying anammox sludge in the presence of elevated salt levels. The process was shown to gradual adaptation to a 3% saline environment by enhancing aeration, but growth of AnAOB was inhibited under a 5% salinity, resulting in the failure of the system [70]. In recent studies, a novel salt-tolerant PD/A process was developed, demonstrating the sustained high activity of both partial denitrification sludge and anammox sludge under a 10 g/L salinity. This innovative approach achieved a remarkable NRE of 90%, offering a promising research direction for effectively treating industrial wastewater containing salt [280, 281, 284]. However, it was observed that a salt concentration of 30 g NaCl/L was the critical limit beyond which the stability of the anammox reactor was compromised [168]. The anammox process is considered an ideal solution for treating industrial wastewater containing high NH_4^+–N concentrations and salinity levels, with the salinity load threshold ranging from 30 to 75 gNaCl/L. Achieving successful treatment involves extended acclimatization periods and the introduction of compatible substances. Ren et al. [200, 201] support this notion, emphasizing the widespread acceptance of anammox as an effective approach for such wastewater treatment scenarios. In summary, anammox-based systems have demonstrated the ability to withstand elevated salinity levels. However, additional investigations are required to understand the mechanisms underlying salt tolerance that influence stability of AnAOB under different salinity conditions.

1.6.4 Coking Wastewater

Coking wastewater is a concentrated and complex wastewater that is notoriously difficult to treat. This type of wastewater is generated by the coke production reaction, coal gas purification, and coke product recovery processes, and it contains high levels of organic matter [134, 139, 270, 333]. Coking wastewater is characterized by the presence of diverse inorganic and organic contaminants, including sulfate, cyanide, polycyclic aromatic hydrocarbons, phenolic compounds, and nitrogen-containing polycyclic compounds [197]. Recently, anammox process has shown promising results in treating coking wastewater with a nitrogen removal rate up to 98% [270]. However, the presence of harmful constituents in the coking wastewater can inhibit the activity of AnAOB. For example, phenol and sulfide are kwon to exhibit inhibitory impacts on the anammox process. The prolonged exposure to 32 mg/L sulfides was reported to inhibit the anammox activity. When thiocyanate (< 50 mg/L) is present in the anammox process, a thiocyanate-based denitrification would compete for NO_2^-–N with anammox process [302]. To address these challenges, researchers have developed various integrated processes to treat coking wastewater. For example, Zhou et al. [333] used a single-stage system that integrated PN, denitrification and anammox to treat phenol-stressed coking wastewater and achieved NRE of 89.8% phenol and 97.5% for COD. Due to the existence of toxic compounds in the coking wastewater, it needs researchers to experiment with different approaches and models, which making treatment process difficult. In addition, specific bacterial strains can be selected based on different sensitivity to different toxic compounds. For instance, it has been reported that NOB is more susceptible than AOB, suggesting that thiocyanate can be used to achieve stable nitridation [270]. While anammox has demonstrated effectiveness in treating coking wastewater, the presence of toxic compounds remains a challenge for the growth of AnAOB.

1.6.5 Pharmaceutical Wastewater

Likewise, pharmaceutical wastewater also poses a significant challenge to the anammox process due to its high levels of ammonia, organic contaminants, and the presence of antibiotics [8, 204]. Among all of the pharmaceutical wastewater components, antibiotics have the most significant impact on the treatment efficiency of the anammox process [270]. High concentrations of tetracyclines such as oxytetracycline and chlortetracycline, have an irreversible effect on AnAOB, with oxytetracycline showing a more pronounced disruptive effect on anammox granules compared to chloramphenicol [62]. However, Zhang et al. [323, 326] reported that AnAOB was not caused a considerably inhibited by 1 μg/L of erythromycin, and instead, it protected the cells and inactivated the antibiotic by inducing the genes of antibiotic resistance (ermB and mphA), thereby protecting the AnAOB cells and rendering the antibiotic ineffective. The acclimatization of AnAOB can develop antibiotic resistance through the EPS secretion as protective mechanism [323, 326]. Furthermore, Zhang et al. [313–325, 317–319] demonstrated the successful treatment of pharmaceutical wastewater using a PD/A SBBR reactor system, achieving an NRR of 80.8%, and an anammox contribution rate of 83.6%. However, the high COD/NH_4^+–N in pharmaceutical wastewater remains an obstacle for anammox treatment. In the treatment of ascorbic digestion-treated vitamin B2 wastewater using anammox, researchers achieved a NLR of 0.35 kg/(m^3 d) and obtained the highest TNRE of 75% [176]. However, despite recent efforts to overcome anammox inhibitions in the treatment of pharmaceutical wastewater, the effective treatment of pharmaceutical wastewater using anammox remains a significant obstacle.

1.6.6 Urban Sewage

Given the rapid progress of urbanization, effective treatment of urban sewage and the harnessing of energy have emerged as pivotal areas of investigation [235, 270]. Anammox process application is anticipated to keep WWTP to attain energy self-sufficiency [112, 270]. At Garmerwolde wastewater treatment plant in The Netherlands (Fig. 1.24) two anammox granule reactors are operating, with conditions of more than 25 °C and more than 0.1 g/L nitrogen concentrations [232]. However, it has been illustrated that the anammox process would have a lesser nitrogen removal efficiency with operated at lower temperatures (< 25 °C) [71]. Mainstream urban sewage presents challenges such as diluted NH_4^+–N levels, fluctuating nitrogen loads, and temperature variations corresponding to different seasons. Even the Ca. *brocadia*, a notable AnAOB, takes a long time to adapt to lower temperatures environment, and its proliferation requires a higher nitrogen consumption for sustainable NRE [71]. Therefore, to successful implement the anammox process in the mainstream urban sewage treatment, additional investigation is required to explore methods that enable

Fig. 1.24 Two reactors at the wastewater treatment plant Garmerwolde in The Netherlands (treating wastewater of 235,000 persons) [235]

efficient operation at a lower temperatures and nitrogen concentration, is need to facilitate energy production in WWTP.

Up to now, the achievements of the anammox process have been obtained with the wastewaters having a high NH_4^+–N concentrations and a C/N ratio > 0.5 [171]. However, implementing anammox for mainstream applications, for instance municipal wastewaters treatment, with a low C/N relies on the influent concentration of organic substances [34]. To avoid adverse effects on AnAOB caused by organic matter, municipal wastewater is often pretreated before the implementation of the anammox process, such as the PN/A process [209]. Hu et al. [90] also implemented the integrated PN/A process, where the SBR successfully acclimated to the low temperature of 12 °C within a short period of 10 days. The system-maintained stability at this temperature without nitrite accumulation remaining stably under this temperature with no accumulation of nitrite, and achieved a NRE of over 90% over more than 300 days operation. In another study, it was showed that a laboratory scale anammox process could acquire an NRR of 6.61 kg/m^3·d at 9.1 °C with gradual cooling and acclimation, or by introducing a low-temperature protective agent such as glycine betaine. Gao et al. [65] observed that combining an anoxic–oxic (AO) process with a PD/A process could effectively maintain stable treatment efficiency in municipal wastewater treatment, even under varying temperatures ranging from 26.8 to 13.1 °C. However, during the initial treatment phase, the excessive production of NO_3^-–N by AnAOB and NOB can result in the accumulation of NO_3^-–N, due to the limited availability of carbon sources for removing NO_3^-–N in the later stages [184]. Although AnAOB has been found in various WWTPs, directly enriching AnAOB is challenging because of its sluggish growth rate and susceptibility to environmental factors. Furthermore, the NH_4^+–N environment in municipal wastewaters hinders the proliferation of AnAOB [71]. Currently, research has shown promising results for implementing the anammox process at ambient temperature or a reduced temperature, which could help WWTPs in achieving energy self-sufficiency. Nonetheless, the competitive mechanism between AOB and NOB under low temperature and low substrate concentration conditions in practical applications remain not fully understood. Therefore, further research strategies are need to explore the utilization of anammox process in municipal wastewater treatment.

1.6.7 Other Wastewaters

Swine wastewater is known to have a high concentration of NH_4^+–N (1120 mg/L) [15, 16, 23, 330, 331]. To address this, PNA processes such as one or two step SBR PNA process and simultaneous PN, anammox and denitrification process have been widely used. In a particular investigation, a solitary-stage biofilm reactor was implemented to couple anammox with intracellular carbon assimilation, leading to significant removal of nitrogen and phosphorus elements. This process demonstrated efficient performance even with a low C/N ratio [15, 16, 23]. Meanwhile, for semiconductor wastewater treatment, Li et al. [138, 145, 146] established integrated anammox and sulfur autotrophic reactors that attained a 3.1 kg/(m^3 d) NRR in anammox process. In research, a novel approach was introduced to convert the conventional nitrification–denitrification process into a PN/A-denitrification process. This transformation was accomplished through selectively inhibiting nitrite oxidation using a high bicarbonate concentration. This innovative method allowed for the promotion of PN while facilitating denitrification (Toyama et al. 2011). The CANON process was employed to treat high NH_4^+–N concentration in photoelectric industrial wastewater. They utilized an 18 L SBR and maintained a basicity level of 850 mg/L $CaCO_3$ through the addition of $NaHCO_3$. With the 3636 mg/L NH_4^+–N in influent, they achieved an impressive NH_4^+–N removal rate of 98% and a NLR of 0.909 kg/m^3 d [42]. Furthermore, sulfate can be transformed into hydrogen sulfide by sulfate-reducing bacteria, which can strongly inhibit the activity of AnAOB. Hence, anammox is typically not suitable for direct remediation of monosodium glutamate wastewater but may be used for subsequent treatment.

In summary, we have explored the successful implementation of anammox processes in diverse wastewater treatment scenarios, encompassing both small-scale laboratory experiments and larger-scale pilot studies (in Sects. 3.1 and 1.6). However, wastewater is a complex mixture containing a variety of contaminants, such as heavy metal ions, antibiotics, cyanide, phenols, and organic nitrogen. For instance, landfill leachate has relatively high concentrations heavy metal ions. Coking and petrochemical wastewater, as well as tannery wastewater, contain various toxic compounds such as cyanide, tar, phenols, organic nitrogen, and heavy metal ions. Inhibitors such as biological inhibitors, sulfide, and phosphates are also present in wastewater generated from seafood processing, tanning, papermaking, fermenting alcohol, aquaculture, fecal sewage, municipal wastewater, chemical fertilizer, and pharmaceutical production. These pollutants pose difficulties in utilizing anammox technology to treat industrial wastewater with high levels of NH_4^+–N. Although several factors inhibiting AnAOB have been identified, the inhibitory thresholds of these inhibitors differ based on different strains or experimental conditions, and their reversibility is also controversial. The combined impact of these inhibitors on AnAOB has not been established, and there is limited research on regulatory strategies associated with it. Furthermore, there is no final consensus on the selection between integrated and separate process configurations for wastewater with varying water characteristics, and operational data to support practical proof is still lacking.

1.7 New Discovery in Anammox Bacteria

The AnAOB has a critical effect on the nitrogen removal in wastewater and industrial effluents. Recent research has revealed new insights into the diversity and metabolic capabilities of AnAOB, including the discovery of new species and metabolic pathways. These findings discoveries have significant ramifications for the optimization of anammox-based bioreactors and the devising approaches for treating nitrogen-rich wastewaters, as well as for understanding the ecology and evolution of AnAOB in natural environments.

1.7.1 New Discovery on the Metabolism via AnAOB

AnAOB convert NH_4^+–N and NO_2^-–N into N_2 through a chemolithoautotrophic process, relying exclusively on CO_2/HCO_3 solely as carbon sources. This autotrophic nature restricts their competitiveness against heterotrophic microorganisms in both natural ecosystems and engineered setups. Recent environmental metagenomic studies have provided evidence of the ability of AnAOB to utilize short-chain fatty acids as a carbon source. This finding has been further supported by experimental investigations conducted on highly enriched cultures.

A microorganism's mixotrophic lifestyle is a selective advantage in nature, and such capability is especially important to species that are critical to global nutrient cycles [225]. Versatile lifestyles have been discovered in microbial cells because of the development of observation power [83], which expanded our knowledge on microbial functionality and niche. The discovery of an organotrophic lifestyle by AnAOB is often challenged by interference from heterotrophic microorganisms that co-exist in any anammox consortium. Direct observation of organotrophic processes on a single-cell level could deliver profound information about the actual physiology AnAOB. However, such studies are scarce. In recent years, nanometer-scale secondary ion mass spectrometry (NanoSIMS) has enabled to gain an in-depth view of microbiology, for example, to track microbial interactions and to study microbial metabolic activities [122, 187]. In a study, using a NanoSIMS to visualize and quantify the incorporation of ^{13}C-labelled acetate and propionate by a highly enriched anammox consortium that was dominated by Candidatus Jettenia asiatica [225]. This microorganism has not yet been reported relating to organotrophy. The results that were based on 279 scans of 4885 cells strongly indicated that the anammox consortium incorporated acetate and propionate at a rate even higher than the uptake rate of bicarbonate. In a mainstream anammox system, organic matter present in wastewater plays a key role in governing the balance between AnAOB and heterotrophic microorganisms such as denitrifying bacteria. Although it has been accepted that AnAOB are protected by heterotrophs when organic carbon is present [140], this study confirms that AnAOB are capable of dealing with organic matters by their own. The niche of AnAOB in engineered ecosystems can therefore be much broader than general assumed. It also reminds us that their actual contribution to the global nitrogen turnover warrants more exploration.

Specifically, the enrichment of *J. asiatica* was carried out in an expanded granular sludge bed (EGSB) reactor that was fed with a standard medium for cultivating AnAOB. No organic matter was added to the EGSB. Red-colored granule-like aggregates were formed in the EGSB, and the size of granules increased to and stabilized at 1–2 mm during the cultivation (Fig. 1.25a). Some granules contained interstitial voids (Fig. 1.26) that might serve as water or gas channels, benefitting to mass transportation [163]. Fluorescence in situ hybridization (FISH) analysis confirmed that the microbial consortium was dominated by anammox bacteria with a mean relative abundance of 96.0% (Fig. 1.25b), identified as *J. asiatica* by 16S rRNA sequencing (GenBank: KJ002641.1). A classic anammox cell architecture and compartmentalization were confirmed by TEM (Fig. 1.25c), clearly showing anammoxosome as the central organelle [236, 237]. The ammonium-oxidation capacity of the EGSB increased by approximately ten times after a long-term operation (Figs. 1.25b and 1.26). Meanwhile, the qPCR analysis showed that the *hzo* gene copy numbers of the EGSB biomass increased by over five orders of magnitude (Fig. 1.25d). The *hzo* gene encodes for the enzyme hydrazine oxidoreductase converting hydrazine to N_2 [135], which is a unique metabolic process specific for AnAOB [118]. The ratio of the consumed NO_2^-–N to the consumed NH_4^+–N stabilized in the range of 1.30–1.37, while the ratio of the produced NO_3^-–N to the consumed NH_4^+–N was consistently in the range of 0.20–0.25, both of which agreed with known anammox stoichiometry.

In conclusion, *J. asiatica* was proven to be able to take-up acetate and propionate in the study. NanoSIMS analysis of an inorganic/organic incubated, highly enriched anammox granular consortium, reveals efficient incorporation of acetate and propionate. Although it is still an enigma how acetate and propionate are metabolized and how energy electrons flow in the cell, the results offer essential insight to the versatility of AnAOB. Here, the shown organotrophic, but unlikely a heterotrophic, lifestyle of anammox bacteria strongly supports a promising niche of AnAOB in a mainstream anammox process where organic matter is present. Follow-up studies about how such versatility could enhance the growth of AnAOB are needed.

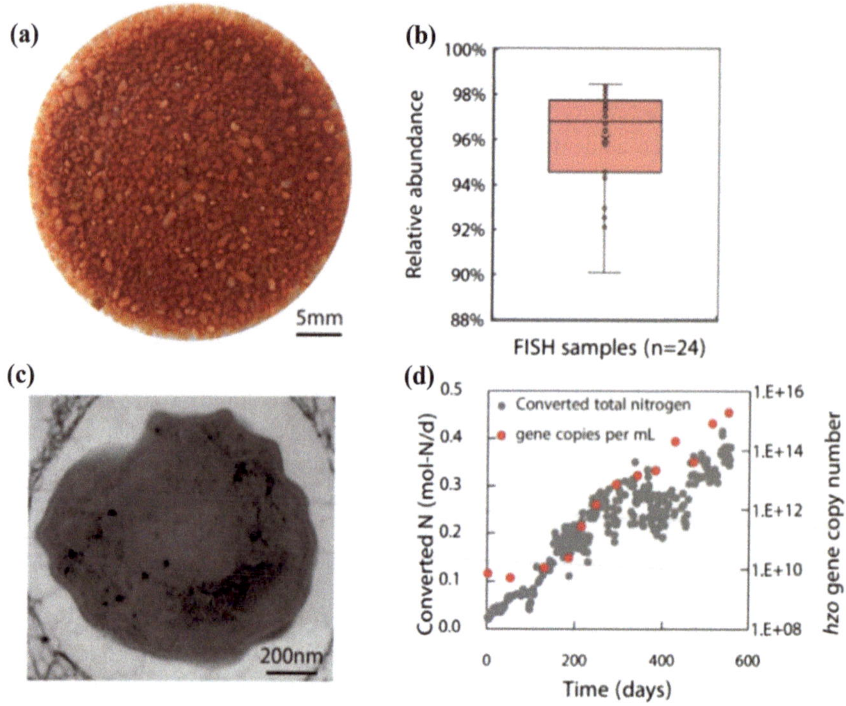

Fig. 1.25 Enrichment and confirmation of AnAOB. **a** A photo of the enriched red-color anammox granules, which is a typical color of anammox consortia/granules. **b** Shows the relative abundance of AnAOB in the community calculated based on the FISH images of twenty-four different samples. **c** Shows a TEM image of an anammox bacterium in the anammox granule. **d** The changes of the total converted nitrogen and the number of AnAOB during the 500-day enrichment

1.7.2 New Discovery of the Genus Anammox

In the past twenty years, there has been significant advancement in understanding the fundamental mechanism of AnAOB. However, there are still unresolved questions regarding their survival strategies. In a recent study, Yang et al. [291] revealed a novel AnAOB genus in a biofilm reactor that treats landfill leachate. The research team also utilized metagenomic analysis to assemble the near-complete genome of the newly identified anammox bacterium, provisionally designated as "*Ca. Loosdrechtia aerotolerans*". Remarkably, "*Ca. Loosdrechtia aerotolerans*" contains all the genes necessary for the core metabolism of anammox. As anticipated, the fundamental genes that are essential for anammox were the most highly transcribed genes, indicating that "*Ca. Loosdrechtia aerotolerans*" has a potential for high nitrogen removal in a biofilm system [291].

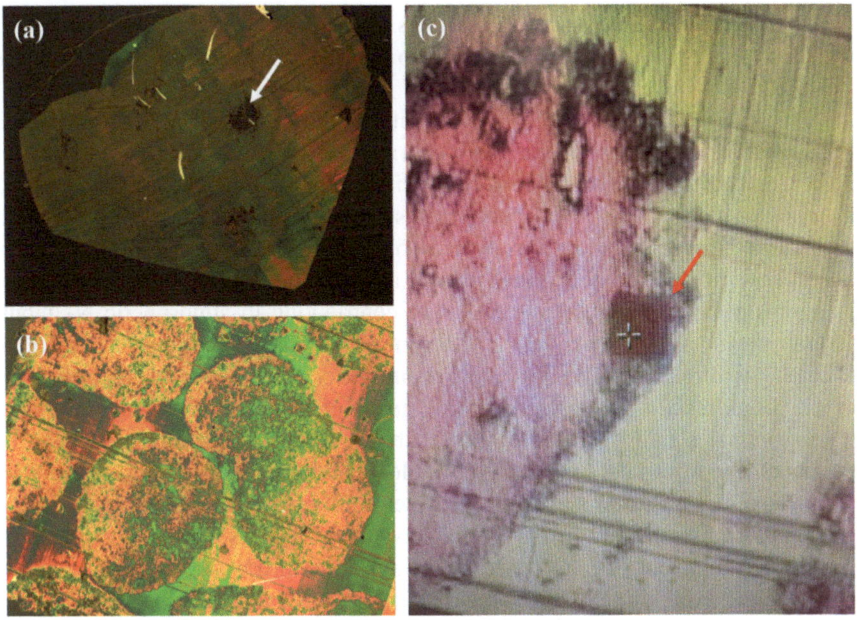

Fig. 1.26 The anammox granule microtome sections for the NanoSIMS analysis. **a** and **b** Ultrathin sections of anammox granules used for NanoSIMS scanning. **c** The blurred area pointed by the red arrow shows a Cs^+ ion beamed area after NanoSIMS scanning

In addition to investigating the core anammox metabolisms, the metabolic versatility of *Ca. Loosdrechtia aerotolerans* in adapting to adverse and variable environmental conditions was also studied [291]. The species possessed and expressed genes associated with DNRA, which could be utilized to regenerate ammonium and support their metabolism by coupling with oxidation of small organic acids. Furthermore, *Ca. Loosdrechtia aerotolerans* exhibited a distinct ferredoxin-dependent NO_3^-–N reductase, a feature not observed in any other anammox genome. This unique characteristic could potentially provide a competitive edge in the process of NO_3^-–N assimilation. Like numerous other microorganisms, *Ca. Loosdrechtia aerotolerans* also possessed and expressed superoxide dismutase and catalase genes that contribute to its ability to withstand oxidative stress. Interestingly, the species and four *Ca. Brocadia* genomes also contained bilirubin oxidase, which likely plays a role in oxygen resistance under fluctuating oxygen concentrations. The activity of bilirubin oxidase was confirmed using isolated recombinant proteins expressed in a foreign host [291]. A survey of oxygen-active proteins in AnAOB uncovered novel oxygen protection mechanisms, such as the novel cbb3-type cytochrome c oxidase and bifunctional catalase-peroxidase. The DO defense systems, including the cbb3-type cytochrome c oxidase and bifunctional catalase-peroxidase, could provide a competitive edge to *Ca. Kuenenia* and *Ca. Scalindua*, which regularly encounter varying oxygen levels. Lastly, the decreased transcription of the *nir* gene suggests that

"*Ca. Loosdrechtia aerotolerans*" likely relies extensively on the HAO-like protein for the conversion of NO_2^-–N to NO, similar to observations made in K. stuttgartiensis [59, 291].

A Research indicated that AnAOB exhibit a versatile lifestyle, enabling them to use various substrates to utilize a wide range of substrates for their survival in diverse and changing environments. In addition, these microorganisms have evolved a variety of mechanisms to safeguard themselves against unpredictable changes in their surroundings, including variations in oxygen levels. They exhibit these adaptations in various ecosystems, both natural and man-made, where they play a role in removing NH_4^+–N from wastewater or contributing to atmospheric N_2 input. Overall, the investigation of a full-scale anammox biofilm system presents a valuable example for nitrogen removal through anammox processes. The recent identification of *Ca. Loosdrechtia aerotolerans,* a new genus of AnAOB exhibiting significant transcriptional activity and metabolic flexibility, enhances our comprehension of the phylogenetic variety, ecological physiology, and niche adaptability of AnAOB within large-scale wastewater treatment systems. Moreover, the DO tolerance strategies employed by distinct AnAOB provide valuable insights for the functioning of wastewater treatment systems based on the anammox process. Future research can prioritize the exploration of enrichment strategies and the distinctive physiological adaptability of novel anammox organisms. This can lead to the enhancement of nitrogen removal from wastewater through the optimization of bioreactor configurations for the implementation of the anammox process. In summary, the discovery of a novel growth mode in AnAOB, which is capable of acetate and propionate, has revealed the metabolic mechanisms of these bacteria for small organic acids, advancing our understanding of the ecological niche of traditional AnAOB. In addition, the identification of a novel genus of AnAOB in full-scale wastewater treatment plants is a milestone in the development and application of AnAOB. Although research on this new strain is limited, it has the potential to greatly expand the methods for enriching AnAOB and enriching the practical application scenarios of anammox technology.

1.8 Prospects

In the last decade, there have been notable breakthroughs in the comprehension of anammox processes and AnAOB. Despite these advances, there remain numerous challenges to the application of anammox-mediated process and the enrichment of AnAOB. Nonetheless, with ongoing research and development, it is only a matter of time before the anammox process can be widely applied in practice. As we move forward with anammox-based processes, there are several factors that should be considered for future studies:

1. AnAOB enrichment has been studied in various types of reactors, primarily on an experimental scale. However, the efficacy of different reactor types for AnAOB enrichment in treating real sewage remains unclear due to the complexity and instability of actual sewage water quality. Therefore, further research is needed in this area.
2. The addition of carriers can effectively enrich AnAOB. The above-mentioned newly proposed carrier with four ideal properties has great potential for enriching AnAOB in large-scale WWTPs. In future research, greater attention should be given to carrier modification, preparation, and the mechanism for enriching AnAOB. This will aid in advancing the practical application of the mainstream anammox-based process.
3. The current research on enriching AnAOB has primarily focused on the enhancement mechanism through adding individual external compounds. Whereas, it is crucial to investigate the collaborative enhancement of diverse external compounds. While numerous investigations have examined the impacts of adding external substances on AnAOB, additional research is required to comprehend the nitrogen conversion pathway, microbial composition, and alterations in functional gene expression.
4. The main factors affecting the enrichment of AnAOB include substrates, salts, organic compounds, inorganic compounds, antibiotics, heavy metals, microplastics and NPs. The mechanism by which these factors affect the anammox process is still not fully understood. Future research on anammox should focus on the metabolic pathway of AnAOB to these pollutants, and should explore the changes of metabolic pathway of single strain. Meanwhile, the role of ARB and MRB in the AnAOB remains unclear. It is essential to examine the impacts of inhibitory substances and microbial community variations on the anammox process and develop relevant management approaches in future studies.
5. The full-scale implementation of the sidestream anammox process has been widespread. However, the issues of a prolonged initialization period and subpar nitrogen removal effectiveness in treating sidestream wastewater have yet to be adequately addressed. Furthermore, the mainstream anammox process faces challenges such as limited stability, elevated C/N ratio, inadequate retention of anammox bacteria, and the challenge of suppressing NOB activity. While the technology has been applied in laboratory settings, its mainstream applications have been limited. To enhance the practical applications of anammox, it is crucial to thoroughly investigate the impact of various water quality obstacles and formulate effective regulation strategies under strict reaction conditions. Future research efforts should focus on the advancement of a mixed microbial consortium integrated process utilizing anammox, alongside the exploration of a hybrid approach that incorporates anammox as an intermediary biological module. Furthermore, it is imperative to explore the prolonged, immediate, and cumulative impacts of various sources of water quality inhibitory factors on each constituent of the integrated anammox systems.

6. Although it was confirmed that the capability of anammox bacteria J. asiatica incorporating acetate and propionate, it definitely requires further experimental studies to explain the following questions. Is such organotrophic capability only specific to J. asiatica, or is it a versatility to all anammox species? Will short-chain fatty acids other than acetate and propionate also be incorporated under similar conditions? What are the evolutionary, physiological and ecological reasons behind a detour metabolism of J. asiatica's organotrophic lifestyle? Is it feasible to engineer truly heterotrophic anammox bacteria through the introduction of production pathways and the modification of autotrophic systems by genetic engineering?
7. Currently, research on anammox biofilm and granular systems has been growing more rapidly than that on floc systems, as these two methods offer higher nitrogen removal performance. However, the use of granular systems may not be suitable for full-scale municipal wastewater treatment due to engineering challenges and wastewater quantity limitations. Biofilm-based systems have been proven effective for AnAOB enrichment, but their widespread adoption has been hindered by cost considerations. Therefore, it is important to prioritize future research on cost optimization of biofilm-based anammox for the successful deployment of this technology on a large scale. Additionally, the limited understanding of membrane fouling and how to alleviate it presents significant challenges in biofilm-based anammox processes.
8. Although a new type of AnAOB has been discovered, the underlying metabolic mechanism of this strain is still unclear. Furthermore, the tolerance, application scenarios, and optimal culture parameters of this strain have not been fully elucidated. Therefore, additional research is required to determine the actual cultivation effect and potential for large-scale application of this new strain. The development of anammox technology based on this novel strain may become a new research hotspot in the future.

Ultimately, this chapter offers a thorough examination of the multifaceted engineering aspects of anammox-based processes, aiming to achieve improved energy efficiency and environmentally sustainable solutions. Hence, it will serve as an indispensable reference for the practical implementation of anammox-based systems.

References

1. Adams M, Xie J, Chang Y, Kabore AWJ, Chen C (2021) Start-up of anammox systems with different biochar amendment: process characteristics and microbial community. Sci Total Environ 790:148242. https://doi.org/10.1016/j.scitotenv.2021.148242
2. Adams M, Xie J, Kabore AWJ, Chang Y, Xie J, Guo M, Chen C (2022) Research advances in anammox granular sludge: a review. Crit Rev Environ Sci Technol 52:631–674. https://doi.org/10.1080/10643389.2020.1831358
3. Ahmad HA, Ni S, Ahmad S, Zhang J, Ali M, Ngo HH, Guo W, Tan Z, Wang Q (2020) Gel immobilization: a strategy to improve the performance of anaerobic ammonium oxidation

(anammox) bacteria for nitrogen-rich wastewater treatment. Bioresour Technol 313:123642. https://doi.org/10.1016/j.biortech.2020.123642
4. Aklujkar M, Coppi MV, Leang C, Kim BC, Chavan MA, Perpetua LA, Giloteaux L, Liu A, Holmes DE (2013) Proteins involved in electron transfer to Fe(III) and Mn(IV) oxides by geobacter sulfurreducens and geobacter uraniireducens. Microbiology (NY) 159:515–535. https://doi.org/10.1099/mic.0.064089-0
5. Al-Hazmi HE, Hassan GK, Maktabifard M, Grubba D, Majtacz J, Mąkinia J (2022) Integrating conventional nitrogen removal with anammox in wastewater treatment systems: microbial metabolism, sustainability and challenges. Environ Res 215:114432. https://doi.org/10.1016/j.envres.2022.114432
6. Al-Hazmi HE, Lu X, Grubba D, Majtacz J, Badawi M, Mąkinia J (2023) Sustainable nitrogen removal in anammox-mediated systems: microbial metabolic pathways, operational conditions and mathematical modelling. Sci Total Environ 868:161633. https://doi.org/10.1016/j.scitotenv.2023.161633
7. Ali M, Oshiki M, Rathnayake L, Ishii S, Satoh H, Okabe S (2015) Rapid and successful start-up of anammox process by immobilizing the minimal quantity of biomass in PVA-SA gel beads. Water Res 79:147–157. https://doi.org/10.1016/j.watres.2015.04.024
8. Ali SS, Sabry R, Ali HM, Gadallah H, Mansor ES, Abdallah H, Shalaby M, Shaban AM (2023) Integrated system of anoxic/activated sludge and ultrafiltration membrane for zero liquid discharge of pharmaceutical industrial wastewater treatment. J Environ Chem Eng 11:109068. https://doi.org/10.1016/j.jece.2022.109068
9. Arrojo B, Mosquera-Corral A, Campos JL, Méndez R (2006) Effects of mechanical stress on anammox granules in a sequencing batch reactor (SBR). J Biotechnol 123:453–463. https://doi.org/10.1016/j.jbiotec.2005.12.023
10. Bettazzi E, Caffaz S, Vannini C, Lubello C (2010) Nitrite inhibition and intermediates effects on anammox bacteria: a batch-scale experimental study. Process Biochem 45:573–580. https://doi.org/10.1016/j.procbio.2009.12.003
11. Biddle J, Jefferson B, Stephenson T, Hassard F, Cartmell E (2015) Rotating biological contactors for wastewater treatment—a review. Trans Inst Chem Eng. Process Saf Environ Prot, Part B
12. Carvajal-Arroyo JM, Puyol D, Li G, Swartwout A, Sierra-Álvarez R, Field JA (2014) Starved anammox cells are less resistant to NO_2^- inhibition. Water Res 65:170–176. https://doi.org/10.1016/j.watres.2014.07.023
13. Carvajal-Arroyo JM, Sun W, Sierra-Alvarez R, Field JA (2013) Inhibition of anaerobic ammonium oxidizing (anammox) enrichment cultures by substrates, metabolites and common wastewater constituents. Chemosphere 91:22–27. https://doi.org/10.1016/j.chemosphere.2012.11.025
14. Chen Y, Guo G, Li Y (2022) A review on upgrading of the anammox-based nitrogen removal processes: performance, stability, and control strategies. Bioresour Technol 364:127992. https://doi.org/10.1016/j.biortech.2022.127992
15. Chen Y, Jia F, Liu Y, Yu W, Cai W, Zhang X, He H, Yao H (2021) The effects of Fe(III) and Fe(II) on anammox process and the Fe–N metabolism. Chemosphere 285:131322. https://doi.org/10.1016/j.chemosphere.2021.131322
16. Chen C, Jiang Y, Zou X, Guo M, Liu H, Cui M, Zhang TC (2021) Insight into the influence of particle sizes on characteristics and microbial community in the anammox granular sludge. J Water Process Eng 39:101883. https://doi.org/10.1016/j.jwpe.2020.101883
17. Chen H, Jin RC (2017) Summary of the preservation techniques and the evolution of the anammox bacteria characteristics during preservation. Appl Microbiol Biotechnol 101:4349–4362
18. Chen YP, Li S, Fang F, Guo JS, Zhang Q, Gao X (2012) Effect of inorganic carbon on the completely autotrophic nitrogen removal over nitrite (CANON) process in a sequencing batch biofilm reactor. Environ Technol 33:2611–2617
19. Chen G, Li J, Wang Y, Deng H, Zhang Y, Zeng J (2016) Novel anammox reactor start-up method using immobilized particles as biocatalyst and its kinetic characteristics. Desalin Water Treat 57:17291–17299

20. Chen C, Wang Y, Ying J, Guo M, Cui M (2020) Effects of organic-matter-induced short-term stresses on performance and population dynamics of anammox systems. J Environ Eng (New York)
21. Chen H, Yu J, Jia X, Jin R (2014) Enhancement of anammox performance by Cu(II), Ni(II) and Fe(III) supplementation. Chemosphere 117:610–616. https://doi.org/10.1016/j.chemosphere.2014.09.047
22. Chen Y, Zhao Z, Liu H, Ma Y, An F, Huang J, Shao Z (2020) Achieving stable two-stage mainstream partial-nitrification/anammox (PN/A) operation via intermittent aeration. Chemosphere 245:125650. https://doi.org/10.1016/j.chemosphere.2019.125650
23. Chen Y, Zheng R, Sui Q, Ritigala T, Wei Y, Cheng X, Ren J, Yu D, Chen M, Wang T (2021) Coupling anammox with denitrification in a full-scale combined biological nitrogen removal process for swine wastewater treatment. Bioresour Technol 329:124906. https://doi.org/10.1016/j.biortech.2021.124906
24. Chen J, Zheng P, Yu Y, Tang C, Mahmood Q (2010) Promoting sludge quantity and activity results in high loading rates in anammox UBF. Bioresour Technol 101:2700–2705. https://doi.org/10.1016/j.biortech.2009.11.085
25. Cheng B, Du J, Bao J, Tufail H, Xu T, Zhang Y, Mao Q, Faheem M (2022) Insight into enrichment of anammox bacteria by a polyurethane sponge carrier coupled with iron-carbon micro-electrolysis under no strict anaerobic condition. Bioresour Technol 347
26. Cheng J, Qiao L, Xu W, Qian Y, Ge Y, Xia T, Li Y (2022) Nickel (ii) effects on anammox reaction: reactor performance, dehydrogenase, sludge morphology and microbial community changes. Environ Technol 43:4227–4236. https://doi.org/10.1080/09593330.2021.1946165
27. Cho K, Choi M, Jeong D, Lee S, Bae H (2017) Comparison of inoculum sources for long-term process performance and fate of ANAMMOX bacteria niche in poly(vinyl alcohol)/sodium alginate gel beads. Chemosphere 185:394–402. https://doi.org/10.1016/j.chemosphere.2017.06.123
28. Cho S, Jung MK, Ju DJ, Lee YH, Cho K, Okabe S (2017) For R2-submission to "Environmental Technology" as a original research paper Anammox biomass carrying efficiency of polyethylene non-woven sheets as a carrier material. Environ Technol 1
29. Cho S, Kambey C, Nguyen V (2020) Performance of anammox processes for wastewater treatment: a critical review on effects of operational conditions and environmental stresses. Water (Basel) 12:20. https://doi.org/10.3390/w12010020
30. Choi D, Cho K, Jung J (2019) Optimization of nitrogen removal performance in a single-stage SBR based on partial nitritation and ANAMMOX. Water Res 162:105–114. https://doi.org/10.1016/j.watres.2019.06.044
31. Choi M, Cho K, Lee S, Chung Y, Park J, Bae H (2018) Effective seeding strategy using flat type poly (vinyl alcohol) cryogel for anammox enrichment. Chemosphere 205:88–97. https://doi.org/10.1016/j.chemosphere.2018.04.055
32. Choi D, Lee D, Yun W, Jung Y (2017) Addition of a foaming agent to improve N_2 gas permeability of PVA/alginate carriers for deammonification process. Int J Hydrogen Energy
33. Chou WP, Tseng SK, Ho ACM (2012) Anaerobic ammonium oxidation improvement via a novel capsule bioreactor. Environ Technol 33
34. Cui H, Zhang L, Zhang Q, Li X, Huang Y, Peng Y (2021) Advanced nitrogen removal from low C/N municipal wastewater by combining partial nitrification-anammox and endogenous partial denitrification-anammox (PN/A-EPD/A) process in a single-stage reactor. Bioresour Technol 339:125501. https://doi.org/10.1016/j.biortech.2021.125501
35. Dale OR, Tobias CR, Song B (2009) Biogeographical distribution of diverse anaerobic ammonium oxidizing (anammox) bacteria in Cape Fear River Estuary. Environ Microbiol 11:1194–1207. https://doi.org/10.1111/j.1462-2920.2008.01850.x
36. Dan NH, Le Luu T (2021) High organic removal of landfill leachate using a continuous flow sequencing batch biofilm reactor (CF-SBBR) with different biocarriers. Sci Total Environ 787:147680. https://doi.org/10.1016/j.scitotenv.2021.147680
37. Dapena-Mora A, Campos JL, Mosquera-Corral A, Jetten MSM, Méndez R (2004) Stability of the ANAMMOX process in a gas-lift reactor and a SBR. J Biotechnol 110:159–170. https://doi.org/10.1016/j.jbiotec.2004.02.005

38. Dapena-Mora A, Van Hulle SWH, Campos JL, Méndez R, Vanrolleghem PA, Jetten M (2004) Enrichment of Anammox biomass from municipal activated sludge: experimental and modelling results. J Chem Tech Biotech
39. Dapena-Mora A, Vázquez-Padín JR, Campos JL, Mosquera-Corral A, Jetten MSM, Méndez R (2010) Monitoring the stability of an anammox reactor under high salinity conditions. Biochem Eng J 51:167–171. https://doi.org/10.1016/j.bej.2010.06.014
40. Daverey A, Chen Y, Dutta K, Huang Y, Lin J (2015) Start-up of simultaneous partial nitrification, anammox and denitrification (SNAD) process in sequencing batch biofilm reactor using novel biomass carriers. Bioresour Technol 190:480–486
41. Daverey A, Chen YC, Sung S, Lin JG (2014) Effect of zinc on anammox activity and performance of simultaneous partial nitrification, anammox and denitrification (SNAD) process. Bioresour Technol 165:105–110
42. Daverey A, Su S, Huang Y, Chen S, Sung S, Lin J (2013) Partial nitrification and anammox process: a method for high strength optoelectronic industrial wastewater treatment. Water Res 47:2929–2937. https://doi.org/10.1016/j.watres.2013.01.028
43. Deng W, Wang L, Cheng L, Yang W, Gao D (2022) Nitrogen removal from mature landfill leachate via anammox based processes: a review. Sustainability 14:995. https://doi.org/10.3390/su14020995
44. Dong Z, Sun T (2007) A potential new process for improving nitrogen removal in constructed wetlands—promoting coexistence of partial-nitrification and ANAMMOX. Ecol Eng 31:69–78. https://doi.org/10.1016/j.ecoleng.2007.04.009
45. Drews A (2010) Membrane fouling in membrane bioreactors—characterisation, contradictions, cause and cures. J Membr Sci 363:1–28. https://doi.org/10.1016/j.memsci.2010.06.046
46. Du R, Cao S, Zhang H, Peng Y (2020) Formation of partial-denitrification (PD) granular sludge from low-strength nitrate wastewater: the influence of loading rates. J Hazard Mater 384:121273. https://doi.org/10.1016/j.jhazmat.2019.121273
47. Du R, Peng Y, Cao S, Wang S, Niu M (2016) Characteristic of nitrous oxide production in partial denitrification process with high nitrite accumulation. Bioresour Technol 203:341–347. https://doi.org/10.1016/j.biortech.2015.12.044
48. Du R, Peng Y, Cao S, Wu C, Weng D, Wang S, He J (2014) Advanced nitrogen removal with simultaneous anammox and denitrification in sequencing batch reactor. Bioresour Technol 162:316–322. https://doi.org/10.1016/j.biortech.2014.03.041
49. Duan X, Zhou J, Qiao S, Xin Y, Tian T, Xu F (2012) Start-up of the anammox process from the conventional activated sludge in a hybrid bioreactor. 环境科学学报(英文版) 24:1083–1090
50. Egli K, Fanger U, Alvarez PJJ, Siegrist H, van der Meer JR, Zehnder AJB (2001) Enrichment and characterization of an anammox bacterium from a rotating biological contactor treating ammonium-rich leachate. Arch Microbiol 175:198–207. https://doi.org/10.1007/s002030100255
51. Elreedy A, Fujii M, Koyama M, Nakasaki K, Tawfik A (2019) Enhanced fermentative hydrogen production from industrial wastewater using mixed culture bacteria incorporated with iron, nickel, and zinc-based nanoparticles. Water Res 151:349–361. https://doi.org/10.1016/j.watres.2018.12.043
52. Elreedy A, Ismail S, Ali M, Ni S, Fujii M, Elsamadony M (2021) Unraveling the capability of graphene nanosheets and γ-Fe_2O_3 nanoparticles to stimulate anammox granular sludge. J Environ Manage 277:111495. https://doi.org/10.1016/j.jenvman.2020.111495
53. Erdim E, Yücesoy Özkan Z, Kurt H, Alpaslan Kocamemi B (2019) Overcoming challenges in mainstream anammox applications: utilization of nanoscale zero valent iron (nZVI). Sci Total Environ 651:3023–3033. https://doi.org/10.1016/j.scitotenv.2018.09.140
54. Fan J, Du R, Liu Q, Li C, Peng Y (2022) Insight into the microbial interactions of anammox and heterotrophic bacteria in different granular sludge systems: effect of size distribution and available organic carbon source. Bioresour Technol 364:128055. https://doi.org/10.1016/j.biortech.2022.128055

55. Feinberg EC, Huff Chester AL, Novak PJ, Hillmyer MA (2021) Porous polyethylene-supported zeolite carriers for improved wastewater deammonification. ACS EST Eng 1:1104–1112. https://doi.org/10.1021/acsestengg.1c00077
56. Feng L, Li J, Ma H, Chen G (2020) Effect of Fe(II) on simultaneous marine anammox and Feammox treating nitrogen-laden saline wastewater under low temperature: enhanced performance and kinetics. Desalination 478:114287. https://doi.org/10.1016/j.desal.2019.114287
57. Feng F, Tang X, Qu C, Lu X, Liu Z, Tang J, Tang C, Chai L (2021) Hydroxylamine addition enhances fast recovery of anammox activity suffering Cr(VI) inhibition. Bioresour Technol 329:124920. https://doi.org/10.1016/j.biortech.2021.124920
58. Fernández I, Vázquez-Padín JR, Mosquera-Corral A, Campos JL, Méndez R (2008) Biofilm and granular systems to improve anammox biomass retention. Biochem Eng J 42:308–313. https://doi.org/10.1016/j.bej.2008.07.011
59. Ferousi C, Lindhoud S, Baymann F, Hester ER, Reimann J, Kartal B (2019) Discovery of a functional, contracted heme-binding motif within a multiheme cytochrome. J Biol Chem 294:16953–16965
60. Fu L, Ding J, Lu YZ, Ding ZW, Zeng RJ (2017) Nitrogen source effects on the denitrifying anaerobic methane oxidation culture and anaerobic ammonium oxidation bacteria enrichment process. Appl Microbiol Biotechnol 101:3895–3906. https://doi.org/10.1007/s00253-017-8163-2
61. Fu J, Wang Y, Yang J, Huang D, Zhang Q, Huang Y, Chen J, Fan N, Jin R (2022) Mitigating the detrimental effects of salt stress on anammox process: a comparison between glycine betaine and mannitol. Sci Total Environ 851:158221. https://doi.org/10.1016/j.scitotenv.2022.158221
62. Fu J, Zhang Q, Huang B, Fan N, Jin R (2021) A review on anammox process for the treatment of antibiotic-containing wastewater: linking effects with corresponding mechanisms. Front Environ Sci Eng 15. https://doi.org/10.1007/s11783-020-1309-y
63. Gamoń F, Banach-Wiśniewska A, Kaur JJ, Cema G, Ziembińska-Buczyńska A (2022) Microbial response of the anammox process to trace antibiotic concentration. J Water Process Eng 46:102607. https://doi.org/10.1016/j.jwpe.2022.102607
64. Gao D, Huang X, Tao Y, Cong Y, Wang X (2015) Sewage treatment by an UAFB–EGSB biosystem with energy recovery and autotrophic nitrogen removal under different temperatures. Bioresour Technol 181:26–31. https://doi.org/10.1016/j.biortech.2015.01.037
65. Gao D, Li Y, Liang H (2022) Biofilm carriers for anaerobic ammonium oxidation: mechanisms, applications, and roles in mainstream systems. Bioresour Technol 353:127115. https://doi.org/10.1016/j.biortech.2022.127115
66. Gao Y, Liu Z, Liu F, Furukawa K (2012) Mechanical shear contributes to granule formation resulting in quick start-up and stability of a hybrid anammox reactor. Biodegradation 23:363–372. https://doi.org/10.1007/s10532-011-9515-8
67. Gao R, Peng Y, Li J, Liu Y, Deng L, Li W, Kao C (2022) Mainstream partial denitrification-anammox (PD/A) for municipal sewage treatment from moderate to low temperature: reactor performance and bacterial structure. Sci Total Environ 806:150267. https://doi.org/10.1016/j.scitotenv.2021.150267
68. Gao D, Wang X, Liang H, Wei Q, Dou Y, Li L (2018) Anaerobic ammonia oxidizing bacteria: ecological distribution, metabolism, and microbial interactions. Front Environ Sci Eng 12. https://doi.org/10.1007/s11783-018-1035-x
69. Gao D, Xiang T (2021) Deammonification process in municipal wastewater treatment: challenges and perspectives. Bioresour Technol 320:124420. https://doi.org/10.1016/j.biortech.2020.124420
70. Ge C, Dong Y, Li H, Li Q, Ni S, Gao B, Xu S, Qiao Z, Ding S (2019) Nitritation-anammox process—a realizable and satisfactory way to remove nitrogen from high saline wastewater. Bioresour Technol 275:86–93. https://doi.org/10.1016/j.biortech.2018.12.032
71. Gong X, Wang B, Qiao X, Gong Q, Liu X, Peng Y (2020) Performance of the anammox process treating low-strength municipal wastewater under low temperatures: effect of undulating seasonal temperature variation. Bioresour Technol 312:123590. https://doi.org/10.1016/j.biortech.2020.123590

72. Gonzalez-Martinez A, Rodriguez-Sanchez A, Garcia-Ruiz MJ, Muñoz-Palazon B, Cortes-Lorenzo C, Osorio F, Vahala R (2016) Performance and bacterial community dynamics of a CANON bioreactor acclimated from high to low operational temperatures. Chem Eng J 287:557–567. https://doi.org/10.1016/j.cej.2015.11.081
73. Gottshall EY, Bryson SJ, Cogert KI, Landreau M, Sedlacek CJ, Stahl DA, Daims H, Winkler M (2021) Sustained nitrogen loss in a symbiotic association of comammox nitrospira and anammox bacteria. Water Res 202:117426. https://doi.org/10.1016/j.watres.2021.117426
74. Gu W, Wang L, Liu Y, Liang P, Zhang X, Li Y, Huang X (2020) Anammox bacteria enrichment and denitrification in moving bed biofilm reactors packed with different buoyant carriers: performances and mechanisms. Sci Total Environ 719:137277. https://doi.org/10.1016/j.scitotenv.2020.137277
75. Guo B, Chen Y, Lv L, Ahmad HA, Ni S, Ren L, Cui Z, Fang X, Qiao Z, Ding S (2019) Transformation of the zero valent iron dosage effect on anammox after long-term culture: from inhibition to promotion. Process Biochem 78:132–139. https://doi.org/10.1016/j.procbio.2019.01.014
76. Guo M, Jiang Y, Xie J, Cao Q, Zhang Q, Mabruk A, Chen C (2022) Bamboo charcoal addition enhanced the nitrogen removal of anammox granular sludge with COD: performance, physicochemical characteristics and microbial community. J Environ Sci (China) 115:55–64. https://doi.org/10.1016/j.jes.2021.07.010
77. Guo Y, Niu Q, Sugano T, Li Y (2020) Biodegradable organic matter-containing ammonium wastewater treatment through simultaneous partial nitritation, anammox, denitrification and COD oxidization process. Sci Total Environ 714:136740. https://doi.org/10.1016/j.scitotenv.2020.136740
78. Güven D, Dapena A, Kartal B, Schmid MC, Maas B, van de Pas-Schoonen K, Sozen S, Mendez R, Op Den Camp HJM, Jetten MSM, Strous M, Schmidt I (2005) Propionate oxidation by and methanol inhibition of anaerobic ammonium-oxidizing bacteria. Appl Environ Microbiol 71:1066–1071. https://doi.org/10.1128/AEM.71.2.1066-1071.2005
79. Guzmán-Fierro V, Sanhueza J, Arriagada C, Pereira L, Campos V, Gallardo JJ, Roeckel M (2021) The prediction of partial-nitrification-anammox performance in real industrial wastewater based on granular size. J Environ Manage 286:112255. https://doi.org/10.1016/j.jenvman.2021.112255
80. Han YM, Liu ZJ, Xu XF, Huang SB, Liu FX (2016) Study on gas-lift bioreactor in combined process for nitrogen removal from wastewater. Modern Chemical Industry
81. Hanson TE, Campbell BJ, Kalis KM, Campbell MA, Klotz MG (2013) Nitrate ammonification by nautilia profundicola AmH: experimental evidence consistent with a free hydroxylamine intermediate. Front Microbiol 4:180
82. Harb R, Laçin D, Subaşı I, Erguder TH (2021) Denitrifying anaerobic methane oxidation (DAMO) cultures: factors affecting their enrichment, performance and integration with anammox bacteria. J Environ Manage 295:113070. https://doi.org/10.1016/j.jenvman.2021.113070
83. Hawley AK, Brewer HM, Norbeck AD, Paša-Tolić L, Hallam SJ (2014) Metaproteomics reveals differential modes of metabolic coupling among ubiquitous oxygen minimum zone microbes. Proc Natl Acad Sci 111:11395–11400. https://doi.org/10.1073/pnas.1322132111
84. He S, Chen Y, Qin M, Mao Z, Yuan L, Niu Q, Tan X (2018) Effects of temperature on anammox performance and community structure. Bioresour Technol 260:186–195. https://doi.org/10.1016/j.biortech.2018.03.090
85. He Y, Mao H, Makinia J, Drewnowski J, Wu B, Xu J, Xie L, Lu X (2023) Impact of soluble organic matter and particulate organic matter on anammox system: performance, microbial community and N_2O production. J Environ Sci (China) 124:146–155. https://doi.org/10.1016/j.jes.2021.11.007
86. He S, Yang W, Li W, Zhang Y, Qin M, Mao Z (2019) Impacts of salt shocking and the selection of a suitable reversal agent on anammox. Sci Total Environ 692:602–612. https://doi.org/10.1016/j.scitotenv.2019.07.215

87. Hendrickx TLG, Wang Y, Kampman C, Zeeman G, Temmink H, Buisman CJN (2012) Autotrophic nitrogen removal from low strength waste water at low temperature. Water Res 46:2187–2193. https://doi.org/10.1016/j.watres.2012.01.037
88. Hu Q, Kang D, Wang R, Ding A, Abbas G, Zhang M, Qiu L, Lu H, Lu H, Zheng P (2018) Characterization of oligotrophic AnAOB culture: morphological, physiological, and ecological features. Appl Microbiol Biotechnol 102:995–1003. https://doi.org/10.1007/s00253-017-8587-8
89. Hu Y, Li N, Jiang J, Xu Y, Luo X, Cao J (2022) Simultaneous feammox and anammox process facilitated by activated carbon as an electron shuttle for autotrophic biological nitrogen removal. Front Environ Sci Eng 16. https://doi.org/10.1007/s11783-021-1498-z
90. Hu Z, Lotti T, de Kreuk M, Kleerebezem R, van Loosdrecht M, Kruit J, Jetten MSM, Kartal B (2013) Nitrogen removal by a nitritation-anammox bioreactor at low temperature. Appl Environ Microbiol 79:2807–2812. https://doi.org/10.1128/AEM.03987-12
91. Huang D, Fu J, Li Z, Fan N, Jin R (2022) Inhibition of wastewater pollutants on the anammox process: a review. Sci Total Environ 803:150009. https://doi.org/10.1016/j.scitotenv.2021.150009
92. Huang X, Gao D, Peng S, Tao Y (2014) Effects of ferrous and manganese ions on anammox process in sequencing batch biofilm reactors. J Environ Sci (China) 26:1034–1039. https://doi.org/10.1016/S1001-0742(13)60531-8
93. Huang X, Gao D, Tao Y, Wang X (2014) C2/C3 fatty acid stress on anammox consortia dominated by Candidatus Jettenia asiatica. Chem Eng J 253:402–407. https://doi.org/10.1016/j.cej.2014.05.055
94. Huang Z, Wang Y, Jiang L, Xu B, Wang Y, Zhao H, Zhou W (2018) Mechanism and performance of a self-flocculating marine bacterium in saline wastewater treatment. Chem Eng J 334:732–740. https://doi.org/10.1016/j.cej.2017.10.076
95. Huff Chester AL, Eum K, Tsapatsis M, Hillmyer MA, Novak PJ (2021) Enhanced nitrogen removal and anammox bacteria retention with zeolite-coated membrane in simulated mainstream wastewater. Environ Sci Technol Lett 8:468–473. https://doi.org/10.1021/acs.estlett.1c00154
96. Ibrahim M, Yusof N, Mohd Yusoff MZ, Hassan MA (2016) Enrichment of anaerobic ammonium oxidation (anammox) bacteria for short start-up of the anammox process: a review. Desalin Water Treat 57:13958–13978
97. Isaka K, Kimura Y, Matsuura M, Osaka T, Tsuneda S (2017) First full-scale nitritation-anammox plant using gel entrapment technology for ammonia plant effluent. Biochem Eng J 122:115–122
98. Isaka K, Nitta S, Osaka T, Tsuneda S (2022) Effect of inorganic carbon limitation on the nitrogen removal performance of the single-stage reactor containing anammox and nitritation gel carriers. J Biosci Bioeng 133:70–75. https://doi.org/10.1016/j.jbiosc.2021.09.014
99. Isaka K, Sumino T, Tsuneda S (2007) High nitrogen removal performance at moderately low temperature utilizing anaerobic ammonium oxidation reactions. J Biosci Bioeng 103:486–490. https://doi.org/10.1263/jbb.103.486
100. Isanta E, Bezerra T, Fernández I, Suárez-Ojeda ME, Pérez J, Carrera J (2015) Microbial community shifts on an anammox reactor after a temperature shock using 454-pyrosequencing analysis. Bioresour Technol 181:207–213. https://doi.org/10.1016/j.biortech.2015.01.064
101. Iskander SM, Zhao R, Pathak A, Gupta A, Pruden A, Novak JT, He Z (2018) A review of landfill leachate induced ultraviolet quenching substances: sources, characteristics, and treatment. Water Res 145:297–311. https://doi.org/10.1016/j.watres.2018.08.035
102. Jetten MS, Niftrik L, Strous M, Kartal B, Keltjens JT, Op DCH (2009) Biochemistry and molecular biology of anammox bacteria. Crit Rev Biochem Mol Biol 44:65–84. https://doi.org/10.1080/10409230902722783
103. Ji J, Peng Y, Li X, Zhang Q, Liu X (2020) A novel partial nitrification-synchronous anammox and endogenous partial denitrification (PN-SAEPD) process for advanced nitrogen removal from municipal wastewater at ambient temperatures. Water Res 175:115690. https://doi.org/10.1016/j.watres.2020.115690

104. Jiang H, Liu G, Ma Y, Xu X, Chen J, Yang Y, Liu X, Wang H (2018) A pilot-scale study on start-up and stable operation of mainstream partial nitrification-anammox biofilter process based on online pH-DO linkage control. Chem Eng J 350:1035–1042. https://doi.org/10.1016/j.cej.2018.06.007
105. Jiang H, Wang Z, Ren S, Qiu J, Li X, Peng Y (2021) Culturing sludge fermentation liquid-driven partial denitrification in two-stage anammox process to realize advanced nitrogen removal from mature landfill leachate. J Hazard Mater 415:125568. https://doi.org/10.1016/j.jhazmat.2021.125568
106. Jiang H, Wang Z, Ren S, Qiu J, Zhang Q, Li X, Peng Y (2021) Enrichment and retention of key functional bacteria of partial denitrification-anammox (PD/A) process via cell immobilization: a novel strategy for fast PD/A application. Bioresour Technol 326:124744. https://doi.org/10.1016/j.biortech.2021.124744
107. Jiang H, Yang P, Wang Z, Ren S, Qiu J, Liang H, Peng Y, Li X, Zhang Q (2021) Efficient and advanced nitrogen removal from mature landfill leachate via combining nitritation and denitritation with anammox in a single sequencing batch biofilm reactor. Bioresour Technol 333:125138. https://doi.org/10.1016/j.biortech.2021.125138
108. Jiang T, Zhang H, Qiang H, Yang F, Xu X, Du H (2013) Start-up of the anammox process and membrane fouling analysis in a novel rotating membrane bioreactor. Desalination 311:46–53. https://doi.org/10.1016/j.desal.2012.10.031
109. Jin R, Ma C, Mahmood Q, Yang G, Zheng P (2011) Anammox in a UASB reactor treating saline wastewater. Process Saf Environ Prot 89:342–348. https://doi.org/10.1016/j.psep.2011.05.001
110. Jin R, Ma C, Yu J (2013) Performance of an anammox UASB reactor at high load and low ambient temperature. Chem Eng J 232:17–25. https://doi.org/10.1016/j.cej.2013.07.059
111. Jin R, Xing B, Yu J, Qin T, Chen S (2013) The importance of the substrate ratio in the operation of the anammox process in upflow biofilter. Ecol Eng 53:130–137. https://doi.org/10.1016/j.ecoleng.2012.12.027
112. Jin R, Yang G, Yu J, Zheng P (2012) The inhibition of the anammox process: a review. Chem Eng J 197:67–79. https://doi.org/10.1016/j.cej.2012.05.014
113. Jin R, Zheng P, Hu A, Mahmood Q, Hu B, Jilani G (2008) Performance comparison of two anammox reactors: SBR and UBF. Chem Eng J 138:224–230. https://doi.org/10.1016/j.cej.2007.06.038
114. Jin R, Zheng P, Mahmood Q, Hu B (2007) Osmotic stress on nitrification in an airlift bioreactor. J Hazard Mater 146:148–154. https://doi.org/10.1016/j.jhazmat.2006.12.004
115. Jo Y, Cho K, Choi H, Lee C (2020) Treatment of low-strength ammonia wastewater by single-stage partial nitritation and anammox using upflow dual-bed gel-carrier reactor (UDGR). Bioresour Technol 304:123023. https://doi.org/10.1016/j.biortech.2020.123023
116. Kartal B, Keltjens JT (2016) Anammox biochemistry: a tale of heme c proteins. Trends Biochem Sci 41:998–1011. https://doi.org/10.1016/j.tibs.2016.08.015
117. Kartal B, Koleva M, Arsov R, van der Star W, Jetten MSM, Strous M (2006) Adaptation of a freshwater anammox population to high salinity wastewater. J Biotechnol 126:546–553. https://doi.org/10.1016/j.jbiotec.2006.05.012
118. Kartal B, Maalcke WJ, Almeida N, Cirpus I, Gloerich J, Geerts W, Camp H, Harhangi HR, Janssen-Megens EM, Francoijs KJ (2011) Molecular mechanism of anaerobic ammonium oxidation. Nature Publishing Group
119. Kartal B, Niftrik LV, Keltjens JT, Camp H, Jetten M (2012) Anammox-growth physiology, cell biology, and metabolism. Adv Microb Physiol 60:211–262
120. Kimura Y, Isaka K (2014) Evaluation of inhibitory effects of heavy metals on anaerobic ammonium oxidation (anammox) by continuous feeding tests. Appl Microbiol Biotechnol 98:6965–6972
121. Kimura Y, Isaka K, Kazama F (2011) Effects of inorganic carbon limitation on anaerobic ammonium oxidation (anammox) activity. Bioresour Technol 102:4390–4394. https://doi.org/10.1016/j.biortech.2010.12.101

122. Kopf SH, Mcglynn SE, Green-Saxena A, Guan Y, Orphan VJ (2015) Heavy water and (15) N labelling with NanoSIMS analysis reveals growth rate-dependent metabolic heterogeneity in chemostats. Environ Microbiol 17
123. Kosugi Y, Matsuura N, Honda R, Yamashita T, Yamamoto-Ikemoto R (2022) Effects of organic carbon and sulfide on the anammox reaction in the anoxic column in the SRDAPN process for treating high-strength wastewater. J Environ Manage 307:114459. https://doi.org/10.1016/j.jenvman.2022.114459
124. Kouba V, Hůrková K, Navrátilová K, Kok D, Benáková A, Laureni M, Vodičková P, Podzimek T, Lipovová P, van Niftrik L, Hajšlová J, van Loosdrecht MCM, Weissbrodt DG, Bartáček J (2022) Effect of temperature on the compositions of ladderane lipids in globally surveyed anammox populations. Sci Total Environ 830:154715. https://doi.org/10.1016/j.scitotenv.2022.154715
125. Kowalski MS, Devlin T, di Biase A, Basu S, Oleszkiewicz JA (2019) Accelerated start-up of a partial nitritation-anammox moving bed biofilm reactor. Biochem Eng J 145:83–89. https://doi.org/10.1016/j.bej.2019.02.015
126. Kowalski MS, Devlin TR, Oleszkiewicz JA (2019) Attachment of anaerobic ammonium-oxidizing bacteria to augmented carrier material. Environ Technol 40:576–583. https://doi.org/10.1080/09593330.2017.1397770
127. Kuenen JG (2008) Anammox bacteria: from discovery to application. Nat Rev Microbiol 6:320–326
128. Kumwimba MN, Lotti T, Enel ES, Li X, Suanon F (2020) Anammox-based processes: how far have we come and what work remains a review by bibliometric analysis. Chemosphere.https://doi.org/10.1016/j.chemosphere.2019.124627
129. Lackner S, Gilbert EM, Vlaeminck SE, Joss A, Horn H, van Loosdrecht MCM (2014) Full-scale partial nitritation/anammox experiences—an application survey. Water Res 55:292–303. https://doi.org/10.1016/j.watres.2014.02.032
130. Langone M, Jia Y, Haaijer S, Camp H, Andreottola G (2014) Coexistence of nitrifying, anammox and denitrifying bacteria in a sequencing batch reactor. Front Microbiol 5
131. Laureni M, Weissbrodt DG, Szivák I, Robin O, Nielsen JL, Morgenroth E, Joss A (2015) Activity and growth of anammox biomass on aerobically pre-treated municipal wastewater. Water Res 80:325–336. https://doi.org/10.1016/j.watres.2015.04.026
132. Laureni M, Weissbrodt DG, Villez K, Robin O, de Jonge N, Rosenthal A, Wells G, Nielsen JL, Morgenroth E, Joss A (2019) Biomass segregation between biofilm and flocs improves the control of nitrite-oxidizing bacteria in mainstream partial nitritation and anammox processes. Water Res 154:104–116. https://doi.org/10.1016/j.watres.2018.12.051
133. Li J, Chen X, Liu W, Tao Y (2020) Biostimulation of a marine anammox bacteria-dominated bioprocess by Co(II) to treat nitrogen-rich, saline wastewater. Sci Total Environ 749:141489. https://doi.org/10.1016/j.scitotenv.2020.141489
134. Li J, Feng L, Qiang Z, Dong H, Wang D (2018) Enhanced performance and kinetics of marine anammox bacteria (MAB) treating nitrogen-rich saline wastewater with Mn(II) and Ni(II) addition. Bioresour Technol 249:1085–1091. https://doi.org/10.1016/j.biortech.2017.10.101
135. Li M, Hong Y, Cao H, Gu J (2011) Mangrove trees affect the community structure and distribution of anammox bacteria at an anthropogenic-polluted mangrove in the Pearl River Delta reflected by 16S rRNA and hydrazine oxidoreductase (HZO) encoding gene analyses. Ecotoxicology 20:1780–1790. https://doi.org/10.1007/s10646-011-0711-4
136. Li Y, Liang H, Yang W, Cheng L, Cao J, Wang P, Gao D (2022) Enhanced nitrogen removal in mainstream deammonification systems at ambient temperature by novel modified carriers and differentiation of microbial community transformation. Bioresour Technol 366:128158. https://doi.org/10.1016/j.biortech.2022.128158
137. Li Z, Peng Y, Gao H (2020) A novel strategy for accelerating the recovery of a Fe(II)-inhibited anammox reactor by intermittent addition of betaine: performance, kinetics and statistical analysis. Chemosphere 251:126362. https://doi.org/10.1016/j.chemosphere.2020.126362

138. Li J, Peng Y, Zhang L, Liu J, Wang X, Gao R, Pang L, Zhou Y (2019) Quantify the contribution of anammox for enhanced nitrogen removal through metagenomic analysis and mass balance in an anoxic moving bed biofilm reactor. Water Res 160:178–187. https://doi.org/10.1016/j.watres.2019.05.070
139. Li J, Qi P, Qiang Z, Dong H, Gao D, Wang D (2018) Is anammox a promising treatment process for nitrogen removal from nitrogen-rich saline wastewater? Bioresour Technol 270:722–731. https://doi.org/10.1016/j.biortech.2018.08.115
140. Li X, Sun S, Badgley BD, Sung S, Zhang H, He Z (2016) Nitrogen removal by granular nitritation–anammox in an upflow membrane-aerated biofilm reactor. Water Res 94:23–31. https://doi.org/10.1016/j.watres.2016.02.031
141. Li J, Tabassum S (2022) Effect of zero-valent iron and iron-carbon particles on the denitrification efficiency of anammox process under gradual cooling conditions. Cleaner Eng Technol 8:100477. https://doi.org/10.1016/j.clet.2022.100477
142. Li H, Tao W (2017) Efficient ammonia removal in recirculating vertical flow constructed wetlands: complementary roles of anammox and denitrification in simultaneous nitritation, anammox and denitrification process. Chem Eng J 317:972–979. https://doi.org/10.1016/j.cej.2017.02.143
143. Li Z, Xu X, Shao B, Zhang S, Yang F (2014) Anammox granules formation and performance in a submerged anaerobic membrane bioreactor. Chem Eng J 254:9–16. https://doi.org/10.1016/j.cej.2014.04.068
144. Li Z, Xu X, Xu X, Yang F, Zhang S (2015) Sustainable operation of submerged anammox membrane bioreactor with recycling biogas sparging for alleviating membrane fouling. Chemosphere 140:106–113. https://doi.org/10.1016/j.chemosphere.2014.08.059
145. Li X, Yuan Y, Huang Y, Bi Z (2019) Simultaneous removal of ammonia and nitrate by coupled S0-driven autotrophic denitrification and Anammox process in fluorine-containing semiconductor wastewater. Sci Total Environ 661:235–242. https://doi.org/10.1016/j.scitotenv.2019.01.164
146. Li N, Zeng W, Guo Y, Li C, Ma C, Peng Y (2019) Nitrogen-associated niche characteristics and bacterial community estimated by 15N-DNA-stable isotope probing in one-stage partial nitritation/anammox process with different ammonium loading. J Environ Manage 247:603–612. https://doi.org/10.1016/j.jenvman.2019.06.110
147. Li YLAL (2011) Macroscale and microscale analysis of anammox in anaerobic rotating biological contactor. J Environ Sci (China)
148. Liang L, Luo J, Xiao X, Wang J, Hong M, Deng C, Li Y, Liu J (2022) Granular activated carbon promoting re-granulation of anammox-hydroxyapatite granules for stable nitrogen removal at low phosphate concentration. Sci Total Environ 805:150359. https://doi.org/10.1016/j.scitotenv.2021.150359
149. Lin L, Pratt S, Crick O, Xia J, Duan H, Ye L (2021) Salinity effect on freshwater anammox bacteria: Ionic stress and ion composition. Water Res 188:116432. https://doi.org/10.1016/j.watres.2020.116432
150. Liu L, Ji M, Wang F, Tian Z, Wang T, Wang S, Wang S, Yan Z (2020) Insight into the short-term effect of fulvic acid on nitrogen removal performance and N-acylated-L-homoserine lactones (AHLs) release in the anammox system. Sci Total Environ 704:135285. https://doi.org/10.1016/j.scitotenv.2019.135285
151. Liu T, Lawluvy Y, Shi Y, Ighalo JO, He Y, Zhang Y, Yap P (2022) Adsorption of cadmium and lead from aqueous solution using modified biochar: a review. J Environ Chem Eng 10:106502. https://doi.org/10.1016/j.jece.2021.106502
152. Liu Y, Li Y, Pan B, Zhang X, Zhang H, Steinberg CEW, Qiu H, Vijver MG, Peijnenburg WJGM (2021) Application of low dosage of copper oxide and zinc oxide nanoparticles boosts bacterial and fungal communities in soil. Sci Total Environ 757:143807. https://doi.org/10.1016/j.scitotenv.2020.143807
153. Liu Y, Ngo HH, Guo W, Peng L, Wang D, Ni B (2019) The roles of free ammonia (FA) in biological wastewater treatment processes: a review. Environ Int 123:10–19. https://doi.org/10.1016/j.envint.2018.11.039

154. Liu T, Tian R, Li Q, Wu N, Quan X (2021) Strengthened attachment of anammox bacteria on iron-based modified carrier and its effects on anammox performance in integrated floating-film activated sludge (IFFAS) process. Sci Total Environ 787:147679
155. Liu L, Wang F, Xu S, Yan Z, Ji M (2022) Long-term effect of fulvic acid amendment on the anammox biofilm system at 15 °C: performance, microbial community and metagenomics analysis. Bioresour Technol 344:126234. https://doi.org/10.1016/j.biortech.2021.126234
156. Liu Y, Xu W, Bao L, Li Y, Liu S, Zhang Q, Alessi DS, Konhauser KO, Zhao H (2019) Cell surface characterization and trace metal adsorptive properties of anaerobic ammonium-oxidizing (anammox) consortia. Chemosphere 221:11–20. https://doi.org/10.1016/j.chemosphere.2019.01.025
157. Liu C, Yu D, Wang Y, Chen G, Tang P, Huang S (2020) A novel control strategy for the partial nitrification and anammox process (PN/A) of immobilized particles: using salinity as a factor. Bioresour Technol 302:122864. https://doi.org/10.1016/j.biortech.2020.122864
158. Liu W, Zeng Z, Chen A, Zeng G, Xiao R, Guo Z, Yi F, Huang Z, He K, Hu L (2018) Toxicity effects of silver nanoparticles on the freshwater bivalve *Corbicula fluminea*. J Environ Chem Eng 6:4236–4244. https://doi.org/10.1016/j.jece.2018.06.032
159. Liu S, Zhang Z, Ni J (2013) Effects of Ca^{2+} on activity restoration of the damaged anammox consortium. Bioresour Technol 143:315–321. https://doi.org/10.1016/j.biortech.2013.06.005
160. Lodha T, Narvekar S, Karodi P (2021) Classification of uncultivated anammox bacteria and Candidatus Uabimicrobium into new classes and provisional nomenclature as Candidatus Brocadiia classis nov. and Candidatus Uabimicrobiia classis nov. of the phylum Planctomycetes and novel family Candidatus Scalinduaceae fam. nov to accommodate the genus Candidatus Scalindua. Syst Appl Microbiol 44:126272. https://doi.org/10.1016/j.syapm.2021.126272
161. Lu G, Ma Y, Zang L, Sun Y, Yu F, Xue R (2021) Effects of granular activated carbon and Fe-modified granular activated carbon on anammox process start-up. Rsc Adv 11:10625–10634. https://doi.org/10.1039/D1RA00384D
162. Lu C, Yuan C, Zhu T, Wang Y (2021) Effect of humic acid on the single-stage nitrogen removal using anammox and partial nitritation (SNAP) process: performance and bacterial communities. J Environ Chem Eng 9:106680. https://doi.org/10.1016/j.jece.2021.106680
163. Lu H, Zheng P, Ji Q, Zhang H, Ji J, Wang L, Ding S, Chen T, Zhang J, Tang C, Chen J (2012) The structure, density and settlability of anammox granular sludge in high-rate reactors. Bioresour Technol 123:312–317. https://doi.org/10.1016/j.biortech.2012.07.003
164. Lu Y, Ma L, Liang Y, Shan B, Chang J (2018) Enhancing nitrogen removal performancein a bioreactor using immobilized anaerobicammonium oxidation sludge by polyvinylalcohol-sodium alginate (PVA-SA). Pol J Environ Stud 27:773–778. https://doi.org/10.15244/pjoes/75805
165. Luo H, Zeng Y, Cheng Y, He D, Pan X (2020) Recent advances in municipal landfill leachate: a review focusing on its characteristics, treatment, and toxicity assessment. Sci Total Environ 703:135468. https://doi.org/10.1016/j.scitotenv.2019.135468
166. Lv Y, Pan J, Huo T, Zhao Y, Liu S (2019) Enhanced microbial metabolism in one stage partial nitritation-anammox system treating low strength wastewater by novel composite carrier. Water Res 163:114872. https://doi.org/10.1016/j.watres.2019.114872
167. Ma M, Cao Q, Mabruk A, Xie J, Wu P, Liu W, Chen C (2022) Promotion of nitrogen removal and microbial enrichment on anammox by exogenous substance addition: a critical review. J Water Process Eng 49:103096. https://doi.org/10.1016/j.jwpe.2022.103096
168. Ma C, Jin R, Yang G, Yu J, Xing B, Zhang Q (2012) Impacts of transient salinity shock loads on anammox process performance. Bioresour Technol 112:124–130. https://doi.org/10.1016/j.biortech.2012.02.122
169. Ma B, Peng Y, Zhang S, Wang J, Gan Y, Chang J, Wang S, Wang S, Zhu G (2013) Performance of anammox UASB reactor treating low strength wastewater under moderate and low temperatures. Bioresour Technol 129:606–611. https://doi.org/10.1016/j.biortech.2012.11.025

170. Ma W, Ren Z, Yu L, Wu X, Yao Y, Zhang J, Guo J, Fan N, Jin R (2021) Deciphering the response of anammox process to heavy metal and antibiotic stress: arsenic enhances the permeability of extracellular polymeric substance and aggravates the inhibition of sulfamethoxazole. Chem Eng J 426:130815. https://doi.org/10.1016/j.cej.2021.130815
171. Ma B, Wang S, Cao S, Miao Y, Jia F, Du R, Peng Y (2016) Biological nitrogen removal from sewage via anammox: recent advances. Bioresour Technol 200:981–990. https://doi.org/10.1016/j.biortech.2015.10.074
172. Ma J, Wang K, Gong H, Yuan Q, Yang M, He C, Shi C, San E (2020) Integrating floc, aggregate and carrier to reap high-quality anammox biofilm. Bioresour Technol 309:123325. https://doi.org/10.1016/j.biortech.2020.123325
173. Ma X, Yan Y, Wang W, Guo J, Wang Y (2020) Metatranscriptomic analysis of adaptive response of anammox bacteria Candidatus 'Kuenenia stuttgartiensis' to Zn(II) exposure. Chemosphere 246:125682. https://doi.org/10.1016/j.chemosphere.2019.125682
174. Ma C, Zeng W, Li J, Meng Q, Peng Y (2023) Metabolomic pathway regulation to achieve optimal control of inorganic carbon in anammox process. Sci Total Environ 855:158653. https://doi.org/10.1016/j.scitotenv.2022.158653
175. Magrí A, Vanotti MB, Szögi AA (2012) Anammox sludge immobilized in polyvinyl alcohol (PVA) cryogel carriers. Bioresour Technol 114:231–240. https://doi.org/10.1016/j.biortech.2012.03.077
176. Mai W, Hu T, Li C, Wu R, Chen J, Shao Y, Liang J, Wei Y (2020) Effective nitrogen removal of wastewater from vitamin B2 production by a potential anammox process. J Water Process Eng 37:101515. https://doi.org/10.1016/j.jwpe.2020.101515
177. Malovanyy A, Trela J, Plaza E (2015) Mainstream wastewater treatment in integrated fixed film activated sludge (IFAS) reactor by partial nitritation/anammox process. Bioresour Technol 198:478–487. https://doi.org/10.1016/j.biortech.2015.08.123
178. Mason SA, Garneau D, Sutton R, Chu Y, Ehmann K, Barnes J, Fink P, Papazissimos D, Rogers DL (2016) Microplastic pollution is widely detected in US municipal wastewater treatment plant effluent. Environ Pollut 218:1045–1054. https://doi.org/10.1016/j.envpol.2016.08.056
179. Meibom A, Sleep NH, Chamberlain CP, Coleman RG, Frei R, Hren MT, Wooden JL (2002) Re–Os isotopic evidence for long-lived heterogeneity and equilibration processes in the Earth's upper mantle. Nature 419:705–708. https://doi.org/10.1038/nature01067
180. Meng F, Chae S, Drews A, Kraume M, Shin H, Yang F (2009) Recent advances in membrane bioreactors (MBRs): membrane fouling and membrane material. Water Res 43:1489–1512. https://doi.org/10.1016/j.watres.2008.12.044
181. Meng Y, Sheng B, Meng F (2019) Changes in nitrogen removal and microbiota of anammox biofilm reactors under tetracycline stress at environmentally and industrially relevant concentrations. Sci Total Environ 668:379–388. https://doi.org/10.1016/j.scitotenv.2019.02.389
182. Meng F, Su G, Hu Y, Lu H, Huang L, Chen G (2014) Improving nitrogen removal in an ANAMMOX reactor using a permeable reactive biobarrier. Water Res 58:82–91. https://doi.org/10.1016/j.watres.2014.03.049
183. Miao Y, Liao R, Zhang X, Wang Y, Wang Z, Shi P, Liu B, Li A (2015) Metagenomic insights into Cr(VI) effect on microbial communities and functional genes of an expanded granular sludge bed reactor treating high-nitrate wastewater. Water Res 76:43–52. https://doi.org/10.1016/j.watres.2015.02.042
184. Miao Y, Peng Y, Zhang L, Li B, Li X, Wu L, Wang S (2018) Partial nitrification-anammox (PNA) treating sewage with intermittent aeration mode: effect of influent C/N ratios. Chem Eng J 334:664–672. https://doi.org/10.1016/j.cej.2017.10.072
185. Morgan-Sagastume F (2018) Biofilm development, activity and the modification of carrier material surface properties in moving-bed biofilm reactors (MBBRs) for wastewater treatment. Crit Rev Environ Sci Technol 1–32
186. Ni B, Yuan Z (2013) A model-based assessment of nitric oxide and nitrous oxide production in membrane-aerated autotrophic nitrogen removal biofilm systems. J Membr Sci 428:163–171. https://doi.org/10.1016/j.memsci.2012.10.049

187. Niculina Musat FMPK, Pett-Ridge J (2016) Tracking microbial interactions with NanoSIMS. Curr Opin Biotechnol 114:121. https://doi.org/10.1016/j.clinmicnews.2015.01.008
188. Niftrik LV, Jetten M (2012) Anaerobic ammonium-oxidizing bacteria: unique microorganisms with exceptional properties. Microbiol Mol Biol Rev 76
189. O'Laughlin J, McElligott K (2009) Forest policy and economics. For Policy Econ 11:535–536. https://doi.org/10.1016/j.forpol.2009.07.001
190. Peng Z, Gao D, Xiang T, Wang X (2019) Achieving stable and efficient single-stage deammonification using plug flow reactor. Environ Sci Pollut Res Int 26:28031–28039
191. Phan TN, Van Truong TT, Ha NB, Nguyen PD, Bui XT, Dang BT, Doan VT, Park J, Guo W, Ngo HH (2017) High rate nitrogen removal by ANAMMOX internal circulation reactor (IC) for old landfill leachate treatment. Bioresour Technol 234:281–288. https://doi.org/10.1016/j.biortech.2017.02.117
192. Pijuan M, Ribera-Guardia A, Balcázar JL, Micó MM, de la Torre T (2020) Effect of COD on mainstream anammox: evaluation of process performance, granule morphology and nitrous oxide production. Sci Total Environ 712:136372. https://doi.org/10.1016/j.scitotenv.2019.136372
193. Qi P, Li J, Dong H, Wang D, Bo Y (2018) Performance of anammox process treating nitrogen-rich saline wastewater: kinetics and nitrite inhibition. J Clean Prod 199:493–502. https://doi.org/10.1016/j.jclepro.2018.07.201
194. Qiao S, Tian T, Zhou J (2014) Effects of quinoid redox mediators on the activity of anammox biomass. Bioresour Technol 152:116–123. https://doi.org/10.1016/j.biortech.2013.11.003
195. Qin Y, Cao Y, Ren J, Wang T, Han B (2017) Effect of glucose on nitrogen removal and microbial community in anammox-denitrification system. Bioresour Technol 244:33–39. https://doi.org/10.1016/j.biortech.2017.07.124
196. Qin Y, Han B, Cao Y, Wang T (2017) Impact of substrate concentration on anammox-UBF reactors start-up. Bioresour Technol 239:422–429. https://doi.org/10.1016/j.biortech.2017.04.126
197. Qin Z, Wei C, Wei T, Li Z, Pang Z, Luo P, Feng C, Qiu G, Wei C, Wu H, Peng Y, Jiang C, Preis S (2022) Evolution of biochemical processes in coking wastewater treatment: a combined evaluation of material and energy efficiencies and secondary pollution. Sci Total Environ 807:151072. https://doi.org/10.1016/j.scitotenv.2021.151072
198. Qin Y, Zhou S (2009) Enrichment and molecular diversity of anammox bacteria in uasb reactor. Environ Prot Eng 35:17–26
199. Qiu S, Li Z, Hu Y, Shi L, Liu R, Shi L, Chen L, Zhan X (2021) What's the best way to achieve successful mainstream partial nitritation-anammox application? Crit Rev Environ Sci Technol 51:1045–1077. https://doi.org/10.1080/10643389.2020.1745015
200. Ren S, Wang Z, Jiang H, Li X, Zhang Q, Peng Y (2022) Efficient nitrogen removal from mature landfill leachate in a step feed continuous plug-flow system based on one-stage anammox process. Bioresour Technol 347:126676. https://doi.org/10.1016/j.biortech.2022.126676
201. Ren Z, Wang H, Zhang L, Du X, Huang B, Jin R (2022) A review of anammox-based nitrogen removal technology: from microbial diversity to engineering applications. Bioresour Technol 363:127896. https://doi.org/10.1016/j.biortech.2022.127896
202. Sabine Marie P, Pümpel T, Markt R, Murthy S, Bott C, Wett B (2015) Comparative evaluation of multiple methods to quantify and characterise granular anammox biomass. Water Res 68:194–205. https://doi.org/10.1016/j.watres.2014.10.005
203. Sari T, Can S, Akgul D (2020) Assessment of Anammox process against acute and long-term exposure of ZnO nanoparticles. Sci Total Environ 727:138603. https://doi.org/10.1016/j.scitotenv.2020.138603
204. Scaria J, Nidheesh PV (2023) Pre-treatment of real pharmaceutical wastewater by heterogeneous Fenton and persulfate oxidation processes. Environ Res 217:114786. https://doi.org/10.1016/j.envres.2022.114786
205. Schmid MC, Maas B, Dapena A, Pas-Schoonen K, Strous M (2005) Biomarkers for in situ detection of anaerobic ammonium-oxidizing (anammox) bacteria. Appl Environ Microbiol 71:1677–1684

206. Schwartz-Narbonne R, Schaeffer P, Hopmans EC, Schenesse M, Charlton EA, Jones DM, Sinninghe Damsté JS, Farhan Ul Haque M, Jetten MSM, Lengger SK, Murrell JC, Normand P, Nuijten GHL, Talbot HM, Rush D (2020) A unique bacteriohopanetetrol stereoisomer of marine anammox. Org Geochem 143:103994. https://doi.org/10.1016/j.orggeochem.2020.103994
207. Shi Z, Hu H, Shen Y, Xu J, Shi M, Jin R (2017) Long-term effects of oxytetracycline (OTC) on the granule-based anammox: process performance and occurrence of antibiotic resistance genes. Biochem Eng J 127:110–118. https://doi.org/10.1016/j.bej.2017.08.009
208. Siegrist H, Salzgeber D, Eugster J, Joss A (2008) Anammox brings WWTP closer to energy autarky due to increased biogas production and reduced aeration energy for N-removal. Water Sci Technol 57:383–388
209. Silveira NC, Oliveira GHD, Damianovic MHRZ, Foresti E (2021) Two-stage partial nitrification-anammox process for nitrogen removal from slaughterhouse wastewater: evaluation of the nitrogen loading rate and microbial community analysis. J Environ Manage 296:113214. https://doi.org/10.1016/j.jenvman.2021.113214
210. Sindhu L, Niu K, Liu X, Ni S, Fang X (2021) Effect of Fe^{2+} addition on anammox consortia, nitrogen removal performance and functional genes analysis during start-up of anammox process. J Water Process Eng 43:102251. https://doi.org/10.1016/j.jwpe.2021.102251
211. Smith JA, Nevin KP, Lovley DR (2015) Syntrophic growth via quinone-mediated interspecies electron transfer. Front Microbiol 6. https://doi.org/10.3389/fmicb.2015.00121
212. Soler-Jofra A, Laureni M, Warmerdam M, Pérez J, van Loosdrecht MCM (2020) Hydroxylamine metabolism of Ca Kuenenia stuttgartiensis. Water Res 184:116188. https://doi.org/10.1016/j.watres.2020.116188
213. Star WRLV, Miclea AI, Dongen UGJM, Muyzer G, Picioreanu C, Loosdrecht MCMV (2010) The membrane bioreactor: a novel tool to grow anammox bacteria as free cells. Biotechnol Bioeng 101:286–294
214. Strous M, Gerven EV, Ping Z, Kuenen JG, Jetten M (1997) Ammonium removal from concentrated waste streams with the anaerobic ammonium oxidation (anammox) process in different reactor configurations. Water Res 31:1955–1962
215. Strous M, Heijnen JJ, Kuenen JG, Jetten MSM (1998) The sequencing batch reactor as a powerful tool for the study of slowly growing anaerobic ammonium-oxidizing microorganisms. Appl Microbiol Biotechnol 50:589–596. https://doi.org/10.1007/s002530051340
216. Strous M, Kuenen JG, Jetten MSM (1999) Key physiology of anaerobic ammonium oxidation. American Society for Microbiology
217. Su JF, Li GQ, Wen Q, Xue L, Chen CL, Huang TL (2020) Highly efficient nitrate and phosphorus removal and adsorption of tetracycline by precipitation in a chitosan/polyvinyl alcohol immobilized bioreactor. Bioprocess Biosyst Eng 43:1761–1771
218. Su B, Liu Q, Liang H, Zhou X, Zhang Y, Liu G, Qiao Z (2022) Simultaneous partial nitrification, anammox, and denitrification in an upflow microaerobic membrane bioreactor treating middle concentration of ammonia nitrogen wastewater with low COD/TN ratio. Chemosphere 295:133832. https://doi.org/10.1016/j.chemosphere.2022.133832
219. Su Y, Yang H, Wang X, Liu X (2022) Response of microbial succession of Anammox granular sludge (AnGS) and essential abundance under salty stress and temperature reduction. J Environ Chem Eng 10:106834. https://doi.org/10.1016/j.jece.2021.106834
220. Sui Q, Wang Y, Wang H, Yue W, Chen Y, Yu D, Chen M, Wei Y (2020) Roles of hydroxylamine and hydrazine in the in-situ recovery of one-stage partial nitritation-anammox process: characteristics and mechanisms. Sci Total Environ 707:135648. https://doi.org/10.1016/j.scitotenv.2019.135648
221. Tan C, Liu Y, Xu H, Cui D, Zuo J, Li J, Ji Y, Qiu S, Yao L, Chen Y (2018) Effects of tourmaline on nitrogen removal performance and biofilm structures in the sequencing batch biofilm reactor. 环境科学学报:英文版 127–135
222. Tang L, Su C, Chen Y, Xian Y, Hui X, Ye Z, Chen M, Zhu F, Zhong H (2021) Influence of biodegradable polybutylene succinate and non-biodegradable polyvinyl chloride microplastics on anammox sludge: performance evaluation, suppression effect and metagenomic analysis. J Hazard Mater 401:123337. https://doi.org/10.1016/j.jhazmat.2020.123337

223. Tang C, Zheng P, Wang C, Mahmood Q, Zhang J, Chen X, Zhang L, Chen J (2011) Performance of high-loaded ANAMMOX UASB reactors containing granular sludge. Water Res 45:135–144. https://doi.org/10.1016/j.watres.2010.08.018
224. Tao Y, Gao D, Fu Y, Wu W, Ren N (2012) Impact of reactor configuration on anammox process start-up: MBR versus SBR. Bioresour Technol 104:73–80. https://doi.org/10.1016/j.biortech.2011.10.052
225. Tao Y, Huang X, Gao D, Wang X, Chen C, Liang H, van Loosdrecht MCM (2019) NanoSIMS reveals unusual enrichment of acetate and propionate by an anammox consortium dominated by Jettenia asiatica. Water Res 159:223–232. https://doi.org/10.1016/j.watres.2019.05.006
226. Tatara M, Ishikawa S, Ueno Y (2017) Continuous nitrogen removal by a single-stage reactor packed with ring-laced string medium. J Biosci Bioeng 124
227. Tian X, Schopf A, Amaral-Stewart B, Christensson M, Morgan-Sagastume F, Vincent S, Delatolla R (2020) Anammox attachment and biofilm development on surface-modified carriers with planktonic- and biofilm-based inoculation. Bioresour Technol 317:124030. https://doi.org/10.1016/j.biortech.2020.124030
228. Tian X, Schopf A, Amaral-Stewart B, Christensson M, Morgan-Sagastume F, St-Pierre JP, Vincent S, Mercier E, Zhang X, Delatolla R (2022) Carrier surface modification for enhanced attachment and growth of anammox biofilm. Sci Total Environ 811
229. Toyama K, Nakamura T, Kataoka K, Yasuda O, Fukuda M, Tokutomi Y, Dong Y, Ogawa H, Kim-Mitsuyama S (2011) Telmisartan protects against diabetic vascular complications in a mouse model of obesity and type 2 diabetes, partially through peroxisome proliferator activated receptor-γ-dependent activity. Biochem Biophys Res Commun 410:508–513. https://doi.org/10.1016/j.bbrc.2011.06.012
230. Tran HT, Park YJ, Cho MK, Kim DJ, Ann DH (2006) Anaerobic ammonium oxidation process in an upflow anaerobic sludge blanket reactor with granular sludge selected from an anaerobic digestor. Biotechnol Bioprocess Eng 11:199–204
231. Vale G, Mehennaoui K, Cambier S, Libralato G, Jomini S, Domingos RF (2016) Manufactured nanoparticles in the aquatic environment-biochemical responses on freshwater organisms: a critical overview. Aquat Toxicol 170:162–174. https://doi.org/10.1016/j.aquatox.2015.11.019
232. van der Star WRL, Abma WR, Blommers D, Mulder J, Tokutomi T, Strous M, Picioreanu C, van Loosdrecht MCM (2007) Startup of reactors for anoxic ammonium oxidation: experiences from the first full-scale anammox reactor in Rotterdam. Water Res 41:4149–4163. https://doi.org/10.1016/j.watres.2007.03.044
233. van der Star WRL, Miclea AI, van Dongen UGJM, Muyzer G, Picioreanu C, van Loosdrecht MCM (2008) The membrane bioreactor: a novel tool to grow anammox bacteria as free cells. Biotechnol Bioeng 101:286–294.https://doi.org/10.1002/bit.21891
234. van Dongen U, Jetten MS, van Loosdrecht MC (2001) The SHARON-anammox process for treatment of ammonium rich wastewater. Water Sci Technol 44:153–160
235. Van Loosdrecht MCM, Damir B (2014) Anticipating the next century of wastewater treatment. Science 344:1451–1452. https://doi.org/10.1126/science.1255819
236. Van Niftrik L, Geerts WJC, Van Donselaar EG, Humbel BM, Webb RI, Fuerst JA, Verkleij AJ, Jetten MSM, Strous M (2008) Linking ultrastructure and function in four genera of anaerobic ammonium-oxidizing bacteria: cell plan, glycogen storage, and localization of cytochrome c proteins. J Bacteriol 190:708–717
237. van Niftrik L, Geerts WJC, van Donselaar EG, Humbel BM, Yakushevska A, Verkleij AJ, Jetten MSM, Strous M (2008) Combined structural and chemical analysis of the anammoxosome: a membrane-bounded intracytoplasmic compartment in anammox bacteria. J Struct Biol 161:401–410.https://doi.org/10.1016/j.jsb.2007.05.005
238. Van DGAA, De Bruijn P, Robertson LA, Jetten M, Kuenen JG (1996) Autotrophic growth of anaerobic ammonium-oxidizing micro-organisms in a fluidized bed reactor. Microbiol 142:2187–2196
239. Vera L, González E, Díaz O, Delgado S (2014) Application of a backwashing strategy based on transmembrane pressure set-point in a tertiary submerged membrane bioreactor. J Membr Sci 470:504–512. https://doi.org/10.1016/j.memsci.2014.07.069

240. Veuillet F, Lacroix S, Bausseron A, Gonidec E, Lemaire R (2014) Integrated fixed-film activated sludge ANITA (TM) Mox process—a new perspective for advanced nitrogen removal. Water Sci Technol 69:915–922
241. Waki M, Yasuda T, Fukumoto Y, Kuroda K, Suzuki K (2013) Effect of electron donors on anammox coupling with nitrate reduction for removing nitrogen from nitrate and ammonium. Bioresour Technol 130:592–598. https://doi.org/10.1016/j.biortech.2012.12.101
242. Wang H, Chen H, Liu S, Li L, Yang D, Dai X (2023) Inhibition and recovery of UASB-anammox process: performance characteristics and microbial community dynamics in response to substrate concentration and dissolved oxygen exposure. J Water Process Eng 51:103472. https://doi.org/10.1016/j.jwpe.2022.103472
243. Wang X, Gao D (2018) The transformation from anammox granules to deammonification granules in micro-aerobic system by facilitating indigenous ammonia oxidizing bacteria. Bioresour Technol 250:439–448. https://doi.org/10.1016/j.biortech.2017.11.046
244. Wang J, Jing K (2005) The characteristics of anaerobic ammonium oxidation (ANAMMOX) by granular sludge from an EGSB reactor. Process Biochem 40:1973–1978
245. Wang Y, Li B, Li Y, Chen X (2021) Research progress on enhancing the performance of autotrophic nitrogen removal systems using microbial immobilization technology. Sci Total Environ 774:145136. https://doi.org/10.1016/j.scitotenv.2021.145136
246. Wang W, Li D, Li S, Zeng H, Zhang J (2021) Characteristics and formation mechanism of hollow anammox granular sludge in low-strength ammonia sewage treatment. Chem Eng J 421:127766. https://doi.org/10.1016/j.cej.2020.127766
247. Wang S, Li J, Wang C, Ma J, Li Z, Zheng Z, Zhang J (2022) Reaction of the anammox granules to various antibiotics and operating the anammox coupled denitrifying reactor for oxytetracycline wastetwater treatment. Bioresour Technol 348:126756. https://doi.org/10.1016/j.biortech.2022.126756
248. Wang J, Liang J (2021) Insight: high intensity and activity carrier granular sludge cultured using polyvinyl alcohol/chitosan and polyvinyl alcohol/chitosan/iron gel beads. Bioresour Technol 326:124778
249. Wang J, Liang J, Ning D, Zhang T, Wang M (2022) A review of biomass immobilization in anammox and partial nitrification/anammox systems: advances, issues, and future perspectives. Sci Total Environ 821:152792. https://doi.org/10.1016/j.scitotenv.2021.152792
250. Wang Z, Liu X, Ni S, Zhuang X, Lee T (2021) Nano zero-valent iron improves anammox activity by promoting the activity of quorum sensing system. Water Res 202:117491. https://doi.org/10.1016/j.watres.2021.117491
251. Wang C, Liu S, Xu X, Zhang C, Wang D, Yang F (2018) Achieving mainstream nitrogen removal through simultaneous partial nitrification, anammox and denitrification process in an integrated fixed film activated sludge reactor. Chemosphere 203:457
252. Wang H, Peng L, Mao N, Geng J, Ren H, Xu K (2021) Effects of Fe^{3+} on microbial communities shifts, functional genes expression and nitrogen transformation during the start-up of anammox process. Bioresour Technol 320:124326. https://doi.org/10.1016/j.biortech.2020.124326
253. Wang Q, Rogers MJ, Ng SS, He J (2021) Fixed nitrogen removal mechanisms associated with sulfur cycling in tropical wetlands. Water Res 189:116619. https://doi.org/10.1016/j.watres.2020.116619
254. Wang D, Wang G, Zhang G, Xu X, Yang F (2013) Using graphene oxide to enhance the activity of anammox bacteria for nitrogen removal. Bioresour Technol 131:527–530. https://doi.org/10.1016/j.biortech.2013.01.099
255. Wang Y, Xie H, Wang D, Wang W (2019) Insight into the response of anammox granule rheological intensity and size evolution to decreasing temperature and influent substrate concentration. Water Res 162:258–268. https://doi.org/10.1016/j.watres.2019.06.060
256. Wang G, Xu X, Zhou L, Wang C, Yang F (2017) A pilot-scale study on the start-up of partial nitrification-anammox process for anaerobic sludge digester liquor treatment. Bioresour Technol 241:181–189. https://doi.org/10.1016/j.biortech.2017.02.125

257. Wang X, Yang H, Liu X, Su Y (2020) Effects of biomass and environmental factors on nitrogen removal performance and community structure of an anammox immobilized filler. Sci Total Environ 710:135258. https://doi.org/10.1016/j.scitotenv.2019.135258
258. Wang T, Zhang H, Yang F (2016) Performance of anammox process and low-oxygen adaptability of anammox biofilms in a FBR with small ring non-woven carriers. Ecol Eng 86:126–134. https://doi.org/10.1016/j.ecoleng.2015.11.025
259. Wang T, Zhang H, Yang F, Li Y, Zhang G (2013) Start-up and long-term operation of the anammox process in a fixed bed reactor (FBR) filled with novel non-woven ring carriers. Chemosphere 91:669–675. https://doi.org/10.1016/j.chemosphere.2013.01.026
260. Wang Z, Zheng M, Duan H, Yuan Z, Hu S (2022) A 20-year journey of partial nitritation and anammox (PN/A): from sidestream toward mainstream. Environ Sci Technol 56:7522–7531. https://doi.org/10.1021/acs.est.1c06107
261. Wang L, Wei F, Lü Y, Wang X, Wang Z (2012) Mechanism and pathways of nitrogen removal in AnRBC Anammox system. J Jiangsu Univ (Nat Sci Ed) 33:586–590
262. Wang Y, Xu L, Wang S, Ye F, Zhu G (2019) Global distribution of anaerobic ammonia oxidation (anammox) bacteria—field surveys in wetland, dryland, groundwater aquifer and snow. Front Microbiol 10.https://doi.org/10.3389/fmicb.2019.02583
263. Wei A, Ma J, Chen J, Zhang Y, Song J, Yu X (2018) Enhanced nitrate removal and high selectivity towards dinitrogen for groundwater remediation using biochar-supported nano zero-valent iron. Chem Eng J 353:595–605. https://doi.org/10.1016/j.cej.2018.07.127
264. Wen J, Tao W, Wang Z, Pei Y (2013) Enhancing simultaneous nitritation and anammox in recirculating biofilters: effects of unsaturated zone depth and alkalinity dissolution of packing materials. J Hazard Mater 244–245:671–680. https://doi.org/10.1016/j.jhazmat.2012.10.058
265. Wu P, Chen Y, Ji X, Liu W, Lv G, Shen Y, Zhou Q (2018) Fast start-up of the cold-anammox process with different inoculums at low temperature (13 °C) in innovative reactor. Bioresour Technol 267:696–703. https://doi.org/10.1016/j.biortech.2018.07.026
266. Wu Y, Chen Z, Wen Q, Fu Q, Bao H (2021) Mechanism concerning the occurrence and removal of antibiotic resistance genes in composting product with ozone post-treatment. Bioresour Technol 321:124433. https://doi.org/10.1016/j.biortech.2020.124433
267. Wu J, Kong Z, Luo Z, Qin Y, Rong C, Wang T, Hanaoka T, Sakemi S, Ito M, Kobayashi S, Kobayashi M, Xu K, Kobayashi T, Kubota K, Li Y (2021) A successful start-up of an anaerobic membrane bioreactor (AnMBR) coupled mainstream partial nitritation-anammox (PN/A) system: a pilot-scale study on in-situ NOB elimination, AnAOB growth kinetics, and mainstream treatment performance. Water Res 207:117783. https://doi.org/10.1016/j.watres.2021.117783
268. Wu D, Li G, Shi Z, Zhang Q, Huang B, Fan N, Jin R (2019) Co-inhibition of salinity and Ni(II) in the anammox-UASB reactor. Sci Total Environ 669:70–82. https://doi.org/10.1016/j.scitotenv.2019.03.130
269. Wu Y, Wang Y, De Costa YG, Tong Z, Cheng JJ, Zhou L, Zhuang W, Yu K (2019) The co-existence of anammox genera in an expanded granular sludge bed reactor with biomass carriers for nitrogen removal. Appl Microbiol Biotechnol 103:1231–1242. https://doi.org/10.1007/s00253-018-9494-3
270. Wu P, Chen J, Garlapati VK, Zhang X, Wani Victor Jenario F, Li X, Liu W, Chen C, Aminabhavi TM, Zhang X (2022) Novel insights into anammox-based processes: a critical review. Chem Eng J 444:136534.https://doi.org/10.1016/j.cej.2022.136534
271. Wuertz S, Okabe S, Hausner M (2004) Microbial communities and their interactions in biofilm systems: an overview. Water Sci Technol
272. Xiang T, Gao D (2019) Comparing two hydrazine addition strategies to stabilize mainstream deammonification: performance and microbial community analysis. Bioresour Technol 289:121710. https://doi.org/10.1016/j.biortech.2019.121710
273. Xiang T, Liang H, Gao D (2022) Effect of exogenous hydrazine on metabolic process of anammox bacteria. J Environ Manage 317:115398. https://doi.org/10.1016/j.jenvman.2022.115398

274. Xiao H, Peng Y, Zhang Q, Liu Y (2021) Pre-anaerobic treatment enhanced partial nitrification start-up coupled with anammox for advanced nitrogen removal from low C/N domestic wastewater. Bioresour Technol 337:125434. https://doi.org/10.1016/j.biortech.2021.125434
275. Xie J, Guo M, Xie J, Chang Y, Mabruk A, Zhang TC, Chen C (2022) COD inhibition alleviation and anammox granular sludge stability improvement by biochar addition. J Clean Prod 345:131167. https://doi.org/10.1016/j.jclepro.2022.131167
276. Xie F, Ma X, Zhao B, Cui Y, Zhang X, Yue X (2020) Promoting the nitrogen removal of anammox process by Fe-C micro-electrolysis. Bioresour Technol 297:122429. https://doi.org/10.1016/j.biortech.2019.122429
277. Xing B, Guo Q, Zhang Z, Zhang J, Wang H, Jin R (2014) Optimization of process performance in a granule-based anaerobic ammonium oxidation (anammox) upflow anaerobic sludge blanket (UASB) reactor. Bioresour Technol 170:404–412. https://doi.org/10.1016/j.biortech.2014.08.026
278. Xing B, Qin T, Chen S, Zhang J, Guo L, Jin R (2013) Performance of the ANAMMOX process using multi- and single-fed upflow anaerobic sludge blanket reactors. Bioresour Technol 149:310–317. https://doi.org/10.1016/j.biortech.2013.09.066
279. Xu J, Cheng Y, Jin R (2020) Long-term effects of Fe_3O_4 NPs on the granule-based anaerobic ammonium oxidation process: performance, sludge characteristics and microbial community. J Hazard Mater 398:122965. https://doi.org/10.1016/j.jhazmat.2020.122965
280. Xu X, Li H, Guo M, Zeng M, Liu W, Wu N, Liang J, Cao J (2022) Deciphering performance and potential mechanism of anammox-based nitrogen removal process responding to nanoparticulate and ionic forms of different heavy metals through big data analysis. Sep Purif Technol 301:122044. https://doi.org/10.1016/j.seppur.2022.122044
281. Xu J, Li C, Shen Y, Zhu N (2022) Anaerobic ammonium oxidation (anammox) promoted by pyrogenic biochar: deciphering the interaction with extracellular polymeric substances (EPS). Sci Total Environ 802:149884. https://doi.org/10.1016/j.scitotenv.2021.149884
282. Xu J, Li C, Zhu N, Shen Y, Yuan H (2021) Particle size-dependent behavior of redox-active biochar to promote anaerobic ammonium oxidation (anammox). Chem Eng J 410:127925. https://doi.org/10.1016/j.cej.2020.127925
283. Xu J, Wu X, Zhu N, Shen Y, Yuan H (2020) Anammox process dosed with biochars for enhanced nitrogen removal: role of surface functional groups. Sci Total Environ 748:141367. https://doi.org/10.1016/j.scitotenv.2020.141367
284. Xu A, Yu D, Qiu Y, Chen G, Tian Y, Wang Y (2022) A novel process of salt tolerance partial denitrification and anammox (ST-PDA) for treating saline wastewater. Bioresour Technol 345:126472. https://doi.org/10.1016/j.biortech.2021.126472
285. Xue Y, Ma H, Li Y (2023) Anammox-based granulation cycle for sustainable granular sludge biotechnology from mechanisms to strategies: a critical review. Water Res 228:119353. https://doi.org/10.1016/j.watres.2022.119353
286. Yan Z, Shen L, Jiao L, Tang R (2022) Effect of Fe (II) on nitrogen removal of anammox under organic matter inhibition. J Water Process Eng 46:102632. https://doi.org/10.1016/j.jwpe.2022.102632
287. Yan Y, Wang Y, Wang W, Zhou S, Wang J, Guo J (2019) Comparison of short-term dosing ferrous ion and nanoscale zero-valent iron for rapid recovery of anammox activity from dissolved oxygen inhibition. Water Res 153:284–294. https://doi.org/10.1016/j.watres.2019.01.029
288. Yang X, Chen Y, Guo F, Liu X, Su X, He Q (2020) Metagenomic analysis of the biotoxicity of titanium dioxide nanoparticles to microbial nitrogen transformation in constructed wetlands. J Hazard Mater 384:121376. https://doi.org/10.1016/j.jhazmat.2019.121376
289. Yang XL, Jiang Q, Song HL, Gu TT, Xia MQ (2015) Selection and application of agricultural wastes as solid carbon sources and biofilm carriers in MBR. J Hazard Mater 283:186–192
290. Yang Y, Li Y, Gu Z, Lu F, Xia S, Hermanowicz S (2019) Quick start-up and stable operation of a one-stage deammonification reactor with a low quantity of AOB and ANAMMOX biomass. Sci Total Environ 654:933–941. https://doi.org/10.1016/j.scitotenv.2018.11.069

291. Yang Y, Lu Z, Azari M, Kartal B, Du H, Cai M, Herbold CW, Ding X, Denecke M, Li X, Li M, Gu J (2022) Discovery of a new genus of anaerobic ammonium oxidizing bacteria with a mechanism for oxygen tolerance. Water Res 226:119165. https://doi.org/10.1016/j.watres.2022.119165
292. Yang S, Peng Y, Zhang S, Han X, Li J, Zhang L (2021) Carrier type induces anammox biofilm structure and the nitrogen removal pathway: demonstration in a full-scale partial nitritation/anammox process. Bioresour Technol 334:125249. https://doi.org/10.1016/j.biortech.2021.125249
293. Yang S, Peng Y, Zhang L, Zhang Q, Li J, Wang X (2019) Autotrophic nitrogen removal in an integrated fixed-biofilm activated sludge (IFAS) reactor: anammox bacteria enriched in the flocs have been overlooked. Bioresour Technol 288:121512. https://doi.org/10.1016/j.biortech.2019.121512
294. Yang S, Xu S, Mohammed A, Guo B, Vincent S, Ashbolt NJ, Liu Y (2019) Anammox reactor optimization for the treatment of ammonium rich digestate lagoon supernatant - Step feeding mitigates nitrite inhibition. Int Biodeterior Biodegradation 143:104733. https://doi.org/10.1016/j.ibiod.2019.104733
295. Yang J, Zhang L, Fukuzaki Y, Hira D, Furukawa K (2010) High-rate nitrogen removal by the anammox process with a sufficient inorganic carbon source. Bioresour Technol 101:9471–9478. https://doi.org/10.1016/j.biortech.2010.07.087
296. Yao Z, Wang F, Wang C, Xu H, Jiang H (2019) Anaerobic ammonium oxidation coupled to ferric iron reduction in the sediment of a eutrophic lake. Environ Sci Pollut Res Int 26:15084–15094. https://doi.org/10.1007/s11356-019-04907-7
297. Yao R, Yuan Q, Wang K (2020) Effective inhibition prevention strategy for the enrichment of anammox bacteria with low concentrations of substrates at 25 °C. J Water Process Eng 37:101514. https://doi.org/10.1016/j.jwpe.2020.101514
298. Yeshi C, Loosdrecht M, Daigger GT (2017) Mainstream partial nitritation-anammox in municipal wastewater treatment: status, bottlenecks, and further studies. Appl Microbiol Biotechnol 年101卷4期:1365–1383
299. Yi Y, Yong H, HuiPing D (2011) Effect of salt on anammox process. Procedia Environ Sci 10:2036–2041. https://doi.org/10.1016/j.proenv.2011.09.319
300. Yin S, Li J, Dong H, Qiang Z (2019) Enhanced nitrogen removal through marine anammox bacteria (MAB) treating nitrogen-rich saline wastewater with Fe(III) addition: nitrogen shock loading and community structure. Bioresour Technol 287:121405. https://doi.org/10.1016/j.biortech.2019.121405
301. Yin X, Qiao S, Yu C, Tian T, Zhou J (2015) Effects of reduced graphene oxide on the activities of anammox biomass and key enzymes. Chem Eng J 276:106–112. https://doi.org/10.1016/j.cej.2015.04.073
302. Yu X, Nishimura F, Hidaka T (2020) Anammox reactor exposure to thiocyanate: long-term performance and microbial community dynamics. Bioresour Technol 317:123960. https://doi.org/10.1016/j.biortech.2020.123960
303. Yu H, Li J, Dong H, Qiang Z (2019) Nitrogen removal performance of marine anammox bacteria treating nitrogen-rich saline wastewater under different inorganic carbon doses: High inorganic carbon tolerance and carbonate crystal formation. Bioresour Technol 288:121565. https://doi.org/10.1016/j.biortech.2019.121565
304. Yu HQ, Tay JH, Fang HHP (2001) The roles of calcium in sludge granulation during uasb reactor start-up. Water Res 35:1052–1060. https://doi.org/10.1016/S0043-1354(00)00345-6
305. Zekker I, Artemchuk O, Rikmann E, Ohimai K, Dhar Bhowmick G, Madhao Ghangrekar M, Burlakovs J, Tenno T (2021) Start-up of anammox sbr from non-specific inoculum and process acceleration methods by hydrazine. Water (Basel) 13:350. https://doi.org/10.3390/w13030350
306. Zhang X, Chen J, Li J (2020) The removal of microplastics in the wastewater treatment process and their potential impact on anaerobic digestion due to pollutants association. Chemosphere 251:126360. https://doi.org/10.1016/j.chemosphere.2020.126360

307. Zhang Q, Chen H, Liu J, Yang B, Ni W, Jin R (2014) The robustness of ANAMMOX process under the transient oxytetracycline (OTC) shock. Bioresour Technol 153:39–46. https://doi.org/10.1016/j.biortech.2013.11.053
308. Zhang X, Chen Z, Ma Y, Zhao J, Chen T, Fu H, Zhai H (2018) Acute and persistent toxicity of Cd(II) to the microbial community of anammox process. Bioresour Technol 261:453–457. https://doi.org/10.1016/j.biortech.2018.04.055
309. Zhang X, Chen Z, Ma Y, Zhou Y, Zhao S, Wang L, Zhai H (2018) Influence of elevated Zn (II) on anammox system: microbial variation and zinc tolerance. Bioresour Technol 251:108–113. https://doi.org/10.1016/j.biortech.2017.12.035
310. Zhang X, Chen Z, Zhou Y, Ma Y, Ma C, Li Y, Liang Y, Jia J (2019) Impacts of the heavy metals Cu (II), Zn (II) and Fe (II) on an anammox system treating synthetic wastewater in low ammonia nitrogen and low temperature: Fe (II) makes a difference. Sci Total Environ 648:798–804. https://doi.org/10.1016/j.scitotenv.2018.08.206
311. Zhang Q, Fu J, Wu Q, Chen J, Fan N, Huang B, Jin R (2021) Build the expressway for the salt-tolerant anammox process: acclimation strategy tells the story. J Clean Prod 278:123921. https://doi.org/10.1016/j.jclepro.2020.123921
312. Zhang Q, Ji X, Tian G, Jin R (2021) Evolution of microbial community and antibiotic resistance genes in anammox process stressed by oxytetracycline and copper. Bioresour Technol 319:124106. https://doi.org/10.1016/j.biortech.2020.124106
313. Zhang Z, Li B, Li N, Sardar MF, Song T, Zhu C, Lv X, Li H (2019) Effects of UV disinfection on phenotypes and genotypes of antibiotic-resistant bacteria in secondary effluent from a municipal wastewater treatment plant. Water Res 157:546–554. https://doi.org/10.1016/j.watres.2019.03.079
314. Zhang Q, Lin J, Kong Z, Zhang Y (2022) A critical review of exogenous additives for improving the anammox process. Sci Total Environ 833:155074. https://doi.org/10.1016/j.scitotenv.2022.155074
315. Zhang J, Peng Y, Li X, Du R (2022) Feasibility of partial-denitrification/anammox for pharmaceutical wastewater treatment in a hybrid biofilm reactor. Water Res 208:117856. https://doi.org/10.1016/j.watres.2021.117856
316. Zhang J, Qu Y, Qi Q, Zhang P, Zhang Y, Tong YW, He Y (2020) The bio-chemical cycle of iron and the function induced by ZVI addition in anaerobic digestion: a review. Water Res 186:116405. https://doi.org/10.1016/j.watres.2020.116405
317. Zhang L, Wang Y, Hao S, Dou Q, Lan S, Peng Y (2022) Anammox-synchronous zero-valent iron oxidation promoting synergistic nitrogen and phosphorus removal from wastewater. Bioresour Technol 347:126365. https://doi.org/10.1016/j.biortech.2021.126365
318. Zhang B, Wang J, Huang JJ, Razaqpur AG, Han X, Fan L (2022) Promotion of anammox process by different graphene-based materials: roles of particle size and oxidation degree. Sci Total Environ 831:154816. https://doi.org/10.1016/j.scitotenv.2022.154816
319. Zhang X, Wei D, Zhang H, He Y, Zhang S, Dai J, Wen X (2022) Comprehensive analysis of the impacts of iron-based nanoparticles and ions on anammox process. Biochem Eng J 180:108371. https://doi.org/10.1016/j.bej.2022.108371
320. Zhang X, Xia Y, Wang C, Li J, Wu P, Ma L, Wang Y, Wang Y, Da F, Liu W, Xu L (2020) Enhancement of nitrite production via addition of hydroxylamine to partial denitrification (PD) biomass: functional genes dynamics and enzymatic activities. Bioresour Technol 318:124274. https://doi.org/10.1016/j.biortech.2020.124274
321. Zhang Q, Yang G, Sun K, Tian G, Jin R (2018) Insights into the effects of bio-augmentation on the granule-based anammox process under continuous oxytetracycline stress: performance and microflora structure. Chem Eng J 348:503–513. https://doi.org/10.1016/j.cej.2018.04.204
322. Zhang L, Yue Q, Yang K, Zhao P, Gao B (2018) Enhanced phosphorus and ciprofloxacin removal in a modified BAF system by configuring Fe-C micro electrolysis: investigation on pollutants removal and degradation mechanisms. J Hazard Mater 342:705–714. https://doi.org/10.1016/j.jhazmat.2017.09.010
323. Zhang Z, Zhang Q, Guo Q, Chen Q, Jiang X, Jin R (2015) Anaerobic ammonium-oxidizing bacteria gain antibiotic resistance during long-term acclimatization. Bioresour Technol 192:756–764. https://doi.org/10.1016/j.biortech.2015.06.044

324. Zhang ZZ, Zhang QQ, Xu JJ, Shi ZJ, Guo Q, Jiang XY, Wang HZ, Chen GH, Jin RC (2016) Long-term effects of heavy metals and antibiotics on granule-based anammox process: granule property and performance evolution. Appl Microbiol Biotechnol 100:2417–2427
325. Zhang F, Peng Y, Wang Z (2022) An innovative process for mature landfill leachate and waste activated sludge simultaneous treatment based on partial nitrification, in situ fermentation, and anammox (PNFA). Environ Sci Technol 56
326. Zhang Y, Zhang X, Zhang W (2015) Research advances in anammox granular sludge. In: International conference on industrial electronics and applications
327. Zhao Y, Feng Y, Li J, Guo Y, Chen L, Liu S (2018) Insight into aggregation capacity of anammox consortia during reactor start-up. Environ Sci Technol 52:3685–3695
328. Zhao K, Wang Y, Gao L, Chai B, Zhuang P, Kou X (2022) Anammox in a biofilter reactor to treat wastewater of high strength nitrogen. J Water Process Eng 49:103169. https://doi.org/10.1016/j.jwpe.2022.103169
329. Zhao Y, Liu D, Huang W, Tran NH (2019) Insights into biofilm carriers for biological wastewater treatment processes: current state-of-the-art, challenges, and opportunities. Bioresour Technol 288
330. Zheng Z, Huang S, Bian W, Liang D, Wang X, Zhang K, Ma X, Li J (2019) Enhanced nitrogen removal of the simultaneous partial nitrification, anammox and denitrification (SNAD) biofilm reactor for treating mainstream wastewater under low dissolved oxygen (DO) concentration. Bioresour Technol 283:213–220. https://doi.org/10.1016/j.biortech.2019.01.148
331. Zheng H, Wu X, Zou G, Zhou T, Liu Y, Ruan R (2019) Cultivation of Chlorella vulgaris in manure-free piggery wastewater with high-strength ammonium for nutrients removal and biomass production: effect of ammonium concentration, carbon/nitrogen ratio and pH. Bioresour Technol 273:203–211. https://doi.org/10.1016/j.biortech.2018.11.019
332. Zheng B, Zhang L, Guo J, Zhang S, Yang A, Peng Y (2016) Suspended sludge and biofilm shaped different anammox communities in two pilot-scale one-stage anammox reactors. Bioresour Technol 211:273–279. https://doi.org/10.1016/j.biortech.2016.03.049
333. Zhou X, Wang G, Yin Z, Chen J, Song J, Liu Y (2020) Performance and microbial community in a single-stage simultaneous carbon oxidation, partial nitritation, denitritation and anammox system treating synthetic coking wastewater under the stress of phenol. Chemosphere 243:125382. https://doi.org/10.1016/j.chemosphere.2019.125382
334. Zhou J, Zhang X, Zhuang J, Xu T, Liu Y, Li W (2022) Saline short-term shock and rapid recovery on anammox performance. Chemosphere 307:135687. https://doi.org/10.1016/j.chemosphere.2022.135687
335. Zhou S, Zhu Y, Yan Y, Wang W, Wang Y (2019) Deciphering extracellular antibiotic resistance genes (eARGs) in activated sludge by metagenome. Water Res 161:610–620. https://doi.org/10.1016/j.watres.2019.06.048
336. Zhu GL, Hu YY, Wang QR (2009) Nitrogen removal performance of anaerobic ammonia oxidation co-culture immobilized in different gel carriers. Water Sci Technol J Int Assoc Water Pollut Res 59:2379
337. Zhu W, Li J, Wang B, Chen G (2020) Enhancement of pollutants removal from saline wastewater through simultaneous anammox and denitrification (SAD) process with glycine betaine addition. Bioresour Technol 315:123784. https://doi.org/10.1016/j.biortech.2020.123784
338. Zhu G, Wang S, Ma B, Wang X, Zhou J, Zhao S, Liu R (2018) Anammox granular sludge in low-ammonium sewage treatment: not bigger size driving better performance. Water Res 142:147–158
339. Zhu GL, Yan J, Hu YY (2014) Anaerobic ammonium oxidation in polyvinyl alcohol and sodium alginate immobilized biomass system: a potential tool to maintain anammox biomass in application. Water Sci Technol J Int Assoc Water Pollut Res 69:718–726
340. Zhu W, Li J, Dong H, Wang D, Zhang P (2017) Effect of influent substrate ratio on anammox granular sludge: performance and kinetics. Biodegradation

Chapter 2
Aerobic Granular Sludge

2.1 Introduction

Aerobic granular sludge (AGS) is a self-immobilizing microbiota consisting mainly of aerobic and parthenogenic bacteria, unlike anaerobic granular methanogenic sludge. Aerobic granular technology is considered a promising technology for wastewater treatment, but it is not yet applied on a large scale. AGS are mainly in sequencing batch reactors (SBR) for water culture under force-selective pressure [28].

Research on AGS technology in wastewater treatment has mainly been at the laboratory scale and to a lesser extent at the pilot scale. The first pilot plant for AGS technology was launched in September 2003 in Ede, The Netherlands. The plant consists of two parallel bioreactors, each 6 m high and 0.6 m in diameter. The footprint is only 20–30% of the traditional process, the residual sludge volume is small, and less oxygen supply is required, saving about 30% of energy consumption. The pilot SBR system was set up at the Hefei Zhu zhuan jing Wastewater Treatment Plant (WWTP) in China, and AGS (granules size of 0.2–0.8 mm) was successfully cultivated, obtaining good effluent quality. In the following years, South Africa, America, Europe and Australia have designed, built, and operated WWTPs with AGS technology, realizing the practical application of AGS technology.

In 2005, the first workshop on AGS was organized by the International Water Association (IWA) at the Technical University of Munich, Germany, and provided a clear definition of AGS: composed of granules believed to be microbial aggregates, does not coagulate under low hydrodynamic shear and settles significantly faster than activated sludge flocs [21].

After a long period of laboratory research, AGS has been found to have several advantages compared with traditional flocculated activated sludge. First, it has superior settling performance due to the high density of AGS and its inability to coalesce, resulting in better settling speed and stronger impact load resistance. Second, AGS has the ability to adsorb heavy metals and toxic substances due to its large specific

G. Dawen and M. Nabi, *Novel Approaches Towards Wastewater Treatment*,
Springer Water, https://doi.org/10.1007/978-3-031-55189-5_2

surface area and granules surface pore density, as well as the ability of microorganisms to secrete extracellular polymers with high viscosity. The structure of AGS is divided into an anaerobic zone, an anoxic zone, and an aerobic zone from the inside out. This structural advantage gives it the ability to complete simultaneous nitrification and denitrification, and because the anaerobic and aerobic zones are combined, it meets the conditions for simultaneous denitrification and phosphorus removal [74, 141].

At the same time, the shortcomings of AGS will limit its promotion in industry to a certain extent. AGS may have problems such as decreased stability after long-term operation or storage, as well as destabilization and eventual disintegration during long-term operation and storage. In the operation of AGS, when the AGS granules size is too large, the mass transfer of the substrate is severely restricted, and cellular hydrolysis and death occur inside the granules, leading to granule disintegration. On the other hand, fast-growing heterotrophic bacteria and overgrown filamentous bacteria can cause AGS to gradually change into a filamentous structure, leading to its instability. Additionally, under certain special circumstances, the loss of activity of the strains in the granular sludge can cause some strains in the granular sludge to lose their ability to produce polysaccharides (PS) and proteins (PN), which can also lead to the disintegration of the granular sludge.

According to the statistical heat map (Fig. 2.1), it has been found that popular research contents include extracellular polymers, sludge, microorganisms, nitrogen removal, denitrification, etc. In terms of time, the research trend in recent years is gradually shifting towards new directions such as carbon emission and storage, as well as reactivation.

This chapter provides a theoretical basis for the development of AGS technology through a detailed explanation of the characteristics of AGS, influencing factors, hypothesis of the formation mechanism, factors affecting the strength and stability

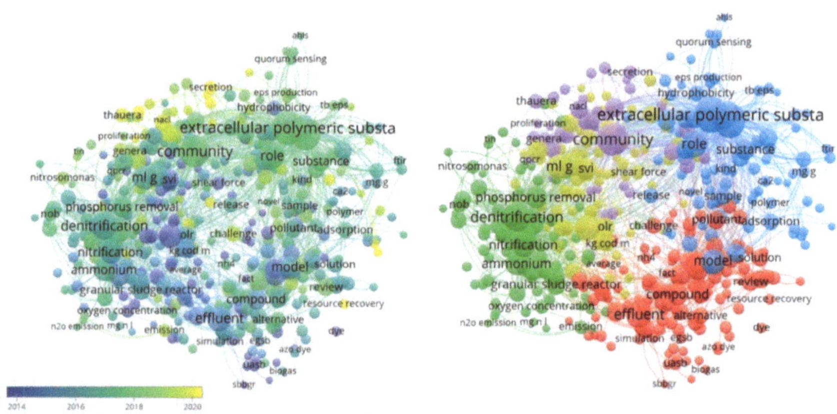

Fig. 2.1 Multiple overlapping visualization map of AGS from Web of Science. The size of the nodes indicates the number of keywords, and the shade of the color indicates the proximity of the year

of granules, rapid formation/initiation technology of AGS, and the combination of AGS in wastewater and membrane technology.

2.2 Physical and Chemical Characteristics of Aerobic Granular Sludge

This section summarizes the characteristics of AGS and the factors that influence them. These characteristics include physical and chemical properties, microbial community, and others. Influencing factors such as substrate, temperature, organic loading rate (OLR), starvation time, and dissolved oxygen (DO) also play a crucial role in determining AGS's properties. The granules properties of AGS, including good settling properties, specific gravity, and density, are responsible for its handling properties and high mechanical strength. Additionally, the granular properties of AGS, which include physical properties, chemical properties, and microbial community, contribute to its unique characteristics, such as its resistance to sludge expansion, ability to withstand high organic loads, and integration of different types of microorganisms (aerobic, anaerobic, and facultative microorganisms).

2.2.1 Physical Characteristics

2.2.1.1 Settleability

The settling capacity of sludge is related to the retention capacity of the material in the reactor and its solid–liquid separation capacity. The settling speed of AGS can reach 18–90 m/s, and in some cases, even up to 130 m/s; Meanwhile, the sludge volume index (SVI) of AGS is generally lower than 80 ml/g [28]. The settling performance of AGS determines the efficiency of solid–liquid separation in the reactor. As AGS exists in dense granular form, its SVI is usually around 30 ml/g (12.6–64.5 ml/g), and the SVI of ordinary activated sludge is 100–150 ml/g, which is much lower than the SVI value of ordinary activated sludge. The sedimentation rate of AGS is related to its granules size, generally 30–70 m/h, while the sedimentation rate of traditional activated sludge is around 8–10 m/h [30]. Thus, AGS exhibits good sedimentation performance. The increase of settling rate can not only shorten the sedimentation time and reduce the volume of the sedimentation tank, but also enhance the concentration of microorganisms in the reactor, and the increase of microbial concentration can obtain a higher pollutant degradation rate. The relationship between granules size and settling performance. The increase of settling rate can not only shorten the sedimentation time and reduce the volume of the sedimentation tank, but also enhance the concentration of microorganisms in the reactor, leading to a higher pollutant degradation rate. The relationship between granules size and settling performance can be determined by Stokes' law, which states that settling velocity is proportional to the square of the granules radius. Strubbe et al. [88] found that AGS settling performance is not only a physical characteristic but also a limiting factor in its ability to remove nitrogen, compared to activated sludge, which had a five times higher settling velocity. Specific gravity and moisture content are also important factors that affect AGS settling performance and solid–liquid separation efficiency.

2.2.1.2 Specific Gravity and Moisture Content

AGS has compact structure and high density, which is conducive to microbial residence and solid–liquid separation. The density of AGS can be determined by utilizing 4,6-diamidino-2-phenylindole hydrochloride and fluorescence image processing techniques. The density of AGS can be estimated through the application of 4,6-diamidino-2-phenylindole hydrochloride and fluorescence image processing. The density of AGS is higher than that of the activated sludge. Density is often used to characterize the compactness of granules and assess their stability [63].

The proportion of AGS is generally about 1.004–1.008, and the moisture content is generally 97–98%, which is lower than ordinary activated sludge (moisture content above 99%). Therefore, the amount of AGS used is at least half less than that of ordinary activated sludge.

2.2.1.3 Mechanical Strength

The mechanical strength of AGS is an important parameter characterizing the structural of AGS and its ability to withstand high abrasion and hydraulic shear forces. Granules that contain higher levels of Ca^{2+} exhibit higher compressive strength than those with lower levels. The mechanical strength of the AGS is critical to the stability of the granules and can be expressed by expressed by determining the integrity factor (IC). To determine the IC value, the sludge mixture is sheared for a certain time using a vortex mixer. Then, the supernatant is decanted, and the suspended solids (SS) are determined. The IC value is calculated using the following formula:

$$\mathrm{IC_t} = (1 - \frac{SS_t}{SS_0}) \times 100\%$$

SS_0 denotes the initial total amount of SS, while SS_t represents the quantity of SS present in the supernatant after t minutes of the test.

The fractal dimension of the AGS image can also reveal the correlation between the compactness and stability of AGS [2]. The higher the value of the fractal dimension, the greater the stability and compactness of the granular structure. Furthermore, the strength of AGS can be visually analyzed by comparing the rates of complete disintegration and dispersion under the influence of ultrasonic fragmentation [109].

2.2.1.4 Shape and Size

Compared with flocculated sludge, AGS has a clear outline and appears spherical or elliptical, with granules sizes ranging from 0.2 to 16 mm. The laser scanning granules size analyzer can measure smaller size granules, while larger size granules are measured computationally using image analysis software. Granules size affects some of the basic physical properties of AGS. In general, the settling properties, density, and biomass of the granules can increase as granules size increases, but sometimes an increase in granules size is accompanied by a decrease in granules strength and specific surface hydrophobicity [63]. The morphological coefficient of the granular sludge system was stabilized at 0.45. AGS has better settling capacity, density, and mechanical strength when the granules size is less than 4 mm. However, if the size of granules is larger than 4 mm, these characteristics deteriorate.

2.2.1.5 Porosity

Porosity is related to substrate transport and DO transfer, and pore wall blockage, restricted substrate diffusion and elevated DO will adversely affect the bioactivity of aerobic granules. Porosity has a major impact on controlling sludge biological activity. The porosity of AGS can be analyzed by size exclusion chromatography, where porosity decreases with increasing granules size [63]. The porosity of AGS ranges from 68 to 93%. Studies have shown that there is a positive correlation between the biological activity of AGS and porosity [99, 152] (Table 2.1).

Table 2.1 Comparison of the physical characteristics of different sludge

Sludge type and characteristic	AGS	Activated sludge	Anaerobic granules
Settling	10–90 m/h	2–10 m/h	< 20 m/h
Size	0.2–5.0 mm	< 0.2 mm	0.2–2.0 mm
Degree of compactness and density	High	Low	High
Shape	Regular and spherical	Irregular and filamentous	Regular and spherical
Sludge volume index			
5 min	30–60 mL/g	–	–
30 min	30–60 mL/g	110–160 mL/g	–
Redox microenvironments	Granules consist of unique aerobic, anoxic, and anaerobic microbial layers	Minimum feasibility for anaerobic zones	Facultative and obligate anaerobic
References	[22, 35]		

2.2.2 Chemical Properties

2.2.2.1 Cell Surface Hydrophobicity

Thermodynamic theory indicates that an excess increase in cell surface hydrophobicity decreases the Gibbs free energy in cells, thus promoting cell aggregation. The surface hydrophobicity of AGS is much higher than that of flocculated sludge, and the average contact angles of flocculated sludge and aerobic granules are 35° and 46.3°, respectively [150]. In the SBR reactor, the cell surface hydrophobicity increased from 43.2 to 78.2% when the sludge changed from flocculated to granular state compared to flocculated sludge, and the cell surface hydrophobicity of AGS increased by 2.5–3.5 times [150].

2.2.2.2 Extracellular Polymeric Substance (EPS)

EPS, are organic polymers produced by microorganisms, primarily bacteria, under specific environmental conditions. These substances form a three-dimensional matrix in which bacteria can be embedded. The main components of EPS include PN, PS, humic acids, nucleic acids, lipids, and various other substances.

The different forms and components of EPS, including soluble EPS (S-EPS), loosely bound EPS (LB-EPS), tightly bound EPS (TB-EPS), PN and PS, play distinct roles in the process of granulation. The content of TB-EPS accounted for 72.81–93.11% of the total EPS [79]. PS also contributes significantly to granules structure formation and gelling properties as well as PN to its stability, hydrophobicity and surface cell charge.

The EPS of AGS is divided into biodegradable and non-biodegradable, with most of the biodegradable EPS located in the core of AGS and the non-biodegradable EPS located mainly in the shell of the AGS. Therefore, only the non-biodegradable EPS produced by AGS contribute to the structural stability of AGS [119]. The application of EPS may affect granules formation by increasing the location of contact between the active bacteria and the suspended granules [8] (Fig. 2.2).

The physical and chemical characteristics are the most basic and proven elements to understand AGS. These characteristics, which reflect some of the unique properties of AGS, allow us to better explore its environmental applications. In addition, we can further investigate the factors that affect the performance of AGS with respect to its characteristics, so that we can better improve or regulate the environment that is more favorable to the application of AGS.

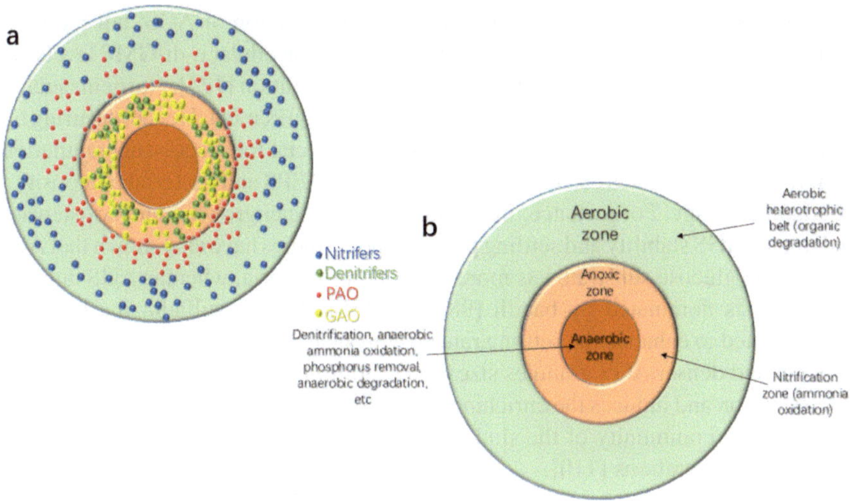

Fig. 2.2 Composition of aerobic granules across varying do concentrations

2.3 Factors Influencing Aerobic Granular Sludge

The granulation process of sludge can be divided into three stages: acclimation, granulation and maturation. Initially, granulation starts in the reactor in the form of mycelial balls and begins to grow at an accelerated rate, known as "granulation onset". The corresponding period from the start of operation to "pellet formation" is referred to as the sludge acclimation phase. The point of maturation is reached when the initial pellet can grow sufficiently and the biomass concentration remains constant. The pelleting stage corresponds to the period from "pellet initiation" to the point of pellet maturity [115].

Many studies have so far shown that the formation, structure and its stability of AGS can be influenced by various factors, such as the type of seed sludge, the composition of the substrate, EPS, metal ions, selection pressure, temperature, etc. (as shown in Fig. 3.4).

2.3.1 Substrate

The composition of substrates used for culturing AGS also influences the structure and stability of the sludge. The type of organic carbon sources and Carbon–nitrogen ratio (C/N) can have the most significant impact on granulation. High levels of readily available substrate can cause an initial crash in biomass yields, while excessively low concentrations can hinder granulation. An imbalance between carbon and nitrogen due to high levels of easily degradable compounds can lead to toxicity from ammonia or nitrite. Nitrate also does not only promote the formation of AGS, but the long-term effect of nitrate leads to the downregulation of intracellular c-di-GMP levels and reduced EPS secretion, which may lead to cell disintegration [124]. Nutrient, such as phosphate, can limit granulation due to their finicky requirement. Hence, ensuring the correct substrate ratio and nutrient metal content is crucial for successful granulation. Differences in substrates can result in variations in the EPS and bacterial composition of granular sludge. For instance, filamentous bacteria can grow in dairy wastewater and impact AGS stability and settling performance. Studies have also shown that AGS cultured on glucose substrate has more filamentous bacteria, whereas AGS cultured on acetate is dominated by bacilli [98, 139, 138]. Adding sulfadiazine has been demonstrated to enhance the settling rate and physical strength of AGS, possibly due to larger and denser AGS granules sizes. The addition of sulfadiazine also increases EPS secretion and induces the enrichment of resistance genes (Sul-1, Sul-2), altering the microbial community of the sludge and enriching various dominant bacteria at different concentrations [110].

2.3.2 Dissolved Oxygen

The dissolved oxygen (DO) concentration also affects the formation of granules. The mechanism of the effect of DO is that lower DO may lead to sludge expansion, which in turn competes with AGS to affect the formation and stability of AGS. On the other hand, AGS has mass transfer limitations for DO due to its large size and compact structure. Therefore, changes in DO concentration affect the transfer and removal efficiency of substrates in aqueous solution, which in turn affects the structure and stability of AGS. Due to the structure of AGS, nitrifying bacteria grow slowly and cannot successfully compete with heterotrophic bacteria for DO in the outer layer, which leads to a decrease in nitrification efficiency. The variation in DO affects the mass transfer rate in AGS reactors, which in turn affects the oxygen penetration depth of AGS and ultimately the nitrification–denitrification process. Yuan and Gao [136] investigated the effect of different DO concentrations on the biological denitrification process using an AGS sequencing batch reactor (SBR) and concluded that high DO concentrations favored nitrification and increased the volumetric removal of NH_4^+–N, while low DO concentrations favored TN removal but prolonged the duration of nitrification.

2.3.3 Extracellular Polymeric Substance (EPS)

EPS are cellular secretions composed of PN and PS that are associated with intercellular adhesion. Microbial flocculation, sludge settling, and dewaterability are all influenced by EPS. EPS can be classified as biodegradable or non-biodegradable, with the former mostly located in the core region of AGS and the latter mostly in the outer shell of the granules. EPS can also be divided by structure into TB-EPS and LB-EPS. An increased protein mass fraction in TB-EPS promotes the formation of AGS, and some PN in LB-EPS affect the electronegativity and dewatering of sludge. These structures improve the stability and the ability of AGS to cope with toxic and hazardous substances.

Peng et al. [79] conducted a series of studies investigating the properties of three EPS and their constituent PN and PS during AGS granulation and their effect on the aggregation efficiency of AGS. Correlations between the various forms of EPS, their components, and sludge properties during granulation were investigated using Pearson correlation analysis. The results indicate that TB-EPS has a stronger impact on the zeta potential and hydrophobicity of the sludge surface compared to S-EPS and LB-EPS. During the process of granulation, the PN content and relative hydrophobicity of TB-EPS increased from 30.38 ± 0.12 mg/g VSS to 83.99 ± 2.05 mg/g VSS and 37.63%–71.25%, respectively. The contribution of TB-EPS to the flocculation capacity of the settling tank increased from 13.30 to 35.37%. Furthermore, the contribution of PN to the flocculation capacity of the settling tank was 76.93%. The contribution of PN to the flocculation capacity of the sedimentation tank was 76.93%, which is 6.8 times higher than that of PS.

Tay et al. [97] suggested that β-PS in TB-EPS could enhance the adhesion between microorganisms and strengthen the structural stability of AGS. In contrast, Liu et al. [64] suggested that the non-cellular protein core in the sludge provides the backbone structure for the stable operation of AGS. Therefore, an in-depth analysis of the functional roles of different components of EPS can help to promote the development of AGS technology. Some studies have shown that when the β-polysaccharide in the extracellular polymer was hydrolyzed, the AGS fragmented, indicating that it had an effect on the structure and stability of AGS [77]. Wang et al. [116] used inoculated sludge, laboratory cultured AGS and AGS from WWTPs as the objects of their study. They observed and analyzed the EPS formed by Fourier infrared spectroscopy and three-dimensional fluorescence spectroscopy. The results showed that the absorption peak intensities of aromatic amino acids, tryptophan/tyrosine and aspartic acid PN were significantly higher in the EPS of the two AGSs compared with the inoculated sludge. Moreover, the presence of four unique organic components, namely phenylpropylamine, hydroxyurea, diethylenebenzene and hexadecylcyclooctasiloxane, in the EPS of the two AGSs is presumed to help shorten the granulation process of inoculated sludge and improve the structural stability of AGS. The effects of EPS on AGS have been shown to be related to hydrophobicity, cell surface charge, and bridging functions, as explained below [30].

Cell surface hydrophobicity: EPS contains more PN than hydrophilic PS and is therefore more hydrophobic. Moreover, the hydrophobicity of the EPS is influenced by the hydrophobicity of the sludge granules shell, which is stronger than that of the core. Changes in EPS polarity can significantly alter the sludge surface hydrophilicity/hydrophobicity and structural compactness, which in turn affect the dewatering properties of sludge. Specifically, the hydrophobicity of EPS is due to an increase in the content of tyrosine-like PN, particularly those associated with the hydrophobic acid fraction (HPO-A) [39]. The trend of relative hydrophobicity of PS in different forms of EPS was less prominent compared to that of PN. There was an overall increasing trend in the relative hydrophobicity of PN in different forms of EPS. Additionally, compared to TB-EPS, the hydrophobicity of PN and PS in LB-EPS is lower, but there is an overall irregular trend. This can be attributed to the higher content of PN in TB-EPS [79].

Cell surface charge: There is also an influential relationship between EPS and intercellular charge. The electrostatic repulsion between cells increases with the absolute value and magnitude of the zeta potential. The electrostatic repulsion between sludge granules is unfavorable for cell aggregation [79]. EPS can granulate sludge by reducing the cell surface charge and overcoming electrostatic repulsion. The protein/polysaccharide ratio has a strong inverse correlation with the AGS surface charge. Increasing the protein to polysaccharide ratio can reduce the negative charge on the surface of bacteria, thus reducing the electrostatic repulsion between bacteria and further promoting aerobic granulation. In addition, LB-EPS from activated sludge method contains more PS with a large amount of negative charges that are easily bound to water. In contrast, PN contains a large amount of positively charged amino groups that can chelate with metal ions to reduce the negative charge on the cell surface and neutralize the negative charges of phosphate and carboxylic acids in PS and DNA [79].

Bridging function: PS in EPS play a crucial role as binders, helping to bind bacteria together to form microcolonies. Moreover, PS present on the surface of small granules or agglomerates act as bridges, enabling them to capture or link other small granules or agglomerates together to form larger size granules (Fig. 2.3).

2.3.4 Metal Ions

The electric charge on the bacterial surface can promote mutual adsorption and aggregation of AGS. Usually, bacteria have a negative charge on their surface, and cells with the same charge will have difficulty in aggregating due to electrostatic repulsion, which can be reduced by the addition of, for example, calcium ions and magnesium ions. On the one hand, metal ions neutralize the negative charge on the cell surface, which facilitates bacterial aggregation. Secondly, the formation of metal ions precipitates accelerates microbial aggregation. Finally, metal ions form ionic bonds on the granules surface, while metal ions can cross-link with EPS, enhancing the stability of AGS.

Mg^{2+} and Ca^{2+} ions have been reported to impact biofilm formation through their influence on electrostatic interactions and their involvement in essential cellular functions as important cellular cations and enzyme cofactors. Furthermore, Ca^{2+}, Mg^{2+}, and Fe^{2+} ions can bind to negatively charged cells and serve as microbial nuclei [37].

Hao et al. [37] investigated the effect of common metal ions, including Ca^{2+}, Mg^{2+}, Cu^{2+}, Fe^{2+}, and Zn^{2+} on microbial attachment in biological wastewater treatment. The results indicated that aerobic granular activated sludge (AGAS) exhibited

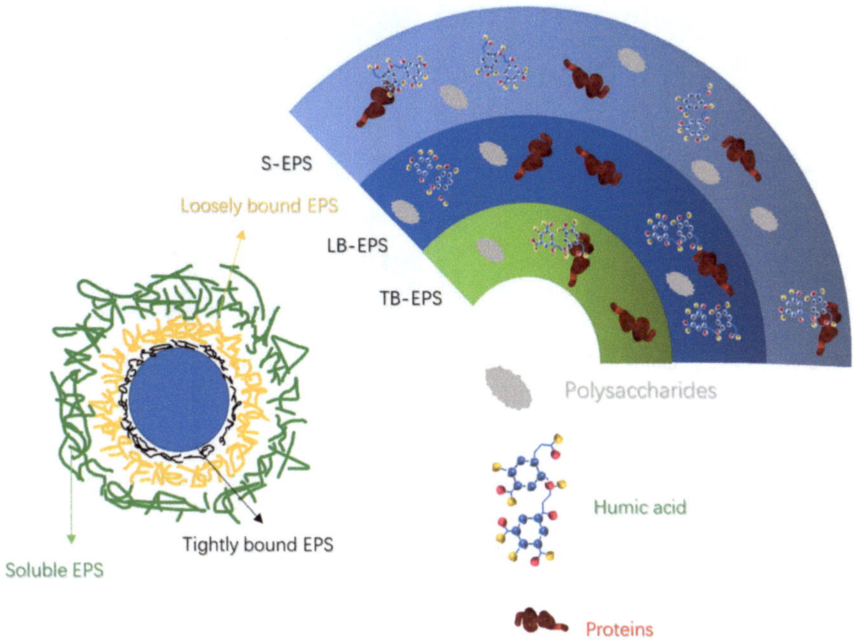

Fig. 2.3 Composition of EPS in AGS

sensitivity to metal ions. Ca^{2+}, Mg^{2+}, and K^{+} enhanced the microbial attachment of AGAS at appropriate concentrations, but Cu^{2+}, Fe^{2+} and Zn^{2+} greatly inhibited the attachment process of AGAS even at very low concentrations.

In addition, Guo et al. [32] looked at the role of calcium ions in AGS reactor start-up in terms of their effects on the physicochemical properties and biological activity of sludge and demonstrated the importance of calcium ions on AGS reactor start-up and the mechanism of their effects on different granules properties. Three sets of reactor parallel reactor experiments were conducted with Ca^{2+} as the variable. The analysis of sludge granulation results and pollutant removal status in the three reactors led to the following conclusions: (1) Ca^{2+} ions can increase the inorganic composition of the sludge to promote granulation; and (2) Ca^{2+} ions can ensure the biological activity of the granules by promoting the enrichment of functional microorganisms and the synthesis of metabolic enzymes (Fig. 2.4).

2.3.5 *Surface Charges*

Surface charge is the electrical charge present at an interface, which creates an electric field that causes granules to interact with each other in a repulsive or attractive manner. Such interactions can lead to the aggregation of colloids. Objects that are

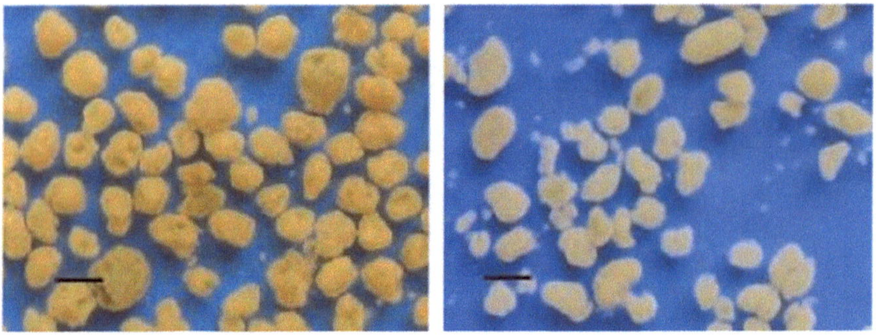

Fig. 2.4 Aerobic granules with Ca^{2+} (left) and Mg^{2+} (right) added (Bar is 2mm) [30]

in fluids are generally charged due to the presence of ions, both positive (cations) and negative (anions), which interact with the surface and become adsorbed onto it. Sludge granulation is an intercellular self-aggregation process that is accompanied by changes in surface properties and surface charge. Among the current theories of microbial aggregation, the widely accepted models are the DLVO model and the polymer bridging model, both of which involve the role of sludge surface charge and electrochemical properties. The extended DLVO theory explains that a reduction in negative charge leads to a reduction in electrostatic repulsion between surfaces, resulting in tightly bound granular sludge [89].

Zeta potential is a crucial parameter for determining surface charge, and it gradually decreases as the negative charge on the surface of granular sludge decreases. In acidic or alkaline conditions, the ionization of chemical groups in EPS can result in colloidal granules to carry a charge. Some literature suggests that positively charged amino groups in the effluent, such as PN, can neutralize the negative charges of hydroxyl and phosphate groups on the surface of the sludge granules, thus reducing the surface potential of the sludge. If the pH of the effluent changes, ionization reactions of the carboxylic acid and amine groups in protein molecules can also be inhibited or promoted, affecting the charge sign on the cell surface. Therefore, the zeta potential in the bilayer on the surface of the granules is strongly related to the pH in the effluent.

In the stern double layer model, counter ions in the adsorption and diffusion layers reach a dynamic equilibrium, and when the concentration or valence of ions inside the solution increases, more counter ions will enter the adsorption layer to establish a new equilibrium. During this process, the potential difference between the sliding surface relative to the charge of the colloidal pollutant in the effluent changes, leading to colloidal movement and, consequently, changes in the zeta potential. According to the DLVO theory, charges attract and repel each other, which makes the compressed double layer of colloids thinner or thicker. The reduction of the zeta potential is favorable for the flocculation or aggregation of the colloids.

By exploring the measurement method and characterization of the zeta potential of AGS, Su et al. [89] found that the zeta potential of fine granules gradually decreased

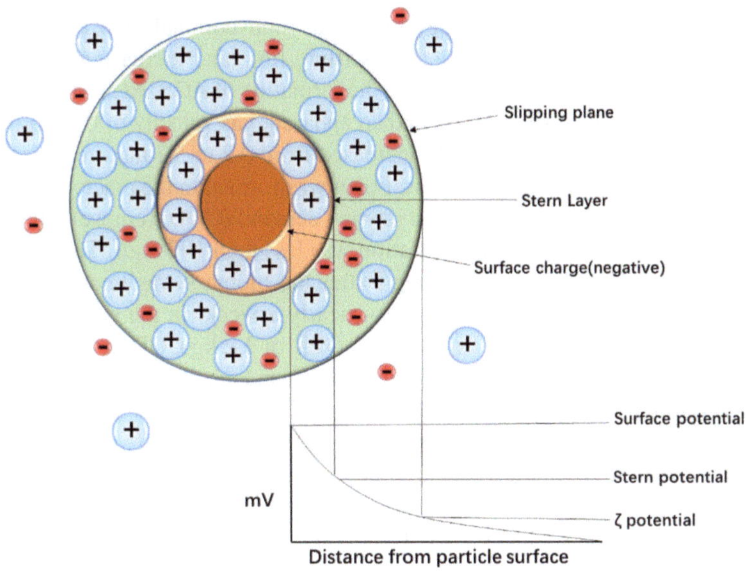

Fig. 2.5 Charge distribution on the AGS surface

from − 19.5 to − 10.0 mV until − 12.0 mV during aerobic granulation. It is thus clear that the surface charge and zeta potential of granular sludge have a degree of influence on the formation and stability of AGS.

Zeta potential calculation formula:

$$\zeta = \mathrm{K}\pi\eta\mathrm{u}/\mathrm{DE}$$

K—constancy related to the shape of the colloidal granules (K = 6 for spherical granules and K = 4 for rod-shaped granules), η—coefficient of viscosity of the liquid (absolute viscosity).

u—the velocity of movement of the colloidal granules relative to the liquid, cm/s, D—the dielectric constant of the liquid, E—electric field, V/m (Fig. 2.5).

2.3.6 Hydraulic Selection Pressure

Hydraulic selection pressure has an important effect on AGS. The selection pressure comprises hydraulic shear force and settling time. A certain amount of hydraulic shear can promote the structural stability of AGS. The selection pressure and shortened settling time can wash away organisms that are difficult to settle during AGS granulation, and increasing OLR can ensure the growth of new organisms. Hydraulic shear can also screen bacteria on the biofilm. It has been shown that an updraft velocity higher than 1.2 cm/s is beneficial to the formation of AGS [43].

Wang et al. [115] used seed sludge which from a conventional (WWTP) to cultivate AGS in an SBR reactor. Shortening the settling time and increasing the selective pressure generated by the OLR promoted the formation of AGS. At an HRT of 15.4 h, an OLR of about 1.8 kg COD/m^3 d, and an apparent gas velocity of 0.0035 m/s, the minimum settling velocity gradually increased from 0.2 to 5.7 m/h, resulting in the flushing out of poorly settling biomass. Foam was formed in the reactor after 1–2 d. In the first stage, the average chemical oxygen demand (COD) removal rate reached 91.8% and the effluent COD concentration was less than 100 mg/l. The effluent SS decreased from 4200 mg/l to 2100 mg/l after the biomass was treated with foam and was stabilized at about 2100 mg/l for nearly a week. In the second stage, the operation was maintained for 8 weeks at an OLR of 3.2 kg COD/m^3 d and an HRT of 9 h. The COD removal rate reached about 86% and the effluent SS concentration decreased from 2100 mg/l to less than 1000 mg/l. On day 67, when the initial granular sludge appeared, the COD removal rate dropped rapidly to 73%, but rebounded to 89% after 2 weeks. When the initial granules were formed, most of the poorly settled filamentous and flocculated sludge was flushed out due to the OLR. In addition, the growth rate of granular sludge was slower than that of flocculated sludge due to diffusion limitations. Thereafter, the biomass concentration decreases rapidly at the beginning of granular emergence. Over time, sludge with good settling properties accumulates, leading to an increase in biomass concentration, an increase in removal efficiency, and a return to normal. When the initial granular sludge appeared for 2 weeks and the COD removal rate reached more than 88%, the OLR of the Round 3 reactor reached 4.8 kg COD/m^3-d, the HRT decreased to 5.9 h, and the apparent gas velocity was maintained at 0.0175 m/s. Although the minimum settling velocity increased significantly from 10.8 to 32.4 m/h, the average COD removal rate and effluent COD during this period remained stable. The effluent SS concentration decreased rapidly from 1000 to 400 mg/l. The effluent SS concentration was still high at the end of this run because most of the flocculated sludge and small size granules in the reactor with settling velocity less than the minimum settling velocity were easily washed out.

2.3.7 *Settling Time*

The settling time allows the sludge to be screened, keeping the sludge with good performance and removing the sludge with poor performance. At the same time, shorter settling time can select cells with strong surface hydrophobicity and improve extracellular polymer to promote cell aggregation. The granules size increases with the shortening of settling time. And when settling time is 5–10 min, larger granules sludge will be formed, and settling time is greater than 15 min granules sludge is difficult to form. The short settling time allows the overly dispersed flocs and granules to be washed away and helps to shorten the reactor cycle time, which can improve the treatment efficiency to a certain extent.

Yin et al. [134] investigated the effect of settling time on the instability of AGS under nitrogen deficient conditions. The instability of AGS under nitrogen deficiency was overcome by shortening the settling time. The results showed that the model could fit the granules diameter growth well when the settling time was shortened to 5 min, 2 min and 1 min in nitrogen deficiency, with the mean granules size increasing to 324.7 ± 2.8, 400.1 ± 2.9 and 548.9 ± 3.9 μm, respectively. Demonstrating that the short settling time enhanced granulation under nitrogen deficiency conditions. In contrast, lack of nitrogen resulted in unstable AGS with a broken, irregular and loose structure and poor settling properties. After shortening the settling time, large granular sludge with regular structure and good settling properties appeared in the reactor. In terms of microbial communities, the dominant genus of *Microbacterium* shifted from the *Micropruina* (24.5 and 25.9%) to the unclassified *Microbacterium* family (22.0%) as the settling time was shortened. *Lysobacteria* increased from 1.0% to 2.6%, which may be related to cell lysis at short settling times. *Norank_f* of the anaerobic Streptomyces family increased from 2.7% to 6.4%. Large filamentous bacteria decreased by 1.5%, 0.5% and 0.4% respectively. The short settling time leads to the loss of filamentous bacteria due to their poor settling performance; another aspect, the short settling time enhances the conversion of COD to PS, which may reduce the competitive advantage of filamentous bacteria. In addition, when the settling time was shortened from 5 to 1 min, the abundance of red bacteria decreased from 1.0 to 0.6%.

It can be seen that settling time can influence the stability of AGS to some extent by affecting EPS, physical scrubbing and microbial communities, so controlling settling time is one of the most important strategies for the formation of stable AGS.

2.3.8 Temperature

Temperature has an effect on the rate of oxygen transfer in water, which in turn affects the formation of AGS. At 8 °C, sludge aggregation occurs when aeration is stopped, but the granules are unstable. Starting from 20 °C and gradually lowering the temperature to 15 and 8 °C, the formed granules exhibited increased stability. At 30 °C, the AGS formed during incubation showed a more compact structure, improved settling properties, and higher biological activity.

Temperature can affect the density and viscosity of the medium. Differences in temperature can cause molecular movement to be affected, thus affecting the settling and stability of AGS. Winkler et al. [123] explored the effect of temperature on the settling performance of AGS by determining the settling velocity of granules at different temperatures (5–40 °C). The results were consistent with general physical mass and heat transfer theorems. The results of this study, suggest that the corresponding slow settling of small granules due to lower temperatures at reduced water viscosity and increased water density may be a key reason for the difficulty in starting the granular sludge reactor. In addition, experiments have been designed to investigate the effect of starting AGS reactors at low to medium temperatures on the performance of granular sludge. de Kreuk et al. [21] investigated the biological performance of AGS under several temperature conditions, including at 8 °C and at 25 °C, and then gradually cooling down. Experiments showed that the granular sludge reactor produced irregular granules when started at 8 °C and that once aeration was stopped, the granules aggregated, leading to severe biomass loss and erratic operation. A start-up temperature of 20 °C and lowering to 15 and 8 °C had no effect on the stability of the pellets and biomass was easily maintained in the system. Analysis of the data and discussion revealed that the activity of the outer layer of the pellet was reduced at lower temperatures, and the depth of oxygen penetration could be increased, resulting in a larger volume of aerobic biomass, compensating for the reduced activity of individual organisms. As a result, the denitrification capacity of granular sludge becomes weaker at lower temperatures, making nitrogen removal less efficient.

The denitrification performance and denitrification and phosphorus removal performance of AGS changed at low temperatures. Bao et al. [12] performed SBAR at 10 °C to study the performance of nitrification process. The results showed that the AGS reactor had a good ability to remove COD, N and P at the same time. When the initial concentrations of COD, NH_4^+–N and PO_4^{3-}–P were 560, 56 and 22.4 mg/l, respectively, the removal rates of COD, NH_4^+–N and PO_4^{3-}–P are 89.5%, 98.9% and 95.5%, respectively. The concentrations of NO_2^-–N and NO_3^-–N in the effluent were lower than 3.4 and 0.5 mg/l, respectively. The efficiency of synchronous nitrification and denitrification was 83.3%. When the initial concentrations of COD and NH_4^+–N were 560–1200 mg/l and 56 mg/l, respectively, the removal rates of COD and NH_4^+–N were 93.0% and 77.45%, respectively. The concentrations of NO_2^-–N and NO_3^-–N in the first phase were 0.4 mg/l, in the second phase they were 11.8 mg/l and 0.4 mg/l, and the third phase was 21.0 mg/l and 0.5 mg/l respectively. These results showed that at this temperature, the granular sludge reactor achieved good removal performance.

Ab Halim et al. [1] conducted research to investigate the performance of AGS cultivation at high temperatures with regards to organic matter and nutrient removal. They used three SBR reactors with the same influent load to start up and explore the removal of COD, phosphate, and ammonia at various high temperatures ranging from 30 to 50 °C. The study revealed that the SBR reactor could successfully cultivate AGS at these high temperatures, indicating that the granulation process can be achieved even in such conditions. However, it should be noted that high temperatures may have an impact on granule characteristics and treatment efficiency.

2.3.9 Seed Sludge

Seed sludge characteristics are known to have a significant impact on the granulation and treatment performance of AGS processes. Kosar et al. [52] compared different seed sludge characteristics on granulation and AGS system treatment performance by using a mixture of sludge from conventional activated sludge process (Phase I) alone and from conventional activated sludge process and efficient activated sludge process (Phase II) in an AGS system. The results showed that despite the larger granules size of the sludge inoculated in the second stage, the granules size of the AGS from the first stage (124 μm) was instead larger than that of the AGS from the second stage (86 μm) after stabilization operation, indicating that the seed sludge had an effect on the anoxic/anaerobic volume in the granular matter and thus on denitrification and phosphate accumulation.

2.3.10 pH

pH also has a significant effect on AGS. Fluctuations in pH can alter the cell surface charge and thus influence AGS flocculation. The population structure and the rate of granules growth can also be influenced to some extent by pH. Most microorganisms cannot survive in highly acidic or alkaline environments, and the morphology and structure of AGS can deteriorate as microorganisms lose their ability to self-aggregate [62]. At a pH of 3.0, AGS is dominated by fungi. Fungi and EPS are uniformly dispersed in the AGS, and the granules structure is loose and easily broken by aeration, despite the short granulation time and small granules size. At pH 8.1, AGS is dominated by bacteria and is distributed in the outer layer of the pellets. Although the structure is dense and the granules size is large, the granulation process takes a longer time. The results suggest that the pH value can be controlled to select the dominant population in the sludge granulation process [132]. Jiang et al. [50] investigated the resistance and resilience of AGS reactors for simultaneous removal of aniline and nitrogen under pH shock. The AGS maintained its structure and sedimentation under both acidic and neutral conditions, but the alkaline environment negatively affected the stability of the AGS. The pH change resulted in over 99% and 88% removal of aniline and COD, respectively, and the specific removal rate with pH change was higher than that of other biological processes. The nitrification reaction was most active in a neutral environment, and denitrification efficiency was slightly higher in an acidic environment. Moreover, the authors found that the alkaline conditions had a significant effect on the sedimentation performance of AGS by comparing the SVI of AGS at different pHs, and this effect was not reversible in the short term (Fig. 2.6).

In summary, AGS in the natural environment is in a complex environment and some factors, both external and its own, can affect its own performance (ability to remove pollutants, its own structural stability, adaptability to the environment, etc.). By understanding the influencing factors, we can improve the living environment of AGS in a targeted manner and provide a better living environment for AGS. At the same time, we can change certain factors to get more specific results.

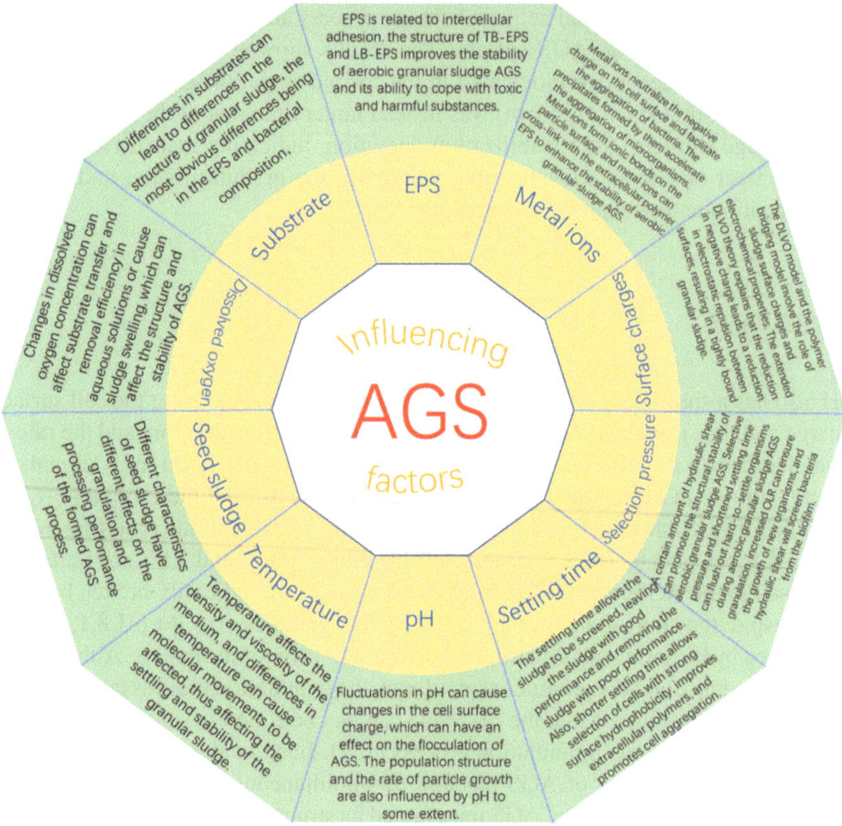

Fig. 2.6 Factors influencing AGS

2.4 Formation Mechanisms of Aerobic Granular Sludge

2.4.1 Self-condensation Hypothesis

The self-condensation hypothesis suggests that under certain conditions, microorganisms can coalesce automatically. The formation of AGS is a self-condensation process of microorganisms under the action of electrostatic repulsion and hydrodynamic forces. The interactions occurring within and between cells, and adsorption occurs between bacteria of different genera, as well as between PN. These interactions cause the microorganisms to aggregate and form a tight, three-dimensional structure with a regular shape. Later, under the influence of extracellular polymers (EPS), cell surface hydrophobicity, hydraulic shear, and mutual adhesion between cells, granular structures with regular shapes gradually form (Fig. 2.7).

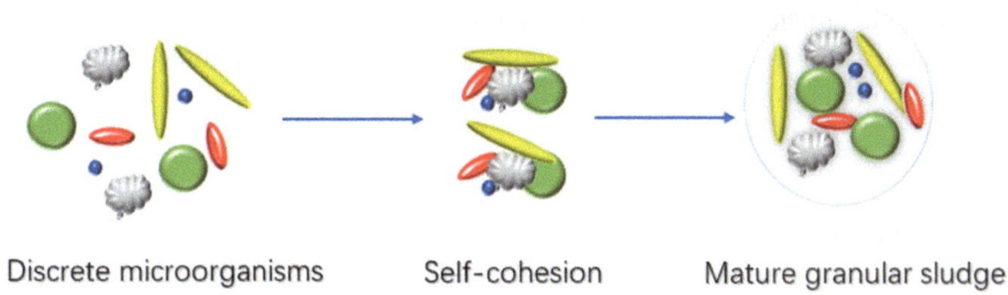

Fig. 2.7 Self-condensation hypothesis

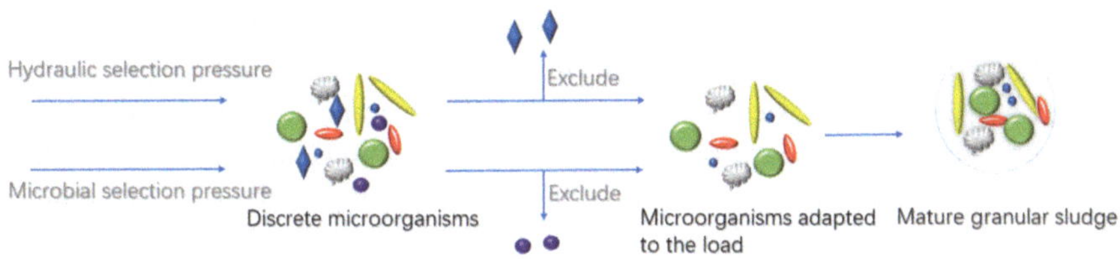

Fig. 2.8 Selective pressure-driven hypothesis

2.4.2 Selective Pressure-Driven Hypothesis

This hypothesis suggests that the nature of aerobic granulation is a cellular self-immobilization process driven by selection pressure. Selection pressures can be divided into hydraulic and microbial selection pressures. Hydraulic selection pressure excludes poorly performing sludge from the reaction system by controlling the properties that affect its structural characteristics and hydraulic conditions, and microbial selection pressure screens out microorganisms adapted to that load by varying the nutrient load in the system (Fig. 2.8).

2.4.3 Extracellular Polymeric Substance Hypothesis

EPS is a substance composed of PS and PN secreted with the cell surface, which facilitates intercellular adhesion, bridging and aggregation, and can affect the hydrophobicity and charge of the cell surface. EPS also contains nucleic acids, phospholipids and humic acids, which are influenced by a variety of factors, such as cell growth secretions, substances absorbed by the external environment, etc. [86].

Individual bacteria are negatively charged, and EPS production produces coalescence by changing the electrostatic repulsion between cells. First, the bacteria come into contact with each other; second, the cells in contact with each other are subjected to external forces to form stable aggregates; then, the bacterial aggregates gradually form structurally complex microbial aggregates through colony growth and EPS secretion, etc.; finally, the microbial structure is continuously adjusted by the action of hydraulic shear to form a stable granular sludge system with a three-dimensional microstructure [140].

2.4.4 *Nuclear Induction Hypothesis*

This hypothesis suggests that AGS is formed by an induced nucleus that triggers microbial aggregation, resulting in the formation of microbial cohesion, which eventually develops into mature AGS through the action of other factors. The theory claims that microorganisms attach to the inorganic sediment of the granules core as a core to grow and multiply in the form of granules. Adding substances such as Mg^{2+}, Ca^{2+}, etc. to the wastewater system can enhance granules formation during the incubation process. These cations can also react with anions like CO_3^{2-} and PO_4^{3-} present in the wastewater, forming precipitates and acting as the core of granules [36]. The inorganic precipitates in the mature AGS formed by this theory contain calcium carbonate, calcium and iron phosphates and hydroxides, hydroxyapatite and some unknown precipitates [107] (Fig. 2.9).

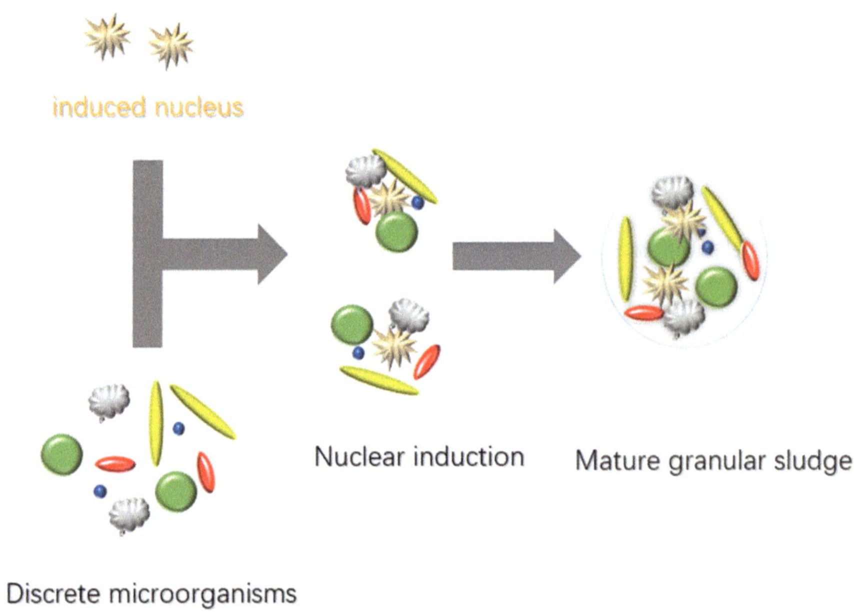

Fig. 2.9 Nuclear induction hypothesis

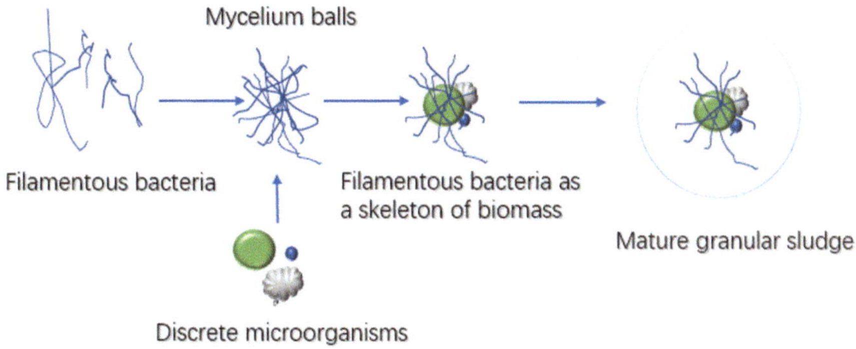

Fig. 2.10 Filamentous bacteria hypothesis

2.4.5 Filamentous Bacteria Hypothesis

Tay et al. [97] found in his study that the sludge inoculated with AGS was dominated by filamentous bacteria. Later, other studies found that the formation of AGS is also accompanied by the development of filamentous bacteria, which shows that filamentous bacteria play an important role in the formation of AGS by promoting intercellular aggregation and bacterial coalescence to form AGS.

The fungal species in seed sludge first form rapidly settling mycelial spheres and dominate the initiation phase, acting as anchors for individual bacterial cells. These "fungal granules" are therefore assumed to be precursors of the true aerobic granules [114] (Fig. 2.10).

2.4.6 Signaling Molecule Hypothesis

This hypothesis suggests that the mutual cohesion of cells is related to "intercellular communication". Under the action of signal molecules, cells within a certain range gather towards a common target, forming microbial clusters. Eventually, under the combined action of other conditions, these clusters can further aggregate and develop into AGS.

Quorum sensing (QS) is a communication system used by bacteria to coordinate their behavior in a population. It involves the secretion and detection of small signal molecules called autoinducers. Bacteria release these autoinducers into their surrounding environment, and when the concentration of these molecules reaches a certain threshold, the bacteria can detect them and respond accordingly [25]. N-acylhomoserine lactones (AHLs) are the major autoinducers in AGS-system microorganisms and play an important role in the formation and stabilization of AGS [48]. Zhang et al. [146] found that the environment with high AHL content can enhance effective communication between cells, increase biomass density, and

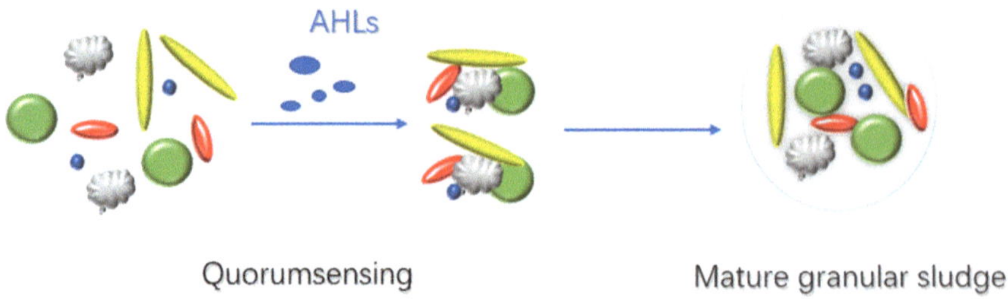

Fig. 2.11 Signaling molecule hypothesis

provide more biological energy to synthesize AHLs and induce bacterial aggregation. AHLs regulate the production of tryptophan and aromatic protein substances, which contribute to an increase in relative hydrophobicity and the maintenance of system stability (Fig. 2.11).

2.4.7 *Surface Hydrophobicity Theory*

The enhancement of the hydrophobicity of the cell surface reduces the Gibbs free energy between cells, which promotes the formation of dense structures and detachment of microbial aggregates from the water surface. As a result, cell surface hydrophobicity is considered to be an important driver of microbial self-aggregation [66].

2.4.8 *Cations Flocculation Theory*

Cations have the ability to compress the double electric layer or facilitate microbial aggregation by forming adsorption bridges among bacteria or flocs. This process reduces electrostatic repulsion and enhances the van der Waals forces between microbial surfaces, thereby promoting the aggregation of AGS [122].

Wan et al. [111] proposed a rational model of bacterial AGS formation, including four stages of core formation, microaggregation, micro-granules formation and maturation. That is, calcium precipitation is first formed as the core of granules; then adherent bacteria attach to the calcium precipitation and grow and multiply to form microaggregates; due to the extracellular PS secreted by adherent bacteria in the aggregates, other bacteria gradually attach to the surface and multiply by selective adsorption, forming microgranules visible to the naked eye; microgranules gradually grow to become mature AGS.

2.4.9 Fungal Based Aerobic Granular Sludge Theory

In addition to the above hypothesis on the formation of AGS, other researchers have reported on the formation of fungal AGS. Fungal AGS uses glucose as a carbon source and creates an environment conducive to fungal growth by releasing hydrogen ions through glycolysis and lowering the pH of the surrounding environment to inhibit the growth of other bacteria [107]. The reason for this is that the lower pH is favorable for filamentous fungi to multiply and degrade the substrate. Also, the rate of glucose utilization by bacteria at low pH is less than that of filamentous fungi. Wan et al. [106, 108] proposed a theory of fungal AGS formation, in which filamentous fungi transform from spores to hyphae and then entangle themselves under the action of shear forces, thereby forming monolithic filamentous granules.

2.4.10 Four Stage Theory of Sludge Granulation

Liu and Tay [65] proposed a four-stage theory based on previous studies: Step 1. Physical interactions (hydrodynamic, diffusive, gravitational and thermodynamic forces) trigger inter-bacterial contacts that allow bacteria to attach to the solid surface, Step 2. This process allows bacteria to build a three-dimensional structure to provide a stable environment; Step 3. Microbial self-interaction (EPS growth of cellular cluster metabolic change and genetic competence) allows the aggregated bacteria to mature; Step 4. Microbial aggregates eventually form a certain structured community through hydrodynamic shear.

The above theories each state a different perspective on the hypothesis of the formation of AGS. It can be seen that the process of AGS formation cannot be separated from the contact, aggregation, maturation and stabilization of bacteria/fungi. Most of the above theories suggest that the process of AGS formation requires some kind of material as the core to proliferate through constant inter-attachment. Although there is no clear theory on the formation of AGS. However, there is no doubt that the formation of AGS must be the result of a combination of factors.

In summary, the formation mechanism of AGS is a complex process with multiple factors, and the above theories explain the formation of AGS from different directions. self-condensation hypothesis, EPS hypothesis, nuclear induction hypothesis, filamentous bacteria hypothesis, signaling molecule hypothesis, and theory of the formation of fungal type AGS, The self-condensation hypothesis, EPS hypothesis, nuclear induction hypothesis, filamentous bacteria hypothesis, signaling molecule hypothesis and theory of the formation of fungal type AGS emphasize the role of microorganisms themselves in the formation of AGS. In contrast, the four-stage theory suggests that the formation of AGS is the result of the combined action of external agents and microorganisms themselves. Although the conclusions are not uniform now, the above studies will provide a good theoretical basis for the initiation of AGS.

2.5 Start-Up of Aerobic Granular Sludge System

Despite the significant advantages of AGS, there are still some non-negligible limitations that hinder its large-scale implementation in practical engineering. One of the primary drawbacks is the extended start-up time of AGS and the difficulty in maintaining stable operation during the start-up phase. This issue has been extensively studied by researchers over time.

2.5.1 Shorten the Sedimentation Time Method

Settling time, as one of the important parameters of AGS process, not only affects the formation of AGS, but also has a regulatory effect on the long-term stable operation of AGS. Its essence is to flush out the flocculated sludge with poor settling performance and retain the dense sludge by short settling time. Huang et al. [42] compared the treatment effect and granulation process of AGS of high concentration organic wastewater under different settling time (5, 2, 1 min) conditions and analyzed the physicochemical characteristics of AGS cultivated with different settling time, and the results showed that: when the settling time was 5 min, the granular sludge granules size reached stability around 12 d, and when the settling time was 1 min, the AGS size stability required When the settling time is 1 min, it takes about 17 d to stabilize the granules size of granular sludge. The granules size and settling rate of granular sludge increased to 563.9 μm and 30.5 m/h when settling time was shortened to 1 min, while the SVI and integrity factor decreased to 13.3 ml/g and 0. 031. The content of EPS, protein (PN) and polysaccharide (PS) decreased to 11.3 ml/g when settling time was shortened to 1 min. While PN/PS increased to 2.3, the higher PN/PS may be the reason for the increased stability of granular sludge In [36].

Indeed, the selective sludge discharge method can be influenced by settling time, among other factors. The study conducted by Guo et al. examined the impact of seed sludge properties and the selective sludge discharge method on the formation of AGS in wastewater biological treatment. The small and loose flocs and the large and dense flocs in the raw sludge were separated by settling columns. Granulation experiments were performed using two types of sludge as seed sludge. The daily sludge discharge rate of both reactors was fixed at about 10% of slow settling sludge. The results showed that seed sludge with different structures and settling properties were able to form AGS in the reactors. The key operational parameter during bioreactor start-up is the daily discharge of relatively slow settling biomass in the reactor. PCR-DGGE analysis showed that the bacterial community structure of the mature AGS in the two reactors did not differ significantly. This implies that the continuous discharge of small, slowly settling flocs removes these competitors for substrate uptake from the system, thus making the substrate more accessible to larger, compact flocs and granules. Selective sludge discharge is a strategy employed in wastewater treatment processes to promote the growth and accumulation of denser sludge in the reactor, which can lead to the formation of complete granules [85].

In general, a short settling time helps to start up the AGS system quickly. However, directly using a very short settling time to start the system, there is a risk of failure to start the AGS system while obtaining a better granular sludge. Therefore, a suitable settling time needs to be selected.

2.5.2 *Increase Organic Load Method*

Organic loading also plays an important role in the initiation of AGS, and Li et al. [55] reported that AGS exhibited different morphological and structural characteristics and bacterial species at different OLRs. At high OLR, large and loose AG formed in a short time, while at low OLR, small and compact AG took longer time to form. In terms of species diversity, the lowest species diversity was found in the reactors with high OLR and the highest species diversity was found in the reactors with low OLR. As a fast start-up strategy, the use of high OLR can accelerate the formation of AG. Zhang et al. [145] applied 24 kg COD/m^3-d and AG was formed within 7 h, but disintegrated within two days. When the OLR was reduced to 12 kg COD/m^3-d and subsequently to 6 kg COD/m^3-d, AGS was formed within 24 h and the granules remained stable and reached steady state within 12 d. In the same research area, [97] emphasized the OLR of 12 kg COD/m^3-d during AGS formation. Once the system reaches steady state, the OLR decreases to 6 kg COD/m^3-d and the system reaches steady state. This allows for rapid start-up of the AG reactor and improves the success rate of the reactor [44].

2.5.3 *Ion Addition Method*

Divalent metal ions have a positive effect on palletization. Metal ions, especially calcium and magnesium, are considered to be the main factors in enhancing SBR granulation. The mechanisms of enhanced granulation may include the following: (1) inorganic ions neutralize the negative charge on the bacterial surface and enhance bacterial aggregation; (2) inorganic ions form precipitation forming nuclei, which accelerates microbial aggregation, especially in the case of Ca^{2+}; and (3) inorganic ions form ionic bonds on the granules surface, which act as cationic bridges and promote granules formation. After granulation, calcium also binds to EPS, thus forming EPS-Ca^{2+}-EPS crosslinks to strengthen the granules structure [28].

Gao et al. [28, 30] successfully formed AGS in an SBR reactor using four different granulation enhancement strategies (R1) controlled organic loading and Ca^{2+} supplementation, (R2) short settling time, (R3) extended aerobic starvation time and (R4) increased shear force (Table 2.2). The results showed that rapid granulation could be achieved by using short settling time in the SBR reactor, and the resulting granules had good settling performance, high storage stability, and higher EPS and protein content than those under other conditions. The granules formed by extended aerobic starvation time and enhanced shear have small granules size, dense structure, high physical strength and low K-value. The AGS formed at higher OLRs had larger AGS sizes (Fig. 2.12). In addition, AGS formed under different conditions had similar maximum COD removal rates, but the apparent half-rate constant (K) was related to the average granules size of the granular sludge.

The results for four groups of reactors with different conditions are shown in Fig. 2.12, and the order of granules appearance is R2-R3-R4-R1; and the COD corresponding to each reactor at the time of granules appearance is 0.36, 0.38, 0.41, and 0.35 g COD/g SS^{-d}, respectively (Fig. 2.12c), thus indicating that an organic load of 0.4 g $\pm$ 0.5 COD/g SS^{-d} is favorable for the rapid start-up of AGS.

The EPS in the reactor all increased after the AGS was fully formed, and different methods caused different PS/PN (Fig. 2.15); it may be influenced by the C/N ratio, with more organic carbon sources and more PS, the larger the AGS.

The AGS of each reactor shows clear contours and colors except for the granules size (Fig. 2.13). The AGS size of the granules in the reactor on day 95; higher organic loading promotes larger AGS formation; while high shear (R4) or prolonged starvation (R3) results in the formation of smaller granules size AGS, while shorter settling times form AGS with uniform granules size (Fig. 2.14). The reason is that high shear enhances the collision of granules and makes the AGS surface loose structure detachment; while longer starvation time increases endogenous respiration and makes microbial growth slower [28, 30].

The Fig. and table in this section are from Gao et al. [28, 30] (Fig. 2.15).

Table 2.2 Operational conditions of four SBRs

Condition	Reactors			
	R1	R2	R3	R4
Influent COD (mg/L)	300–1000	500	500	500
COD load (g/(L^{-d}))	0.45–1.5	0.75	0.75	0.75
Added Ca^{2+}(mg/L)	50	0	0	0
Settling time (min)	15	15 and then 1	15	15
Aeration time (min)	280	280	445	280
Total cycle time (min)	300	300	465	300
SRT (days)	15	Various	15	15
Air flow (m^3/h)	0.3	0.3	0.3	0.6

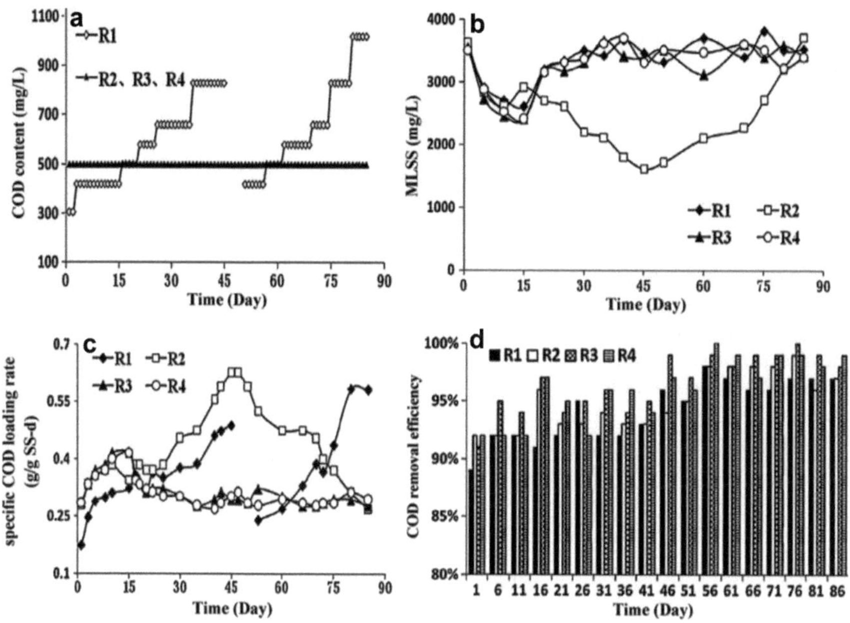

Fig. 2.12 **a** Changes in influent COD concentration. **b** Sludge concentration changes in R1–R4 reactors. **c** Changes in specific COD loading rates in R1, R2, R3 and R4 during granulation process. **d** COD removal efficiencies of SBRs during granulation; (R1: controlled organic loading and Ca^{2+} supplementation, R2: short settling time, R3: extended aerobic starvation time and R4: increased shear force)

2.5.4 Different Seed Sludge

Different seed sludge also affects the initiation of AGS, a study investigated the initiation of AGS by inoculating the SBR reactor with flocculated activated sludge and frozen stored mature granular sludge as inoculated sludge, and the results showed that using frozen stored granular sludge as seed sludge can start the AGS SBR system more quickly, and the structure of the formed AGS. The AGS formed was compact, with clear and smooth contours, mixed liquid suspended solids (MLSS) over 8 g/L, and degraded effluent index [118]. Xin et al. [127] added flocculant-producing denitrifying bacterial TN-14 sludge to a continuous flow reactor treating domestic wastewater based on microbial self-flocculation theory. The bacterial TN-14 sludge acted as the main seed sludge to promote the rapid formation of AGS. To explore the initiation of AGS, the results showed that the use of inoculated flocculant-producing denitrifying bacterial TN-14 sludge could promote the formation of AGS. The mature granular sludge was compact. The SVI_{30} value decreased from 122.62 ml/g to 46.61 ml/g and then remained at 44.28–60.51 ml/g. In this system, the PN content of sludge EPS increased rapidly, while the polysaccharide content did not change much.

Fig. 2.13 Granules formed under different operating conditions (Bar is 2 mm): R1: high OLR and Ca^{2+} addition; R2: short sedimentation time; R3: long starvation period; R4: high shear. The sequence of the average diameter of granules is R1 > R2 > R3 = R4. (R1: controlled organic loading and Ca^{2+} supplementation, R2: short settling time, R3: extended aerobic starvation time and R4: increased shear force)

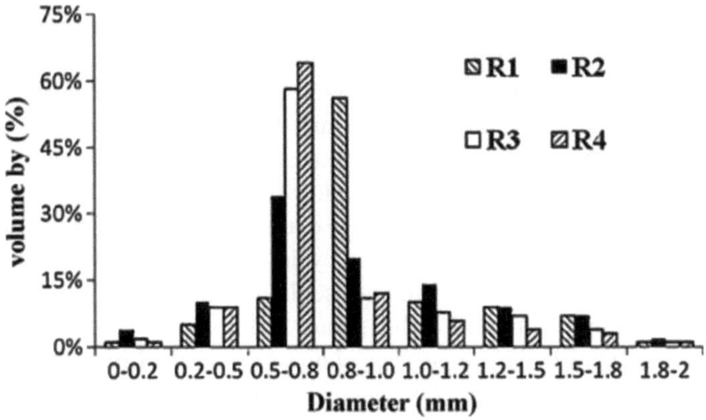

Fig. 2.14 Size distributions of the granules in all reactors on day 95. (R1: controlled organic loading and Ca^{2+} supplementation, R2: short settling time, R3: extended aerobic starvation time and R4: increased shear force)

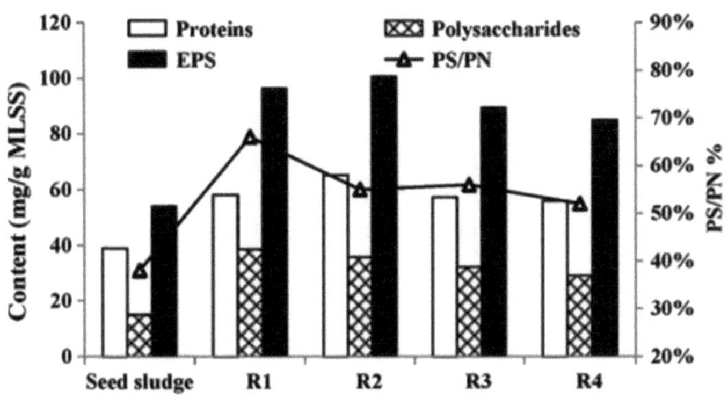

Fig. 2.15 EPS content and the PS/PN of the granules in SBRs and inoculum activated sludge. Granule samples were taken on day 90. (R1: controlled organic loading and Ca^{2+} supplementation, R2: short settling time, R3: extended aerobic starvation time and R4: increased shear force)

2.5.5 Carriers Addition

According to the nucleation theory, the addition of a carrier or substance can act as an initial nucleus for the cells, allowing for faster attachment of microorganisms and granulation of the sludge. Different carriers of different natures have been used to try to accelerate the sludge keratinization process.

Zhang et al. [147], in their study to improve the degradation efficiency of AGS systems for difficult-to-degrade compounds such as pyridine, found that the addition of biochar to the wastewater treatment process resulted in some positive effects on pelletization and sludge characteristics. Tao et al. [96] experimented with granular activated carbon (GAC) and sludge granulation. The results showed that GAC acted as a different role in each of the three stages of granulation to facilitate the sludge granulation process. In the lag phase, GAC provides the sludge with a cell nucleus and promotes sludge morphological regulation. In the early stage, GAC accelerates granulation by reducing the compression caused by inter-granules collisions. Similarly, Li et al. [54] have attempted to add granular or powdered activated carbon to bioreactors to enhance the granulation of slow-growing AOB. The results demonstrated that the addition of GAC to the sludge could greatly accelerate the granulation process and could shorten the granulation cycle from around 6 weeks to less than 3 weeks for the formation of large and fast settling granules.

In addition to pure activated carbon, some scholars have tried to improve the quality of activated carbon by improving it into a new carrier. Lin et al. [61] prepared iron-loaded activated carbon (FAC) to culture granular sludge. The results showed an increase in PN and PS content with the addition of FAC, in addition to an increase in the number of microorganisms and denitrifying bacteria, leading to rapid granulation and removal of nutrients.

In addition to organic carriers/materials, inorganic materials also have an impact in terms of accelerated granulation. Tang et al. [95] attempted to enhance the aerobic granulation process with sodium alginate (SA) as the nucleus (SAN). A laboratory-scale sequential batch experimental study found that the addition of a certain amount of SAN effectively enhanced aerobic palletization and improved biomass retention. In contrast, Liu et al. [67, 68] attempted to add polymerized ferric sulphate (PFS), aluminum sulphate (AS) and diatomaceous earth to enhance aerobic granulation. The results showed that the addition of carriers accelerated the granulation of sludge and that PFS and AS significantly increased the PN content in EPS compared to diatomaceous earth.

2.5.6 Inoculating Special Strains

Bioaugmentation is the addition of specialized microorganisms to a biological treatment system to increase biomass or improve biological activity to achieve certain objectives. Several studies have shown that the addition of relevant microorganisms or bacteria to sludge systems can promote aerobic granulation. For example, during the cultivation of flocculent-enhanced AGS, the inoculated flocculent bacteria can allow AGS to self-aggregate or act as a bridge in a copolymeric colony. Both of these approaches will contribute to the rapid formation of microbial aggregates and accelerate AGS formation [36]. Ivanov et al. [45] successfully obtained AGS with an average diameter of 446 lm in 8 d by using two special strains of bacteria with a high self-aggregation capacity and activated sludge. Quan et al. [82] used the 2,4-D degrading bacterium Bacillus saprophyticus QXH for bioaugmentation and eventually successfully cultured 2,4-D degradable AGS.

In addition to single strains, researchers have also tried co-cultures of multiple flocculating bacteria as inoculum to enhance co-aggregation and accelerate the initiation of sludge granulation.

Wan et al. [106, 108] conducted a study on the granulation process of aggregate-functional bacteria X9, which included Pseudomonas putida X-1, Acinetobacter sp. X-2, Alcaligenes sp. X-3, and Comamonas testosteroni X-4. The researchers observed that after 6 h, granules of approximately 0.6 mm were granulated. Additionally, after 40 h, larger granules measuring 8.2 × 6.5 mm were formed. The reason for this is that the increase in PN forms microgranules, while the cell membrane and extracellular polysaccharide hydrophobicity promote the formation of wickedly, suggesting that combinatorial functional bacteria also favor aerobic granulation. Liang et al. [60] selected NJUST18 and NJUST29 and NJUST32 as inoculate for AGS for aerobic pelletizing. The results showed that the formation period of AGS was reduced to 42 d. This experiment showed that the inoculum is also suitable for refractory organic wastewater.

Similarly, fungi can accelerate sludge granulation, with fungal spores and hyphae acting as a core and skeleton for other microorganisms to adhere to, thus speeding up the aerobic granulation process. Wang et al. [112] made the first attempt to granulate paper wastewater sludge using fungi (white rot fungi: *Coriolus* versicolor and *Phanerochate chrysosporium*) as inocula. The results showed that the forming time of AGS was improved. The reason for this may be due to the combined effect of sludge granulation and lignin degrading enzymes secreted by the white rot fungi.

2.6 Microbiology of Aerobic Granular Sludge System

The microbial community of AGS is extremely diverse and is influenced by factors such as substrate, wastewater composition, inoculants, environment and operation. If AGS is cultured with sewage or similar wastewater, its biological community contains organic matter degrading bacteria, nitrifying and denitrifying bacteria, and anaerobic bacteria (such as anaerobic ammonia-oxidizing (anammox) bacteria and methanogenic bacteria, etc.). Due to their configuration, AGS can consist of three microbial communities, namely heterotrophic aerobic bacteria growing on the outside of the granules, ammonia-oxidizing bacteria (AOB) growing in the middle of the granules and facultative anaerobe growing on the inside of the granules.

The dominant phylum of AGS usually includes Proteobacteria, Actinobacteria, Firmicutes, Bacteroidetes, Acidobacteria, Verrucomivrobia, Chloroflexi, and Saccharibacteria [18]. In general, Proteobacteria are the main phylum in AGS, followed by Actinobacteria, Firmicutes and Bacteroidetes, usually with bacterial composition at the phylum level consistent with the seed sludge at [125].

Mature AGS contains a clear stratification of bacterial community, with filamentous bacteria as support in the interior and bacilli distributed on the surface of the granules. Mature AGS is similar to a ball, with the core dominated by the families Rhodocyclaceae and Moraxellaceae, which are responsible for granulation and other functions of AGS, and the outer layer dominated by the families Flavobacteriaceae, Xanthomonadaceae, Rhodobacteraceae, and Microbacteriaceae. The distribution of bacteria is also influenced by the granules size of granular sludge, and the transfer of

DO and nutrients is affected by the granules size, which can lead to the division of the granular sludge into aerobic, parthenogenic, and anaerobic zones from the outermost to the innermost layers, further resulting in stratification of the bacterial population. AOB and nitrite-oxidizing bacteria (NOB) are most significantly affected by this, with AOB mainly existing on the outside of AGS and NOB usually existing on the inside, while granular sludge with a large granules size facilitates the growth of AOB and inhibits the growth of NOB.

2.6.1 EPS-Producing Bacteria

EPS is a viscous substance containing PN, PS, humic acids and lipids that are involved in the formation of the structure of AGS and the establishment of stability. At the same time, the amount of EPS in AGS far exceeds the amount of EPS in flocculated sludge and biofilms. It has been shown above that EPS has a positive influence on the AGS granulation process.

During pelleting, the abundance of *Zoogloea* may have increased from 2 to 38%, which is consistent with fluctuations in EPS content. Other genera (*Thauera, Meganema, Devosia, Rhodocyclus, Stenotrophomonas*, etc.) also appear to contribute to EPS production, while *Sphingobacteriales and Cytophage* are known to consume EPS. Thus, AGS can be considered as a producer and consumer of EPS, and the ecological balance within the reactor has an important influence on the stability and function of AGS [125].

2.6.2 Denitrifing Bacteria

The abundance of nitrifying bacterial communities in AGS is much higher than in flocculated sludge due to the high retention capacity of slow-growing bacteria. AOB and NOB are the main microorganisms involved in the nitrification and denitrification process. *Nitrosomonas-like* and *Nitrosospira-like* populations is the main AOB in wastewater treatment. However, only *Nitrosomonas-like* was present in the AGS, possibly due to the effect of selection pressure, which allowed *Nitrosospira-like* to be removed. Also, there are many EPS-producing bacteria that are denitrifying bacteria, e.g. the members within the families *Rhodocyclaceae, Xanthomonadaceae, Hyphomicrobiaceae, Comamonadaceae*, and *Methylobacteriaceae* and the genera *Thauera, Zoogloea, Meganema, Devosia*, and *Stenotrophomonas*. And the anaerobic ammonia-oxidising bacteria present in the AGS also contribute to denitrification, such as *Brocadiaceae* [125].

2.6.3 Phosphorus Removing Bacteria

Biological phosphorus removal is mainly achieved by phosphorus accumulating organisms (PAOs), which are present in AGS where anaerobic phosphate release promotes phosphorus precipitation in the core. However, there is competition between glycogen accumulating organisms (GAOs) (*Competibacter*) and PAOs (*Accumulibacter*) during phosphorus removal, which are mostly found in dense AGS. Longer settling times facilitate larger granules size and density, improving phosphorus removal. *Rhodocyclus* and *Accumlibacter*, known to anaerobically synthesize PHAs and aerobically accumulate polyphosphate, have also been Also, the filamentous genus *Thiothrix* in AGS contributed to the accumulation of precipitated phosphate. In addition, the AGS in which *Competibacter* and *Accumulibacter* in synergy can achieve simultaneous denitrification and phosphorus removal. In addition, it has been confirmed that AGS dominated by Accumulibacter, Nitrosomonas, and Nitrospira exhibit excellent performance in both nitrogen and phosphorus removal [125].

2.6.4 Refractory Organic Matter Removing Bacteria

AGS treatment of refractory or toxic substances can also be attributed to the strains of AGS. The presence of *Acinetobacter calcoaceticus* at the core of AGS can degrade phenol. The genus *Thauera* in the Proteobacteria phylum has the ability to degrade p-chloroaniline. Furthermore, *Pseudomonas* and *Comamonas* are capable of degrading aniline, and *Rhodoccus* is capable of degrading 2-fluorophenol. Additionally, Acinetobacter has been identified as being responsible for the degradation of formaldehyde, and *Defluviicoccus* was found to be linked to the degradation of azo dye in AGS [125] (Table 2.3).

The conjoined actions of filamentous or rod-shaped bacteria for skeletal functions, extracellular polysaccharide producing bacteria for adhesion, and both of them together for combined action, are essential for the AGS formation process. Filamentous or rod-like bacteria, such as *Rhodocyclus* and *Hyphomicrobiaceae*, play a crucial role in the initial formation and structural maintenance of granules. During the granule formation process, there is a significant increase in the abundance of *Accumulibacter*, *Nitrosospira*, and *Thauera*, which are closely associated with nutrient removal. Furthermore, EPS (Extracellular Polymeric Substances) producers like *Brachymonas*, *Thauera*, *Zoogloea*, and *Acidovorax* may also contribute to the formation of AGS.

Many of the bacteria known for producing EPS are reported to be denitrifiers, such as those belonging to the Rhodocyclaceae, Xanthomonadaceae, Hyphomicrobiaceae, Comamonadaceae, and Methylobacteriaceae families as well as the genera *Thauera*, *Zoogloea*, *Meganema*, *Devosia*, and *Stenotrophomonas*. Denitrifiers play

Table 2.3 Main functions of the bacterial community in AGS

Function	Family	Genus	Reactor condition/or substrate	Characteristics	References
EPS producers	Rhodocyclaceae	*Thauera*	Laboratory-SBR		[92]
		Zoogloea	Pig manure		[27]
	Methylobacteriaceae	*Meganema*	Laboratory-SBR		[92]
	Hyphomicrobiaceae	*Devosia*	Continuous gas-lift aerobic pellet biomass reactor		[71]
	Xanthomonadaceae	*Stenotrophomonas*	Laboratory-SBR		[92]
	Comamonadaceae	*Brachymonas*	Pig manure		[27]
	Rhodocyclaceae	*Rhodocyclus*	Fresh activated sludge obtained from a local municipal wastewater treatment plant (3.5 g) mixed liquor volatile suspended solids (MLVSS) L^{-1} and sludge volume index (SVI) of 212 mL g^{-1}		[5]
	Flavobacteriaceae	*Flavobacterium*			[9]
Denitrification	Xanthomonadaceae	*Pseudoxanthomo-nas*		Amphitrophy denitrifiers	[9]
		Lysobacter	Continuous gas-lift aerobic pellet biomass reactor laboratory-SBR		[71]
		Stenotrophomonas	Laboratory-SBR		[92]
	Rhodocyclaceae	*Thauera*	Lab-scale SBR	Amphitrophy denitrifiers	[145]
		Zoogloea	Reactor for synthetic wastewater	Amphitrophy denitrifiers	[121]
		Azoarcus	Laboratory-SBR	Heterotrophic denitrifiers	[92]

(continued)

Table 2.3 (continued)

Function	Family	Genus	Reactor condition/or substrate	Characteristics	References
	Thiobacillacaeae	*Thiobacillus*		Autotrophic denitrifiers	[129]
	Rhodobacteraceae	*Paracoccus*	GSBR	Amphitrophy denitrifiers	[113]
	Accumulibacter	–	Lab-scale SBR		[145]
	Hyphomicrobiaceae	*Devosia*	Continuous gas-lift aerobic pellet biomass reactor laboratory-SBR		[71]
	Methylobacteriaceae	*Meganema*	Laboratory-SBR		[92]
	Pseudomonadaceae	*Pseudomonas*	SBR reactor		[131]
	Comamonadaceae	*Comamonas*	SBR reactor		[131]
Nitrification	Nitrosomonadaceae	*Nitrosomonas*	Pig manure		[27]
	Nitrospiraceae	*Nitrospira*	Laboratory-SBR		[92]
	Bradyrhizobiaceae	*Nitrobacter*	Laboratory-SBR		[92]
	Brocadiaceae	–	Laboratory-SBR	Anammox bacteria	[92]
Phosphor removal	Accumulibacter	–			[151]
	Competibacter	–			[121]
Hydrolysis	Flavobacteriaceae	*Flavobacterium*	Laboratory-SBR		[92]
	Sphingobacteriac-eae	–	Laboratory-SBR		[92]
Removal of refractory pollutants	Rhodocyclaceae	*Thauera*	High concentration 4-chloroaniline (4-CLA) wastewater	High-level 4-chloroaniline biodegradation	[153]

(continued)

Table 2.3 (continued)

Function	Family	Genus	Reactor condition/or substrate	Characteristics	References
		Azoarcus		High-molecular substance biodegradation AMP biodegradation	[113, 151]
		Zoogloea	Antibiotics GSBR Synthetic wastewater	PPCPs biodegradation	[113, 149]
	Rhodobacteraceae	*Paracoccus*	GSBR Synthetic wastewater SBR with 4-CP as sole carbon source	Degradation of pyridine Degradation of 4-chlorophenol	[31, 149]
	Moraxellaceae	*Acinetobacter*	Aerobic granules are cultivated by a single bacterial strain, Acinetobacter calcoaceticus, in a sequencing batch reactor (SBR)	o-cresol biodegradation Phenol biodegradation	[4, 46]
	Rhodospirillaceae	*Defluviicoccus*	SBR Synthetic textile wastewater	Azo dye biodegradation	[26]

crucial roles in AGS, and their cumulative relative abundance can account for a significant portion, ranging from 45 to 60%, within aerobic granules. In addition, the role of anammox bacteria in denitrification in the AGS process has been demonstrated. For instance, the presence of *Brocadiaceae*, which belongs to anammox bacteria, has been reported in AGS. The communities of AOB, NOB, anammox bacteria, and denitrifying bacteria may interact and cooperate within AGS, promoting efficient nitrogen removal. However, our understanding of the underlying microbiological mechanisms is currently limited [126].

2.7 Storage of Aerobic Granular Sludge

The research and development of AGS has become more and more widespread in recent years and the use of this technology has gradually become more common. However, the long-term inactivity of AGS or their long-term storage can cause microorganisms to lose their biological activity, which is an obstacle in the development of AGS [29]. As a result, there is a trend towards research into the storage and reactivation of AGS. Storage and reactivation techniques are being investigated in order to achieve a rapid recovery of the physical properties and microbiological activity of AGS after storage.

The realization of effective storage and reactivation technology for AGS offers several advantages as mentioned by Wan et al. [107].

1. AGS storage can function as an external reserve, ensuring continuous and stable operation.
2. The stored granular sludge can serve as seed sludge, enabling rapid start-up operations for reactors or WWTP.
3. The storage process is convenient and allows for easy transportation over long distances, facilitating the commercialization of AGS.

To a certain extent, the cost of AGS technology is reduced, which facilitates the wider diffusion of AGS technology. The physical properties and biological activity of AGS can be affected by different storage methods, including storage temperature, storage substrate, storage time, and the degree of dryness and wetness, due to internal cell hydrolysis, protein hydrolysis, and endogenous respiration [29].

Regarding the reactivation of AGS, the focus is on maintaining the physical properties and degradability of the granular sludge. The reactivation method aims to recover the granules by prioritizing the use of the original culture conditions when drastic environmental changes may affect storage. Attention is also required to details of reactivation, such as setting nutrient concentration gradients, controlling temperature and osmotic pressure during recovery [107].

2.7.1 Temperature

Storage temperature has an important influence on the storage of granular sludge. The higher the storage temperature, the higher the rate of hydrolysis and endogenous respiration, the more severe the loss of microbial activity and therefore storage can be problematic. Low temperatures can inhibit cellular hydrolysis and protein conversion to some extent, effectively maintaining the stability of the granules. At the same time, there are certain difficulties with low temperature storage. Low temperature conditions (4 or − 20 °C) are energy intensive and the stringent requirements lead to increased operating costs. When stored at low temperatures for longer than six months, aerobic granules require long recovery times and there is a risk of instability and reduced removal efficiency of the recovered AGS [29].

Gao et al. [29] investigated the effect of three storage temperatures (− 25, 4 and 25 °C) (Table 2.4) on the long-term storage and reactivation performance of AGS. The results (Table 2.5) showed that a storage temperature of 4 °C was more suitable for maintaining the structural integrity and long-term stability of the granules, while 25 °C was more suitable for maintaining the internal microstructure. After 8 months of storage at different temperatures, the appearance of AGS stored at 4 °C was still brownish yellow spherical or oval with a clear and dense structure. In contrast, the sludge stored at 25 °C had a brown hollow flake shape and irregular shape. As shown in Table 2.5, the structural properties, specific gravity, physical strength, biomass and sedimentation of AGS gradually decreased after storage under different conditions. Granules stored at 4 °C had little change on physical properties. After 8 months storage the integrity coefficient of R2 and R5 was 88.9 and 86.3%, and the specific gravity of R2 and R5 was 1.013 and 1.011 g/cm^3, which were significantly higher than other storage conditions. Along with the storage, the settleability of granules stored at 4 °C was deteriorated as shown from values of settling velocity and SVI. The settling velocity of R3 and R5 was remarkably decreased to 30.67 and 29.15 m/h, which presented granules after such long-term storage could still maintain a good settling ability. Therefore, storage at 4 °C had little influence on granules physical structure, biomass and settleability, which is an optimal way to preserve granules structure and physical properties for granules long-term stability storage. Storage at 25 °C had great impact on granules physical structure and integrity. Because of the freezing and thawing process during the storage, granules in R1 and R4 were serious disintegration with loose and irregular shape. Consequently, the physical properties were gradually deteriorated, and granules with distilled water and glucose solution had little difference. The integrity coefficient of R1 and R4 was 76.3 and 75.5%, which was much lower than the granules stored at 4 °C. All three storage temperatures resulted in reactivation after ten days of storage, with 4 °C showing the best recovery of the granules and room temperature showing the worst recovery. It can be seen that different storage temperatures have different effects on the storage and reactivation of AGS. In addition, fresh AGS were collected by another researcher for low temperature storage (4 °C for six weeks). This was followed by reactivation (at 15 °C). During storage the granules changed color to grey and during reactivation the dehydrogenase activity (DHA) was gradually restored and exceeded the initial amount, leading to an overgrowth of filamentous bacteria. At mid-activation, the AGS became smoother, the settling performance became better, DHA returned to an acceptable range and the nitrification performance improved and stabilized at 85%. This study shows that AGS performance can be fully restored by reactivation over time after low temperature storage. The contents of EPS, protein and PS in the granules under different storage conditions were significantly lower than those in fresh granules before preservation (Figs. 2.16a and 2.17a). The lowest EPS content was found in granules stored at room temperature, and essentially the same in granules stored at 4 and – 25 °C. EPS can form a buffer layer for cells against the harsh external environment and act as a source of carbon and energy during starvation. Storage of granules can be seen as a special state of starvation, in response to which bacteria secrete large amounts of EPS to cope with starvation. During long-term storage, high storage temperature, high cellular activity, high EPS hydrolysis utilization and lowest EPS content in granules stored at room temperature. However, the utilization ratios of biodegradable PS and PN in pellets are different, and usually the utilization of PN is higher, which explains the large differences in PN/PS ratios in pellets under different storage conditions (Figs. 2.16b and 2.17b) [137].

In addition, Campo et al. [15] attempted to study the granules size distribution, EPS composition and activity of AGS after freezing storage (– 20, – 80 °C). The results showed that the granules were reactivated after 4 weeks of storage at – 80 °C. 91.71% of the granules were < 1.0 mm in size and 57.16% were < 1.7 mm in size the granules remained intact after storage but broke up and disintegrated when aerated the EPS, PN and PS were reduced by 53.22%, 51.81% and 41.34% respectively. In contrast, EPS, PN and PS changed relatively little at – 20 °C. Compared to the other temperatures these two storage temperatures showed the greatest ability to oxidize and denitrify ammonia and nitrogen in AGS [130].

Table 2.4 Different storage conditions of aerobic granules [137]

Sludge no	Storage time (months)	Storage media	Storage temperature (°C)
R1	8	Distilled water	– 25
R2	8	Distilled water	4
R3	8	Distilled water	20–26
R4	8	400 mg/L glucose solution	– 25
R5	8	400 mg/L glucose solution	4
R6	8	400 mg/L glucose solution	20–26

Table 2.5 Granules characteristics under different storage conditions

Characteristics	Before storage	After 8 months storage					
		R1	R2	R3	R4	R5	R6
Physical properties integrity coefficient (%)	99.8	76.3	88.9	73.1	75.5	86.3	71.5
Specific gravity (g/cm^3)	1.023	1.01	1.013	0.987	1.009	1.011	0.985
Settling velocity (m/h)	56.87 ± 2.31	26.82 ± 2.76	30.67 ± 2.64	18.49 ± 2.44	26.75 ± 2.87	29.15 ± 2.73	18.28 ± 1.04
VSS/SS (%)	95.71	90.25	93.08	85.24	91.36	94.06	83.94
SVI (ml/g)	28.89	81.17	94.73	166.57	80.68	78.92	158.73
Reactor performance COD removal (%)	99.12	0.13	0.1	0.15	0.12	0.14	0.21
NH_4^+–N removal (%)	97.83	0.08	0.15	0.09	0.14	0.08	0.04
PO_4^{3-}–P removal (%)	98.62	0.13	0.21	0.16	0.19	0.11	0.18
SOUR (mg-O_2/g-SS h^{-1})	36.18	0.05	0.12	0.13	0.17	0.12	0.09

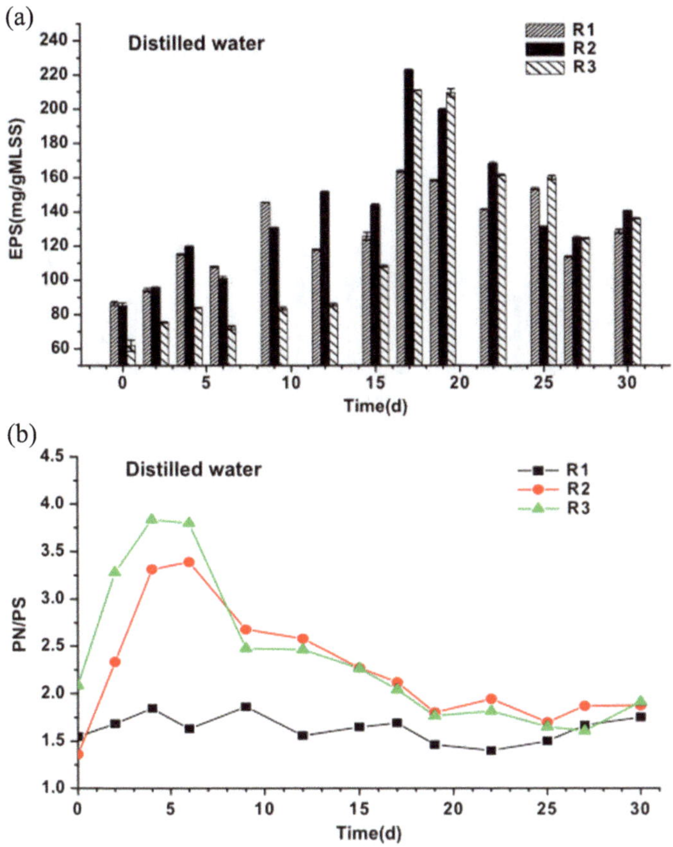

Fig. 2.16 EPS content and PN/PS of the granules. (R1: distilled water, − 25 °C; R2: distilled water, 4 °C; R3: distilled water, 20–26 °C; R4: 400 mg/l glucose solution, − 25 °C; R5: 400 mg/l glucose solution, 4 °C; R6: 400 mg/l glucose solution, 20–26 °C) [137]

2.7.2 Substrate

As the sole carbon source, the substrate present in the storage liquid during the storage process of granular sludge has a direct impact on the hydrolysis and endogenous respiration of the granules. Consequently, it significantly influences the storage efficacy of the granules and their physical and biological characteristics after reactivation [7].

Gao et al. [29] soaked fresh granules in glucose solution and deionised (DI) water and stored them at − 25, 4 °C and room temperature (20–26 °C) for eight months, respectively (Tables 2.4 and 2.5). Performance analysis of the reactivated sludge revealed that storage in a glucose solution at 4 °C was the best option compared to other temperatures. However, storage in deionised water at 4 °C was superior to the glucose solution due to higher performance, sustainable physical properties and simpler solution preparation. Yuan et al. [137] performed reactivation experiments

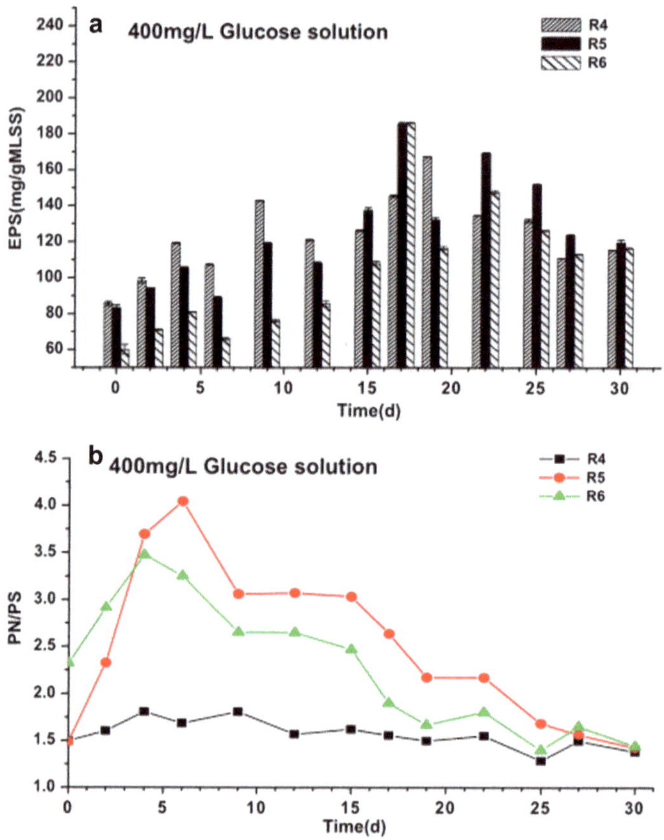

Fig. 2.17 EPS content and the PN/PS of the granules in R4, R5 and R6. (R1: distilled water, − 25 °C; R2: distilled water, 4 °C; R3: distilled water, 20–26 °C; R4: 400 mg/l glucose solution, − 25 °C; R5: 400 mg/l glucose solution, 4 °C; R6: 400 mg/l glucose solution, 20–26)

on fresh granular sludge after washing and soaking in tap water and storing at room temperature for six months. The results showed that the best reactivation of particulate matter was achieved at an OLR of 0.8 kg COD/m^3/d, an ammonia concentration of 15–20 mg/l and an apparent updraft velocity of 2.6 cm/s. SOUR, as a bioactivity factor, indicates the respiratory and catabolic activity of microorganisms and can be used to evaluate the microbial activity and biodegradation efficiency of aerobic pellets. The SOUR of fresh granules before storage was 36.18 mg-O_2/g-SS h^{-1}. Aerobic granules under different storage conditions basically lost their initial metabolic activity after 8 months of storage. As shown in Fig. 2.18, the aerobic pellets in R1-R6 gradually recovered their microbial activity (SOUR) with reactivation. After 9 d of activation, the SOUR of R1-R6 recovered rapidly to 33.17, 36.36, 35.61, 29.32, 35.42 and 33.82 mg-O_2/g SS h^{-1}. The SOUR of R1-R3 increased significantly faster than that of R4-R6. This indicates that distilled water has better

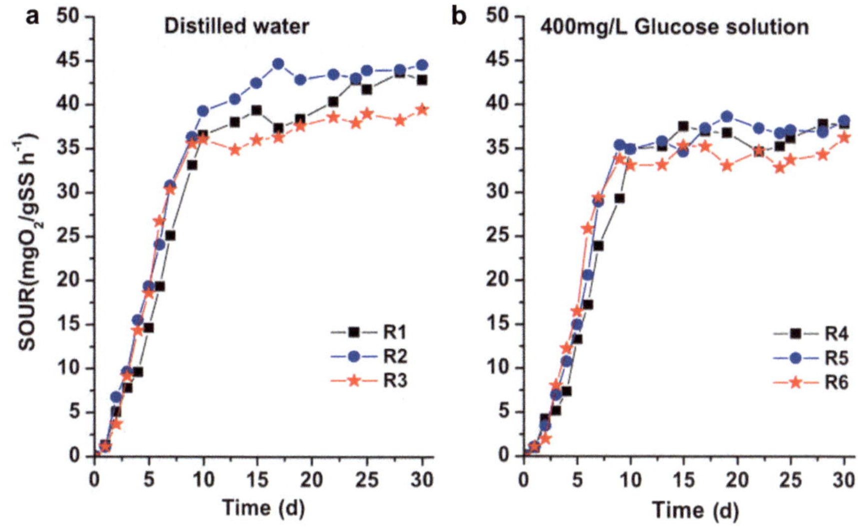

Fig. 2.18 The profiles of SOUR during the reactivation [137] (R1: distilled water, 25 °C; R2: distilled water, 4 °C; R3: distilled water, 20–26 °C; R4: 400 mg/l glucose solution, – 25 °C; R5: 400 mg/l glucose solution, 4 °C; R6: 400 mg/l glucose solution, 20–26 °C)

recovery performance on microbial activity. This indicates that tap water can also be used as a storage medium for the AGS, but that certain conditions are required for reactivation. Wan et al. [106, 108] collected AGS for storage in a 5% NaCl solution at 4 °C for over 180 d and subsequently successfully reactivated the stored AGS and used them in SBR and continuous flow reactors (CFR) under different COD conditions. This means that the reactivation of AGS after salt solution storage can be recovered for long-term operation and that the use of salt solution stored pellets is suitable for long-term storage.

2.7.3 Time

The storage of granular sludge is also limited by time, as it is hoped that the granular sludge can be stored for long periods of time for various applications. However, long-term storage under certain conditions may result in permanent damage to the internal structure of the AGS or a reduction in properties such as physical and biological activity after reactivation, which may have negative impacts on subsequent applications.

Duarte et al. [23] observed a study on AGS with different storage times (short term 40 d and long term 180 d) to discuss the effects of these two storage times on the physical properties and biological activity of the granules. The results showed that the AGS stored for a short time had a strong resistance to breakage and a gentle color change. After reactivation, the AGS stored for 180 d swelled on the first day and the settling properties were severely affected. In terms of biological activity, COD removal recovered within a few d of activation for the short-storage AGS. In contrast, the AGS stored for 180 d required a longer recovery time to reach the condition of the AGS stored for 40 d. The experimental results also revealed that 180 d of inactivity can have a detrimental effect on the polyphosphate organisms.

2.7.4 Dry Storage

Dried AGS storage methods are used to dehydrate fresh AGS by techniques such as air drying, drying in ethanol solutions, microwave drying, etc. [94]. Tanavarotai et al. [93] used the drying technique to store the AGS for one week and one month respectively. The AGS was analyzed by reactivation runs for one day at the end of each storage period. The results indicated that the reactivated AGS became darker in color and smaller in storage during each activation, indicating that there was some loss of AGS as a result of each dry storage. Lv et al. [73] used six methods such as air drying at 25 or 50 °C, freeze drying, acetone or ethanol dehydration and microwave heating to dehydrate the aerobic particulate matter, which was then recovered and characterized. The results showed that the air-drying technique at both temperatures causes minimal damage to the granules structure and that the granules are stable and hard. Drying the granules using ethanol solution provided good recovery performance. The microwave heating technique used the least amount of drying time compared to the other techniques. However, this technique results in the granules losing more strength after reactivation.

In addition, Zhang et al. [148] enclosed AGS in agar for thirty d in dry storage at 4 °C. The changes in the composition of the microbial community during dry storage and reactivation were investigated. The results showed that the main genus of AGS changed from *Zoogloea* (relative abundance of 22.39%), *Thauera* (16.03%) and *Clostridium_sensu_stricto* (11.17%) to *Acidovorax* (26.79%), *Macellibacteroides* (12.83%) and *Pseudoxanthomonas* (5.69%), respectively. Parthenogenic and anaerobic bacteria proliferate and dominate in stored AGS, and their metabolic activity gradually leads to structural damage and deterioration of the AGS properties (Table 2.6).

Table 2.6 The state of AGS under different storage conditions

Conditions	Storage method	Storage time	Conclusion	Reference
Temperature	− 25, 4, 25 °C	Eight months	A storage temperature of 4 °C was more suitable for maintaining the structural integrity and long-term stability of the AGS, while 25 °C was more suitable for maintaining the internal microstructure	[29]
	− 25, − 80 °C	Four weeks	The AGS were reactivated after 4 weeks of storage at − 80 °C. These AGS remained intact after storage but broke up and disintegrated during aeration. 53.22%, 51.81% and 41.34% reduction was observed for EPS, PN and PS, respectively	[15]
Substrate	Soaked fresh pellets in glucose solution and deionised (DI) water	Eight months	Glucose solution as substrate has better storage for AGS; deionized water has good sustainability and operating costs	[29]
	Tap water	Six months	Tap water can also be used as a storage medium for the AGS, but that certain conditions are required for reactivation	[137]
	Sodium chloride solution	Six months	The reactivation of AGS after salt solution storage can be recovered for long-term operation and that the use of salt solution stored AGS is suitable for long-term storage	[106]

(continued)

Table 2.6 (continued)

Conditions	Storage method	Storage time	Conclusion	Reference
Time	Short-term storage (40 d) and long-term storage (180 d)		AGS stored for a short period of time is resistant to breakage and has little color change; it is easier to regain activity; AGS stored for a long period of time has its sedimentation properties destroyed	[23]
Dry storage	Store dry	A week and a month	Reactivated AGS becomes darker, the storage volume becomes smaller, and each dry storage will damage the AGS	[93]
	AGS is dehydrated and dried by air drying, freeze drying, acetone or ethanol dehydration and microwave heating		The air drying technique causes minimal damage to the granules structure, and the AGS are stable and hard. The ethanol solution drying method provides good recovery properties. Microwave heating technology uses the least drying time	[73]
	Agar-wrapped AGS is stored dry	Thirty days	The metabolic activity of parthenogenic and anaerobic bacteria in the stored granules gradually leads to structural damage and deterioration of the granules properties	[148]

2.8 Pollutants Removal by Aerobic Granular Sludge

2.8.1 Organic Pollutant Removal

AGS technology can treat various organic polluted wastewater, such as food industry wastewater, pharmaceutical wastewater, coking wastewater, waste leachate, etc. The above wastewater has high suspended matter content, high chromaticity, odor, complex water quality components, and COD concentration higher than 2000 mg/l.

AGS technology can effectively remove COD when treating the corresponding high concentration organic wastewater. Compared with the traditional activated sludge technology, the AGS technology has low residual sludge production, strong resistance to shock load and certain self-recovery ability.

AGS treatment of high concentration organic wastewater is mainly adsorption and biodegradation. Hamza et al. [33, 34] synthesized organic wastewater by incubating AGS in a small SBR reactor while using sodium acetate as the sole carbon source with the following effluent contents: Sodium Acetate (NaAc)-water, 2930 mg/l; NH_4Cl, 350 mg/l; K_2HPO_4, 30 mg/l; K_2HPO_4, 25 mg/l. After that, the NAAC concentration was adjusted to increase the COD concentration to 7500 ± 600 mg/l to investigate the effect of AGS on the treatment of high concentration organic wastewater. It was shown that the AGS formation resulted in a significant improvement in biomass characteristics and system performance, and the COD removal rate was maintained at more than 98%.

Phenol is one of the major organic pollutants in petrochemical wastewater and can also be detected in various other industrial wastewaters [105]. As the concentration of phenol increases phenol becomes more toxic and the biodegradation of phenol is gradually inhibited. Nitrogen removal may also be reduced by high phenol loadings that disrupt microbial structure II [76]. Ho [41] found experimentally that bacterial communities in AGS can remain relatively stable under high phenol loadings (Table 2.7).

The resistance of AGS to phenol is attributed to its spatial heterogeneity, leading to ecological niche differentiation and an enhanced physiological diversity within the resident microbial community. In addition, AGS functional microorganisms can also positively influence the bioaggregation of granular sludge due to its high propensity to attach to solid surfaces [3, 6]. Phenol biodegrades simultaneously on the surface and in inner layers of AGS. Due to substrate diffusion limitations, phenol degraders with low half-saturation constants (Ks) tend to be confined to the interior of the granules, while those with high Ks tend to be confined near the granules surface [35].

Chlorophenols are representative organic pollutants in agriculture. Chlorophenols are a typical group of aromatic compounds containing both chlorine atoms and hydroxyl groups and are highly toxic and resistant to biodegradation [102]. They can be classified as monochlorophenols, dichlorocresorcinols and polychlorophenols

Table 2.7 Degradation performance of AGS on phenol (excerpted from [41])

Phenol concentration (mg/L)	Degradation time (h)	Degradation rate (g phenol/g VSS/d)
1000	20	3.7
2000	80	3.3
3000	160	2.5
3500	NA	NA
4000	300	19
5000	560	1.6

(including pentachlorophenols, tetrachlorophenols and trichlorophenols). Mesdaghinia et al. [75] confirmed that AGS predominantly utilizes the meta cleavage pathway for the biodegradation of 4-chlorophenol. This pathway generates non-growth-associated energy, which contributes to the resistance against toxicity. Initially, 4-chlorophenol is transformed into 4-chlorocatechol. Subsequently, the meta cleavage of 4-chlorocatechol is facilitated by dioxygenases, resulting in the formation of a compound called 5-chloro-2-hydroxymuconic semialdehyde (CHMS). Concurrently, a yellowish color accumulates as a byproduct of this process. Studies by Khan et al. [51] and [72] speculate that AGS is resistant to the high toxicity of trichlorophenols and pentachlorophenols, respectively. The high microbial activity and biomass retention provided by the laminar structure was hypothesised to play an important role. Chen et al. [19] described a three-layer structure for chlorophenol-degrading AGS: aerobic bacteria were mainly distributed in the surface layer; the middle layer contained a range of syntrophic bacteria, with methanogenic bacteria forming the core. Mesdaghinia et al. [75] asserted that only a limited community of microorganisms, mainly anaerobic bacteria in the inner layer of aerobic granules, were able to metabolize 4-chlorophenol as a substrate. However, the hierarchical structure of the AGS allows for the presentation of different microbial communities, which enables the formation of metabolic networks that facilitate the complete breakdown of hazardous pollutants, leading to their complete mineralization.

Under good culture and operating conditions, AGS technology can effectively treat highly concentrated organic wastewater. In addition, periodic starvation enhances the stability of sludge granules and controls granules size and integrity [33].

2.8.2 *Nitrogen and Phosphorus Removal*

The structure of AGS consists of an anaerobic zone, an anoxic zone, and an aerobic zone arranged in order from inside to outside. Due to its unique structural characteristics, AGS can achieve simultaneous nitrification and denitrification. Furthermore, the alternating aerobic and anaerobic phases facilitates simultaneous denitrification and phosphorus removal. Campo et al. [14] used AGS simultaneous nitrification denitrification to treat domestic wastewater with low carbon to nitrogen ratios. The results showed that the efficiency of simultaneous removal of COD, PO_4^-–P and N was 84%, 96% and 71%, respectively, with effluent water quality possessing attributes of 25.1 mg COD/L, 0.09 mg PO_4^-–P/L, 8.6 mg TIN/L (9.8 mg TN/L) and 22.1 mg TSS/L, respectively. Meanwhile, the average ratio between biodegradable soluble chemically aerobic polyhydroxy fatty acid esters (PHAs) and nitrogen removed was 3.1 g COD/g N, indicating that nitrification/denitrification occurred partly via the nitrite pathway. It has also been shown that volatile fatty acid mixtures containing acetate and propionate can select for enrichment of PAOs relative to GAOs at pH 6.5–7.5 and temperatures of 15–30 °C. GAOs and PAOs will compete for available organic substrates. The dominance of GAOs in AGS is associated with a reduction in phosphorus removal. COD is the limiting condition in wastewater and is required for both N and P removal processes. This limitation can be addressed by enriching and maintaining microorganisms that are capable of simultaneous nitrogen removal and anoxic phosphorus removal. Denitrifying phosphorus-polymerizing bacteria (DPAOs) are a type of microorganisms that utilize nitrate and/or nitrite as electron acceptors while also removing phosphorus. In addition, these microorganisms have the ability to store organic carbon in the form of intracellular polyhydroxyalkanoates (PHAs) and use the stored PHAs as a source of electrons for nitrogen removal. Treatment systems containing DPAOs are therefore energetically efficient because of the proper utilization of COD and less aeration requirements [77].

In addition, Li et al. [57] explored and compared the phosphorus removal performance of granular sludge (R1) and flocculated sludge (R2). Experiments were conducted in two identical SBR in the same anaerobic–aerobic alternating environment, with two daily cyclic operations of about 8 h. The inlet water was mixed anaerobically for 180 min and aeration for 240 min. because of the different settling times due to sludge type, AGS was cultivated by shortening the settling time, which was gradually reduced from 20 to 2 min in the R1 reactor (daily reduction 1.5 min), while the R2 reactor maintained a settling time of 20 min. The feed water is all synthetic wastewater (Table 2.8). The performance of the two reactors during the start-up period is shown in Fig. 2.19. During the initial startup stage, the flocculated sludge system exhibited a phosphorus release phenomenon, albeit with a low phosphorus removal capacity. However, over time, both phosphorus release and phosphorus absorption gradually increased. Eventually, after stabilization, the system achieved an impressive phosphorus removal rate of 80%. In contrast, the AGS was formed by shortening the settling time, so the biomass within the AGS system decreased in the initial stage, making its performance unstable, but the phosphorus removal treatment of the AGS system also reached more than 80% in the anaerobic phase. At day 25, the average granules size of AGS was around 0.5 mm, indicating that the phosphorus-polymerizing microbial granules were well developed and reached steady state. When both reactors reached steady state, it can be seen that the phosphorus removal effect of flocculated sludge is better than that of AGS, and the phosphorus concentration in the effluent of flocculated sludge system is 1 mg/l, while that of AGS system is 2.5 mg/l. However, the removal rate of ammonia nitrogen is better in AGS system (96%) than that of flocculated sludge system (86%); the COD removal rate of both reactors is basically similar (Fig. 2.20). After 40 d, the sludge was taken and analyzed for its physical properties, and the results were obtained, and the MLVSS/MLSS of the two reactors were similar with little difference. While the SVI_{30} of AGS was significantly lower than that of flocculated sludge, the settling rate of AGS was much higher than that of flocculated sludge (Table 2.9). The specific phosphorus release rate and specific phosphorus uptake rate of flocculated sludge were consistently greater than those of AGS, which may imply that the number of microorganisms accumulating phosphorus in the flocculated sludge system was higher than that of the AGS system, indicating that the flocculated sludge system had a higher phosphorus removal rate than the AGS system (Fig. 2.21).

Table 2.8 Synthetic wastewater composition [57]

INF	mg/l	Trace elements	mg/l
$COD(C_6H_{12}O_6)$	300	$FeCl_3$	20
NH_4^+–$N(NH_4Cl)$	15	$CuSO_4$	50
PO_4^{3-}–$P(KH_2PO_4)$	10	$CaCl_2$	100
$MnSO_4{\cdot}H_2O$	50	$CoCl_2{\cdot}6H_2O$	50
KCl	18	$AlCl_3$	15
$NaHCO_3$	250	$ZnSO_4{\cdot}7H_2O$	30
Trace elements	1 ml/L	H_3BO_3	40
		$(NH_4)_6Mo_7O_{24}{\cdot}4H_2O$	30

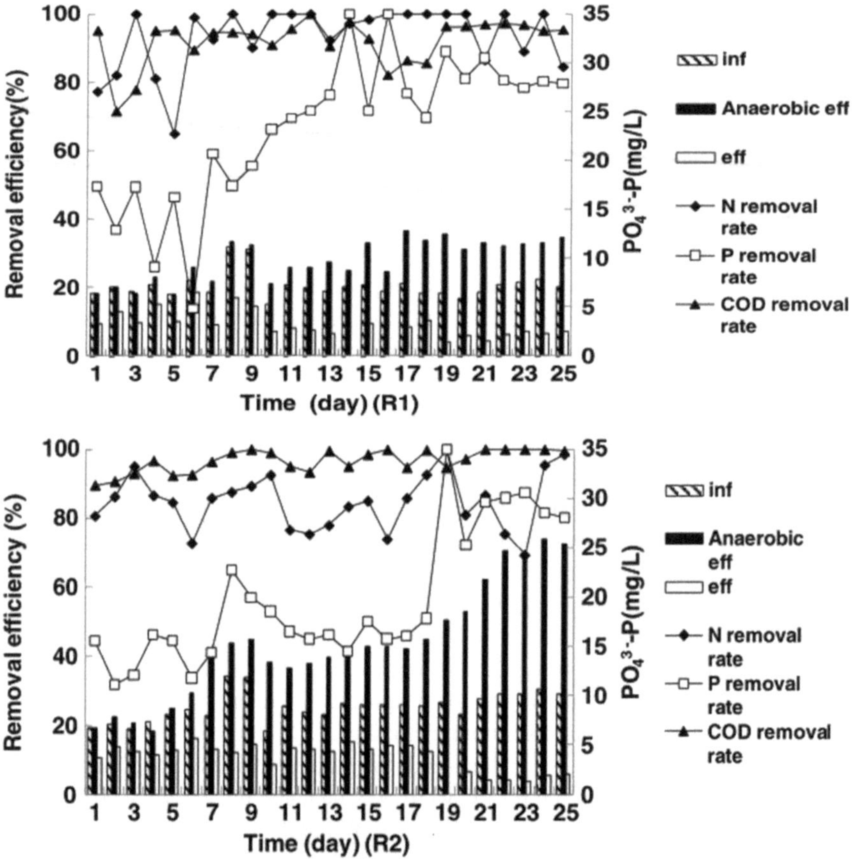

Fig. 2.19 COD, NH_4^+–N, and phosphorus removal during the start-up period (R1: AGS; R2: flocculated sludge) [57]

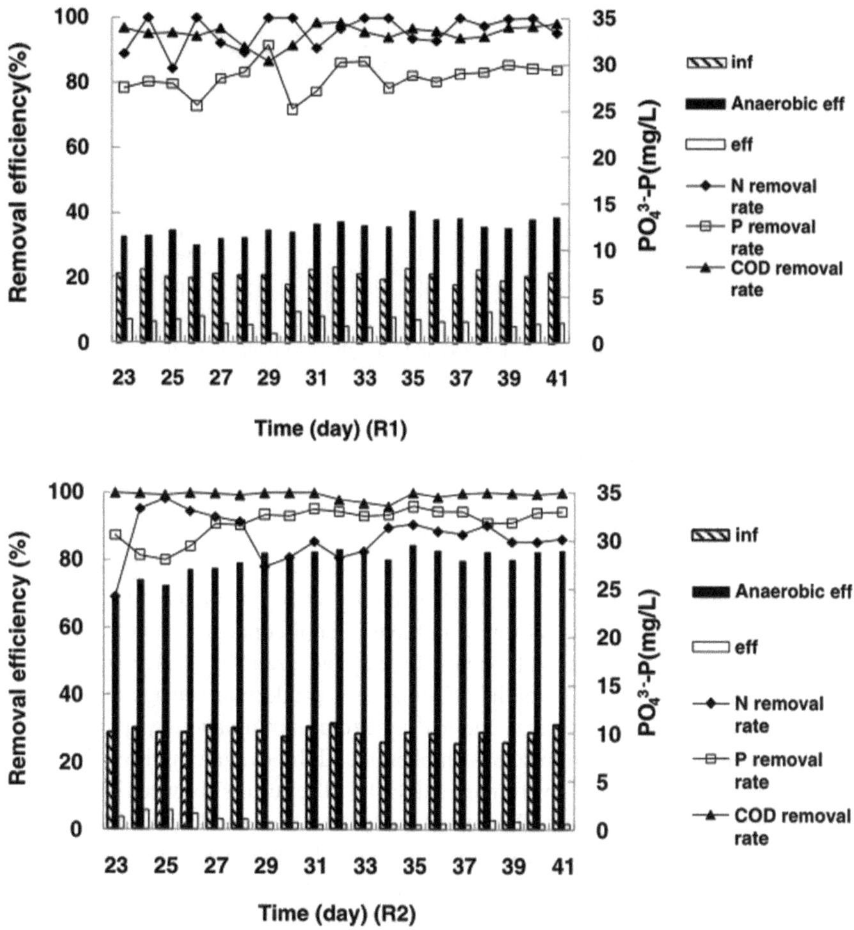

Fig. 2.20 COD, NH_4^+–N, and phosphorus removal during the steady period (R1: AGS; R2: flocculated sludge) [57]

Table 2.9 Characteristics of sludge in the two reactors after 40 d [57]

	AGS	Flocculated sludge
MLVSS/MLSS	99.55	99.44
SVI_{30} (ml/g)	16	44
Setting velocity (m/h)	40–85	5–8
Specific gravity (g/cm^3)	1.045	1.004
Moisture content (%)	85.5	98.7

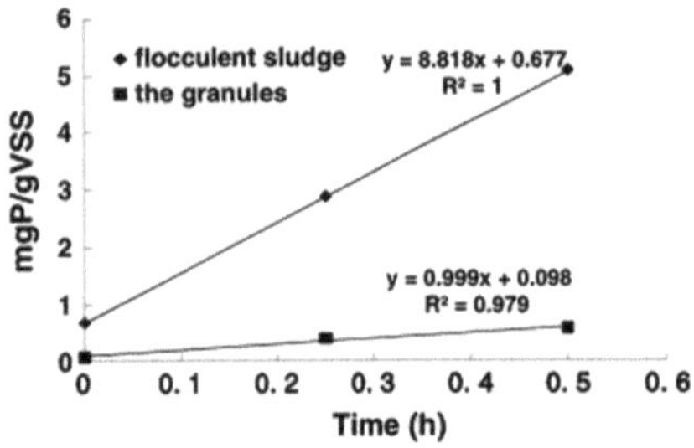

(a)

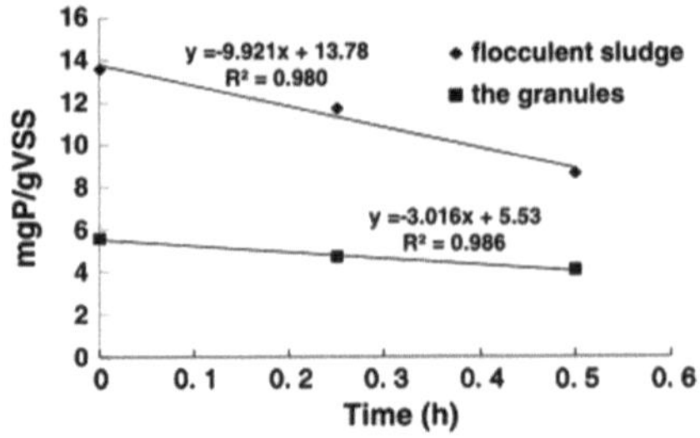

(b)

Fig. 2.21 Specific phosphorus re-lease rate (**a**) and phosphorus uptake rate (**b**) in two reactors at the steady stage [57]

2.8.3 Toxic Contaminant Removal

AGS have been used for the treatment of toxic organic chemicals, especially phenolic compounds. In one study, it was found that the aerobic granular sludge achieved a degradation rate of phenol greater than 1 g phenol/g VSS^{-d} when the influent phenol concentration was 500 mg/l. As for an influent phenol concentration of 1900 mg/l in SBR, the degradation rate of phenol by aerobic granular sludge was 0.53 g phenol/g VSS^{-d} [28]. The potential of AGS to degrade two nitrogenous heterocyclic compounds (NHCs) (thiazoline and pyrazole) simultaneously, pyridine and indole as well as ammonia nitrogen with phenol and thiocyanate in three SBR (R1, R2 and R3) was investigated by Tomar et al. [101]. The results showed that the AGS removal of pyridine and other pollutants (phenol, thiocyanate and ammonia nitrogen) was about 74% and 98%, respectively. It was shown that in the presence of phenolics in the wastewater, aerobic granules maintain structural stability by increasing the production of EPS, especially PS folds and various functional groups, which were characterized by EEM, FTIR and XPS spectroscopy, in response to the increased phenol load. Illumina MiSeq sequencing results showed that the increased phenol load reduced bacterial diversity and abundance and also led to changes in the structure of microbial populations. Significant changes in the composition of microbial populations. Several halophilic bacteria, including Stappia, Luteococcus and Formosa, provided the biological basis for stable and efficient aerobic granules bioproduction and phenol removal. Redundancy analysis (RDA) revealed the critical role of phenol in shaping the relative abundance and genus of key bacteria. This study shows that AGS is suitable for the treatment of low strength wastewater with low salt and phenol [38].

2.8.4 *Heavy Metals Removal*

AGS with large surface area and high activity are considered as a biosorbent for dyes and heavy metals removal. Dyes and pigments are extensively employed in various industries such as textile, tanning, food, paper and pulp, printing, carpet, and mineral processing for coloring purposes. However, if these dyes and pigments are not properly treated before discharge, they can become significant pollutants in the environment. Therefore, appropriate treatment methods are necessary to mitigate their potential adverse effects [28].

The AGS had a high adsorption capacity for Cd^{2+} and Zn^{2+}. The maximum adsorption capacities of Cd^{2+} and Zn^{2+} were 566 and 270 mg/g, respectively. The adsorption efficiency of the granules was related to the concentration of metals and granular sludge. When the AGS MLSS was kept at 100 mg/l, the amount of adsorbed metals was proportional to the metal ion concentration [28].

A study was also conducted to obtain manganese oxidized AGS (Mn-AGS) by continuously adding Mn(II) to the sludge in a SBR. During incubation, the PN content in EPS increased from 80.3 mg/g volatile suspended solids (VSS) to 87.8 mg/g, which may facilitate sludge granulation and heavy metal removal. The sequential batch experiments revealed that Mn-AGS could better oxidize As(III)/Sb(III) to the less toxic As(V)/Sb(V) and reduce Cr(VI) to the relatively harmless Cr(III) compared to the conventional AGS [40]. This shows that AGS has great potential for heavy metal removal from wastewater.

The contents of this section demonstrate that AGS finds extensive application in wastewater and pollutant treatment processes. However, it is important to note that the practical implementation of AGS necessitates some adjustments based on the specific type of wastewater or pollutants being treated. It is essential to achieve the corresponding effect through the targeted regulation of AGS. At the same time, a single AGS system can face more complex situations in practical application, and there may be difficult results to meet the requirements. Therefore, it can be combined with other processing systems to meet the relevant requirements or to improve the overall performance.

2.9 Application of Aerobic Granular Sludge in Wastewater

2.9.1 *Printing and Dyeing Wastewater*

Wastewater generated from the textile industry, particularly during the dyeing, printing, and finishing steps, poses significant challenges for treatment due to its high concentration of pollutants. These pollutants include dyes, surfactants, heavy metal ions, detergents, solvents, and recalcitrant compounds, which are often toxic to microbial biomass and resistant to degradation. As a result, textile wastewaters are often characterized by high concentrations of suspended solids and a low ratio between BOD to COD [69]. With the development of the dye industry, the printing and dyeing industry uses a large number of new auxiliaries and dyestuffs, etc. The content of recalcitrant and toxic organic substances in the printing and dyeing wastewater is getting higher and higher, causing greater harm to the environment, especially the water system [154]. The AGS technology is also useful for the treatment of textile printing and dyeing wastewater.

Lotito et al. [69] investigated the performance of a sequencing batch biofilter granular reactor for the treatment of wastewater from a dyeing plant. Using real dyeing wastewater provided in different stocks. The results showed that COD removal ranged from a minimum of approximately 36% to a maximum of 80% due to fluctuations in feed water concentration. The removal rate of TKN steadily decreases as the feed water COD and organic load increases. In the aforementioned study, the researchers also observed certain challenges related to biomass settling and treatment performance stability under specific conditions. These difficulties were attributed to a gradual increase in OLR within the system. Removal rates ranged from 40 to 90%, with an average value of 70%. Long sludge residence times may have promoted the development of species capable of degrading dyes. Subsequent results of the above mentioned researcher's trials using a granular sludge system SBBGR for the treatment of municipal mixed textile wastewater. Good high average removal rates were obtained (82.1% COD, 94.7% TSS, 87.5% total Kjeldahl nitrogen, 77.1% surfactant) [70]. Zhu et al. [154] used an IC reactor (micro aerobic granules as the main body) to treat printing and dyeing wastewater under micro-oxygen conditions. The results showed that the experiment was feasible when the DO concentration was 0.3–0.5 mg/l, with a 25% increase in mixed water load and a 7% increase in influent load. When the COD load of the mixed wastewater reached 2188.2 mg/l, the components in the printing and dyeing wastewater were toxic and inhibited the growth of the AGS, which also indicated that the micro AGS was limited in treating the printing and dyeing wastewater. The reactor operated best when the influent load was about 4655.1 mg/l, the DO concentration was 0.3–0.5 mg/l and the hydraulic retention time was 6 h. The COD removal rate could reach 85.8% at this time.

2.9.2 Aquaculture Wastewater

In general, livestock wastewater contains high COD, biological oxygen demand (BOD), color, nitrogen, phosphorus and suspended solids. The high levels of phosphorus and nitrogen in discharged livestock wastewater lead to eutrophication of the receiving water body. Therefore, specific treatment methods are required to treat farm wastewater. In the past, small amounts of livestock wastewater were discharged without appropriate technology, which seriously polluted the environment. The higher the discharge and frequency of untreated livestock, the greater the negative impact. Biological treatment is the preferred method due to the high cost of reagents and the poor performance of physicochemical methods in removing soluble organic matter. AGS is a biological treatment technology for livestock wastewater that has been successfully cultivated in SBR [78]. Othman et al. [78] treated livestock wastewater by incubating AGS in a laboratory-scale SBR. Untreated livestock wastewater containing urine, manure and washing water collected from cattle farms was fed into the SBR for treatment. AGS (diameter: 0.1 to 4.1 mm, SVI 42 ml/g; MLSS of 10.3 g l^{-1}) achieved maximum removal rates of 74, 73 and 70% for COD, total nitrogen and total phosphorus from livestock wastewater. This shows that there is also a vast scope for the development of AGS in the treatment of livestock wastewater.

In addition to livestock farming wastewater, with the development of the aquaculture industry, aquaculture wastewater has become a major area of farming wastewater. In aquaculture, where the water requirements are high, nitrogenous compounds are the main water pollutants that are toxic to fish. The mortality of rainbow trout increases when ammonium concentrations exceed 2.3 mg NH_4^+–N/L. Nitrite is even more harmful than ammonium, with concentrations ranging from 0.14 to 0.15 mg NO_2–N/L, resulting in approximately 50% mortality of rainbow trout. When water use is very high, as in intensive rainbow trout farms, the resulting HRT is often not long enough to eliminate the nitrogen pollutant. Technologies that can achieve good removal performance at short HRTs are therefore of particular interest in this sector [84].

2.9.3 Chemical Wastewater

The wastewater discharged from chemical plants is collectively known as chemical wastewater. Chemical wastewater is a highly diverse and toxic mixture that poses significant challenges in terms of purification. It is known to accumulate and have a detrimental impact on living organisms, as well as possess strong oxygen-consuming properties that negatively affect the quality of water bodies. Industries that utilize inorganic minerals as the fundamental chemical raw materials to produce acids, bases, and salts, generate inorganic chemical wastewater. During the production process, water primarily serves as a cooling agent, but the resulting discharged wastewater contains substantial quantities of acids, bases, salts, and suspended matter, with occasional inclusions of sulphides and other toxic substances. Due to its strong oxygen-consuming nature, this substance is highly toxic and, since the majority of its constituents are synthetic organic compounds, it is highly polluting and not easily decomposed.

Carrera et al. [16] installed a, pilot-scale SBR at a fish cannery to cultivate AGS for the treatment of salt-containing chemical wastewater generated. The results show that the system treats either low to medium or high concentrations of wastewater, and a pilot-scale SBR with a working volume of 3 m^3 was installed at the fish cannery to cultivate AGS to treat the resulting wastewater. In addition, Qi et al. [81] investigated the feasibility of AGS for treating glycol industrial wastewater by running a continuous flow reactor for 80 d. The results showed that after 40 d of domestication, the COD degradation rate of 2000 $\pm$ 150 mg/l was maintained at more than 85% and the ethylene glycol (EG) concentration was below the detection limit. At a COD concentration of approximately 2000 mg/l, the removal of EG industrial wastewater by AGS was stable. However, at higher concentrations (> 9000 mg/l), the toxicity of EG leads to the decomposition of AGS. This study confirms that AGS can treat industrial wastewater containing EG.

2.9.4 Pharmaceutical Wastewater

The worldwide pharmaceutical industry has been growing in recent years in line with the increasing demand for medicines for human health.

Large-scale pharmaceutical effluent discharges are linked to the sub-chronic or chronic toxicity of aquatic ecosystems and humans. This is primarily due to the extensive pollution of water bodies and the subsequent accumulation of pharmaceutical residues in the environment and food chain. Pharmaceutical wastewater is characterized by elevated concentrations of recalcitrant organic pollutants, significant levels of waste solvents, substantial quantities of inorganic salts, and various residual drugs [49]. Pharmaceutical wastewater is known to pose significant challenges for conventional biological treatment technologies due to its high COD and complex composition, which often includes highly toxic antibiotics [47]. The dense and powerful microbial structure, the ability to withstand high organic loads, and the resistance to toxicity make the AGS process a great potential for the treatment of pharmaceutical wastewater.

Amorim et al. [10] added simulated wastewater containing a mixture of chiral drugs (CPs) such as alprenolol, bisoprolol, metoprolol, propranolol, venlafaxine, salbutamol, fluoxetine and desmethylfluoxetine to an AGS-SBR for 28 d at a concentration of 1.3 mg/l each. The experimental results showed that AGS was the main mechanism for CPs removal. Phosphorus removal was negatively affected by the dosing of CPs, with phosphorus release decreasing during the anaerobic period, but effluent phosphorus levels remaining almost constant throughout the operation. The loading of CPs influenced the two-step nitrification process, with a greater impact on the reduction of nitrite to nitrate. Although the COD removal capacity changed during the CPs loading, but the treated effluent quality always met the discharge standard. This experiment confirms that the AGS process holds great promise for the treatment of pharmaceutical wastewater. In addition, some researchers have improved the ability of AGS to treat antibiotics in pharmaceutical wastewater by modifying it. Ji et al. [47] constructed AGS using *Aspergillus niger* filamentous spheres as a biological carrier and used it for the treatment of actual pharmaceutical wastewater containing β-lactam antibiotics (*A. niger*-Bla system). The study's findings demonstrated that the *A. niger*-Bla system had a substantial positive impact on antibiotic removal (> 60%) and overall reactor performance. The modified *A. niger*-Bla AGS system effectively mitigated the negative effects of antibiotics while preserving the richness and diversity of the microbial community within the AGS. As a result, the system's overall performance improved, and the stability of the AGS was maintained.

2.9.5 Domestic Sewage

Domestic wastewater is the wastewater discharged from residential life and occupies a major part of the wastewater treatment system. Domestic sewage primarily contains organic matter (such as PN, carbohydrates, fats, urea, ammonia nitrogen, etc.) and a significant quantity of pathogenic microorganisms (including parasitic eggs and intestinal infectious viruses). Various methods exist for the treatment of domestic wastewater and the AGS method is equally applicable to domestic wastewater.

Riccardo Campo et al. [14] used AGS to treat real domestic wastewater with a low C/N ratio in a simultaneous nitrification denitrification phosphorus removal trial. The test results show that nitrification/denitrification occurs partly via the nitrite pathway and that the AGS process has a low organic load, while having a high C/N/P removal rate (84/71/96%). Li et al. [56, 58] investigated the feasibility of the simultaneous nitrification, endogenous denitrification, and phosphorus (P) removal aerobic granular sequence batch reactor (SNEDPR-AGSBR) process for treating low C/N ratios (3.3–5.0) domestic wastewater using a single-stage anaerobic/aerobic/anoxic sequencing batch reactor. Preliminary results showed that the SNEDPR-AGSBR process rapidly formed mature AGS in 38 d. The long-term operation of the SNEDPR-AGSBR process in treating domestic wastewater, featuring a C/N ratio of 3.9, has demonstrated stable NH_4^+–N (100%), total N (84.3%), and P (91.8%) removal performances. These experimental results indicated that the AGS technology has excellent potential for treating low C/N domestic wastewater, and there are extensive opportunities for further development of this technology.

In addition, the coupling of the AGS process with other processes in the treatment of domestic wastewater gives also a development direction. Kosar et al. [53] coupled the High Efficiency Activated Sludge process with the AGS process for improving the energy recovery of municipal wastewater and reducing the particulate load of the AGS process. By adding municipal wastewater to the coupled process, the effluent quality was analyzed after a period of operation. The results were that the effluent concentrations of total suspended solids (TSS), COD, ammonia nitrogen (NH_4^+–N) and total phosphorus (TP) were less than 10 mg/l, 60 mg/l, 0.4 mg/l and 1.3 mg/l respectively at each stage, indicating that the system was effective in treating domestic wastewater.

2.9.6 Pulp and Paper Wastewater (PPW)

The pulping process usually consists of three stages, each of which generates different pollutants that enter the waste stream. Indeed, the characteristics of wastewater generated in the paper industry can vary depending on several factors, including the specific processes involved, the type of wood used, the technology employed, the internal circulation of wastewater, and the volume of water utilized [103]. As a water-intensive industry, the pulp and paper industry accounts for 42% of the

3 billion tons of industrial wastewater worldwide [100]. Pulp and paper mill wastewater typically has high concentrations of COD and suspended solids and contains a range of organic compounds that are difficult to treat, such as tar acids, chlorinated phenols, and organohalogen compounds [13, 104]. Lignin and tannin are prominent and challenging-to-treat organic compounds found extensively in wastewater from the pulp and paper industry. Their degradation is particularly difficult. However, there have been significant advancements in biological treatment methods and the identification of functional microorganisms that have proven effective in the biodegradation of lignin and tannin [11]. Vashi et al. [104] developed AGS by feeding paper mill wastewater and obtained high removal rates of COD (79%) and tannin/lignin (54%). The main mechanism for tannin and lignin removal was biosorption. As AGS is a biosorbent, EPS provides electrostatic force and tannins and lignin are adsorbed on the granules surface and then biodegraded [104].

In addition to tannins and lignin, pulp and paper wastewater also contains a group of organic compounds that are typically resistant. These compounds are known as adsorbable organic halides (AOX), which include chlorophenols, guaiacols, furans, and dioxins.

The introduction of AOX into wastewater can greatly increase the toxicity and extensibility of pulp and paper wastewater due to the inevitable introduction of chlorine and chlorides in various pulp and paper processes [35].

Carucci et al. [17] experimentally found that Pulp and paper wastewater (PPW) contains a wide variety of chlorinated compounds due to the chlorine bleaching process, AGS can treat these toxic compounds without interfering with biological activity. Chen et al. [19] demonstrated that PCP, a common chemical in wood and wood products, can be readily degraded by aerobic granules because they form a dense community of symbiotic anti-toxic bacteria. Wang et al. [117] used glucose as a co-substrate to degrade 2,4-DCP in an aerobic pellet reactor, the results showed that the presence of glucose allowed for the regeneration of NADH, which further promoted the transformation of 2,4-DCP and facilitated granulation. The pulping product contained high doses of 2,4-dichlorophenol, which could be degraded with the aid of AGS technology with glucose as co-substrate. Farooqi and Basheer [24] biodegraded AOX and COD using aerobic granules in a pilot-scale reactor operating for 780 d; 90% of the AOX in the feed water was completely degraded. The results of the study showed the feasibility of AGS in a real industrial setting, where AOX can be reduced significantly in a large reactor. AGS is perfectly suited to treat PPW because of its high tolerance to AOX.

2.10 Combination Technologies Based on Aerobic Granular Sludge

Numerous cases of AGS WWTPs at home and abroad have shown that AGS technology has a broad prospect in the development of wastewater treatment technology. On this basis, some scholars have developed the combination of AGS technology with membrane biotechnology, microbial fuel cells and other processes to provide technical support with practical application value for urban domestic wastewater treatment.

2.10.1 Aerobic Granular Sludge and Membrane Bioreactor

AGS and membrane bioreactor (MBR) processes have achieved significant success in the treatment of high concentration organic wastewater. The combination of the processes was developed to solve the problems arising from their individual operation. The AGS effluent requires further subsequent treatment to meet the effluent standard, while the MBR process increases the operating cost due to membrane contamination problems. The aerobic granular sludge membrane bioreactor (AGS-MBR) compensates for the shortcomings and inherits the advantages of both processes; it exhibits strong adaptability in treating wastewater and shows stable simultaneous denitrification and phosphorus removal performance with less fouling, resulting in good water quality. The rigid structure of AGS provides attachment points for bacteria, reducing the attachment of bacteria and microbial EPS on the membrane surface and in the membrane pores, thus reducing membrane contamination [138].

Zhang et al. [148], developed and optimized the AGS-MBR process, explored the effect of influent C/N on the treatment effect of the process, and investigated membrane module contamination and membrane flux recovery. It was concluded that: the AGS-MBR process was successfully started with domestic wastewater and artificial water distribution as raw water, and the AGS of 2 mm was cultivated to obtain good treatment results. Changes in the influent C/N ratio were used to simulate the load shock in wastewater treatment plant operation, and overall process operation was stable. Physical cleaning + chemical cleaning resulted in 99% membrane flux recovery.

AGS-MBR is also an effective treatment technology for high-concentration organic wastewater. The special spatial structure of AGS screens out functional bacteria, such as nitrifying and denitrifying bacteria, and copes with toxic and harmful external environments to ensure stable reactor operation and high pollutant removal rates. AGS-MBR is effective at reducing membrane contamination and extending reactor operation time, although it cannot achieve zero contamination. AGMBR contamination mainly results from membrane pore clogging, and filtration performance of the mixture can be improved by changing the nature of AGS and adjusting the granules size [138] (Fig. 2.22).

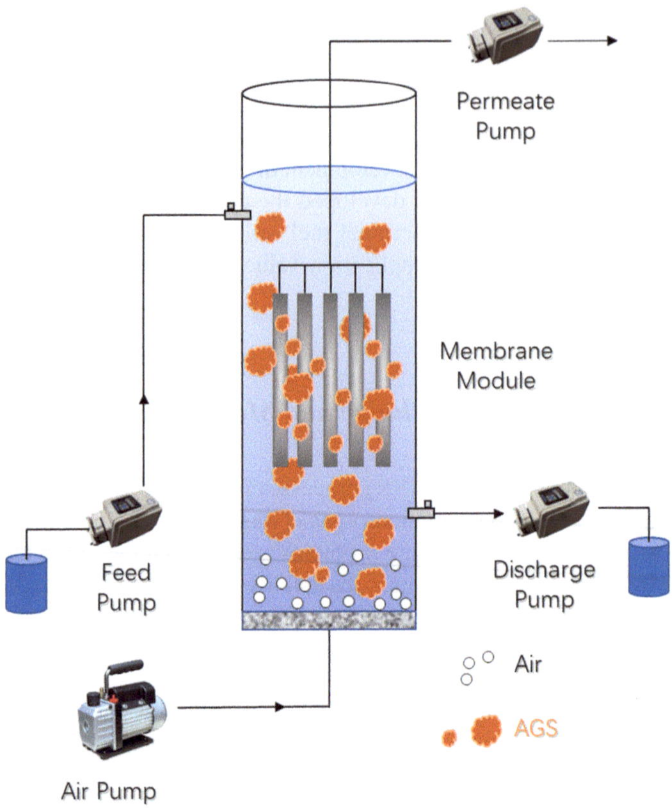

Fig. 2.22 Schematic diagram of AGS-MBR system

2.10.2 Aerobic Granular Sludge and Microbial Fuel Cells

By effectively coupling the microbial fuel cell and AGS, a double simple MFC reactor is designed, in which the anode is inoculated with anaerobic sludge and the cathode with AGS. The reactor makes full use of the advantages of both MFC and AGS to improve the decontamination performance and electricity production performance, and is used to treat hard-to-degrade wastewater—waste leachate.

Zhang et al. [148], formed the MFC-AGS process by coupling AGS with microbial fuel cells to explore its decontamination effect and electricity production performance, and optimize the operating conditions. The following conclusions were drawn: the system transmission voltage signal increased at the beginning of reactor operation, and the high voltage could reach 157 mV, and the highest voltage rose to 250 mV and 459 mV after substrate replacement, respectively, and the higher voltage output could be maintained for a longer period of time, HRT could affect the contact time between microorganisms and waste leachate in MFC and thus the removal efficiency, and the optimal HRT setting was found to be 72 h; in the circuit The best HRT setting was found to be 72 h. Connecting solar panels in series in the circuit will form a built-in electric field, which will improve the electricity production performance of the MFC and promote photoelectron transfer, thus improving the degradation efficiency. Changing the light time of the solar panel solves the thermodynamic barrier problem in the electron transfer process and improves the electricity production performance of the system; at the level of genus, *Trichococcus* in AGS is significantly affected by the external electric field and forms electrotropism among each other. The additional *Saccharibacteria* in anaerobic sludge prefer to degrade complex organic compounds and difficult pollutants after the external voltage is applied, thus improving the performance of the MFC system in degrading pollutants.

While the combined process of MFC and activated sludge method is commonly employed for organic matter removal and recovery of electrical energy, it still faces the challenge of high nitrate emissions. This issue arises due to the limited availability of a carbon source in the feedwater, which is essential for denitrification. Zhao et al. (2022), established a polybutylene glycol succinate (PBS) assembled MFC in an AGS bioreactor for simultaneous promotion of nitrogen removal and electricity generation. The total inorganic nitrogen (TIN) removal rate and COD removal rate of the AGS-MFC group were increased to 84.3 ± 2.6% and 93.5 ± 0.5%, respectively. The average output voltage of the MFC module was 223.7 mV and the maximum power density was 59.6 mW/m^2. The relative abundance of *Thauera*-associated denitrifying bacteria was highest in both bioreactors by high-throughput sequencing analysis, 20.0% and 31.4%, respectively. The relative abundance of *nitrosomonas*-associated AOB was higher in AGS-MFC (1.8%) than in AGS-Control (1.1%). In the MFC module, the anode biofilm was dominated by *Thauera* (16.2%) with denitrification and power generation under PBS enhancement. The MFC-AGS process has likewise seen some research progress in the treatment of toxic wastewater, Cheng et al. [20], assembled AGSon the surface of MFC anode to form an MFC-AGS system with excellent performance for the treatment of epoxy activated diluent (ERD) wastewater. Compared with MFC-AGS inoculated with domestic wastewater, MFC-AGS successfully reduced the start-up time from 13 to 7 d. Improved toxicity tolerance and COD removal (77.8 vs. 63.6%).

2.10.3 Aerobic Granular Sludge Coupled Microalgae System

The advantages of AGS technology are undeniable, but the AGS process has the disadvantage of low structural stability and slow biomass growth over long periods of operation. Algae, on the other hand, have a high growth rate and can effectively adsorb heavy metals from wastewater, but they have poor settling properties. To address the characteristics of AGS and algae, researchers have proposed a new AGS process, algae-bacterial AGS, which combines the advantages of both algae and AGS. It has been reported that algae can grow naturally on AGS to form algae-bacterial granules consortia, which exhibit excellent and consistent nutrient removal performance (90–99% COD removal, 60–70% total N removal, and 65–90% total phosphorus removal) with a low SVI. At the same time the algal–bacterial AGS is able to adapt to low carbon environments (COD/N = 1) and maintain good granules stability. Therefore, the algae-AGS process has great potential for development [133].

Compared to conventional AGS, algae-AGS is more promising because algae can absorb and take up inorganic nutrients (N and P) and release oxygen for metabolism by aerobic bacteria [67]. Furthermore, the development of the algae-bacterial AGS system aligns with the concept of carbon neutral for municipal wastewater treatment technology. The algae produce oxygen, which supports the bacterial activity through photosynthesis, thus significantly reducing the energy input from external aeration. Li et al. [59] found that at a short hydraulic retention time (HRT) (6 h), phototrophs coexisting in algal–bacterial AGS were able to photosynthesize enough O_2 for nitrification and aerobic phosphorus uptake. On the other hand, microalgae can indirectly induce polyphosphate hydrolysis in the granules under strong light conditions and grow on the nutrients in the granules [58]. Meanwhile the carbon dioxide produced by the bacterial oxidation process facilitates the growth of the algae, thus significantly reducing greenhouse gas emissions [142].

The researchers carried out a characterization of microalgae activated granules by adding Chlorella spp. and Fritillaria spp. as seed microalgae. The results showed that algal–bacterial granular sludge could be successfully developed using specific seed algae and could be maintained in the granules at a relative abundance of over 98%. And the algae-bacterial granule system can achieve 50 and 36% removal of total nitrogen and phosphorus respectively, which is better than the AGS system (33 and 25%). Mature algal granules had a dense structure and excellent settling performance. The algal AGS reached maturity between 20 and 40 d and reached an average diameter of 0.29 mm, while SVI_{30} was stable at < 50 mL/g. Microbiological studies have demonstrated that algal–bacterial granules can be rapidly formed in a photo-sequencing batch reactor. These granules exhibit compact structures, excellent settling ability, and resistance to temperature fluctuations. This is attributed to the presence of certain microbial classes, including Betaproteobacteria, Gammaproteobacteria, and Flavobacterium, as well as algae classes such as Chlorophyta, Trebuxiophyceae, and Bacillariophyceae [80].

The removal of Cr(VI) by algal–bacterial AGS as a biosorbent was first reported in a study by Yang et al. [133] compared with conventional AGS in terms of biosorption performance. Sequential batch experiments showed that the biosorption of Cr(VI) by algal–bacterial AGS was highly pH dependent, with the maximum biosorption of Cr(VI) at pH 2 (51.0 mg g^{-1}). The higher biosorption capacity and better granules stability of Alcaligenes AGS compared to conventional bacterial AGS suggest that Alcaligenes AGS can be used as a biomaterial for the removal and recovery of Cr(VI) in the treatment of Cr(VI)-containing wastewater. In addition, the biosorption of Pb(II) by algal–bacterial AGS (A-B AGS) was studied for the first time by Wang et al. [114, 120]. The results showed that the maximum adsorption of Pb (II) by A-B AGS was 72.4 mg at pH 6.0, and about 66% of the adsorbed Pb (II) was accumulated in A-B AGS microbial cells. The results of this study provides a theoretical basis for the large-scale application of A-B AGS in the biological treatment of lead-containing wastewater.

At the same time there have been additional developments of algae-bacterial AGS systems by researchers. Su et al. [90] developed a filamentous algae-bacterial symbiotic granules system in a photo sequential batch reactor. A comparative investigation was conducted to evaluate the characteristics and performance of a well-established algae-bacterial pellet system in comparison to an activated sludge system operating under identical conditions. Dynamic synergy tests showed that the COD removal rates were all greater than 90% under non-aeration conditions. However, the effect of the addition of AHLs (5, 50 and 200 nmol/L) was negligible. Under static conditions, $10^{(-9)}$ nmol/L AHLs significantly modulated the algal symbiosis and < 50 nmol/L AHLs increased COD and NH^{+}–N removal by up to 90%. In addition, the addition of AHLs increased PS secretion, but had little effect on protein secretion. AHLs ($10^{(-9)}$ mol/L) modulated the operational status of the algal–bacterial symbiosis and altered the composition and content of EPS in the algal–bacterial system.

2.10.4 Aerobic Granular Sludge and Continuous-Flow Reactors

AGS technology has been applied in WWTPs all over the world, and more than 80 WWTPs worldwide have already applied AGS technology well in SBR. And compared with traditional activated WWTPs, AGS WWTPs have the advantages of low surface area, low energy consumption and low operating costs [88]. However, there are only 10 or so continuous flow WWTPs that have successfully applied AGS, and its promotion still needs to be done. Although SBR is easier to cultivate AGS, it suffers from the drawbacks of small treatment capacity and low equipment utilization. On the other hand, filamentous aerobic granular sludge (FAGS) has many desirable properties. In SBR, selective pressure based on settling velocity is not effective in separating FAGS from water, which limits the development of FAGS [87]. Nowadays, WWTPs are facing increasing loading rates due to the increase in living standards and the improvement of the pipeline system. Replacing or expanding existing WWTPs will face expensive costs. AGS WWTPs with SBR as the main model will face great challenges [88, 135]. The AGS technology in SBR mode usually has the following cycles: (1) influent; (2) aeration; (3) granules settling and mud-water separation; and (4) discharge of effluent. This model usually uses a short settling time to select AGS with better settling performance and flocculated material is flushed and discharged in the final stage. Such a model will limit the development of future WWTPs [83]. In contrast, AGS continuous-flow reactors (CFRs) is a more suitable model for WWTPs.

Recently, more relevant studies have become abundant due to the fact that AGS CFRs have more advantages than SBRs. CFRs are easier to build, operate and maintain. In addition, the constant flow rate allows the treatment of larger volumes of wastewater and the compactness of the technology [83]. According to the preliminary results of the laboratory-scale study conducted by Rosa-Masegosa et al. [83]. It was found that a laboratory-scale configuration (AGS-CFR) can serve as a viable model for future studies aiming to implement anaerobic granular sludge with carbon capture and utilization at full scale. This is due to the simple design of the laboratory-scale setup and the excellent results achieved during the study. Strubbe et al. [88] investigated the effectiveness of aerobic granular sludge in continuous WWTPs using modeling and simulation methods and concluded that AGS replacing activated sludge can significantly increase the maximum treatment capacity of continuous WWTPs; the use of relatively small growing granules is recommended for the implementation of AGS technology in continuous WWTPs to avoid affecting settling performance and diffusion limitations.

A review of research highlights on continuous flow aerobic granular sludge by Xu et al. [128] found that the process has made substantial progress in treating complex wastewater or enhancing its multi-pollutant removal. Sun et al. [91] explored the feasibility of achieving successful aerobic granulation in continuous flow infrastructures such as modern WWTPs, concluding that aerobic granulation can be achieved in plug-flow reactors (PFRs) between 10 and 22.5 °C; Feast/famine and selection pressure can be achieved in CFRs. In contrast, Zou et al. [155] investigated the formation of AGS using a continuous-flow reactor with two-zone sedimentation tank (CFR-TST) and successfully formed AGS in the CFR-TST (mean granules size 105 μm and SVI of about 26 mL/g).

2.11 Prospects

This chapter provides a comprehensive review of the representative properties of AGS. It begins by introducing the physical and chemical properties of AGS. The factors influencing AGS, the formation mechanism, and the fast start-up method of AGS are also summarized in this chapter. By systematically understanding the AGS mechanism, readers will be better equipped to apply the relevant theories to practical applications. Furthermore, this knowledge will enable them to effectively apply and maintain the high performance of AGS systems.

When considering the formation mechanism and basic properties of AGS, one can grasp its favorable characteristics compared to flocculated sludge. However, it is important to acknowledge that AGS is limited by various influencing factors. For instance, different substrates, temperatures, and additives have varying effects on AGS, necessitating further research to bridge these gaps. There are still no uniform results on the formation mechanism of AGS, and there are still some controversies, such as filamentous bacteria hypothesis, induced nucleation hypothesis, and population induction hypothesis, which are all partial summaries of the current situation and do not fully explain the formation mechanism of AGS in aerobic granular sludge, which adversely affects the development of this technology. In addition, the growth of AGS can be made faster and more cost effective by further understanding of the AGS formation mechanism. Therefore, the mechanism of sludge granulation remains the focus and difficulty of granular sludge research, and there is still a need to try different conditions to improve the relevant AGS formation process to facilitate better engineering application of AGS. In addition, the factors affecting the stabilization process of granulation and maintaining stability are another major challenge for AGS technology. Multiple microbial communities co-exist during long-term operation, and after a long period of competition, the dominance of microorganisms may be affected. How to maintain the dominant position and stable existence of AGS in the competition of complex microbial environment should be considered in terms of mechanism and influencing factors. On the other hand, information about the diversity and abundance of microbial populations and interactions between microbial populations can be fully understood through cutting-edge technologies such as Metagenomics. Better research on granulation mechanism and function from microbial ecology.

The improvement of AGS stability and the rapid start of AGS are complementary to each other, and the rapid start of AGS can also ensure its relative stability, and the two complement each other and will become the trend of AGS development in the future. The study of rapid start-up and storage and reactivation of AGS is still a small part of the research, but it does not affect this to become a relevant hot spot for AGS research, and the study of these two aspects will help to use AGS in practical engineering applications. Co-exist and subsequent treatment is still required to remove the flocculated sludge. The rapid start-up of AGS and its storage and reactivation are currently described based on specific aspects that are more relevant, such as shortening settling time, increasing organic load, and using additives for start-up, and varying wet and dry environments, different temperatures, and substrates for storage and reactivation. However, real-world applications of AGS are far more complex and unpredictable than controlled experimental environments. To improve AGS performance in practical scenarios, further experiments under diverse and complex conditions are necessary. Additionally, it is crucial to investigate methods to reduce the generation of flocculated sludge during AGS start-up, which would help minimize subsequent treatment costs. By gaining a deeper understanding through extensive research, we can develop more effective strategies to ensure rapid start-up, storage, and reactivation of AGS, leading to its successful application and enhanced performance in real-world engineering and life situations.

In terms of AGS applications, the treatment of common wastewater and pollutants (e.g., chemical wastewater, medical wastewater, heavy metals, nitrogen and phosphorus, etc.) and related combined processes (e.g., AGS-MBR, MFC-AGS, AGS coupled with microalgae system, etc.) have been summarized respectively. However, with the development and progress of society, different types of wastewater are gradually increasing, and at the same time the types of pollutants in wastewater are also increasing and changing, and some emerging pollutants that are trace but contain high toxicity (e.g., endocrine disruptors (EDCs), perfluorinated compounds (PFCs), brominated flame retardants (BRPs), etc.) have become a hot topic for pollution treatment in today's society; therefore, in the future research, it is important to apply AGS to more types of wastewater and apply AGS to the degradation of emerging pollutants. Meanwhile, due to the good treatment efficiency and outstanding economic benefits, the combination process of AGS should have a broad development prospect in engineering applications, and more combination processes should be tried and optimized for the existing combination processes. More combination processes and optimize the existing ones, so as to complement the shortcomings and establish an efficient and low-cost treatment process.

References

1. Ab Halim MH, Anuar AN, Jamal N, Azmi SI, Ujang Z et al (2016) Influence of high temperature on the performance of aerobic granular sludge in biological treatment of wastewater. J Environ Manage 184:271–280. https://doi.org/10.1016/j.jenvman.2016.09.079
2. Abdullah N, Yuzir A, Curtis TP, Yahya A, Ujang Z (2013) Characterization of aerobic granular sludge treating high strength agro-based wastewater at different volumetric loadings. Biores Technol 127:181–187. https://doi.org/10.1016/j.biortech.2012.09.047
3. Adav SS, Chen M, Lee D, Ren N (2007) Degradation of phenol by Acinetobacter strain isolated from aerobic granules. Chemosphere 67(8):1566–1572. https://doi.org/10.1016/j.chemosphere.2006.11.067
4. Adav SS, Lee D (2008) Single-culture aerobic granules with Acinetobacter calcoaceticus. Appl Microbiol Biotechnol 78(3):551–557. https://doi.org/10.1007/s00253-007-1325-x
5. Adav SS, Lee D, Lai J (2009) Aerobic granulation in sequencing batch reactors at different settling times. Biores Technol 100(21):5359–5361. https://doi.org/10.1016/j.biortech.2009.05.058
6. Adav SS, Lee D, Ren NQ (2007) Biodegradation of pyridine using aerobic granules in the presence of phenol. Water Res 41(13):2903–2910. https://doi.org/10.1016/j.watres.2007.03.038
7. Adav SS, Lee DJ, Tay JH (2007) Activity and structure of stored aerobic granules. Environ Technol 28(11):1227–1235. https://doi.org/10.1080/09593332808618883
8. Amin Vieira Da Costa NP, Libardi N, Ribeiro Da Costa RH (2022) How can the addition of extracellular polymeric substances (EPS)-based bioflocculant affect aerobic granular sludge (AGS)? J Environ Manage 310:114807. https://doi.org/10.1016/j.jenvman.2022.114807
9. Amorim CL, Alves M, Castro PML, Henriques I (2018) Bacterial community dynamics within an aerobic granular sludge reactor treating wastewater loaded with pharmaceuticals. Ecotoxicol Environ Saf 147:905–912. https://doi.org/10.1016/j.ecoenv.2017.09.060
10. Amorim CL, Moreira IS, Ribeiro AR, Santos LHML, Delerue-Matos C, Tiritan ME et al (2016) Treatment of a simulated wastewater amended with a chiral pharmaceuticals mixture by an aerobic granular sludge sequencing batch reactor. Int Biodeterior Biodegradation 115:277–285. https://doi.org/10.1016/j.ibiod.2016.09.009
11. Asina F, Brzonova I, Kozliak E, Kubátová A, Ji Y (2017) Microbial treatment of industrial lignin: successes, problems and challenges. Renew Sustain Energy Rev 77:1179–1205. https://doi.org/10.1016/j.rser.2017.03.098
12. Bao RL, Yan XJ, Yu SL (2011) Performance of denitrifying phosphorus removal by aerobic granular sludge at low temperature. In: Sun D, Sung WP, Chen R (eds) Frontiers of green building, materials and civil engineering, PTS 1–8 (71–78, pp. 2966). International conference on green building, materials and civil engineering (GBMCE 2011). Reprinted
13. Cai F, Lei L, Li Y (2019) Different bioreactors for treating secondary effluent from recycled paper mill. Sci Total Environ 667:49–56. https://doi.org/10.1016/j.scitotenv.2019.02.377
14. Campo R, Sguanci S, Caffaz S, Mazzoli L, Ramazzotti M, Lubello C et al (2020) Efficient carbon, nitrogen and phosphorus removal from low C/N real domestic wastewater with aerobic granular sludge. Bioresour Technol 305:122961. https://doi.org/10.1016/j.biortech.2020.122961
15. Campo R, Vassallo A, Rabbeni G, Arancio W, Gallo G et al (2021) Reactivation of aerobic granular sludge for the treatment of industrial shipboard slop wastewater: effects of long-term

storage on granules structure, biofilm activity and microbial community. J Water Process Eng 42. https://doi.org/10.1016/j.jwpe.2021.102101
16. Carrera P, Casero-Díaz T, Castro-Barros CM, Méndez R, Val Del Río A et al (2021) Features of aerobic granular sludge formation treating fluctuating industrial saline wastewater at pilot scale. J Environ Manage 296:113135. https://doi.org/10.1016/j.jenvman.2021.113135
17. Carucci A, Milia S, De Gioannis G, Piredda M (2009) Acetate-fed aerobic granular sludge for the degradation of 4-chlorophenol. J Hazard Mater 166(1):483–490. https://doi.org/10.1016/j.jhazmat.2008.11.049
18. Chen Y, Chen Y, Li C, Yuan SC (2021) Structural characteristics and stability of aerobic granular sludge. Ind Water Treat 41(10):28–35. https://doi.org/10.19965/j.cnki.iwt.2020-0814
19. Chen Y, Hsiau S, Lai S, Chyou Y, Li H et al (2009) Filtration of dust particulates with a moving granular bed filter. J Hazard Mater 171(1):987–994. https://doi.org/10.1016/j.jhazmat.2009.06.103
20. Cheng K, Hu J, Hou H, Liu B, Chen Q, Pan K et al (2016) Aerobic granular sludge inoculated microbial fuel cells for enhanced epoxy reactive diluent wastewater treatment. Bioresour Technol 229
21. de Kreuk MK, Pronk M, van Loosdrecht MCM (2005) Formation of aerobic granules and conversion processes in an aerobic granular sludge reactor at moderate and low temperatures. Water Res 39(18):4476–4484. https://doi.org/10.1016/j.watres.2005.08.031
22. de Sousa Rollemberg SL, Mendes Barros AR, Milen Firmino PI, Bezerra Dos Santos A (2018) Aerobic granular sludge: cultivation parameters and removal mechanisms. Biores Technol 270:678–688. https://doi.org/10.1016/j.biortech.2018.08.130
23. Duarte KLS, Castellanos RM, Costa RC, Mahler CF, Bassin JP (2021) Start-up of an aerobic granular sludge system from stored granules: evaluating the impact of storage period on biomass activity and stability and the effect of temperature on nitrification and phosphorus removal rates. J Environ Manage 296:113200. https://doi.org/10.1016/j.jenvman.2021.113200
24. Farooqi IH, Basheer F (2017) Treatment of adsorbable organic halide (AOX) from pulp and paper industry wastewater using aerobic granules in pilot scale SBR. J Water Process Eng 19:60–66. https://doi.org/10.1016/j.jwpe.2017.07.005
25. Feng H, Ding Y, Wang M, Zhou G, Zheng X, He H et al (2014) Where are signal molecules likely to be located in anaerobic granular sludge? Water Res 50:1–9. https://doi.org/10.1016/j.watres.2013.11.021
26. Franca RDG, Vieira A, Mata AMT, Carvalho GS, Pinheiro HM et al (2015) Effect of an azo dye on the performance of an aerobic granular sludge sequencing batch reactor treating a simulated textile wastewater. Water Res 85:327–336. https://doi.org/10.1016/j.watres.2015.08.043
27. Fra-Vázquez A, Morales N, Figueroa M, Val Del Río A, Regueiro L, Campos JL et al (2016) Bacterial community dynamics in long-term operation of a pilot plant using aerobic granular sludge to treat pig slurry. Biotechnol Prog 32(5):1212–1221. https://doi.org/10.1002/btpr.2314
28. Gao DW, Liu L, Liang H, Wu WM (2011) Aerobic granular sludge: characterization, mechanism of granulation and application to wastewater treatment. Crit Rev Biotechnol 31(2):137–152. https://doi.org/10.3109/07388551.2010.497961
29. Gao DW, Yuan XJ, Liang H (2012) Reactivation performance of aerobic granules under different storage strategies. Water Res 46(10):3315–3322. https://doi.org/10.1016/j.watres.2012.03.045
30. Gao D, Liu L, Liang H, Wu W (2011) Comparison of four enhancement strategies for aerobic granulation in sequencing batch reactors. J Hazard Mater 186(1):320–327. https://doi.org/10.1016/j.jhazmat.2010.11.006
31. Gómez-Acata S, Vital-Jácome M, Pérez-Sandoval MV, Navarro-Noya YE, Thalasso F, Luna-Guido M et al (2018) Microbial community structure in aerobic and fluffy granules formed in a sequencing batch reactor supplied with 4-chlorophenol at different settling times. J Hazard Mater 342:606–616. https://doi.org/10.1016/j.jhazmat.2017.08.073

32. Guo Y, Zhang B, Feng S, Wang D, Li J et al (2022) Unveiling significance of Ca^{2+} ion for start-up of aerobic granular sludge reactor by distinguishing its effects on physicochemical property and bioactivity of sludge. Environ Res 212:113299. https://doi.org/10.1016/j.envres.2022.113299
33. Hamza RA, Iorhemen OT, Zaghloul MS, Tay JH (2018) Rapid formation and characterization of aerobic granules in pilot-scale sequential batch reactor for high-strength organic wastewater treatment. J Water Process Eng 22:27–33. https://doi.org/10.1016/j.jwpe.2018.01.002
34. Hamza RA, Sheng Z, Iorhemen OT, Zaghloul MS, Tay JH (2018) Impact of food-to-microorganisms ratio on the stability of aerobic granular sludge treating high-strength organic wastewater. Water Res 147:287–298. https://doi.org/10.1016/j.watres.2018.09.061
35. Hamza R, Rabii A, Ezzahraoui F, Morgan G, Iorhemen OT (2022) A review of the state of development of aerobic granular sludge technology over the last 20 years: full-scale applications and resource recovery. Case Stud Chem Environ Eng 5:100173. https://doi.org/10.1016/j.cscee.2021.100173
36. Han X, Jin Y, Yu J (2022) Rapid formation of aerobic granular sludge by bioaugmentation technology: a review. Chem Eng J 437:134971. https://doi.org/10.1016/j.cej.2022.134971
37. Hao W, Li Y, Lv J, Chen L, Zhu J (2016) The biological effect of metal ions on the granulation of aerobic granular activated sludge. J Environ Sci 44:252–259. https://doi.org/10.1016/j.jes.2015.10.031
38. He Q, Xie Z, Fu Z, Wang H, Chen L, Gao S et al (2021) Effects of phenol on extracellular polymeric substances and microbial communities from aerobic granular sludge treating low strength and salinity wastewater. Sci Total Environ 752:141785. https://doi.org/10.1016/j.scitotenv.2020.141785
39. He S, Feng L, Zhao W, Li J, Zhao Q et al (2023) Composition and molecular structure analysis of hydrophilic/hydrophobic extracellular polymeric substances (EPS) with impacts on sludge dewaterability. Chem Eng J 462:142234. https://doi.org/10.1016/j.cej.2023.142234
40. He Z, Wei Z, Zhang Q, Zou J, Pan X (2019) Metal oxyanion removal from wastewater using manganese-oxidizing aerobic granular sludge. Chemosphere 236:124353. https://doi.org/10.1016/j.chemosphere.2019.124353
41. Ho KCYL (2010) Degrading high-strength phenol using aerobic granular sludge. Appl Microbiol Biotechnol 85:2009–2015. https://doi.org/10.1007/s00253-009-2321-0
42. Huang S, Zhao Y, Liu S, Yunjun Y (2011) A study on the stability of the water treatment process in the United States of America (Coleoptera, Staphylinidae) (2022) Effect of settling time on the characteristics of aerobic granular sludge from highly concentrated organic wastewater. Paper presented at the 2022 annual scientific and technical conference of the chinese society for environmental science—environmental engineering technology innovation and application session, Nanchang, Jiangxi, China
43. Iorhemen OT, Hamza RA, Zaghloul MS, Tay JH (2018) Simultaneous organics and nutrients removal in side-stream aerobic granular sludge membrane bioreactor (AGMBR). J Water Process Eng 21:127–132. https://doi.org/10.1016/j.jwpe.2017.12.009
44. Iorhemen OT, Liu Y (2021) Effect of feeding strategy and organic loading rate on the formation and stability of aerobic granular sludge. J Water Process Eng 39:101709. https://doi.org/10.1016/j.jwpe.2020.101709
45. Ivanov V, Wang XH, Tay S, Tay JH (2006) Bioaugmentation and enhanced formation of microbial granules used in aerobic wastewater treatment. Appl Microbiol Biotechnol 70(3):374–381. https://doi.org/10.1007/s00253-005-0088-5
46. Jemaat Z, Suárez-Ojeda ME, Pérez J, Carrera J (2014) Partial nitritation and o-cresol removal with aerobic granular biomass in a continuous airlift reactor. Water Res 48:354–362. https://doi.org/10.1016/j.watres.2013.09.048
47. Ji J, Gao T, Salama E, El-Dalatony MM, Peng L, Gong Y et al (2021) Using *Aspergillus niger* whole-cell biocatalyst mycelial aerobic granular sludge to treat pharmaceutical wastewater containing β-lactam antibiotics. Chem Eng J 412:128665. https://doi.org/10.1016/j.cej.2021.128665

48. Ji Y, Yu H, Cao R, Xu X, Zhu L (2022) Promoting the granulation process of aerobic sludge via a sustainable strategy of effluent reflux in view of AHLs-mediated quorum sensing. J Environ Manage 303:114091. https://doi.org/10.1016/j.jenvman.2021.114091
49. Jiang Y, Shi X, Ng HY (2021) Aerobic granular sludge systems for treating hypersaline pharmaceutical wastewater: start-up, long-term performances and metabolic function. J Hazard Mater 412:125229. https://doi.org/10.1016/j.jhazmat.2021.125229
50. Jiang Y, Yang K, Shang Y, Zhang HN, Wei L et al (2019) Response and recovery of aerobic granular sludge to pH shock for simultaneous removal of aniline and nitrogen. Chemosphere 221:366–374. https://doi.org/10.1016/j.chemosphere.2018.12.207
51. Khan MZ, Mondal PK, Sabir S, Tare V (2011) Degradation pathway, toxicity and kinetics of 2,4,6-trichlorophenol with different co-substrate by aerobic granules in SBR. Biores Technol 102(13):7016–7021. https://doi.org/10.1016/j.biortech.2011.04.057
52. Kosar S, Isik O, Akdag Y, Gulhan H, Koyuncu I, Ozgun H et al (2022) Impact of seed sludge characteristics on granulation and performance of aerobic granular sludge process. J Cleaner Prod 363:132424. https://doi.org/10.1016/j.jclepro.2022.132424
53. Kosar S, Isik O, Cicekalan B, Gulhan H, Cingoz S, Yoruk M et al (2023) Coupling high-rate activated sludge process with aerobic granular sludge process for sustainable municipal wastewater treatment. J Environ Manage 325:116549. https://doi.org/10.1016/j.jenvman.2022.116549
54. Li AJ, Li XY, Yu HQ (2013) Aerobic sludge granulation facilitated by activated carbon for partial nitrification treatment of ammonia-rich wastewater. Chem Eng J 218:253–259. https://doi.org/10.1016/j.cej.2012.12.044
55. Li A, Yang S, Li X, Gu J (2008) Microbial population dynamics during aerobic sludge granulation at different organic loading rates. Water Res 42(13):3552–3560. https://doi.org/10.1016/j.watres.2008.05.005
56. Li D, Guo W, Liang D, Zhang J, Li J, Li P et al (2022) Rapid start-up and advanced nutrient removal of simultaneous nitrification, endogenous denitrification and phosphorus removal aerobic granular sequence batch reactor for treating low C/N domestic wastewater. Environ Res 212:113464. https://doi.org/10.1016/j.envres.2022.113464
57. Li X, Gao DW, Liang H, Liu L, Fu Y (2012) Phosphorus removal characteristics of granular and flocculent sludge in SBR. Appl Microbiol Biotechnol 94(1):231–236. https://doi.org/10.1007/s00253-011-3593-8
58. Li Z, Wang J, Chen X, Lei Z, Yuan T, Shimizu K et al (2022) Insight into aerobic phosphorus removal from wastewater in algal-bacterial aerobic granular sludge system. Bioresour Technol 352:127104. https://doi.org/10.1016/j.biortech.2022.127104
59. Li Z, Wang J, Liu J, Chen X, Lei Z, Yuan T et al (2023) Highly efficient carbon assimilation and nitrogen/phosphorus removal facilitated by photosynthetic O_2 from algal-bacterial aerobic granular sludge under controlled DO/pH operation. Water Res 238:120025. https://doi.org/10.1016/j.watres.2023.120025
60. Liang J, Li W, Zhang HL, Jiang XB, Wang LJ, Liu XD et al (2018) Coaggregation mechanism of pyridine-degrading strains for the acceleration of the aerobic granulation process. Chem Eng J 338:176–183. https://doi.org/10.1016/j.cej.2018.01.029
61. Lin DX, Li XZ, Hou MX, Chen YL, Zeng J et al (2022) Aerobic granular sludge cultivated from Fe-loaded activated carbon as carrier working low-strength wastewater conditions by bioreactor. Chemosphere 306. https://doi.org/10.1016/j.chemosphere.2022.135532
62. Liu S, Zhou M, Daigger GT, Huang J, Song G (2023) Granule formation mechanism, key influencing factors, and resource recycling in aerobic granular sludge (AGS) wastewater treatment: a review. J Environ Manage 338:117771. https://doi.org/10.1016/j.jenvman.2023.117771
63. Liu X, Lee D (2022) Chapter 7—aerobic granular sludge processes. In: Bui X, Nguyen DD, Nguyen P, Ngo HH, Pandey A (eds) Current developments in biotechnology and bioengineering. Elsevier, pp 193–225
64. Liu Y, Shu FY, Tay JH (2002) Aerobic granules novel zinc biosorbent. Lett Appl Microbiol 35:548–551

65. Liu Y, Tay J (2002) The essential role of hydrodynamic shear force in the formation of biofilm and granular sludge. Water Res 36(7):1653–1665. https://doi.org/10.1016/S0043-1354(01)00379-7
66. Liu Y, Tay JH (2004) State of the art of biogranulation technology for wastewater treatment. Biotechnol Adv 22:533–563
67. Liu Z, Ning F, Hou Y, Zhang D, Yang R, Wang J et al (2022) Deciphering the effect of algae sources on the formation of algal-bacterial granular sludge: endogenous versus exogenous algae. J Cleaner Prod 363:132468. https://doi.org/10.1016/j.jclepro.2022.132468
68. Liu Z, Zhang XH, Zhang SM, Qi H, Hou YW, Gao M et al (2022) A comparison between exogenous carriers enhanced aerobic granulation under low organic loading in the aspect of sludge characteristics, extracellular polymeric substances and microbial communities. Bioresource Technol 346. https://doi.org/10.1016/j.biortech.2021.126567
69. Lotito AM, Fratino U, Mancini A, Bergna G, Di Iaconi C (2012) Effective aerobic granular sludge treatment of a real dyeing textile wastewater. Int Biodeterior Biodegradation 69:62–68. https://doi.org/10.1016/j.ibiod.2012.01.004
70. Lotito AM, De Sanctis M, Di Iaconi C, Bergna G (2014) Textile wastewater treatment: aerobic granular sludge vs activated sludge systems. Water Res 54:337–346. https://doi.org/10.1016/j.watres.2014.01.055
71. Luo J, Hao T, Wei L, Mackey HR, Lin Z et al (2014) Impact of influent COD/N ratio on disintegration of aerobic granular sludge. Water Res 62:127–135. https://doi.org/10.1016/j.watres.2014.05.037
72. Lv Y, Chen Y, Song W, Hu Y (2014) Enhanced selection of micro-aerobic pentachlorophenol degrading granular sludge. J Hazard Mater 280:134–142. https://doi.org/10.1016/j.jhazmat.2014.07.067
73. Lv Y, Wan CL, Lee DJ, Liu X, Zhang Y et al (2018) Recovery of dehydrated aerobic granules: a comparison. Bioresour Technol 267:769–773. https://doi.org/10.1016/j.biortech.2018.08.013
74. McCarty PL (2018) What is the best biological process for nitrogen removal: when and why? Environ Sci Technol 52(7):3835–3841. https://doi.org/10.1021/acs.est.7b05832
75. Mesdaghinia A, Ghahremani MH, Nabizadeh R, Nasseri S, Rafiee M (2017) Role of CODPCP/CODTotal ratio on p-chlorophenol toxicity towards aerobic granular sludge. J Ind Eng Chem 54:440–446. https://doi.org/10.1016/j.jiec.2017.06.025
76. Muñoz-Palazon B, Rodriguez-Sanchez A, Hurtado-Martinez M, de Castro IM, Juarez-Jimenez B, Gonzalez-Martinez A et al (2019) Performance and microbial community structure of an aerobic granular sludge system at different phenolic acid concentrations. J Hazard Mater 376:58–67. https://doi.org/10.1016/j.jhazmat.2019.05.015
77. Nancharaiah YV, Reddy GKK (2018) Aerobic granular sludge technology: mechanisms of granulation and biotechnological applications. Biores Technol 247:1128–1143. https://doi.org/10.1016/j.biortech.2017.09.131
78. Othman I, Anuar AN, Ujang Z, Rosman NH, Harun H et al (2013) Livestock wastewater treatment using aerobic granular sludge. Bioresour Technol 133:630–634. https://doi.org/10.1016/j.biortech.2013.01.149
79. Peng T, Wang Y, Wang J, Fang F, Yan P et al (2022) Effect of different forms and components of EPS on sludge aggregation during granulation process of aerobic granular sludge. Chemosphere 303:135116. https://doi.org/10.1016/j.chemosphere.2022.135116
80. Purba L, Ibiyeye HT, Yuzir A, Mohamad SE, Iwamoto K et al (2020) Various applications of aerobic granular sludge: a review. Environ Technol Innovation 20. https://doi.org/10.1016/j.eti.2020.101045
81. Qi K, Li Z, Zhang C, Tan X, Wan C, Liu X et al (2020) Biodegradation of real industrial wastewater containing ethylene glycol by using aerobic granular sludge in a continuous-flow reactor: performance and resistance mechanism. Biochem Eng J 161:107711. https://doi.org/10.1016/j.bej.2020.107711
82. Quan XC, Ma JY, Xiong WC, Wang XR (2015) Bioaugmentation of half-matured granular sludge with special microbial culture promoted establishment of 2,4-dichlorophenoxyacetic acid degrading aerobic granules. Bioprocess Biosyst Eng 38(6):1081–1090. https://doi.org/10.1007/s00449-014-1350-y

83. Rosa-Masegosa A, Muñoz-Palazon B, Gorrasi S, Fenice M, Gonzalez-Martinez A et al (2023) Description of new single-chamber continuous-flow reactors of aerobic granular sludge: technical and biological study. J Environ Chem Eng 11(3):109938. https://doi.org/10.1016/j.jece.2023.109938
84. Santorio S, Couto AT, Amorim CL, Val Del Rio A, Arregui L, Mosquera-Corral A et al (2021) Sequencing versus continuous granular sludge reactor for the treatment of freshwater aquaculture effluents. Water Res 201:117293. https://doi.org/10.1016/j.watres.2021.117293
85. Sheng GP, Li AJ, Li XY, Yu HQ (2010) Effects of seed sludge properties and selective biomass discharge on aerobic sludge granulation. Chem Eng J 160(1):108–114. https://doi.org/10.1016/j.cej.2010.03.017
86. Sheng G, Yu H, Li X (2010) Extracellular polymeric substances (EPS) of microbial aggregates in biological wastewater treatment systems: a review. Biotechnol Adv 28(6):882–894. https://doi.org/10.1016/j.biotechadv.2010.08.001
87. Song T, Zhang X, Li J (2022) Aerobic granular sludge with filamentous bacteria immobilized by string carriers to treat simulated municipal wastewater in a continuous flow reactor. Biores Technol 363:127917. https://doi.org/10.1016/j.biortech.2022.127917
88. Strubbe L, Pennewaerde M, Baeten JE, Volcke EIP (2022) Continuous aerobic granular sludge plants: better settling versus diffusion limitation. Chem Eng J 428:131427. https://doi.org/10.1016/j.cej.2021.131427
89. Su B, Qu Z, Song Y, Jia L, Zhu J (2014) Investigation of measurement methods and characterization of zeta potential for aerobic granular sludge. J Environ Chem Eng 2(2):1142–1147. https://doi.org/10.1016/j.jece.2014.03.006
90. Su KZ, Wu RY, Zhou H, Wang W, Hao TW (2022) Mechanism and regulation of filamentous algal-bacterial symbiosis based on microbiological quorum sensing. Water Environ Res 94(3). https://doi.org/10.1002/wer.10697
91. Sun Y, Angelotti B, Wang Z (2019) Continuous-flow aerobic granulation in plug-flow bioreactors fed with real domestic wastewater. Sci Total Environ 688:762–770. https://doi.org/10.1016/j.scitotenv.2019.06.291
92. Szabó E, Liébana R, Hermansson M, Modin O, Persson F, Wilén BM (2017) Microbial population dynamics and ecosystem functions of anoxic/aerobic granular sludge in sequencing batch reactors operated at different organic loading rates. https://doi.org/10.3389/fmicb.2017.00770
93. Tanavarotai K, Anuar AN, Yuzir A, Halim MHA, Aris AM, Chin KB et al (2023) The storage of dried aerobic granular sludges under ambient condition. J Cleaner Prod 383:135401. https://doi.org/10.1016/j.jclepro.2022.135401
94. Tanavarotai K, Kamyab H, Anuar AN, Khademi T, Yuzir A, Ashokkumar V et al (2022) Storage and reactivation of aerobic granular sludge: a review. FUEL 330. https://doi.org/10.1016/j.fuel.2022.125536
95. Tang YH, Wu QY, Chen Y, Liu Z, Chen Y, Chen RY et al (2022) Addition of sodium alginate as a nucleus shortens granulation of aerobic sludge. Environ Sci—Water Res Technol 8(10:2216–2230. https://doi.org/10.1039/d2ew00310d
96. Tao J, Qin L, Liu XY, Li BL, Chen JN, You J et al (2017) Effect of granular activated carbon on the aerobic granulation of sludge and its mechanism. Bioresour Technol 236:60–67. https://doi.org/10.1016/j.biortech.2017.03.106
97. Tay JH, Liu QS, Liu Y (2001a) The role of cellular polysaccharides in the formation and stability of aerobic granules. Lett Appl Microbiol 33(3):222–226. https://doi.org/10.1046/j.1472-765x.2001.00986.x
98. Tay JH, Liu QS, Liu Y (2001b) Microscopic observation of aerobic granulation in sequential aerobic sludge blanket reactor. J Appl Microbiol 91(1):168–175. https://doi.org/10.1046/j.1365-2672.2001.01374.x
99. Tay JH, Liu QS, Liu Y (2002) Characteristics of aerobic granules grown on glucose and acetate in sequential aerobic sludge blanket reactors. Environ Technol 23:931–936
100. Toczyłowska-Mamińska R (2017) Limits and perspectives of pulp and paper industry wastewater treatment—a review. Renew Sustain Energy Rev 78:764–772. https://doi.org/10.1016/j.rser.2017.05.021

101. Tomar SK, Kumar R, Chakraborty S (2022) Simultaneous biodegradation of pyridine, indole, and ammonium along with phenol and thiocyanate by aerobic granular sludge. J Hazard Mater 422:126861. https://doi.org/10.1016/j.jhazmat.2021.126861
102. Vallecillo A, Garcia-Encina PA, Peña M (1999) Anaerobic biodegradability and toxicity of chlorophenols. Water Sci Technol 40(8):161–168. https://doi.org/10.1016/S0273-1223(99)00622-8
103. Vashi H, Iorhemen OT, Tay JH (2018) Aerobic granulation: a recent development on the biological treatment of pulp and paper wastewater. Environ Technol Innov 9:265–274. https://doi.org/10.1016/j.eti.2017.12.006
104. Vashi H, Iorhemen OT, Tay JH (2019) Extensive studies on the treatment of pulp mill wastewater using aerobic granular sludge (AGS) technology. Chem Eng J 359:1175–1194. https://doi.org/10.1016/j.cej.2018.11.060
105. Veeresh GS, Kumar P, Mehrotra I (2005) Treatment of phenol and cresols in upflow anaerobic sludge blanket (UASB) process: a review. Water Res 39(1):154–170. https://doi.org/10.1016/j.watres.2004.07.028
106. Wan CL, Lee DJ, Yang X, Wang YY, Lin L (2014) Saline storage of aerobic granules and subsequent reactivation. Biores Technol 172:418–422. https://doi.org/10.1016/j.biortech.2014.08.103
107. Wan CL, Fu LY, Li ZW, Liu X, Lin L et al (2022) Formation, application, and storage-reactivation of aerobic granular sludge: a review. J Environ Manage 323. https://doi.org/10.1016/j.jenvman.2022.116302
108. Wan CL, Yang X, Lee DJ, Wang XY, Yang QL et al (2014) Aerobic granulation of aggregating consortium X9 isolated from aerobic granules and role of cyclic di-GMP. Bioresour Technol 152:557–561. https://doi.org/10.1016/j.biortech.2013.11.052
109. Wan C, Sun S, Lee D, Liu X, Wang L, Yang X et al (2013) Partial nitrification using aerobic granules in continuous-flow reactor: rapid startup. Bioresour Technol 142:517–522. https://doi.org/10.1016/j.biortech.2013.04.120
110. Wan XP, Gao MM, Ye MS, Wang YK, Xu H, Wang MY et al (2018) Formation, characteristics and microbial community of aerobic granular sludge in the presence of sulfadiazine at environmentally relevant concentrations. Bioresour Technol 250:486–494. https://doi.org/10.1016/j.biortech.2017.11.071
111. Wan C, Lee D-J, Yang X, Wang Y, Wang X, Liu X (2015) Calcium precipitate induced aerobic granulation. Bioresource Technology 176:32-37. https://doi.org/10.1016/j.biortech.2014.11.008
112. Wang HL, Yu GL, Liu GS, Pan F (2006) A new way to cultivate aerobic granules in the process of papermaking wastewater treatment. Biochem Eng J 28(1):99–103. https://doi.org/10.1016/j.bej.2005.10.002
113. Wang L, Deng S, Wang S, Su H (2017) Analysis of aerobic granules under the toxic effect of ampicillin in sequencing batch reactors: performance and microbial community. J Environ Manage 204:152–159. https://doi.org/10.1016/j.jenvman.2017.08.027
114. Wang L, Liu X, Li Z, Wan C, Zhang Y (2023) Filamentous aerobic granular sludge: a critical review on its cause, impact, control and reuse. J Environ Chem Eng 11(3):110039. https://doi.org/10.1016/j.jece.2023.110039
115. Wang Q, Du GC, Chen J (2004) Aerobic granular sludge cultivated under the selective pressure as a driving force. Process Biochem 39(5):557–563. https://doi.org/10.1016/S0032-9592(03)00128-6
116. Wang S, Huang X, Liu L, Shen Y, Yan P, Chen Y, Guo J, Fang F (2020) Understanding the mechanism in aggregation ability between aerobic and anammox granular sludge from the perspective of exopolysaccharides. Journal of Water Process Engineering 38:101629, ISSN 2214-7144, https://doi.org/10.1016/j.jwpe.2020.101629
117. Wang S, Liu X, Zhang H, Gong W, Sun X et al (2007) Aerobic granulation for 2,4-dichlorophenol biodegradation in a sequencing batch reactor. Chemosphere 69(5):769–775. https://doi.org/10.1016/j.chemosphere.2007.05.026

118. Wang X, Chen Z, Kang J, Zhao X, Shen J et al (2019) The key role of inoculated sludge in fast start-up of sequencing batch reactor for the domestication of aerobic granular sludge. J Environ Sci 78(04):127–136
119. Wang ZW, Liu Y, Tay JH (2005) Distribution of EPS and cell surface hydrophobicity in aerobic granules. Appl Microbiol Biotechnol 69:469–473
120. Wang ZW, Wang HX, Nie Q, Ding Y, Lei ZF, Zhang ZY et al (2023) Pb(II) bioremediation using fresh algal-bacterial aerobic granular sludge and its underlying mechanisms highlighting the role of extracellular polymeric substances. J Hazard Mater 444. https://doi.org/10.1016/j.jhazmat.2022.130452
121. Weissbrodt DG, Shani N, Holliger C (2014) Linking bacterial population dynamics and nutrient removal in the granular sludge biofilm ecosystem engineered for wastewater treatment. FEMS Microbiol Ecol 88(3):579–595. https://doi.org/10.1111/1574-6941.12326
122. Wil´en BM, Jin B, Lant P, (2003) The influence of key chemical constituents in activated sludge on surface and flocculating properties. Water Res 37:2127–2139
123. Winkler M, Bassin JP, Kleerebezem R, van der Lans R, van Loosdrecht M (2012) Temperature and salt effects on settling velocity in granular sludge technology. Water Res 46(12):3897–3902. https://doi.org/10.1016/j.watres.2012.04.034
124. Wu D, Zhao B, Zhang P, An Q (2023) Insight into the effect of nitrate on AGS granulation: granular characteristics, microbial community and metabolomics response. Water Res 236:119949. https://doi.org/10.1016/j.watres.2023.119949
125. Xia JT, Ye L, Ren HQ, Zhang XX (2018) Microbial community structure and function in aerobic granular sludge. Appl Microbiol Biotechnol 102(9):3967–3979. https://doi.org/10.1007/s00253-018-8905-9
126. Xia J, Ye L, Ren H, Zhang X-XJAm, biotechnology (2018) Microbial community structure and function in aerobic granular sludge. 102:3967-3979
127. Xin X, Lu H, Yao L, Leng L, Guan L (2017) Rapid formation of aerobic granular sludge and its mechanism in a continuous-flow bioreactor. Appl Biochem Biotechnol 181(1):424–433. https://doi.org/10.1007/s12010-016-2221-6
128. Xu D, Li J, Liu J, Qu X, Ma H (2022) Advances in continuous flow aerobic granular sludge: a review. Process Saf Environ Prot 163:27–35. https://doi.org/10.1016/j.psep.2022.05.018
129. Xu G, Peng J, Feng C, Fang F, Chen S, Xu Y et al (2015) Evaluation of simultaneous autotrophic and heterotrophic denitrification processes and bacterial community structure analysis. Appl Microbiol Biotechnol 99(15):6527–6536. https://doi.org/10.1007/s00253-015-6532-2
130. Yan L, Guo Y, Zhang X, Yu L, Bao X, Hao G et al (2017) Influence of storage strategies on the reactivation characteristics of shortcut nitrification aerobic granular sludge. Chem Biochem Eng Q 31(1):115–121. https://doi.org/10.15255/CABEQ.2016.960
131. Yan L, Zhang S, Hao G, Zhang X, Ren Y, Wen Y et al (2016) Simultaneous nitrification and denitrification by EPSs in aerobic granular sludge enhanced nitrogen removal of ammonium-nitrogen-rich wastewater. Bioresour Technol 202:101–106. https://doi.org/10.1016/j.biortech.2015.11.088
132. Yang SF, Li XY, Yu HQ (2008) Formation and characterisation of fungal and bacterial granules under different feeding alkalinity and pH conditions. Process Biochem 43(1):8–14. https://doi.org/10.1016/j.procbio.2007.10.008
133. Yang XJ, Zhao ZW, Yu Y, Shimizu K, Zhang ZY, Lei ZF et al (2020) Enhanced biosorption of Cr(VI) from synthetic wastewater using algal-bacterial aerobic granular sludge: batch experiments, kinetics and mechanisms. Sep Purif Technol 251. https://doi.org/10.1016/j.seppur.2020.117323
134. Yin YJ, Liu FY, Wang L, Sun J (2019) Overcoming the instability of aerobic granular sludge under nitrogen deficiency through shortening settling time. Bioresour Technol 289. https://doi.org/10.1016/j.biortech.2019.121620
135. Yu C, Wang K, Zhang K, Fu X, Liu R et al (2023) Pilot study of continuous flow aerobic granular sludge technology for the treatment of low concentration municipal wastewater. J Environ Eng 17(03):713–721

136. Yuan XJ, Gao DW (2010) Effect of dissolved oxygen on nitrogen removal and process control in aerobic granular sludge reactor. J Hazard Mater 178(1–3):1041–1045. https://doi.org/10.1016/j.jhazmat.2010.02.045
137. Yuan XJ, Gao DW, Liang H (2012) Reactivation characteristics of stored aerobic granular sludge using different operational strategies. Appl Microbiol Biotechnol 94(5):1365–1374. https://doi.org/10.1007/s00253-011-3660-1
138. Yuan J-B, Li X-D, Zhang X, Li L, Wu J-F (2019) Research progress of aerobic granular sludge MBR for treatment of high concentration organic wastewater. Water Treat Technol 45(09):1–5. https://doi.org/10.16796/j.cnki.1000-3770.2019.09.001
139. Yuan Q, Gong H, Xi H, Xu H, Jin Z, Ali N et al (2019) Strategies to improve aerobic granular sludge stability and nitrogen removal based on feeding mode and substrate. J Environ Sci 84:144–154. https://doi.org/10.1016/j.jes.2019.04.006
140. Yuan X (2011) Study on long-term storage and activity recovery of aerobic granular sludge. Master, Northeast Forestry University. Retrieved from Available from CNKI
141. Zeng W (2011) Study on N_2O production during nitrogen removal by aerobic granular sludge in SBR method. M.S., Northeast Forestry University. Retrieved from Available from CNKI
142. Zhang B, Wu L, Guo Y, Lens PNL, Shi W (2023) Rapid establishment of algal-bacterial granular sludge system by applying mycelial pellets in a lab-scale photo-reactor under low aeration conditions: performance and mechanism analysis. Environ Pollut 322:121183. https://doi.org/10.1016/j.envpol.2023.121183
143. Zhang J (2019) Construction of microbial fuel cell based on aerobic granular sludge and its application. Master, Jiangsu University of Science and Technology. Retrieved from Available from CNKI
144. Zhang WJ (2019) Aerobic granular sludge SBR and MBR combined process for the treatment of municipal wastewater. Master, Harbin Institute of Technology. Retrieved from Available from CNKI
145. Zhang B, Ji M, Qiu Z, Liu H, Wang J et al (2011) Microbial population dynamics during sludge granulation in an anaerobic–aerobic biological phosphorus removal system. Bioresour Technol 102(3):2474–2480. https://doi.org/10.1016/j.biortech.2010.11.017
146. Zhang B, Li W, Guo Y, Zhang Z, Shi W, Cui F et al (2020) A sustainable strategy for effective regulation of aerobic granulation: augmentation of the signaling molecule content by cultivating AHL-producing strains. Water Res 169:115193. https://doi.org/10.1016/j.watres.2019.115193
147. Zhang DJ, Li W, Hou C, Shen JY, Jiang XB, Sun XY et al (2017) Aerobic granulation accelerated by biochar for the treatment of refractory wastewater. Chem Eng J 314:88–97. https://doi.org/10.1016/j.cej.2016.12.128
148. Zhang L, Long B, Wu J, Cheng Y, Zhang B, Zeng Y et al (2019) Evolution of microbial community during dry storage and recovery of aerobic granular sludge. Heliyon 5(12):e3023. https://doi.org/10.1016/j.heliyon.2019.e03023
149. Zhao X, Chen Z, Wang X, Li J, Shen J et al (2015) Remediation of pharmaceuticals and personal care products using an aerobic granular sludge sequencing bioreactor and microbial community profiling using Solexa sequencing technology analysis. Bioresour Technol 179:104–112. https://doi.org/10.1016/j.biortech.2014.12.002
150. Zheng Y, Yu H, Sheng G (2005) Physical and chemical characteristics of granular activated sludge from a sequencing batch airlift reactor. Process Biochem 40(2):645–650. https://doi.org/10.1016/j.procbio.2004.01.056
151. Zheng X, Lu D, Wang M, Chen W, Zhou G et al (2018) Effect of chromium (VI) on the multiple nitrogen removal pathways and microbial community of aerobic granular sludge. Environ Technol 39(13):1682–1696. https://doi.org/10.1080/09593330.2017.1337230
152. Zheng YM, Yu HQ (2007) Determination of the pore size distribution and porosity of aerobic granules using size-exclusion chromatography. Water Res 41:39–46
153. Zhu L, Dai X, Xu X, Lv M, Cao D (2014) Microbial community analysis for aerobic granular sludge reactor treating high-level 4-chloroaniline wastewater. Int J Environ Sci Technol 11(7):1845–1854. https://doi.org/10.1007/s13762-013-0380-3

154. Zhu LF, Huang DD, Ye C, Wang XM, Wu JY et al (2013) Research on the characteristic of dyeing-printing wastewater treatment with micro-aerobic granular sludge. In: Zhao J, Iranpour R, Li X, Jin B (eds) Advances in environmental technologies, PTS 1–6 (726–731, pp 2757–2760). 2nd international conference on energy and environmental protection (ICEEP 2013). (Reprinted)
155. Zou J, Tao Y, Li J, Wu S, Ni Y (2018) Cultivating aerobic granular sludge in a developed continuous-flow reactor with two-zone sedimentation tank treating real and low-strength wastewater. Biores Technol 247:776–783. https://doi.org/10.1016/j.biortech.2017.09.088

Chapter 3
Anaerobic Membrane Bioreactor

3.1 Introduction

Microorganisms have a propensity to form complex structures known as biofilms on surfaces that provide them with a source of nutrients and water. Biofilm formation is a widespread phenomenon observed in numerous ecosystems and has been a subject of significant interest in the field of microbial ecology for an extended period [40].

The membrane bioreactor (MBR) is an innovative wastewater treatment technology that integrates membrane separation and biological treatment techniques. This technology offers several advantages, including a compact design, efficient treatment capabilities, and high-quality effluent production. By combining the benefits of both membrane separation and biological processes, MBRs have emerged as a promising solution for wastewater treatment. At present, MBR is widely used in municipal wastewater treatment, but also successfully used in the treatment of wastewater with high ammonia nitrogen concentration, pharmaceuticals, municipal wastewater, and landfill leachate.

The concept of anaerobic membrane bioreactor (AnMBR) was first introduced in 1978 by H. E. Grethlein, an American scholar was the first to utilize the external staggered flow membrane filtration technology for treating septic tank effluent, achieving an increase in biomass concentration while removing 85–95% of the BOD. After 2–3 years of development, the advantages of AnMBR systems were well proven and began to emerge in the field of industrial wastewater treatment. Recognizing the value of AnMBR, numerous governments, and private sector began to invest in AnMBR systems, and a U.S. company developed the world's first commercial AnMBR system, called the Membrane Anaerobic Reactor System (MARS), in the 1980s, primarily for high concentration food processing wastewater treatment. These AnMBR systems were primarily based on external staggered flow configurations until the 1990s when the success of submerged aerobic MBRs greatly encouraged

G. Dawen and M. Nabi, *Novel Approaches Towards Wastewater Treatment*,
Springer Water, https://doi.org/10.1007/978-3-031-55189-5_3

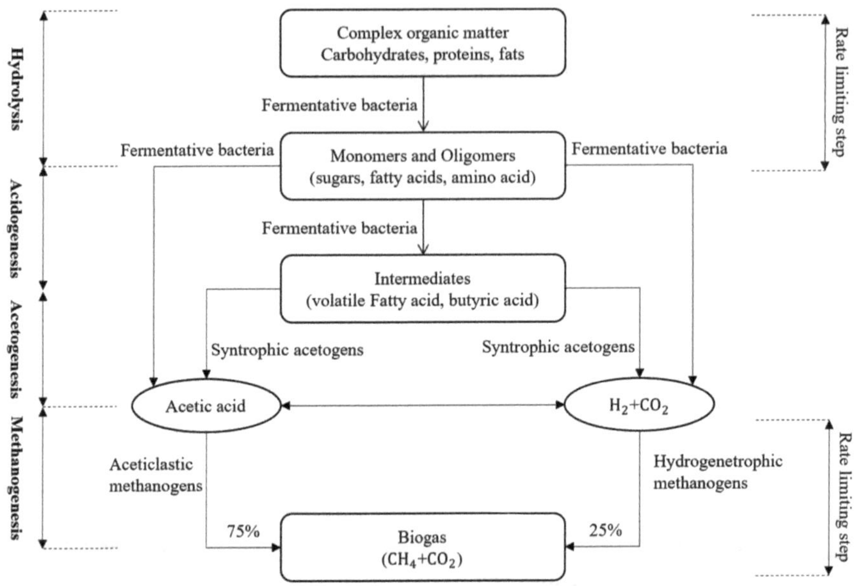

Fig. 3.1 Stages of anaerobic digestion

the exploration of submerged AnMBRs and research began to focus on system performance, filtration characteristics, membrane material development, membrane fouling identification, and characterization, and membrane fouling control (Fig. 3.1).

Despite the extensive research on AnMBR, several challenges still exist. Factors such as membrane performance, wastewater characteristics, sludge properties, operating parameters, and reactor configuration significantly impact the performance of Anaerobic Membrane Bioreactors. Under high organic loading conditions, the anaerobic digestion system generates a considerable amount of volatile fatty acids (VFAs), which reduces the system's ability to produce methane. Additionally, membrane fouling is also one of the major limiting factors in promoting the widespread application of AnMBRs [39]. Three modes, aerobic, anoxic, and anaerobic, are available in the reactor of a MBR. Compared with aerobic MBR technology, anaerobic technology is considered to be a reliable and promising alternative for wastewater treatment. AnMBRs are widely used in industrial and domestic wastewater treatment. In Table 3.1, the differences between the operating processes of partial MBR and AnMBR are presented.

The MBR is an innovative water treatment technology that integrates a membrane separation unit with a biological treatment unit, which is a coupled bioreactor with membrane filtration. MBR has three modes, aerobic, anoxic, and anaerobic. The interception effect of the membrane in MBR effectively reduces the amount of residual sludge and reduces the cost of sludge disposal. In the traditional activated sludge method, controlling hydraulic retention time (HRT) and solid retention time (SRT) is difficult it. Due to the good solid–liquid separation effect of the membrane

Table 3.1 Comparison of AnMBR and MBR in wastewater treatment [99]

Feature	AnMBR	MBR	Reference
Organic loading rate (kg COD/L)	0.25–0.8	0.17–35.5	[94]
Organic removal efficiency (%)	> 95	> 90	[100]
Biomass concentration (g/L)	5–20	10–40	[94]
Hydraulic retention time (h)	4–8	> 8	[17]
Sludge retention time (day)	5–20	> 100	[96]
Operational temperature (°C)	20–30	20–50	[73]

components, efficient screening can make microorganisms stay in the reactor more completely, and effectively improve the purification efficiency of refractory organic matter in wastewater. The effluent from the membrane reactor is not affected by factors such as sludge expansion, so the reactor effluent can be directly reused. Therefore, in the past twenty years, membrane bioreactors have been widely used in domestic and industrial wastewater treatment. AnMBR has a higher removal rate of organic matter, can treat various types of wastewater, and can also produce biogas by converting organic matter, achieving energy recovery, and other advantages. Compared with aerobic MBR technology, anaerobic technology is considered to be a reliable and promising alternative for wastewater treatment due to its advantages [86]. In comparison to MBR, AnMBR is a biological treatment process that synergistically combines anaerobic digestion with physical separation membranes. By integrating anaerobic digestion and physical separation membranes, AnMBR aims to optimize the removal of organic loads and enhance biogas generation (Table 3.1) [31, 57].

The membrane filtration process achieves interception of microorganisms and pollutants, providing a basis for better pollutant biodegradation and energy recovery. AnMBR effectively removes ammonia nitrogen and hard-to-degrade organic matter. The microorganisms in the bioreactor are completely trapped, which benefits slow-proliferating microorganisms like nitrifying bacteria. This enhances their growth state and improves the nitrification efficiency of the system due to their confinement by the membrane. Additionally, the hydrodynamic residence time of refractory organics in the system is extended, leading to improved degradation efficiency. As a result, AnMBR can be applied to treat various types of wastewater, such as domestic, industrial, and municipal wastewater (e.g., waste leachate), as well as solid waste (e.g., food scraps and residual activated sludge). By reducing the HRT while maintaining a relatively high SRT of 50–700 days, AnMBR achieves low suspended solids in the effluent and high chemical oxygen demand (COD) removal [10].

Due to the synergistic benefits of anaerobic digestion and membrane technology, AnMBR has found a wide range of applications in sustainable wastewater treatment, gaining attention from both commercial and research sectors. However, despite its potential, several key obstacles still hinder the widespread use of AnMBR. These challenges include lower throughput, membrane fouling, and higher investment and operational requirements. Among these obstacles, membrane fouling poses a particularly severe challenge in AnMBR compared to aerobic MBR. This is attributed to factors such as the composition of the sludge mixture and specific operating conditions unique to AnMBR. As a result, membrane fouling has emerged as a significant technical barrier, limiting the broad adoption of AnMBR technology. Efforts to overcome these challenges and develop effective fouling mitigation strategies will be crucial in fully realizing the potential of AnMBR for sustainable wastewater treatment solutions [39].

Therefore, the engineering and commercial deployment of AnMBR faces many challenges, including, for example, membrane fouling, the presence of dissolved methane, temperature, pH and alkalinity, and sulfide in the feed water, all of which have different degrees of negative impact on the stable and efficient operation of AnMBR.

Under high OLR conditions, anaerobic digestion systems produce a significant amount of VFAs, which reduce the system's ability to produce methane.

Certain researchers are committed to studying the composition and mechanisms of membrane fouling in order to tackle this problem. The accumulation of extracellular polymeric substances (EPS) and soluble microbial products (SMPs) by microbial communities plays a crucial role in membrane fouling, particularly as the OLR increase [108]. The accumulation, blockage, and adsorption of pollutants on the membrane surface and pores lead to a significant decrease in the efficiency of membrane filtration, resulting in increased operational costs during the initial stages of operation. Therefore, the research on AnMBR mainly focuses on two aspects: one is to enhance digestion (anaerobic unit), and the other is to address membrane fouling [65].

In AnMBR, polymer or ceramic membranes with pore sizes ranging from 0.1 to 0.01 μm prevent the loss of active methane-producing bacteria and volatile suspended solids from bioreactor. This enables higher bioenergy production than traditional anaerobic bioreactors without the need to integrate membrane separation processes [115].

This chapter aims to provide a comprehensive review of relevant studies on AnMBR. It systematically examines the materials utilized in AnMBRs, various system configurations, and their application in treating different types of wastewater, such as high concentration wastewater, pharmaceutical wastewater, municipal wastewater, and landfill leachate. The primary focus is on the removal of pollutants, including COD, N, P, as well as biogas production. Additionally, this chapter explores the impact of different materials on membrane fouling, elucidates membrane pollution mechanisms, and proposes control strategies. By analyzing existing academic research, this chapter emphasizes the role of material properties and their associated mechanisms in mitigating membrane fouling and enhancing anaerobic digestion. Overall, AnMBR is an efficient and low-cost technology for treating organic wastewater, which can be applied in the fields of sewage treatment, industrial wastewater treatment, and other areas, with a wide range of application prospects [54].

3.2 Configuration of AnMBR

AnMBR can be used to treat many types of wastewater, enabling resource recovery and energy recycling in highly concentrated wastewater. The implementation and design of an AnMBR plant rely on the selection of appropriate materials and membrane modules, as well as the specific type of bioreactor employed. In general, three configurations of membrane filtration modules can exist: internally submerged, externally submerged, or side-flow, depending on the installation, location, and membrane operation. In the internal submerged configuration, the membrane unit is submerged inside the anaerobic reactor and is in direct contact with the mixture, and the membrane is pumped to operate to produce the effluent. This implies that the energy requirements are lower compared to an external setup, as the permeate can be obtained through vacuum suction or driven by gravity. In the case of an internal reactor configuration, the generated biogas is commonly compressed and circulated to manage membrane fouling. This is achieved by generating a flow of bubbles along the surface of the membrane [59].

This approach is typically used to treat lower concentrations of wastewater; in the side-flow type, the membrane unit is set outside the anaerobic reactor and the membrane permeates the water under pressure while the mixture is transferred to the membrane unit to produce the permeate. The side-flow filtration mode is known for its utilization of cross-flow to effectively clean the membrane surface. This method requires high-speed recirculation pumps to achieve optimal cross-flow rates, resulting in high effluent fluxes. However, the significant energy consumption associated with the recirculation pumps, potential harm to the sludge structure, and the expensive nature of tubular membranes pose challenges to its widespread adoption [29]. Over time, a combination of these two approaches resulted in the external submersion type, in which the membrane units are submerged and immobilized inside the membrane cell chamber outside the anaerobic reactor, in a two-stage configuration that facilitates membrane cleaning with strong shear forces, thus extending the membrane lifetime. This particular arrangement of the AnMBR facilitates membrane cleaning and maintenance without affecting the microbial community in bioreactor. The intercepted material can be recirculated from the membrane back to the anaerobic zone for further biodegradation of pollutants and further reduction of Mixed Liquor Suspended Solids (MLSS) entering the tank. This method offers the advantage of recirculating the leachate and biogas from the membrane chamber back to the bioreactor. This promotes efficient biodegradation and facilitates exchange between the liquid and gas phases, respectively. Thus, two-stage modulation holds promise for use in scale large-scale AnMBR applications.

There are three common types of membrane: hollow fiber, flat plate, and tubular. However, unlike flat and tubular DM (Direct Membrane) modules, hollow fiber DM modules typically do not utilize coarse pore support materials for assembly. This is likely because DM support materials have lower strength and more complex fabrication procedures, which makes them unsuitable for hollow fiber modules. The performance and efficiency of AnMBR systems are influenced by the type of membrane assemblies employed. Hollow fiber (HF), flat plate (FS), and multi-tube (MT) membranes are widely utilized in AnMBR systems due to their affordability and superior quality. For submerged processes in anaerobic reactors, such as submerged vacuum-driven membranes, HF or FS membranes are commonly employed. Similarly, MT is compatible with side-stream configurations as staggered flow membranes driven by external pressure. Modules and their materials are very important for process performance [31].

With the continuous development and application of wastewater treatment technologies, high-retention AnMBR systems have gradually become a research hotspot.

Anaerobic membrane distillation bioreactors (AnMDBR) and anaerobic osmotic membrane bioreactors (AnOMBR) are the main systems employed to enhance the removal efficiency of pollutants in wastewater deep treatment and promote water resource reuse. These systems combine membrane distillation (MD) and Forward Osmosis (FO) processes with AnMBR. By integrating these technologies, AnMDBR and AnOMBR enable the efficient treatment and reuse of wastewater, thereby ensuring the preservation and utilization of water resources [10].

In comparison to traditional AnMBRs, the filtration process described here operates without the need for any energy input. Gu et al. [46] conducted a study on the treatment of low-strength wastewater using an AnOMBR system, aiming to recover energy in the form of biogas at mesophilic temperatures. Remarkably, they achieved a considerable amount of methane ranging from 0.25 to 0.3 $LCH_4/gCOD_{Removed}$. Similarly, Chen et al. [23] explored the methane recovery capability of an FO-AnMBR system and observed methane production of 0.21 $LCH_4/gCOD_{Removed}$ with a methane content of 65–78%. However, the actual methane yield was only 58% of the theoretical yield due to the presence of dissolved methane in the effluent and low methanogenic activity in a salinity environment.

Therefore, additional research is necessary to examine the activity of methanogens and microbial kinetics, with the goal of improving biogas production in salinity conditions [69].

In AnMBR or AnDMBR, high cross-flow velocity (CFV) in the external configuration can lead to two effects: excessive energy consumption and degradation of sludge properties, such as diminished particle size and microbial activity, can be observed. However, high CFV also helps to effectively control membrane fouling. There is currently no consensus on the membrane configuration in AnDMBR, as most studies have used the submerged configuration. Although only one laboratory-scale study has compared the effects of submerged and side stream bioreactor configurations on treatment and filtration performance, their applicability has not been further evaluated. However, similar to the experience gained from conventional MBR and AnMBR, the submerged configuration is usually preferred due to its low energy consumption and costs. In addition, no additional space is required for membrane installation in the internal submerged configuration, which is different from the side stream configuration in AnDMBR.

In theory, any type of anaerobic bioreactor has the potential to be transformed into an AnMBR by incorporating submerged and sidestream membrane structures.

Therefore, many potential AnMBR configurations may be developed and validated in the future [54] (Fig. 3.2).

AnOMBR, a cutting-edge wastewater treatment technology, integrates forward osmosis and anaerobic treatment. It utilizes a semi-permeable membrane to transport water from the feed solution to the draw solution, driven by the osmotic pressure

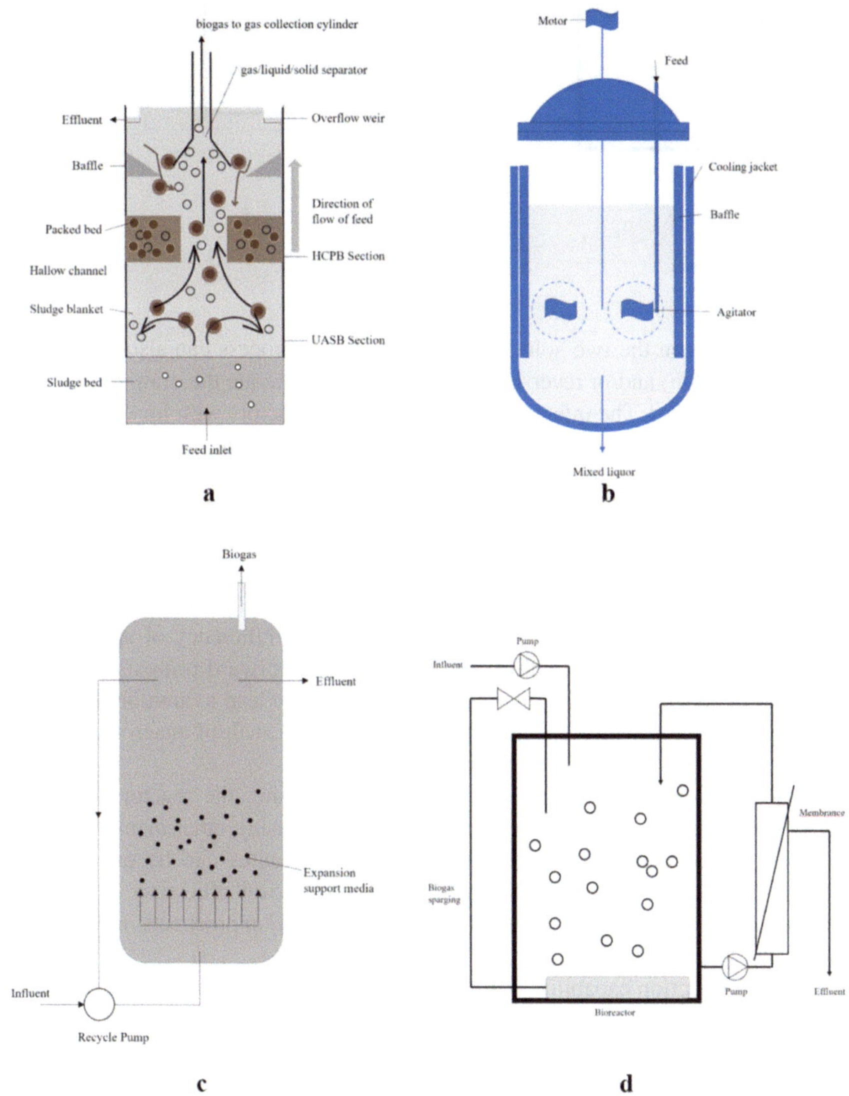

Fig. 3.2 Typical anaerobic bioreactors **a** up-flow anaerobic sludge reactor, **b** continuous stirred-tank reactor, **c** anaerobic fluidized bed reactor and their integration with membrane separation process, **d** side-stream, **e** submerged, **f** external chamber modes. Adopted from Song et al. [99]

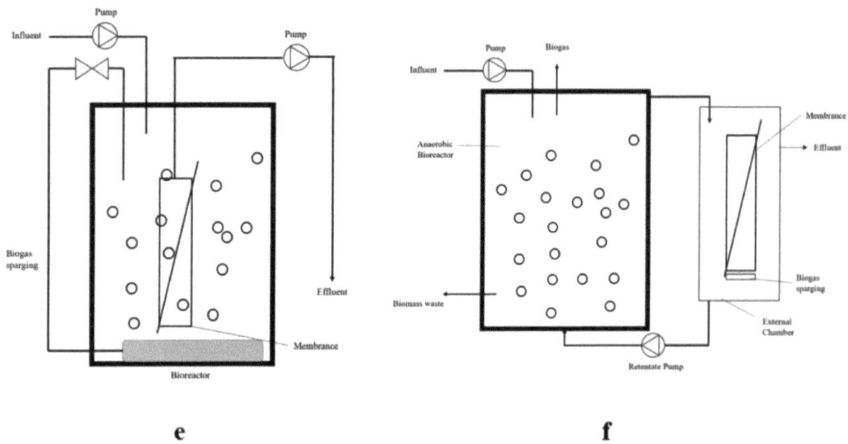

Fig. 3.2 (continued)

difference between the two solutions. The AnOMBR process can also incorporate nanofiltration (NF) and/or reverse osmosis (RO) to regenerate the draw solution and produce clean water. The integration of anaerobic bioreactors with forward osmosis membrane technology has been proposed as a means to reduce energy consumption in wastewater reclamation for pollution control [110].

Due to the dense membrane properties of the semi-permeable forward osmosis membrane and the increased residence time of organic matter in the reactor, AnOMBR has a significantly higher organic matter removal efficiency than traditional AnMBR. This technology has a wide range of applications in the field of wastewater treatment. However, the ammonia removal efficiency of AnOMBR is relatively low compared to traditional AnMBR, as the forward osmosis membrane has a lower capacity for ammonia interception. This may lead to ammonia accumulation in the reactor, which can affect the activity and growth of anaerobic bacteria and thus the overall performance of the reactor.

To address this problem, measures such as increasing the reactor volume of AnOMBR, adjusting the feed ammonia concentration, using other membrane technologies, or combining AnOMBR with other technologies can be taken. These measures can significantly improve the ammonia removal efficiency and overall performance of AnOMBR, further enhancing its application prospects in the field of wastewater treatment [23].

The high rejection capability of forward osmosis membrane in AnOMBR results in an increased retention time of organic matter and nutrients in the reactor, leading to an increase in salinity and membrane fouling, which suppresses microbial activity. Therefore, periodic sludge settling and effluent discharge are commonly considered in AnOMBR to maintain salinity levels in the reactor. This approach can effectively reduce salinity in AnOMBR and keep the biological reactor running smoothly.

Additionally, an integration of porous microfiltration (MF) membrane technology in AnOMBR has been proposed to ensure the removal of soluble salts from the biological reactor. Compared to conventional AnMBR, the permeate from MF membrane in AnOMBR contains less organic matter, which is due to the longer retention time of organic matter in the reactor caused by the high rejection level of forward osmosis membrane. As a result, MF membrane in AnOMBR can more effectively remove salt dissolved in the permeate while maintaining the good operational status of the biological reactor.

In summary, AnOMBR technology can improve wastewater treatment efficiency and quality by integrating forward osmosis and MF membranes. Periodic sludge settling and effluent discharge, as well as the use of MF membrane, are all key factors in maintaining the stable operation of AnOMBR [111]. Therefore, the permeate from MF and forward osmosis can be reused simultaneously and may be applicable for different purposes.

In recent years, the feasibility of using AnMBR for agricultural irrigation of treated wastewater has been studied. While this method can recover nutrients from wastewater, the microorganisms, viruses, and chemicals present in the wastewater may pose potential risks to crops and soil. Therefore, proper disinfection and treatment of AnMBR-treated wastewater are necessary to ensure safe use. In addition, the energy cost of AnMBR is still a challenge. Although anaerobic digestion can produce biogas and be used to provide some of the energy required for AnMBR, other energy sources are still needed to maintain system operation. Therefore, methods to reduce the operating cost of AnMBR, such as using solar energy or other renewable energy sources, need to be explored [16].

Compatibility between anaerobic bioreactors and membrane filtration units is crucial for the operation and maintenance of AnMBR. Different combinations of bioreactors and membrane materials can affect pollutant removal efficiency. Literature suggests that the completely stirred tank reactor and the up-flow anaerobic sludge blanket (UASB) are the two most commonly used bioreactors for the AnMBR treatment of municipal wastewater. Completely stirred tank reactor is widely used in various small-scale and medium-scale explorations in AnMBR because of its simple construction and operation, but its effluent has a high concentration of suspended solids and is prone to membrane fouling. The UASB coupled with a membrane configuration is relatively advanced because the UASB achieves separation of HRT and SRT to some extent, and the biomass at the bottom of the reactor can be retained, resulting in better wastewater treatment. Additionally, the addition of membrane devices can further improve the water quality of the effluent, reduce suspended solids concentration, reduce membrane fouling, ensure long-term membrane operation, and make the process more stable. AnMBR can be operated in a sideways or immersed mode [39].

Gao et al. [39] investigated the feasibility of using an integrated activated carbon anaerobic fluidized bed membrane bioreactor (IAFMBR) for treating domestic wastewater. The performance of the laboratory-scale IAFMBR was tested with different HRT, including COD removal, biogas production, and membrane fouling. When the hydraulic retention time was 6 h, a high COD removal rate was achieved, and further reduction of the hydraulic retention time significantly affected the COD removal capacity and efficiency. Accumulation of more VFA at shorter hydraulic retention times resulted in a decrease in specific methane production rate, and vice versa. The addition of activated carbon helped to control membrane fouling and achieve longer operating cycles between backwash and cleaning. As the amount of activated carbon increased, the protein in the cake layer improved the membrane filtration performance (Fig. 3.3).

The key distinction between the two bioreactor configurations lies in the contact between the membrane and the active sludge. In the Continuous Stirred-Tank Reactor (CSTR) model, the membrane directly interacts with a substantial volume of active sludge. On the other hand, in the UASB model, the membrane filtration occurs with the supernatant, which contains fewer suspended solids. Additionally, the UASB model has less severe membrane fouling problems than the completely stirred tank reactor model, as the membrane in the completely stirred tank reactor model comes into direct contact with the sludge. However, in the CSTR, the substrate in the reactor is homogeneously mixed, which is necessary for digestion and thus biogas production. The choice of reactor model can have a notable impact on the formation and composition of microbial communities within the AnMBR system. Although the removal rates of COD in the two reactors are similar, both exceeding 90%, it was

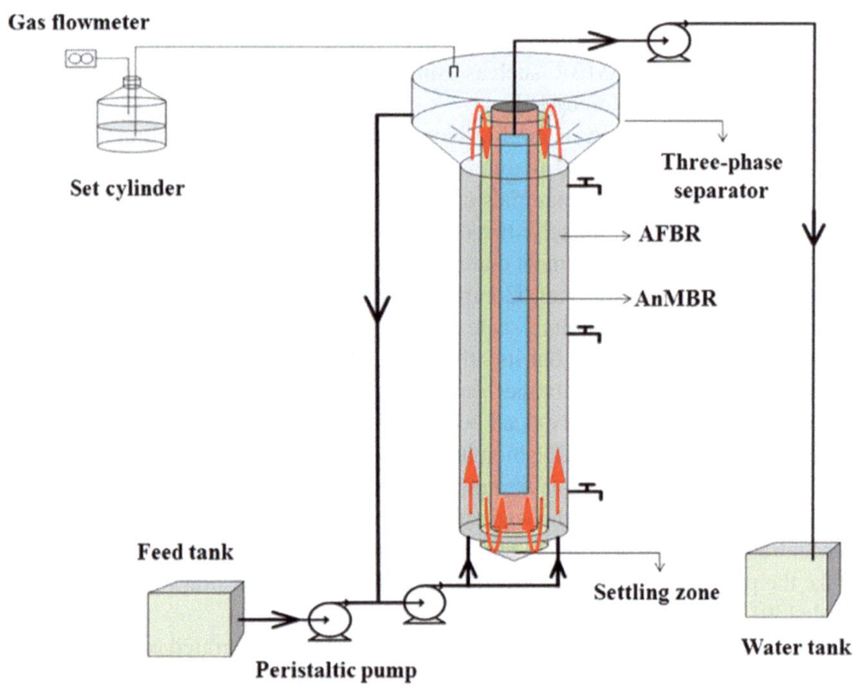

Fig. 3.3 Schematic diagram of integrated anaerobic fluidized bed membrane bioreactor (IAFMBR) [39]

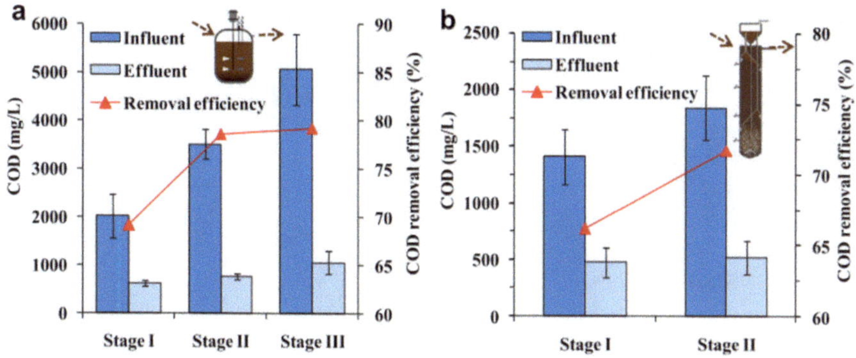

Fig. 3.4 Performances of the CSTR and internal circulation reactor (IC) during the start-up stage. **a** COD removal efficiency of the CSTR, **b** COD removal efficiency of the IC [83]

found that the captured methane in the completely stirred tank reactor was 11–18% higher [31] (Fig. 3.4).

As research progresses, reactors such as Anaerobic Fluidized Bed Reactor (AFBR) and Expanded Granular Sludge Bed reactors (EGSB) are gradually being used in the construction of AnMBR. These reactors have advantages such as higher treatment efficiency, lower energy consumption, smaller footprint, and lower operating costs. AFBR has good suspension and high microbial attachment rates, which can effectively treat high-concentration and complex organic wastewater. EGSB, on the other hand, can better treat low-concentration organic wastewater (Fig. 3.5; Table 3.2).

Several factors have been shown to impact the growth characteristics of biofilms. However, there is limited understanding of how community biodiversity, in response to environmental changes, reflects the growth status of biofilms. Recent research has highlighted the significant connection between community evenness and functional stability, suggesting that community evenness plays a vital role in biofilm growth. In this study, we examined the growth rates of biofilms under varying levels of evenness by manipulating community evenness in a mesocosm experiment, following the principle of selective stresses [40].

In summary, these new types of biological reactors provide more options and possibilities for the construction of AnMBR and bring new impetus to the development of wastewater treatment technology.

3.3 Membrane Material

The membrane module holds great importance within the AnMBR system, as it directly influences the fouling rate of the membrane and the overall performance of the reactor. Factors including membrane materials, pore sizes, surface charge, surface

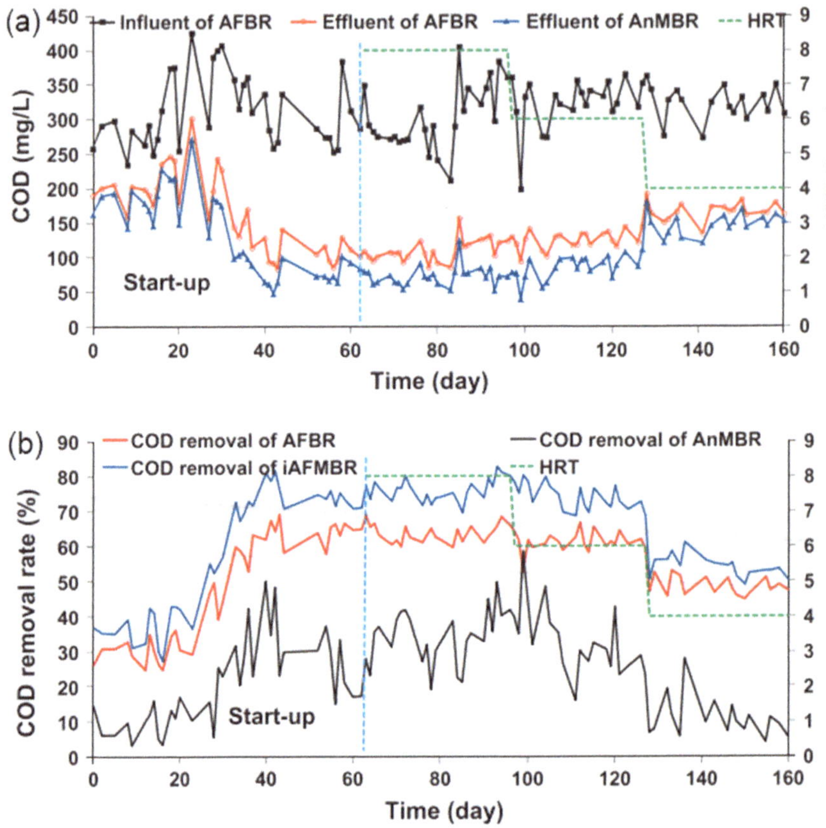

Fig. 3.5 The COD removal efficiency at different HRTs, **a** COD concentrations of influent and effluent of AFBR and the effluent of AnMBR, **b** COD removal efficiency of AFBR, AnMBR and IAFMBR [39]

roughness, hydrophobicity, design/configurations, and other elements all contribute significantly to these aspects [127].

Membranes are commonly made of polymers, ceramics, and metals, each exhibiting unique performance and efficiency characteristics. Among these materials, polymeric membranes are extensively employed in membrane separation processes due to their affordability and versatile applications. In contrast, metal and ceramic membranes are more costly and have narrower fields of application. In AnMBR systems, various membrane materials are utilized, including polyamide (PA), polyvinylidene fluoride (PVDF), polysulfone (PSF), polypropylene (PP), polyethylene (PE), and polyethersulfone (PES)-based polymer membranes [78]. The main reasons why organic polymer membranes are suitable for anaerobic membrane bioreactors are their low cost, high packing density, and promising commercial applications [127].

Table 3.2 The operational conditions of the deammonification process

Phase	Period	HRT	Temperature	Internal recirculation radio	DO	Inf. NH_4^+-N	Inf. COD	NLR
1–1	0–15	18	35.4 ± 1.0	1:3–1:6	< 0.06	125.8 ± 3.1	54.9 ± 3.0	168.2 ± 4.7
1–2	16–72	10–12	–	–	–	69.8 ± 2.8	62.6 ± 4.0	143.1 ± 11.8
II	73–103	7.2–12	22.6 ± 2.0	1:6–1:15	< 0.06	73.2 ± 2.0	58.1 ± 4.2	157.3 ± 2.7
III-1	104–115	6.5–7.2	25.5 ± 1.2	1:15–1:24	< 0.06	42.7 ± 1.6	76.2 ± 1.4	150.1 ± 1.1
III-2	116–140	5.1–7.2	–	–	–	–	93.2 ± 10.5	

Generally, membranes are made of ceramic, metal and polymeric materials while ceramic membranes are generally made of aluminum oxide (Al_2O_3), silicon dioxide (SiO_2) and titanium dioxide (TiO_2). At this stage, polymeric membranes are widely used in the construction and operation of AnMBR due to their lower fabrication cost, with PVDF being the most widely used. However, membrane fouling problems lead to higher operating costs. Ceramic membranes have attracted much attention in recent years because of their higher treatment efficiency and better contamination resistance compared to polymeric membranes. However, high fabrication cost is the bottleneck for commercial application of ceramic membranes. To reduce the cost, natural mineral materials such as chlorite, dolomite, and kaolin are also used in the production of ceramic membranes. In addition, to further reduce the cost, some high porosity materials have been applied in the exploration of AnMBR construction. In practical applications, membranes of different materials are fabricated in the form of hollow fiber membranes, flat plate membranes and tubular membranes respectively for better coupling with various types of anaerobic reactors (Fig. 3.6). The choice of materials plays a crucial role in determining the performance and efficiency of membrane-based water treatment systems. Polymeric membranes offer a cost-effective and suitable alternative to metal and ceramic membranes, which are often expensive and less compatible for such processes. This creates significant opportunities for the utilization of polymeric membranes in membrane treatment processes.

The DM module can be divided into two types: plate and tubular, both of which have been applied in MBRs with different membrane configurations (immersed and sidestream). Although various commercial membrane components can be selected and applied in AnMBR, there are no available DM components for purchase, and the components used in research are all homemade. Although a few studies have used tubular DM modules to develop MBRs, this may be due to the difficulty of assembling tubular modules. Tubular DM modules differ from plate DM modules in that they feature double-sided large inner and outer diameter filters assembled on the exterior of the support skeleton. This design sets them apart from traditional tubular ultrafiltration (UF)/MF membranes. Plate DM modules, on the other hand, are commonly composed of a frame, inner support layer, and outer support layer. The frame serves to hold the inner and outer support layers together, with the inner support layer requiring high strength and chemical stability, making stainless steel the primary choice. The outer support layer utilizes supporting materials to provide support to the DM layer, enabling effective solid–liquid separation in both aerobic and anaerobic DMBRs. Previous studies have shown that plate modules are suitable for immersed and AnMBR configurations, but not for sidestream modules, and the plate DM modules that can be used in sidestream configurations are very different from those observed in AnMBRs, but the reasons are still unclear. Tubular modules are also mainly used in immersed and MBR configurations and are not suitable for use in sidestream and MBR constructions with their existing structures [54] (Table 3.3).

Biological membrane carriers can be broadly categorized into two types: inorganic-based carriers and organic-based carriers. Inorganic-based carriers,

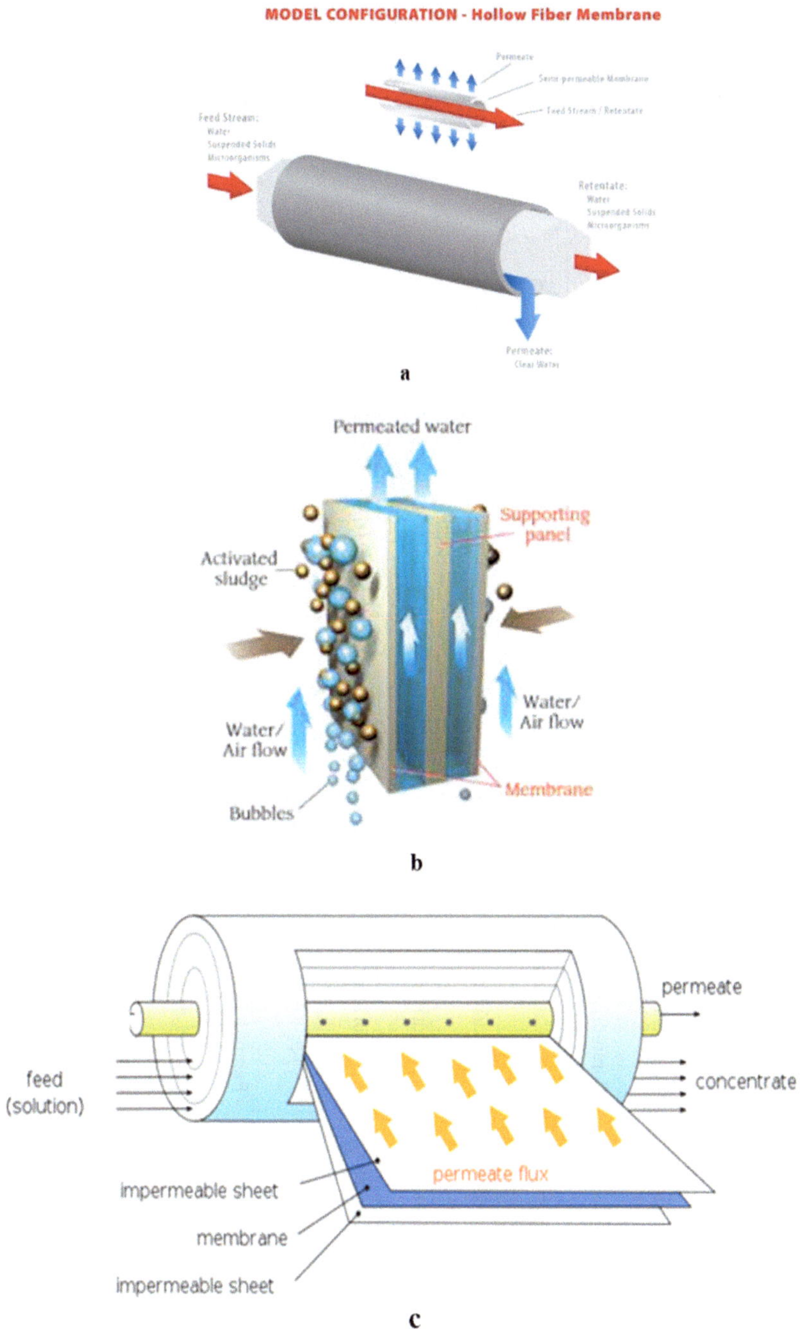

Fig. 3.6 **a** Hollow fiber membranes, **b** flat plate membranes, **c** tubular membranes

Table 3.3 DM filtration performance of the AnDMBRs with different configurations

Reactor configuration	Module configuration	Filtration mode	Support material	Pore size (μm)	Filtration area (m^2)	Operational cycle (d)	Reference
CSTR/ submerged	Tubular	Gravity-driven	Non-woven	~ 10	0.14	5	[119]
UASB/ submerged	Tubular	Pump-driven	Non-woven	0.64	0.98	20–30	[8]
UASB/ submerged	Flat-sheet	Pump-driven	Dacron mesh	61	–	90	[125]
CSTR/ submerged	Tubular	Gravity-driven	Non-woven	–	0.11	12–15	[126]
UASB/ submerged	Flat-sheet	Pump-driven	Dacron mesh	40	–	15–25	[118]
CSTR/ side-stream	Flat-sheet	Pump-driven	Nylon mesh	200	0.004	70	[6]
CSTR/ submerged	Flat-sheet	Pump-driven	Monofilament woven	10a	0.018	50–80	[33]
CSTR/ side-stream	Flat-sheet	Pump-driven	Nylon mesh	200	0.004	130	[5]
CSTR/ side-stream	Flat-sheet	Pump-driven	Monofilament woven fabric	10	0.014	50	[32]
CSTR/ submerged	Tubular	Gravity-driven	Silk	100	0.0314	15	[67]
CSTR/ submerged	Flat-sheet	Pump-driven	Monofilament woven fabric	10	0.014	80	[34]

(continued)

Table 3.3 (continued)

Reactor configuration	Module configuration	Filtration mode	Support material	Pore size (μm)	Filtration area (m^2)	Operational cycle (d)	Reference
CSTR/ side-stream	Flat-sheet	Pump-driven	Monofilament woven	10	0.014	80	[34]
CSTR/ submerged	Flat-sheet	Gravity-driven	Nylon mesh	50	0.04	6	[101]

including volcanic rocks and furnace slag, find common applications in biological filters and artificial wetland media. On the other hand, inert organic materials such as polymers are commercially utilized in full-scale wastewater treatment plants (WWTP) owing to their advantageous features such as large specific surface area, excellent mechanical strength, and low density. In contrast, organic carriers like biochar and corn cobs exhibit a larger surface area and possess a porous surface structure, distinguishing them from polymer-based organic carriers. The formation and development of biofilms typically go through four stages, namely initial biomass attachment, biomass development, biofilm maturation, and biofilm aging and detachment. During the initial stage, the physical and chemical characteristics of the carrier surface, such as surface roughness and ζ potential, significantly impact the biomass attachment by microorganisms. Higher carrier roughness aids in the fixation of microbial tissue and the development of biofilms. Due to their porous structure, inorganic and natural organic-based carriers are typically rougher than inert organic-based carriers, thus being more favorable for microbial cell immobilization. Additionally, hydrophilicity/hydrophobicity (also known as wetting) has been shown to affect cell adhesion and the interaction between the carrier surface and bacterial cells. Water contact angle is commonly measured to evaluate the wettability of carrier surfaces and bacterial cells. The current mainstream view is that a higher surface hydrophilicity is beneficial for the initial bacterial cell attachment to the carrier surface [43].

As the biofilm thickens and matures, it is important to focus on preserving the biomass. Separation of aging biomass is crucial for maintaining the activity of the biofilm. However, excessive separation of biomass from the carrier surface can reduce the biomass attached to the AnMBR and therefore decrease system performance. Therefore, it is very important to protect the AnMBR from hydraulic shear stress. Although the formation of the biofilm is influenced by various factors, such as operational parameters and influent quality, different carriers have a significant impact on the formation time of the biofilm due to their unique characteristics. Compared to gel carriers, common inorganic and organic carriers have a clear disadvantage in anaerobic ammonia oxidation systems' initial biofilm formation stage. It can be inferred that the biofilm formed on the surface of inorganic-based carriers is relatively thin. In terms of large pore structure, most inorganic-based carriers have no advantage. In contrast, inert organic carriers (such as polyurethane sponges) tend to grow thicker biofilms. Therefore, under oxygen injection, a thicker anaerobic biofilm layer can be obtained on inert organic carriers, providing a better growth environment and larger growth space for AnMBR [43].

Currently, the application of ceramic membranes at the laboratory scale is relatively limited. These membranes are typically made of alumina and have pore sizes of 0.05–0.5 μm. Under extreme conditions such as high organic loads, high temperature, and high pressure, AnMBR integrated with ceramic membranes is an ideal choice for treating short HRTS wastewater without generating excessive sludge. The average temperature resistance of ceramic membranes can reach 350 °C, the pressure resistance can reach 4 bar, and the pH range can be 0–14, so they can withstand corrosive chemicals such as acids, bases, and cleaning agents.

In a study conducted by Jeong et al. [58] a laboratory-scale AnMBR system employing ceramic flat sheet membranes demonstrated effective treatment of domestic wastewater supplemented with food waste recycling wastewater. The system achieved impressive COD and total organic carbon (TOC) removal efficiencies of 98.3% and 97.9%, respectively. The system's target HRT was set at 12 h, with a corresponding OLR of 3.0 kgCOD/m^3. Notably, the average methane production rate during operation was measured at 0.21 LCH_4/gCOD. The ceramic flat sheet membrane employed in this study featured a double-layer structure, with a coating on the porous support layer capable of absorbing the permeating agent. This design allowed the target flux to be sustained for approximately 200 days without the need for chemical cleaning. Additionally, the ceramic membrane exhibited substantial elimination of biological polymers and low molecular weight substances, which are primary precursors of fouling. The use of ceramic membranes offered the advantage of reduced clogging frequency due to their high hydrophilicity, enhancing fouling resistance. Moreover, the presence of a greater number of negative charges on ceramic membranes proved effective in mitigating fouling, as it facilitated strong electrostatic repulsion between the membrane surface and fouling agents.

3.4 AnMBR Combined Process and Internal Filling Material

3.4.1 AnMBR Combined Process

Novel bioreactor configurations or rational process integration exhibit high potential for reducing energy demand and improving performance, with the potential to promote wastewater treatment plants towards net energy operation and resource recovery [52].

Microbial electrolytic cell (MEC) coupled AnMBR technology is a promising technology that can alleviate the membrane fouling problem during AnMBR operation by using the MEC cathode to produce hydrogen to scour pollutants in situ and enhance the electrostatic repulsion between pollutants and the membrane surface [15].

MEC is a technology that uses microorganisms as catalysts in electrochemical reactions to convert organic matter in wastewater into electricity or other valuable products. However, the single MEC technology has some problems such as low wastewater treatment efficiency and unstable effluent quality, which limit its application in the field of wastewater treatment. In recent years, a coupled technology called MEC-AnMBR has been proposed to solve these problems [22].

MEC-AnMBR technology combines the advantages of MEC and AnMBR technologies. AnMBR technology is an anaerobic membrane bioreactor that can efficiently remove organic matter while maintaining stable treatment performance. By combining MEC technology with AnMBR technology, organic matter can be converted into electricity or other valuable products through MEC technology, while AnMBR technology can efficiently remove organic matter, thereby improving the treatment efficiency and effluent quality of the entire system. This technology can achieve efficient wastewater treatment and produce clean energy and valuable by-products, which has broad application prospects.

Therefore, MEC-AnMBR coupled technology is increasingly attracting attention as a complementary technology. With the continuous development and improvement of technology, it is believed that MEC-AnMBR technology will play an increasingly important role in the future of wastewater treatment, creating a cleaner and more sustainable living environment for us.

In wastewater treatment, the removal efficiency of organic matter is an important indicator for evaluating the treatment efficiency of wastewater treatment methods, which is often characterized by the COD removal rate. AnMBR process has a high removal efficiency of organic matter due to its high sludge concentration and the filtering effect of membrane materials. However, research has found that by coupling with MEC technology, the removal efficiency of AnMBR for organic pollutants can be further enhanced. The introduction of MEC-AnMBR technology can promote the growth of microorganisms in the system, enhance microbial metabolic processes, and improve microbial diversity and abundance. In the study by Ding et al. [30] although the enhancement effect of MEC technology on the filtering effect of membrane components was not significant, it significantly improved the biodegradation effect, thereby increasing the biodegradation efficiency by 8.4%. Therefore, MEC-AnMBR technology can not only improve the efficiency of wastewater treatment, but also promote the growth and metabolism of microorganisms and produce more valuable by-products, which has a wider range of application prospects.

The integration of MEC and AnMBR can improve the efficiency of wastewater treatment, especially in terms of organic matter removal. The membrane in this combined system functions as a cathode, performing hydrogen evolution reactions, and also serves as a separation layer for filtering the treated water. Compared to traditional AnMBR, MEC coupled with AnMBR has higher organic matter removal efficiency due to MEC's ability to enhance biodegradation and the direct anode oxidation occurring at the anode surface under the driving force of oxidants generated by electron transfer reactions, as well as indirect oxidation in the anode liquid. In addition to producing high-quality permeate, the MEC coupled with AnMBR system also has a trend towards less membrane fouling and lower energy consumption, making it a promising technology for wastewater treatment in the future [30, 104].

Integrating MEC with AnMBR systems offers a promising solution to address key challenges in wastewater treatment. This integration enables effective removal of nitrogen and phosphorus nutrients, methane recovery, and reduction of dissolved methane concentration in AnMBR permeate. In the hybrid system, the anode and cathode of the MEC generate potentials that facilitate the extraction of ammonium and phosphate from the bulk solution of the AnMBR. This enables the recovery of these compounds in separate solutions. This process effectively removes charged ions from the AnMBR solution, leading to a reduction in osmotic pressure drop, improved water flux, and reduced fouling potential. Additionally, the hybrid system helps maintain a low concentration of dissolved methane (< 2.5 ml/L) in the permeate, while keeping the chemical oxygen demand (COD) concentration at a low level (< 20 mg/L) [51].

In certain scenarios, such as when operating conditions are challenging or when AnMBR treatment is subjected to high shock loads, the treated water may not meet the standards required for water reuse. Additionally, achieving the goal of water reuse necessitates the production of high-quality water, which may be challenging to accomplish solely through AnMBR treatment in terms of removing ammonia and phosphate to levels that align with specific reuse standards. Hence, the integration of advanced post-treatment technologies becomes crucial. These technologies include activated carbon adsorption, FO/FOMBR (forward osmosis/forward osmotic membrane bioreactor), MEC, MD, and activated sludge processes (comprising anaerobic, anoxic, and aerobic processes). By integrating these techniques with AnMBR, it becomes possible to produce wastewater of higher quality suitable for reuse [88]. Nevertheless, it is essential to assess the economic sustainability of this two-stage AnMBR approach for water reuse, considering the cost and efficiency of the post-treatment stage.

In recent years, many researchers have been devoted to developing a new type of wastewater regeneration platform aimed at achieving environmentally sustainable wastewater reuse while recovering nutrients. Anaerobic membrane bioreactor (AMBR) has been proven effective in removing organic compounds from various sources of wastewater. However, due to the dominance of anaerobic bacterial biodegradation, the nutrient removal efficiency is still limited. Therefore, the permeate generated by AnMBR typically contains high levels of nitrogen and phosphorus. Additionally, the biogas produced by AnMBR contains approximately 20% to 40% carbon dioxide, which is essential for the growth of algae. Algae can utilize these nutrients and inorganic carbon to produce value-added products or generate energy [25]. Therefore, algal photobioreactors or algal bacteria photobioreactors are a good choice as post-treatment for AnMBR processes.

Algae photobioreactors have demonstrated promising results in removing ammonia and phosphates from AnMBR permeate, with reported studies showing removal rates ranging from 34 to 67% for ammonia and over 75% for phosphates. Notably, these impressive removal efficiencies have been observed even in challenging conditions, including the presence of antibiotics. Algae photobioreactors offer a potential solution for nutrient removal in AnMBR systems, highlighting their effectiveness and versatility in treating wastewater [89]. In addition, research has found that the removal rate of ammonia by algal cells is closely related to temperature and light intensity. These results have been verified by experimental data and mathematical modeling. The results of the study show that different types of algae, such as *Chlorella vulgaris, Scenedesmus obliquus, Chlorella miniata* and *Brachionus plicatilis,* have no significant effect on the nutrient removal rate, and all algae have high removal rates of phosphorus and nitrogen. However, mixed algae have been proposed to further improve nutrient removal in AnMBR post-treatment. Nevertheless, it is necessary to further separate algal cells from the treated water. In addition, the economic and energy demands of algae photobioreactors in the ANMBR process need further analysis.

Researchers have recently been investigating the application of purple phototrophic bacteria and algae, communities in membrane photobioreactors. The primary objective of these studies is to minimize the footprint and operational expenses related to algae separation. Optimizing operational conditions is vital for achieving optimal nutrient recovery and enhancing water quality treatment. Previous research has demonstrated the significant impact of various operational parameters on nutrient recovery and the quality of treated water in membrane photobioreactors. These parameters encompass solar radiation intensity, algae concentration, temperature, nutrient loading rate, algae culture recirculation ratio, and algae biomass retention time. Notably, temperature and solar irradiance are key factors influencing algae biomass productivity and nutrient recovery efficiency. Furthermore, the nitrogen to phosphorus ratio of influent water is a critical determinant of phosphorus recovery rate [107]. This design and operation of mixed algal culture AnMBR systems, facilitating the cost-effective and environmentally friendly recovery of wastewater [115, 121].

Under anaerobic conditions in a mixed algae-bacteria system, ammonia can be utilized by algae or autotrophic bacteria. Phosphorus is removed through a synergistic interaction between heterotrophic bacteria, autotrophic organisms, algae, and phosphorus-accumulating organisms due to the activity of algae and phosphorus-accumulating organisms. This combination can achieve nutrient recovery and wastewater regeneration economically [115, 121].

It has been reported that nutrients produced by ANMBR can be extracted by direct concentration through MD or FO membranes. In addition, nutrients can also be extracted from AnMBR biomass suspension by charging in microbial recovery cells (MRC), but this may lead to the formation of struvite. Due to the organic loading rate in AnMBR typically ranging from 0.43 to 0.9 kgCOD/(kgVSSday), with corresponding SRTs ranging from 50 days to infinity, energy can be recovered from wastewater. Previous studies have shown that controlling operational parameters of AnMBR, such as methane production activity and temperature, is key to achieving maximum biogas production [54].

During denitrification, the conversion of nitrate and nitrite to nitrogen gas (N_2) takes place, while methane (CH_4) is oxidized to carbon dioxide (CO_2). This concept is achieved by integrating AnNMBR with the anaerobic ammonia oxidation process. By concurrently removing nitrate and dissolved methane, the implementation of nitrite-dependent anaerobic methane oxidation (N-DAMO) can mitigate the environmental impact of ANMBR effluent. However, the growth rate of N-DAMO bacteria is inherently slow, which considerably reduces the overall reaction rate. To enhance the activity and utilization of N-DAMO microorganisms for nitrite, membrane bioreactors can be employed to fully retain N-DAMO microorganisms or adjust the partial pressure of methane, thus achieving higher methane utilization efficiency [18].

The application of membrane separation technology has significantly advanced AnMBR in the field of wastewater treatment. Numerous studies have explored the utilization of AnMBR in this context. For instance, Ho [50] conducted research using a laboratory-scale CSTR equipped with a side-stream membrane. The system was designed for the anaerobic treatment of composite urban wastewater with a COD concentration ranging from 500 to 1000 mg/L at a temperature of 25 °C. The study achieved an impressive 97% COD removal efficiency and demonstrated an infinite SRT, highlighting the effectiveness of the AnMBR approach [105]. Xie et al. [117] achieved COD removal rates of 93–99% in a 3.5 L AnMBR laboratory scale system with an HRT of 11–26 h. For other studies with different operating conditions, the calculated COD removal efficiency is about 90%.

In addition to producing biogas, the AnMBR system is considered a viable choice for treating high-concentration wastewater, such as leachate, allowing separate control of HRT and SRT. An anaerobic filter (AF) removes approximately 21.5% of the chemical oxygen demand (COD), while the AnMBR system achieves a removal efficiency of approximately 90.4% when treating the same leachate. Laboratory-scale AnMBR testing has also demonstrated consistent and stable permeate consistency during continuous operation. Many researchers have summarized the application of AnMBR technology in landfill leachate treatment under different operating conditions. Enzyme hydrolysis processes are recommended to be carried out at medium to high temperatures (37–55 °C). Trzcinski and Stuckey achieved COD removal efficiencies of 90 and 79–95% in a laboratory-scale CSTR with an SRT of 300 days and HRT of 1 day, and an SRT of 30 days and HRT of 1.5 days, respectively [105].

3.4.2 AnMBR Internal Filling Material

AnMBR is an advanced wastewater treatment technology that can simultaneously remove organic matter and nutrients such as nitrogen and phosphorus. The type and properties of the internal packing materials in AnMBR have a significant impact on the system's treatment efficiency and stability. This article will introduce the research status and development trends of internal packing materials in AnMBR and explore the advantages and disadvantages of different packing materials.

Internal packing materials in AnMBR are mainly used to increase the microbial growth area in the bioreactor, promote the decomposition and removal of organic matter. Traditional packing materials mainly include polypropylene (PP), polyester (PE), and foam plastics, which have good corrosion resistance and mechanical strength. However, their smooth surface makes it difficult for microorganisms to adhere and grow, which can easily cause packing blockage and an increase in reactor pressure.

To overcome these shortcomings, some new types of packing materials have emerged in recent years, such as bio-ceramic particles, cellulose-based materials, and ceramic particles. Bio-ceramic particles have a good pore structure and a large specific surface area, which can provide more bio-adhesive surface area. They also have good corrosion resistance and mechanical strength, and have good application prospects. Cellulose-based materials have good biocompatibility and biodegradability, which can promote microbial growth and metabolic activity. However, their mechanical strength and corrosion resistance are poor. Ceramic particles have high hardness and corrosion resistance, can withstand high operating pressure and temperature, but have higher costs and production difficulties.

In addition to the type of filling material, the shape and size of the filling material also have an important impact on the performance of AnMBR. In recent years, some new types of filling shapes, such as cubes, hexagons, and spheres, have appeared. These shapes have a larger surface area and better fluidity, which can improve the efficiency and stability of the reactor. At the same time, the size of the filling material also needs to be selected appropriately. Small filling materials are easy to cause blockage, while large filling materials can affect the fluidity of the liquid.

In general, the type and properties of filling materials inside AnMBR are very important for the processing efficiency and stability of the system. The surface characteristics, pore structure, and material composition of the filling material can all affect the attachment and growth of microorganisms, thereby affecting the stability and performance of AnMBR. Therefore, different types of filling materials need to be selected for different sewage treatment requirements [93].

In addition, the self-cleaning ability of filling materials is also an important consideration factor. Some new types of filling materials, such as nanomaterials with self-cleaning function and bioceramic particles, can reduce sludge attachment and accumulation to a certain extent, thereby reducing the operating costs and maintenance difficulties of the system.

In recent years, some researchers have also tried to combine filling materials with membrane technology to develop a new type of AnMBR filling membrane reactor. This reactor can not only increase the bio-attachment area and stability but also use the screening effect of the membrane to achieve better solid–liquid separation, further improving the processing efficiency and reliability of the system.

In summary, the type, shape, size, and material composition of filling materials inside AnMBR all have important impacts on the processing efficiency, stability, and cost of the system. In the future, further research and development of more excellent filling materials and structures are needed to provide better technical support for achieving more efficient, stable, and reliable AnMBR sewage treatment systems.

3.5 Factors Affecting the Performance of AnMBR

The performance of AnMBR systems is influenced by both the biological treatment and filtration processes. Understanding the interplay between these two factors is crucial for comprehending the entire AnMBR process. Biological phenomena occurring within the reactor medium and the associated increase in transmembrane pressure, which indicates a decline in flux, are two key factors that need to be considered [21]. The subsequent sections will delve into the impact of operational parameters on treatment efficiency and membrane performance.

The current anaerobic treatment system, AnMBR, is used for municipal wastewater treatment and its advantages such as high treatment efficiency and good effluent quality are key to its wide application. However, the application of AnMBR in China is still hampered by several factors including membrane fouling, salinity accumulation, and temperature, and more research is needed to investigate potential problems to make AnMBR a viable and sustainable solution.

The main mechanisms involved in the removal of pollutants in wastewater treatment plants include biotic and abiotic transformations, as well as adsorption through sludge. In practical terms, various operating parameters such as HRT, SRT, pH, temperature, and organic loading have a significant impact on biotransformation processes. These parameters play a crucial role in stimulating microbial community diversity, enhancing microbial activity, and influencing metabolic pathways within the treatment system.

The operation of AnMBR is regulated based on the conditions of several related parameters in a complex network. In particular, HRT and SRT are two key parameters in the design of anaerobic bioreactors. HRT and SRT are very important, where HRT refers to the water residence time in the reactor and SRT refers to the biomass retention time in the reactor. Literature studies have shown that the separation of HRT and SRT is crucial to improve the wastewater treatment efficiency of anaerobic bioreactors. In different scales of bioreactors, CSTR and UASB are two commonly used designs. Although the construction and operation of CSTR reactors are relatively simple, the effluent suspended solids (SS) is high, and the membrane fouling is more severe, resulting in relatively lower operating efficiency. In contrast, the HRT and SRT of UASB have achieved a certain degree of separation by retaining biomass at the bottom, thereby achieving better wastewater treatment efficiency. In addition, the addition of membrane devices has further improved the effluent quality of UASB and ensured longer membrane operating time, making the process more stable [39].

AnMBR has been widely used in urban wastewater treatment field, and its efficient organic matter removal and methane production capacity have been confirmed in many studies. Although many studies have explored the treatment performance of AnMBR under long HRT conditions, there are few studies on the treatment performance and effluent organic matter composition characteristics under short HRT conditions. SRT is another key factor that affects the performance of AnMBR, especially under high OLR conditions. Short SRT may cause membrane fouling, while excessively long SRT may increase sludge viscosity and membrane surface resistance, thereby affecting system performance. The unique advantage of HRT and SRT separation in AnMBR enables it to operate at high OLR, so the regulation of SRT is crucial for the optimization of AnMBR system. The experiment conducted by Sanguanpak et al. [92] showed that the change in pH value has an impact on the properties of the mixed solution. An increase in the pH value of the mixed solution caused a significant increase in the pH value in the solution.

For the treatment of urban wastewater using AnMBR technology, the urban wastewater can be pre-concentrated by using the biological adsorption-precipitation process (BSP) or various membrane-based methods such as FO. Recent studies have shown that a new type of BSP technology can concentrate domestic wastewater in a short HRT of 60–120 min, enhancing the biological adsorption process while maintaining an infinite SRT in the system, thereby minimizing sludge treatment and disposal costs. A shorter hydraulic retention time can help organic matter adsorb on the surface of microorganisms, resulting in concentrated organic matter effluent. Research on these emerging technologies will help optimize the application of AnMBR technology in urban wastewater treatment [10].

The use of membrane separation technology has brought significant advancements to the field of wastewater treatment, particularly in the context of AnMBR. Extensive research has been dedicated to evaluating the effectiveness of AnMBR for wastewater treatment. In a study conducted by Jeong et al. [58]. A laboratory-scale AnMBR system employing ceramic plate membranes exhibited impressive performance in treating a combination of domestic wastewater and food waste recycling wastewater. The system achieved remarkable removal rates, with COD removal reaching 98.3% and TOC removal reaching 97.9%. The system operated with a targeted HRT of 12 h and an OLR of 3.0 kgCOD/m^3. Notably, the system exhibited an average methane production rate of 0.21 LCH_4/gCOD during operation. Ho and Sung [50] used a laboratory-scale continuously CSTR with side-stream membranes to treat composite urban wastewater containing 500–1000 CODmg/L at 25 °C), achieving a COD removal rate of 97% and an infinite SRT. Xie et al. [117] used a 3.5 L laboratory-scale AnMBR system (37 ± 1 °C) and achieved COD removal rates of 93–99% at HRTs of 11–26 h. For other research examples with different operating conditions, the calculated COD removal efficiency was approximately 90%. In addition to using biogas, the AnMBR system is considered a reasonable choice for treating high-concentration wastewater such as leachate, allowing for the control of HRT and SRT, respectively. AF removed approximately 21.5% of COD, while the AnMBR system removed approximately 90.4% of COD when treating the same leachate. Laboratory-scale AnMBRs were also tested for consistent and stable permeate consistency during continuous operation. Many researchers have summarized the application of AnMBR technology in leachate treatment under different operating conditions. The hydrolysis process is recommended to occur at medium or high temperatures (37–55 °C). Trzcinski and Stuckey [105] used a laboratory-scale CSTR with an SRT of 300d and an HRT of 1.5d, achieving COD removal rates of 90% and 79–95%, respectively.

3.5.1 Sludge Retention Time

SRT refers to the time that microorganisms stay in an AnMBR reactor and is one of the key parameters that affect AnMBR performance. The impact of SRT on AnMBR reactors is mainly reflected in the following aspects:

Affects microbial community structure and quantity: An increase in SRT will increase the time that microorganisms stay in the reactor, promoting microbial growth and reproduction, and thus increasing the number of microorganisms. In addition, the growth rate and wastewater degradation ability of different microorganisms are also affected by SRT. Therefore, controlling SRT can regulate the microbial community structure and quantity.

Affects wastewater treatment efficiency: Changes in SRT will have an impact on the wastewater treatment efficiency of the AnMBR reactor. Under suitable SRT conditions, microorganisms can fully degrade organic wastewater, improve COD and BOD removal rates, and produce more biogas. However, under too low or too high SRT conditions, wastewater treatment efficiency will be affected and COD and BOD removal rates will decrease.

Affects energy recovery efficiency: Adjusting SRT can optimize biogas production and quality, thereby improving energy recovery efficiency. When SRT is appropriate, methane-producing bacteria can be maximally promoted in growth and reproduction, maximizing biogas production and methane content.

Typically, using longer SRTs and HRTs is beneficial for operating AnMBR systems because it can improve treatment performance, reduce sludge production, and also increase methane recovery rates. Starting up relatively quickly in a temperature-appropriate environment, typically taking 2–4 months, the cost-effective solution to treating wastewater is possible with AnMBR processing in its normal state because of the facility's long sludge retention time. At environmental temperatures, the anaerobic process becomes feasible if the SRT is approximately twice that of the thermophilic condition. For example, Cao et al. [19] noted that the UASB reactor started up quickly when the alkalinity concentration was about 1.5 g/L, with OLRs starting at 2000 g/m^3·d, shortening the start-up time compared to the initial value of 250 g/m^3·d, with influent COD of 2 g/L.

Controlling the HRT and SRT can regulate the biogas production and growth of methane-producing bacteria. Adjusting these two parameters can optimize the microbial population and their growth in the reactor, thereby increasing methane yield and quality. Typically, the methane content in biogas generated by AnMBR (reaches around 80%). However, methane content is influenced by OLR. Recovering methane in AnMBR is challenging because methane in the effluent dissolves in water. It has been reported that about 67% of methane dissolves in the effluent of AnMBR. Additionally, operating temperature significantly affects the solubility of methane. Methane solubility is higher at lower temperatures. For example, methane produced at 15 °C has a solubility of approximately 1.5 times that of 35 °C. Therefore, when controlling methane solubility, the impact of operating temperature must be considered. In normal biogas plants, the methane content in biogas produced is approximately 70%. In biogas from AnMBR, the methane content is typically between 70 and 90%, along with impurities such as nitrogen and carbon dioxide. The content of these impurities affects the quality and application of biogas [66].

Sludge age is a critical parameter that significantly impacts the operation of AnMBRs. It is well recognized that a longer SRT is advantageous for achieving improved COD removal and methane production in specific biological waste scenarios. However, in certain cases, a substantial portion of the removed COD is retained as particulate matter within the AnMBR reactor rather than being fully degraded. It has been established that a longer sludge age and higher biomass concentration are necessary conditions for adequate substrate degradation. Therefore, optimizing the sludge age and biomass concentration is crucial for enhancing the overall performance of AnMBR systems. In a study conducted by Huang et al. [56], the impact of different SRT (30, 60, and infinite days) on the treatment of domestic wastewater was investigated. The results revealed that under an infinite SRT, a maximum specific methane production rate of 0.056 L-CH_4/g-MLVSS/d was observed. However, it is important to note that higher SRT does not always guarantee better overall performance. Both excessively short and excessively long SRTs can have detrimental effects on process stability. A short sludge age not only leads to lower methane production but also results in significant accumulation of SMP and severe membrane fouling. On the other hand, an excessively long SRT accelerates membrane fouling and compromises system stability. While a longer SRT can maintain a higher MLSS concentration, thereby improving organic matter degradation efficiency, extremely high MLSS levels can hasten membrane fouling. Therefore, finding an optimal SRT is crucial to establish a balance between achieving sufficient methane production and avoiding excessive accumulation of SMP or membrane fouling. It is important to carefully consider the specific conditions and requirements of each AnMBR system to determine the appropriate SRT for optimal performance and stability. Moreover, a high SRT may also lead to a decline in permeate flux, limiting the treatment capacity of the bioreactor, despite an increase in suspended solids concentration under a constant HRT [98]. Achieving the optimal SRT is crucial for a stable and efficient AnMBR process. To maintain the desired SRT, periodic removal of digested biomass through manual withdrawal or sludge wasting is necessary. However, the actual SRT value and discharge frequency can vary due to variations in the waste stream and operational condition [127].

3.5.2 Organic Load

Urban wastewater typically has low concentrations of organic matter, which makes it difficult to directly apply anaerobic digestion technology for treatment. There are several challenges associated with implementing anaerobic digestion technology for treating low-concentration wastewater. These challenges include significant capital investments and the issue of methane loss. Nonetheless, it is essential to study the optimal organic load due to its significant influence on the operational efficiency of AnMBR reactors. This efficiency is regulated by various factors, such as operating conditions, microbial community, and influent characteristics.

In an AnMBR system, the importance of controlling the OLR cannot be overlooked, as excessive OLR can lead to system failure, while insufficient OLR fails to maximize its treatment capacity. To modify the OLR, options include increasing the influent flow rate, raising the pollutant concentration in the influent, or reducing the HRT. These measures can be employed to achieve changes in OLR [27].

Increasing the OLR has been found to have positive effects on the removal efficiency of COD and the production of biogas, particularly methane. Research conducted by Hu et al. [55] demonstrated that elevating the OLR from 2 kgCOD/m^3·day to 10 kgCOD/m^3·day resulted in a 600% increase in biogas production and a 625% increase in methane production. These improvements were attributed to the positive influence on the microbial community, as observed by Yee et al. [122]. In an AnMBR treating pig manure, the dominant methanogens in the microbial community shifted from *Methanosaeta* spp. to *Methanoscina* spp. with increasing OLR. Similarly, in an AnMBR treating kitchen waste slurry, *Methanosaeta* was replaced by *Methanoscina* as the primary methane-producing microorganism. These changes in the microbial community potentially enhanced microbial activity, as indicated by Xiao et al. [116]. Specifically, in an AnMBR treating high-strength kitchen waste slurry, the activities of hydrolytic enzymes such as amylase, protease, and lipase increased by at least 1.9-fold as the OLR increased from 4.7 to 9.3 kgCOD/m^3 day [27].

On the contrary, it is worth noting that an increase in the OLR can lead to a decrease in the methane content within the biogas. This phenomenon can be attributed to the heightened production of CO_2 per unit volume of the liquid phase as the OLR increases. Once the liquid phase becomes saturated, CO_2 is released into the gas phase, thereby affecting the methane content in the biogas [85]. The reduction in methane proportion in the biogas could also be attributed to the proliferation of acidogenic bacteria, supported by higher OLR. This acidification and decrease in methane yield are a result of VFA accumulation and inhibition. In the study by Gao et al. [38] increasing the OLR by approximately 60% by increasing the influent COD also led to a slight decrease in biogas production (15%), possibly due to the inhibitory effects of higher OLR.

The capability to operate effectively under varying and elevated OLRs serves as a strong indicator of the stability and feasibility of AnMBR systems in broader applications. This adaptability demonstrates the system's ability to handle fluctuating wastewater conditions and maintain efficient treatment performance, thus showcasing its potential for widespread implementation [7]. Higher OLR improves the economic feasibility of AnMBR by increasing biogas production, reducing physical footprint, and lowering capital costs [27].

For example, Hu et al. [55] studied the effect of organic load on domestic wastewater treatment using two anaerobic dynamic membrane bioreactors under low-temperature conditions. In stages I, II, and III of the experiment, the provided organic loads were 0.88, 1.55, and 3.01 kgCOD/m^3 d, respectively. The experimental results showed that the COD removal rates of the two AnDMBRs increased from 75 and 73% to 92% and 92%, respectively. The observed results were due to the fact that when the organic load was increased, the added substrate including organic matter and trace nutrients were effectively degraded by anaerobic microorganisms.

According to the study by Khan et al. [62], the highest organic load resulted in the lowest COD removal rate (65.1 $\pm$ 2.2%) and as organic load increased, NO^{3-} removal rate also decreased (91.9 $\pm$ 0.5%) during AnMBR treatment of low-concentration wastewater. This could be because VFAs accumulated and caused a drop in pH at higher organic loads, and the contact time between denitrifying microorganisms and wastewater decreased due to a shorter time for wastewater to enter. When the

HRT was 18 and 12 h, the transmembrane pressure (TMP) increase rate accelerated, possibly due to soluble organic matter and nutrient addition intensifying membrane fouling. When the HRT was reduced from 8 to 4 h, the observed biogas production and methane content significantly decreased to 0.23 L/d and 16.9%, respectively, with a higher amount of dissolved methane observed.

3.5.3 *Potential of Hydrogen*

In AnMBR reactors, pH control is crucial for maintaining appropriate microbial communities and normal operation of the reactor. Changes in pH can affect cell membrane stability, microbial metabolite production, and microbial community structure. Therefore, maintaining an appropriate pH range is essential for the stability and efficiency of AnMBR reactors.

PH is typically used as an indicator of hydrogen ion concentration in liquids. For AnMBR reactors, a pH range of 6.5–8.5 is considered an appropriate operating range because microbial communities within this range are the most active and stable. When pH values exceed this range, microbial growth and metabolism may be limited, leading to instability and decreased efficiency of the reactor.

Both anaerobic digestion and aerobic digestion in AnMBR reactors are affected by pH. Under anaerobic conditions, too low or too high pH values can inhibit the growth and metabolism of anaerobic digestion microorganisms, thereby reducing organic load removal efficiency. Changes in pH also affect the growth and metabolism of aerobic microorganisms under aerobic conditions, thereby affecting the stability and efficiency of AnMBR reactors.

Therefore, to maintain an appropriate pH range, a series of measures need to be taken, such as adding alkaline or acidic substances to regulate pH, optimizing influent characteristics, and adjusting influent rates. These measures can help maintain the stability and efficiency of AnMBR reactors, achieving efficient wastewater treatment [96].

3.5.4 *Temperature*

AnMBR is sensitive to environmental changes, especially temperature fluctuations, and a key factor in achieving satisfactory performance is the proper regulation of temperature, as temperature fluctuations can promote dissolution and have a direct impact on the maintenance of biokinetics and chemical equilibrium. When the temperature of the AnMBR decreases, it leads to lower biomethane recovery as methane solubility increases and bacterial activity decreases, which in turn leads to more EPS secretion and even more membrane fouling.

The operational conditions of AnMBR are also one of the important factors that affect its performance. Generally, AnMBR can maintain a stable COD removal rate

(87–97%) in wastewater with a COD concentration of about 500 mg/L. However, at low temperatures and an HRT of 6 h, the COD removal rate significantly decreases to 70–80%. In addition, AnMBR can be operated in mesophilic or thermophilic environments, between 30 to 40 °C and 50 to 60 °C, respectively [53].

Generally, AnMBR can maintain a stable COD removal rate (87–97%) in wastewater with a COD concentration of approximately 500 mg/L. However, at low temperatures and an HRT of 6 h, the COD removal rate significantly decreases to 70–80%. Therefore, treating municipal wastewater (MWW) with anaerobic technology in cold regions remains a challenge. Although the operating temperature range of AnMBR is wider than MWW treatment, its performance is also affected under low-temperature conditions below 20 °C. Due to the insufficient methane production for offsetting the heating costs required to raise the operating temperature to a mesophilic level, this temperature range is deemed unsuitable for AnMBR operation. Instead, it is more appropriate for AnMBR to operate within a suitable temperature range, characterized by low HRT, increased OLR, and the ability to maintain a high biomass concentration and extended SRT. Robles et al. further supported this notion in their study, which involved the treatment of urban wastewater at a temperature range of approximately 25–30 °C. By utilizing an industrial prototype AnMBR for 350 days, they successfully demonstrated the feasibility of AnMBR under such conditions. The results showcased an average COD removal rate of 92%, satisfactory filtration performance, low energy consumption (0.15 kWh/m^3 of wastewater), and the potential for generating nutrient-rich and pathogen-free permeate suitable for bi-fertilizer recovery [86].

In a study conducted by Martinez-Sosa et al. [73] the impact of temperature reduction from 35 °C to 20 °C on AnMBR performance was investigated. The findings revealed a significant increase in the concentration of dissolved COD and Total Suspended Solids (TSS) within the reactor. Additionally, a decrease in CH_4 production and an increase in membrane fouling were observed. These changes can be attributed to the higher solubility of organic compounds in the effluent under colder temperature conditions. Furthermore, as the temperature decreases, the viscosity of the mixture increases, requiring more energy for effective mixing and pumping. In a study conducted by Peña et al. [81] the impact of uncontrolled seasonal temperature fluctuations (ranging from 10 °C to 28 °C) on AnMBR performance was investigated. The study concluded that the efficiency of COD removal noticeably decreases when the temperature falls below 15 °C. Specifically, when the temperature dropped to 10 °C, methane production was low, and the effluent exhibited higher COD concentrations. Operating AnMBR at low temperatures poses a considerable challenge to process stability. The low temperatures contribute to a significant accumulation of suspended solids in the reactor, exacerbating membrane fouling [31].

Gao et al. [44] investigated the effects of temperature on microbial community and membrane fouling in an anaerobic membrane bioreactor. The COD removal efficiencies were 85%, 93%, and 90% at 15 °C, 25 °C, and 35 °C, respectively. Membrane fouling was aggravated at lower temperatures, and the fouling cycle, microbial activity, and metabolic products were directly related. The size of the Shannon index reflected changes in microbial community diversity. In addition, microbial population diversity increased with increasing temperature. The microbial community Shannon indices were 0.820–1.370, 0.963–1.846, and 2.218–2.995 at 15 °C, 25 °C, and 35 °C, respectively. Finally, the Gini coefficient of the microbial community evenness in the cake layer varied with temperature. Pearson correlation analysis (at a 95% confidence level) showed a positive correlation between microbial community evenness and TMP.

Recovering CH_4 from AnMBR poses a challenge due to its dissolution in the effluent. Approximately 67% of the produced CH_4 gets dissolved in the AnMBR effluent, while the wastewater treatment process achieves a removal rate of 86–88% for COD [123]. The solubility of methane is greatly influenced by the operating temperature. Research indicates that methane produced at a temperature of 15 °C has a solubility approximately 1.5 times higher than that at 35 °C (Fig. 3.7). In a typical biogas plant, the methane content is around 70%. Similarly, biogas generated from AnMBR exhibits methane content in the range of 70–90%, accompanied by other impurities such as nitrogen (0–15%) and carbon dioxide (3–15%).

The influent COD concentration fluctuated between 250 and 450 mg/L (Fig. 3.7). In general, when the temperature increased to 25 °C, the COD removal for AnMBR increased significantly. The COD removal efficiencies was 85.0%, 93.0%, and 90.0% at 15, 25, and 35 °C, respectively. The AnMBR was operated for over 180 days at three different temperatures. The COD removal rate exceeded 85%. The membrane fouling cycle varied with temperature and is summarized in Fig. 3.7. At 15, 25, and 35 °C, the membrane fouling cycles were 8, 15, and 23 days, respectively. Figure 3 illustrates the TMP curve. With a sharp increase in TMP (dP/dt) at 15 °C, the membrane fouling rate reached approximately 4.9 kPa/day. Similar TMP curves were observed in submerged anaerobic bioreactors treating synthetic wastewater and real urban wastewater, with variations corresponding to changes in temperature [113]. At 25 °C,

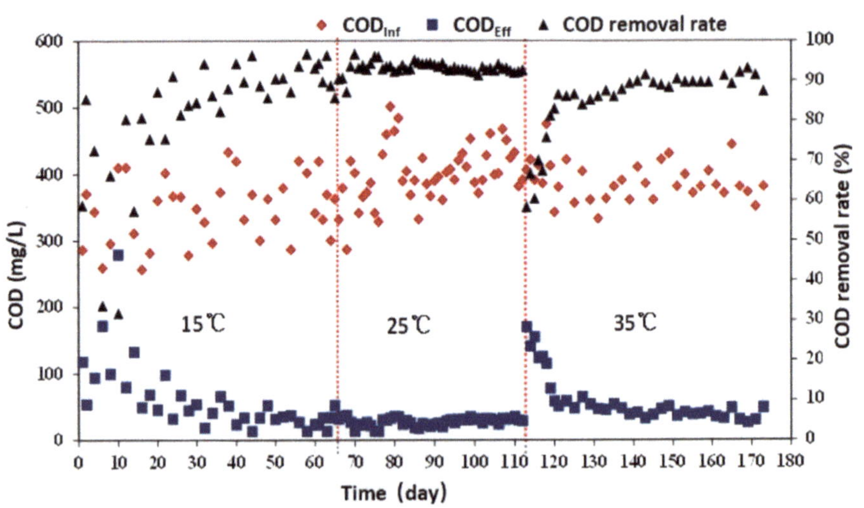

Fig. 3.7 COD removal performance under different temperatures

the first membrane fouling cycle ranged from 1 to 14 days, and the TMP exhibited a rapid increase. We attributed this trend during the initial temperature increase phase to the microbial community not yet adapting to the temperature change. Subsequently, for the next 12 days, the TMP increased gradually. After the 12th day, there was a sudden increase in TMP. At 35 °C, this sudden increase occurred after the 19th day. The results showed that the membrane fouling was getting worse at low temperature conditions, agreeing with some studies [41].

The AnMBR ran for more than 180 days under the three temperatures. The COD removal efficiencies were > 85%. The change of membrane fouling cycle with temperature is summarized in Fig. 3.8. At 15, 25, and 35 °C, the membrane fouling cycles were 8, 15, and 23 d, respectively. Figure 3.8 shows the TMP profiles. With a sharp increase in TMP (dP/dt) at 15 °C, the membrane fouling rate reached to about 4.9 kPa d^{-1}. It has been reported that microbial metabolism products (SMP and EPS) contributed to membrane fouling. Due to temperature differences, SMP and EPS in the reactor were dissimilar in quantity. Such differences could inform about the different membrane fouling rates. Figure 3.8 displays the EPS and SMP changes when the temperature rises. The variation in EPS and SMP concentrations with the TMP are shown in Fig. 3.8. We observed that during TMPs 5–25, the EPS concentrations were maintained under each temperature. Nevertheless, when TMP reached 30 kPa, the EPS concentration peaked, reaching 48.3, 55.3, and 38.3 mg/g, respectively, at 15, 25, and 35 °C. The EPS content decreased at 25 °C but increased at 35 °C, which was consistent with TMP variation [44].

AnMBR systems offer flexibility in terms of operating conditions and can be employed under thermophilic (50–60 °C) or mesophilic (30–40 °C) conditions [45, 73]. However, it is worth noting that psychrophilic conditions (< 20 °C) are generally unsuitable for municipal wastewater treatment. Consequently, anaerobic treatment of municipal wastewater poses a challenge in cold regions, as a substantial amount of energy is required to heat the wastewater to mesophilic conditions. This energy demand presents an obstacle in achieving efficient anaerobic treatment in colder climates.

Operating an AnMBR at low temperatures can give rise to various challenges, such as intensified membrane fouling, sluggish biodegradation of contaminants, and increased solubility of methane in the effluent. At lower temperatures, the hydrolysis of particulate matter into dissolved molecules is limited, resulting in the accumulation of suspended solids within the reactor and a reduction in methanogenic activity. These factors collectively contribute to the negative impacts associated with AnMBR operation under low-temperature conditions. In a study conducted by Martinez-Sosa et al. [73], the impact of temperature reduction from 35 to 20 °C on AnMBR performance was examined. The findings indicated that lowering the temperature resulted in an increase in the total suspended solids content and soluble COD in the bioreactor. This led to severe membrane fouling and a decrease in CH_4 production. The reduced CH_4 production can be attributed to its increased solubility in the effluent at lower temperatures, specifically at 20 °C. Furthermore, as the temperature decreased, the viscosity of the mixed liquor also increased, necessitating more energy for efficient mixing and pumping [99].

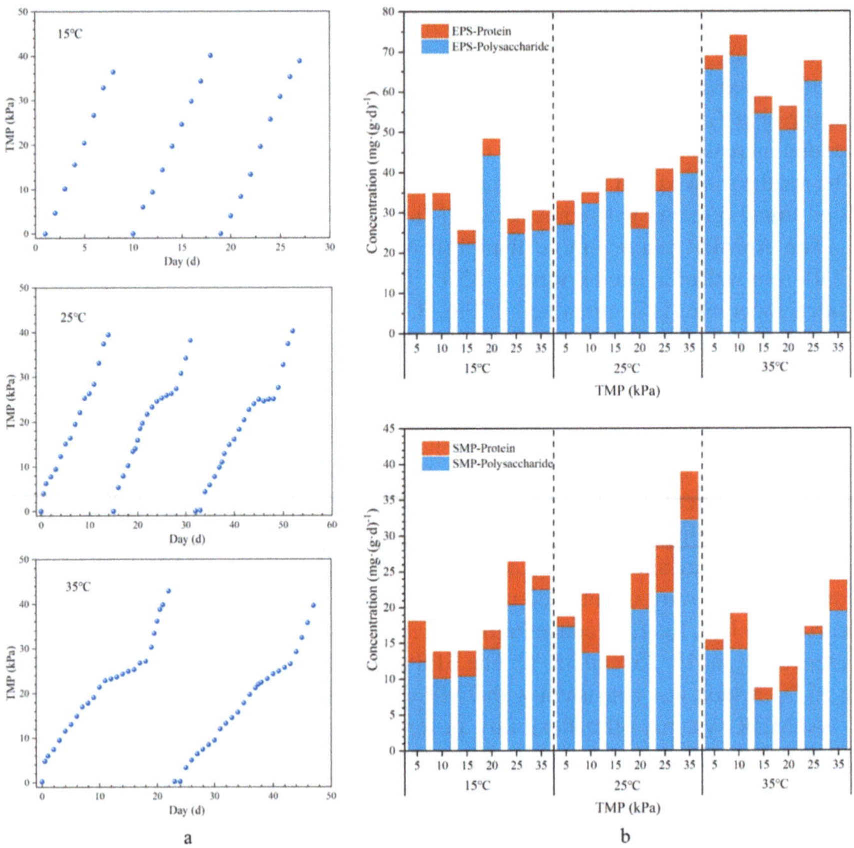

Fig. 3.8 **a** The changes in membrane fouling under different temperatures. **b** EPS and SMP changes under different temperature

3.6 Application of AnMBR in Wastewater Treatment

3.6.1 Pharmaceutical Wastewater

Drugs are compounds used for the prevention, treatment and diagnosis of diseases. Antimicrobial drugs are extensively utilized in the prevention and treatment of infectious diseases in both humans and animals. These medications have played a crucial role in revolutionizing modern life and are indispensable for human medical treatment and animal therapy. They are a large class of compounds with pharmacological and physiological properties, available as prescription drugs, over-the-counter drugs, in their original state or as metabolites, whichever source they come from, and end up in WWTP. In recent years, AnMBR have attracted significant research attention in the treatment of pharmaceuticals in various wastewaters.

Global drug consumption has increased exponentially. Taking human antibiotic consumption as an example, the total amount of drug use and its types have increased by 65.0% from 2000 to 2015. This is mainly attributed to rapid urbanization and higher incidence of infectious diseases. Currently, the COVID-19 pandemic has intensified the use of antibiotics. The consumption of antibiotics for the treatment of bacterial co-infections in COVID-19 patients is high. However, the large consumption of drugs has brought serious environmental problems. As drugs cannot be completely metabolized by organisms, they are excreted in feces and urine, forming mixtures of drugs or their metabolites. The main sources of drugs include pharmaceutical industry, hospitals, wastewater from animal farms and aquaculture, and the disposal of unused and/or expired drugs in cleaning and sewage systems. Drugs eventually reach wastewater treatment plants.

The test results show the presence of drugs in wastewater and their presence in surface water, groundwater and drinking water as well, but the existing conventional activated sludge (ASP) method has insufficient treatment capacity. In this regard, MBR is considered as a promising alternative to ASP and has been widely used for higher and wider range of drug removal. Studies on drug removal by MBR have so far focused on aerobic conditions. Studies on drug removal using AnMBR are very limited, with different microbial communities and redox conditions present to facilitate their removal [109].

Biological transformation is the main mechanism for removing hydrophilic compounds such as ciprofloxacin (CIP), amoxicillin (AMOX), sulfamethoxazole (SMX), and trimethoprim in AnMBR processes. Inhibition of sludge metabolic activity in the AnMBR system, especially under increased salinity, significantly reduces the removal efficiency of these compounds. Therefore, SRT is one of the most critical factors in the AnMBR system. When SRT is at a level of 30 days or higher, it can promote enhanced removal of drugs. In addition, a long SRT enriches slow-growing microorganisms and provides greater microbial population diversity, which helps the biological transformation process to occur [80].

AnMBR technology is regarded as a dependable solution for the removal of pharmaceuticals and sustainable wastewater treatment. Nonetheless, a comprehensive comprehension of the biological transformation mechanism is essential to elucidate the involvement of crucial microbial communities and to determine the fate of pharmaceutical compounds, particularly in the presence of a membrane biofilm layer. To address common negative perceptions of biofouling and to ascertain its beneficial role in pharmaceutical removal, a more focused and targeted study is required. This will aid in gaining a deeper understanding of the processes involved and dispelling any misconceptions related to biofouling in the context of pharmaceutical treatment.

Membrane biofouling remains a significant challenge that limits the widespread implementation of MBRs. However, the fouling layer formed on the membrane surface can have a positive impact on the water quality of the effluent. It has the potential to restrict the permeation of antibiotics and antibiotic-resistant genes (ARGs), thus improving overall water quality.

While there is a limited number of studies investigating its role, it can be concluded that the microbial communities within the biofilm are notably more diverse compared to the suspended biomass. The selection of specific microbial communities may depend on both the surface properties of the membrane and the microbial characteristics themselves. Therefore, a comprehensive characterization of the biofouling layer, including the analysis of EPS and microbial community composition, is necessary to gain a thorough understanding of microbial interactions and their role in the biotransformation of pharmaceuticals within the biofilm.

In recent years, considerable progress has been made in research on pharmaceutical removal and fate in AnMBR. Previous studies have focused primarily on the efficiency and rate of pharmaceutical removal, with limited understanding of the removal mechanisms of different pharmaceuticals and key microbial communities. Therefore, future research should focus on elucidating the major biological transformation pathways and formed metabolic intermediates, and carefully evaluating the potential toxicity and fate of intermediate products in the biological transformation process to comprehensively understand their environmental impact. Understanding the key microbial communities is crucial for designing and operating effective AnMBR systems for the treatment of pharmaceutical wastewater.

To achieve stable operation and efficient pharmaceutical removal in AnMBR systems, optimization of reactor configuration and implementation of effective operational strategies are essential. These strategies aim to maintain a dynamic balance between biofoul growth and degradation. Recently, research has also focused on prediction and modeling techniques to effectively control membrane fouling in AnMBR systems. This innovative approach provides a valuable strategy for optimizing operational conditions and enhancing the overall performance of AnMBR systems in pharmaceutical treatment [80].

Li et al. [63] investigated the impact of benzothiazole on the operational performance and microbial community structure of an integrated anaerobic fluidized bed membrane bioreactor treating benzothiazole wastewater (with a benzothiazole concentration range of 1–50 mg/L). The addition of benzothiazole had an adverse effect on the accumulation of volatile fatty acids (increasing from 10.86 to 57.83 mg/L) and membrane fouling (reducing the service period from 5.9 days to 5.3 days). The removal efficiency of benzothiazole was 96.0%. Biological degradation was the main pathway for benzothiazole removal.

The variations in effluent COD concentrations can be observed in different phases (Fig. 3.9). After 58 days of operation, the COD removal efficiency of AFBR was significantly maintained above 93% (effluent COD below 230 mg/L), which suggested the completion of phase I. At the beginning of phase II, the influent benzothiazole concentration was low (1 mg/L), and its inhibition to microbial activity was not obvious, so that the COD removal efficiency (approximately 93% in the AFBR) was similar to that of phase I. The COD removal efficiency of the AFBR decreased continuously from 89.1% to 80.9% when the influent benzothiazole concentration increased from 20 mg/L to 40 mg/L. With 50 mg/L benzothiazole, COD efficiency of the AFBR increased to 84.8%. In phase III, the COD removal efficiency of the AFBR increased further to 87.3%, which indicated the reactor had entered the stable phase of adaptation (Fig. 3.9) [63].

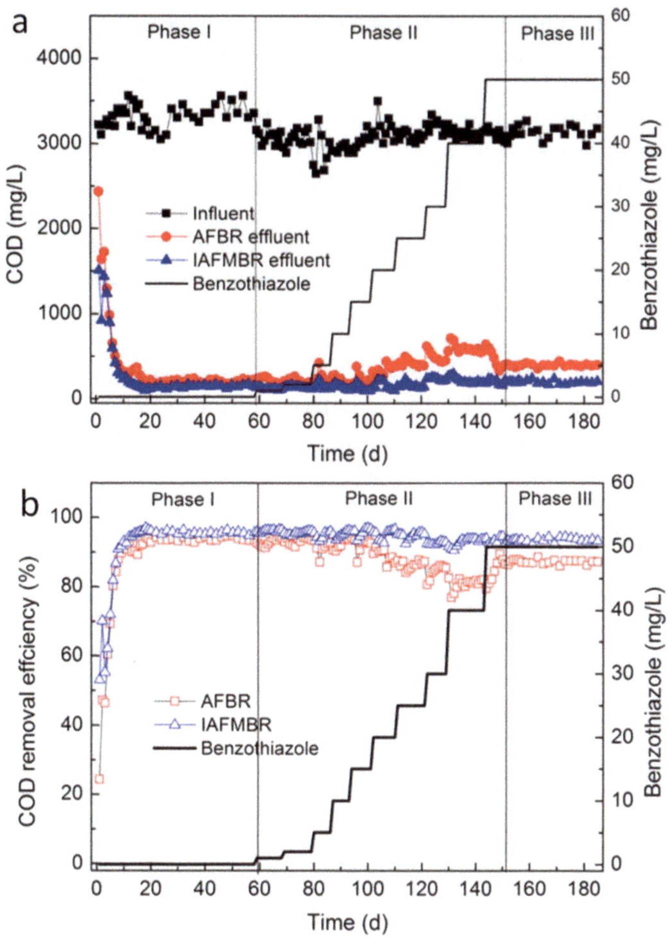

Fig. 3.9 COD removal performance in different reactor operation phases, **a** variations of COD concentration, **b** variations of COD removal efficiency

Meanwhile, Li et al. [63] found that the presence of benzothiazole increased the effluent COD of an anaerobic/anoxic/oxic sequencing batch reactor from 210 mg/L to 397 mg/L. This result is consistent with the finding that when the influent concentration of tetracycline reached 8, the COD removal efficiency decreased from over 96% to only 9%. In activated sludge systems, COD removal mainly depends on the metabolic function of bacteria. The presence of antibiotics severely inhibits bacterial growth, resulting in deteriorating effluent COD.

A plausible explanation is that the high biomass density in the biological treatment of antibiotic-contaminated wastewater makes bacteria more prone to acquiring resistance and antibiotic resistance genes more likely to proliferate. Aydin et al. [13]

also made similar findings when using real-time polymerase chain reaction technology to detect antibiotics in wastewater. They found that the addition of BTH led to an increase in VFA accumulation (from 30.21 mg/L COD to 143.47 mg/L COD) when the concentrations of sulfamethoxazole (10–15 mg/L) and tetracycline (0.5–1.5 mg/L) increased. Aquino and Stuckey [9] found that the toxic effects in anaerobic systems usually manifest in the accumulation of intermediate VFAs, which can cause an increase in wastewater COD. Although the accumulation of VFAs is evident, these substances only account for 1% of the COD. The remaining increase in effluent COD may be due to SMP, which respond to environmental stress, such as that caused by toxic compounds. Moreover, SMP constitutes most of the effluent COD.

Li et al. [63] demonstrated the feasibility of treating high-concentration BTH wastewater with an integrated anaerobic fluidized membrane bioreactor (IAFMBR). The results showed that BTH had an adverse effect on VFA accumulation and membrane fouling. However, it did not affect the COD removal efficiency, methane percentage, or methane production rate of the IAFMBR. More importantly, a high removal efficiency of BTH was achieved. The biodegradation capacity of BTH was significantly enhanced, and its biodegradation kinetics followed a first-order kinetic model. The dominant phylum was Firmicutes, and the dominant genera were Bacillus and Clostridium, *Methanosaeta* was the dominant genus in archaea.

3.6.2 Municipal Wastewater

Stable and abundant MWW is increasingly recognized as a valuable resource for the generation of clean water, bioenergy, and nutrients. However, the effective recovery of bioenergy and biological resources from MWW presents challenges in conventional aerobic and anaerobic biotechnologies due to technological limitations. AnMBR, a technology that combines the benefits of anaerobic digestion and membrane filtration, holds significant potential in overcoming these limitations.

MWW typically possesses a low organic strength, making direct anaerobic treatment impractical. The application of anaerobic digestion for low-strength wastewater treatment faces several obstacles, including the need for substantial capital investment, substantial methane loss in the wastewater, and low efficiency in methane production.

By integrating anaerobic digestion with membrane filtration, AnMBR offers a promising solution to address these challenges. The membrane component helps in retaining biomass and facilitating the separation of solids, while the anaerobic digestion process efficiently treats low-strength wastewater. This combination enhances organic matter degradation, minimizes methane loss, and maximizes methane production efficiency.

Overall, AnMBR presents a compelling approach for the treatment of low-strength municipal wastewater, providing opportunities for sustainable bioenergy generation and resource recovery.

To improve the efficiency of wastewater treatment systems, various technologies have been implemented to treat low organic concentration wastewater. These technologies include co-digestion of urban wastewater, concentration of wastewater flow through the bioadsorption-settling process (BSP) to enhance organic concentration, and pre-concentration of urban wastewater using membrane-based methods such as FO. A new type of BSP can concentrate domestic wastewater in a short HRT of 60–120 min, enhancing the bioadsorption process while maintaining an infinite SRT, thereby minimizing the cost of sludge treatment and disposal. The shorter hydraulic retention time facilitates the adsorption of organic matter onto the microbial surface, resulting in concentrated organic matter effluent [10].

The characteristics of MWW are affected by various factors, including economic conditions, residents' living habits, seasons, and climate, resulting in large material fluctuations and wide spatial and temporal distributions. The concentration of organic matter (expressed as Total COD) in real MWW ranges from 200–900 mg/L, with the most common concentration being 400–500 mg/L. Particulate COD in real MWW accounts for 44–84% of Total COD and belongs to the range of colloidal and suspended particles COD. Therefore, MWW is a complex substrate with a high COD proportion and moderate biodegradation ability. Under similar operating conditions, due to differences in biodegradability, the HRT required for treating RMWW is usually longer than that for treating synthetic MWW, typically 5–20 h [52].

3.6.3 Landfill Leachate

In recent years, the amount of municipal waste generated globally has increased significantly, causing many economic and environmental problems. Landfills in many cities around the world are filled with untreated solid waste almost every day, especially in developing countries. Globally, about 70% of solid waste is disposed of in open dumps or landfills, which may have negative impacts on the environment and health [2]. Sanitary landfill, as a convenient, economical, and widely used waste disposal method, is more popular than traditional methods such as composting and incineration. However, leachate from landfills, which is formed when rainwater or surface water infiltrates the landfill, contains a large amount of waste such as ammonia nitrogen, heavy metals, inorganic salts, and organics, posing a serious threat to the local water and soil environment [77].

One of the issues associated with landfill sites is the generation of leachate. Leachate can contain various insoluble components, such as oil in suspended solid form, which can infiltrate the soil, surface water, and groundwater, causing pollution. Therefore, it is strongly recommended to remove pollutants from leachate before its final discharge. The characteristics of leachate undergo significant changes over time and are influenced by factors such as site moisture content, temperature, and waste composition. The volume of leachate generated primarily depends on environmental factors such as rainfall, permeability, and surface runoff [2].

Biological treatment methods have been found to be very efficient for the treatment of young landfill leachate but not for mature landfill leachate (Fig. 3.10). Anaerobic digestion is an efficient biological treatment method for the removal of organic matter from young landfill leachate, due to its high BOD/COD ratio. However, anaerobic digester systems frequently become acidified due to acidification of the substrate, especially with young landfill leachate [68]. The addition of suitable additives to the anaerobic digestion system can address these problems, significantly enhancing digester system stability, organic degradation, biogas yields, electron transfer and syntrophic metabolism [76].

The quality of waste leachate is primarily influenced by factors such as its age and specific properties, including COD, BOD, pH value, floating solids, and ammonia content. These parameters play a crucial role in determining the overall characteristics and potential environmental impact of the waste leachate. It is reported that fresh leachates younger than 5 years old have the highest COD value (> 10,000) and contain higher organic matter content. Consequently, treating waste leachates early is a promising approach to maximize biogas production and may contribute to a more sustainable process [84]. Waste leachates can be treated using a variety of technologies. One such technology is leachate treatment using a membrane bioreactor. It has been reported that MBR has been used to treat industrial wastewater on an industrial scale, as well as its use for waste leachate treatment [2].

The AnMBR is a highly promising technology for the treatment of permeate, demonstrating significant advantages compared to traditional anaerobic and aerobic techniques. In recent years, there have been extensive studies on the performance of AnMBR in waste leachate treatment. For example, Lin et al. [66] studied the performance of AnMBR in treating general waste water. Xie et al. [118] conducted studies on the use of AnMBR for waste leachate treatment and studied the microbial communities in these systems in detail [74].

The quality of leachate from landfill depends on its generation time, while its quantity is affected by factors such as climate and rainfall. The quality of this liquid is mainly determined by its characteristics such as COD, BOD, pH, suspended solids, and ammonia content. Studies have shown that young leachate generated within 5 years has the highest COD (> 10,000) and higher organic content. Therefore, treating the leachate at an early stage of its formation can efficiently treat young leachate with higher organic content, and produce more biogas, thus maximizing the biogas production [84].

AnMBR is a special biological treatment process that utilizes membrane separation of the activated sludge and the untreated solution under anaerobic conditions to treat organic matter. Some researchers have studied the performance and technology of AnMBR in treating leachate. AnMBR technology can be utilized in conjunction with other processes, such as UF, NF, and RO, or in combination with chemical processes to enhance leachate treatment efficiency. However, relying solely on the MF process is not advisable. Ultrafiltration can be employed to separate organic matter and assess the molecular weight of organic compounds in leachate. Nonetheless, it is important to consider that the use of ultrafiltration may lead to clogging of reverse osmosis membranes.

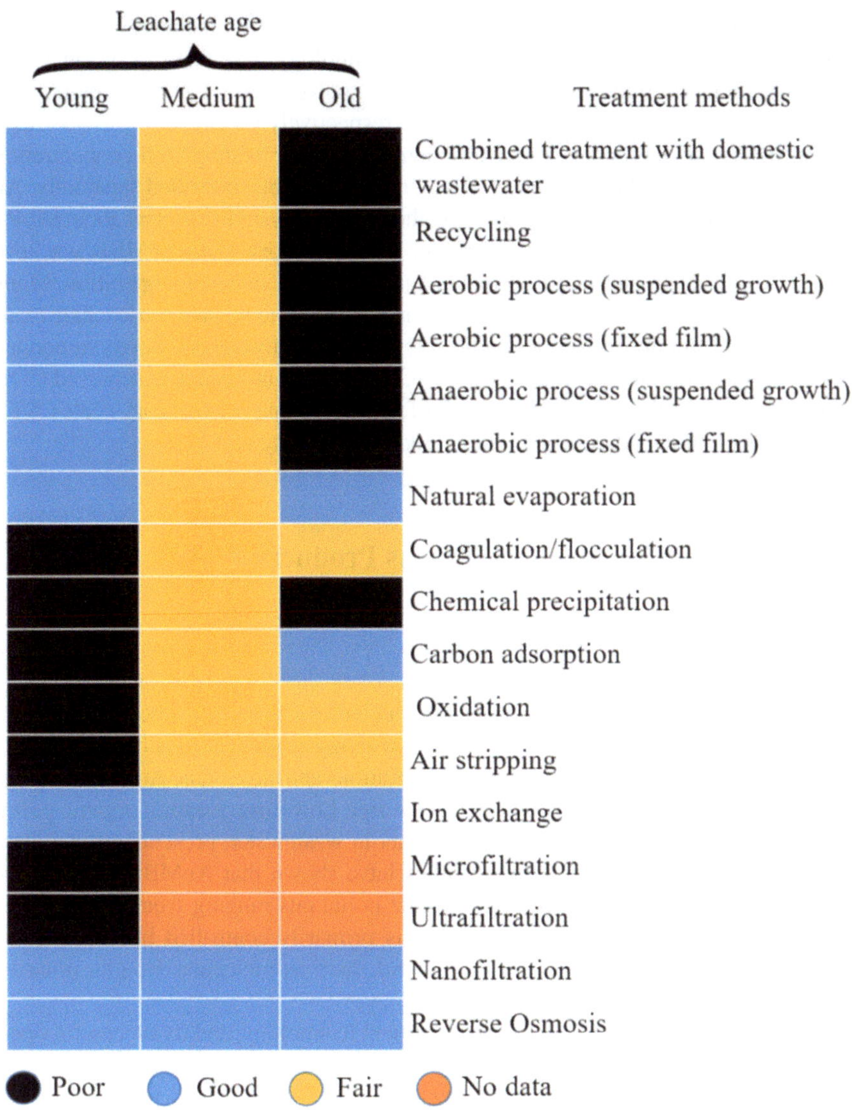

Fig. 3.10 Efficiency of various treatment methods according to the degree of ageing of landfill leachate

NF is a multifunctional method that can improve water quality by limiting the transmission of organic, inorganic, and microbial pollutants. Using NF, the transmembrane pressure is between 6 and 30 bar, with an average speed of 3 m/s, and can remove nearly 50% of ammonia and 60–70% of COD, suitable for any geometric shape (such as flat, tubular, or helical). NF is usually used in conjunction with physical methods. It has been reported that NF is effective in removing difficult-to-treat

COD in leachate, with a removal efficiency of 70–80%. RO technology is another promising and effective leachate treatment technology developed in recent years. RO can efficiently remove heavy metal concentrations and COD parameters, with removal efficiencies of up to 99% and 98%, respectively [2].

AnMBR system is considered a reasonable choice for treating high-concentration wastewater such as leachate, where HRT and SRT can be controlled separately. An AF removed about 21.5% of COD, while the AnMBR system removed about 90.4% of COD when treating the same leachate. The laboratory-scale AnMBR was also tested for consistent and stable permeate quality during continuous operation. Many researchers have summarized the application of AnMBR technology in leachate treatment under different operating conditions. Enzymatic hydrolysis is recommended at moderate or high temperatures (37–55 °C). Trzcinski and Stuckey achieved COD removal rates of 90% and 79–95% (30dSRT, 1.5dHRT) in the laboratory-scale CSTR with SRTs of 300d and HRTs of 1d and 30d, respectively [105].

3.7 Pollutant Removal and Biogas Production

3.7.1 Pollutants Removal

In recent years, AnMBR has been widely studied, particularly in the field of water reuse. AnMBR is an effective method for treating wastewater with high organic content, and as such, numerous pilot demonstrations and large-scale AnMBR systems have been used for treating industrial wastewater. Low-concentration organic pollutants are among the most complex pollutants in wastewater, presenting significant challenges for water reuse. Recent studies have shown that AnMBR can achieve removal rates for low-concentration organic pollutants ranging from negligible to over 90%. The reason for this variability is primarily controlled by the inherent physicochemical properties of the compound, such as molecular weight, polarity, and biodegradability [99].

AnMBR is an effective wastewater treatment technology, and its successful operation mainly depends on the efficiency of organic matter removal. Organic matter is one of the main pollutants in wastewater, and its content and types can affect the efficiency and stability of AnMBR. Organic matter removal involves multiple processes, including biodegradation, membrane fouling, reactor operation, etc. For AnMBR, biodegradation is one of the most critical factors, and its performance in AnMBR is influenced by various factors.

In AnMBR, the removal efficiency of organic matter is influenced by the influent COD concentration. Studies have shown that AnMBR can achieve over 85% COD removal efficiency in a short HRT and ambient temperature, and under appropriate operating conditions, the removal efficiency can be stabilized at over 90%. High

COD influent will lead to excessive microbial growth in the reactor and the production of a large amount of sludge, increasing the risk of membrane fouling and clogging, thereby reducing the efficiency and stability of AnMBR. Therefore, controlling the influent COD concentration is crucial for maintaining the stable operation of AnMBR.

Another factor affecting the removal of organic matter in AnMBR is the quality, type, and composition of the influent organic matter. Studies have shown that AnMBR has different removal efficiencies for different types of organic matter. For example, AnMBR exhibits higher removal efficiency for easily biodegradable organic matter (such as sugars, lactates, etc.), while the removal efficiency for refractory organic matter (such as fatty acids and aromatic compounds) is relatively low. In addition, some specific organic matter, such as antibiotics and antibacterials, can also have a negative impact on the performance of AnMBR.

The operational conditions of the AnMBR reactor play a significant role in the efficiency of organic matter removal. Factors such as HRT, temperature, pH, and DO levels can influence the organic matter removal efficiency in AnMBR. Research has demonstrated that shorter HRT, appropriate temperature, and pH levels can enhance the efficiency of organic matter removal in AnMBR.

Furthermore, the membrane employed in AnMBR plays a crucial role in the process. Membrane fouling and blockage can negatively impact the efficiency of organic matter removal and operational stability in AnMBR. It is essential to address and mitigate membrane fouling issues to ensure consistent performance and optimize organic matter removal in AnMBR systems.

The performance of AnMBR for COD removal was monitored during the whole operation period, and the removal efficiency was compared before and after graphite addition by Nabi et al. [78]. To provide suitable (easy biodegradable) substrate for anaerobes, synthetic wastewater was used. For the sake of microbial acclimatization and to treat maximum proportion of landfill leachate without inhibition of microbiological process, landfill leachate was digested together with synthetic wastewater, and its percentage was gradually increased. Influent COD of landfill leachate was increased gradually from 350 mg/l to 3500 mg/l. The detail of substrate proportion (leachate + synthetic wastewater) is shown in Table 3.4. The proportion of influent leachate was increased step by step, and finally pure leachate (100%) was used in AnMBR. With the stepwise increase of leachate proportion (COD concentration) in the influent, COD removal efficiency dropped and then gradually recovered (Fig. 3.11). It can be inferred that the anaerobes were adversely affected by organic loading shock, and then gradually acclimatized with the complex organics of leachate. The threshold COD concentration of pure leachate was 2000 mg/l for the AnMBR with the removal ability of 65% (day134). The COD removal efficiency at this stage was relatively unsatisfactory, even the influent COD concentration was low (2000 mg/l). Studies showed that composition of landfill leachate is highly heterogeneous and contain recalcitrant COD, and such removal efficiency (65%) is not enough to meet the discharge standard. Graphite was added in AnMBR at day 135. Although the COD concentration was increased (Table 3.1), removal efficiency improved with graphite addition (stage H). The COD removal efficiency improved till the influent COD concentration reached 3000 mg/l, and dropped with the further increase i.e., 3500 mg/l. The highest COD removal of 75% was achieved for 100% landfill leachate, and the remaining COD might be recalcitrant. It can be concluded that the graphite addition enhanced COD removal and the optimum influent COD concentration was 3000 mg/l. Muratçobanoğlu et al. [75] added graphite in the anaerobic co-digestion system of food waste and cow manure, and found that the activities of fermentative bacteria were improved, which significantly enhanced VS removal [77].

AnMBR is particularly well-suited for treating wastewater with a high organic content. It has been successfully demonstrated in several pilot and full-scale AnMBR systems for treating effluents generated by field crop processing (such as sauerkraut, wheat, maize, soybean, and palm oil), dairy processing, and the beverage industry. These applications have showcased the effectiveness of AnMBR in treating wastewater from various organic-rich industries (Table 3.5).

In recent years, the application and research of AnMBR coupled with anaerobic ammonium oxidation (Anammox) for nitrogen removal have been constantly deepened, and some cutting-edge technologies have emerged. For example, the use of electrochcmical oxidation technology can effectively remove sludge and enhance the activity of Anammox microorganisms. The use of high-pressure membrane reactors can increase the pressure in AnMBR reactors, improve the concentration of ammonium nitrogen and the oxygen diffusion rate, and thus improve the nitrogen removal efficiency of the reactor. In addition, some new carrier materials such as nanoparticles and microbial carriers have also been introduced into the AnMBR reactor to improve its stability and denitrification efficiency [48]. The core of the AnMBR coupled with Anammox for nitrogen removal is the direct conversion of ammonium nitrogen to nitrogen gas by Anammox bacteria. Anammox bacteria use nitrite reductase and nitric oxide reductase to reduce NH^{4+} and NO^{2-} to N_2 and H_2O. Due to their relatively low growth rate, Anammox bacteria require control of conditions such as aeration rate,

Table 3.4 Influent composition at different stages

Stages	Leachate COD	Synthetic wastewater COD	Total COD
A	150	–	150
B	–	2000	2000
C	200	1800	2000
D	500	1500	2000
E	800	1200	2000
F	1200	800	2000
G	1200	400	1600
H	2000	–	2000
I	2500	–	2500
J	3000	–	3000
K	3500	–	3500

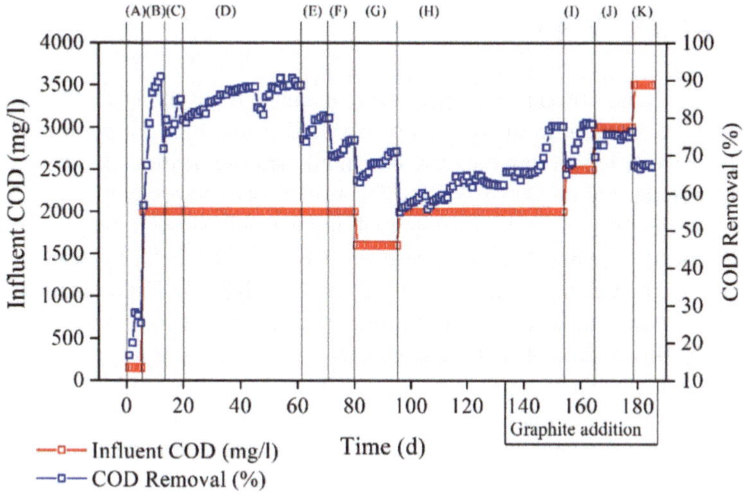

Fig. 3.11 COD removal efficiency of AnMBR

Table 3.5 Examples of AnMBR performance regarding organic removal and methane production

Wastewater	Organic removal (COD, %)	Methane yield (L/kg COD)	Reference
Food wastewater	81–94	136	[49]
Kraft evaporator condensate	97–99	290–310	[64]
Landfill leachate	90	460	[124]
Real municipal	86–88	300	[123]
Pre-concentrated synthetic wastewater	96	223	[26]
Meat packing wastewater	88–98	130–180	[37]
Raw tannery	90	160	[106]
Synthetic wastewater	> 98	290	[17]
Malting wastewater	90.2–94.1	308–345	[70]

temperature, and pH to promote their growth and metabolic activity in the reactor. In addition, suitable substrates (such as acetate and propionate) and carbon sources need to be introduced into the AnMBR reactor to meet the growth and metabolic demands of anaerobic ammonia-oxidizing microorganisms [48]. In summary, these factors are interrelated and interact to affect the efficiency of AnMBR coupled with anammox bacteria. In practical applications, the reactor's operating parameters need to be controlled reasonably to maximize the advantages of this technology, improve reactor performance, and achieve efficient nitrogen removal [52].

3.7.2 Biogas Production

Investigating biogas production, specifically methane recovery, is a crucial area that requires more research in the context of AnMBR technology for wastewater treatment. Effectively recovering methane has significant implications for the economic and commercial viability of AnMBR. The availability of approximately 250 peer-reviewed review papers, with a substantial number published in the past six years, underscores the growing interest and research activity in this field. These publications contribute to the existing knowledge base and highlight the importance of further exploration and advancements in biogas production and methane recovery in AnMBR systems [2]. Biogas is a renewable energy source that can be produced through anaerobic digestion of various types of biomass. The entire process of biogas production not only converts biomass into energy, but also has multiple environmental benefits. Studies have shown that biogas production can reduce carbon emissions, reduce pathogens, and better manage waste odors. Compared to other forms of renewable energy, the biogas production process is considered relatively easy to manage and predict results accurately. In addition, biogas production effectively promotes the concept of a circular economy, which revolves around the basic principles of waste reuse and recycling, driving research on the circular economy. As a result, the circular economy provides an effective solution to resource scarcity and promotes sustainable development of society [1].

AnMBR is a highly attractive technology for both research and industrial sectors due to its ability to integrate the benefits of anaerobic processes and membrane technology. Extensive studies have demonstrated the efficacy of AnMBR in treating diverse wastewater types, including leachate, with notable advantages in energy recovery. It is worth noting that the observed methane production in AnMBR systems typically falls within the range of 0.23–0.33 $LCH_4/gCOD_{removal}$. However, it is important to acknowledge that methane has a high solubility, which often leads to lower methane production compared to the theoretical methane production of 0.382 LCH_4/gCOD at 25 °C [73].

Biogas is the general form of chemical energy conversion in wastewater within AnMBR and consists of CH_4, H_2, CO, N_2 and a small amount of hydrogen sulfide (H_2S). In general methane is the main component of biogas, accounting for 70–90% of the total gas. The recovery of methane is important for both the alleviation of energy demand constraints and the reduction of carbon emissions, and its production and conversion rate are important indicators for evaluating the effectiveness of AnMBR operation. Theoretically, OLR will produce more methane, and increasing the OLR of AnMBR from 1 kgCOD/($m^3 \cdot d$) to 2 kgCOD/($m^3 \cdot d$) increases the system methane production from 0.13 to 0.95 L/d. However, at this stage, it is common that the actual methane production rate is lower than the theoretical value. The inhibition of microbial activity by low temperature, the dissolution of methane, and the presence of inhibitory substances such as sulfate may all contribute to this discrepancy. Under laboratory conditions that do not take into account these inhibitory conditions, AnMBR is able to achieve high methane yields of 0.30–0.35 $LCH_4/g\ COD_{removed}$.

Therefore, the recovery of AnMBR energy in municipal wastewater treatment still has a large potential and room for development, and it is of great importance for the development of the process to further improve the methane yield of AnMBR and achieve efficient recovery of dissolved methane in wastewater in subsequent studies.

Biogas, a renewable energy source derived from the anaerobic digestion of diverse biomass, offers several pathways for utilization. These include cogeneration engines for heat and electricity generation, substitution for natural gas, and utilization in fuel cell technology. It is worth noting that, in terms of cost competitiveness, electricity production from biogas is not on par with other fossil fuel-based alternatives. However, electricity generated from biogas holds potential for sale as green power or qualification for tariff exemptions provided by authorities. In the case of AnMBR, one of the many advantages of the produced biogas is that it can offset the energy demand required for operating wastewater treatment plants. Biogas production is controlled by controlling the number of methane-producing bacteria through controlling HRT and SRT to achieve biogas control. Biogas from AnMBR contains approximately 80% methane, but the methane content is proportional to the organic load rate. Therefore, AnMBR reactors require proper monitoring and management to optimize methane production and wastewater treatment efficiency.

Recovering methane from AnMBR is challenging because the methane in the effluent can dissolve. Approximately 67% of methane dissolves in the effluent of AnMBR, which means measures need to be taken to improve methane recovery. Meanwhile, the removal rate of COD in the wastewater throughout the AnMBR reactor process can reach 86–88%. This means that the AnMBR reactor can not only produce high-quality biogas but also achieve efficient wastewater treatment [123].

The solubility of methane is also significantly affected by the operating temperature. It has been reported that the solubility of methane produced at 15 °C is approximately 1.5 times higher than that at 35 °C. In a typical biogas plant, the CH_4 content is around 70%, while biogas from AnMBR has methane content ranging from 70 to 90% and other impurities such as nitrogen (0–15%) and carbon dioxide (3–15%).

In AnMBR systems operated with synthetic wastewater, it was estimated that removing 1 kg of COD can generate about 2 kWh of energy, which is approximately seven times the energy required to run the system. Dissolving CH_4 in wastewater is dangerous due to methane's greater global warming potential than CO_2. To recover CH_4 from effluent, methods such as forced ventilation aerators, vacuum filling, or bubble covers can be used. Dissolved CH_4 poses a significant challenge in terms of recovery, but membrane contactors have shown promise in addressing this issue, albeit being in the early stages of development. Thermophilic anaerobic membrane bioreactors offer several advantages over other types of anaerobic membrane bioreactors, including shorter reaction times and higher loading rates, resulting in increased biogas production. Given that methane fermentation occurs at a pH range of 6.5–8.5, AnMBR systems typically operate at a neutral pH. However, many studies on AnMBR for industrial wastewater treatment have overlooked the accumulation of dissolved methane in the permeate, which is particularly important in the treatment of dilute wastewater. Moreover, further research is needed to adequately explore the potential of AnMBR for biogas recovery from landfill leachate [52].

The utilization of biogas from AnMBR is predominantly observed on a large scale in urban wastewater treatment plants, while its application on a smaller scale for leachate treatment remains limited. The use of biogas from AnMBR in leachate treatment is still in the early stages, with limited adoption. Empirical results indicate that the methane content of biogas derived from AnMBR using leachate ranges from 70 to 90%, with a methane production rate of 0.34 L/g $COD_{removed}$ [118].

In the case of leachate, it can contain elevated concentrations of organic pollutants, including ammonia, which can lead to biogas with higher levels of contamination. A significant challenge in leachate treatment lies in the presence of refractory compounds whose composition can vary over time. These compounds pose a risk to the overall sustainability of the treatment process [2].

Methanogens are hydrogenotrophic methanogens that utilize H_2 and CO_2 to produce methane. Graphite additives can promote the growth of methanogens, improve the hydrogenotrophic process, and create a more favorable environment for the growth of acetate-degrading methanogens. In anaerobic digestion systems, graphite additives can especially enhance the accumulation of volatile fatty acids (VFA), increase methane production, and ensure stable reactor operation. After studying microbial reactions with the addition of graphite, it was found that Clostridia, *Methanosarcina*, and *Methanosaeta* were significantly enhanced in methane generation. Although graphite additives can increase methane production, excessive addition can lead to the overproduction of hydrogen and VFA in the reactor, thereby reducing reactor stability. Therefore, the optimal amount of graphite additives needs to be optimized for maximum effectiveness [77].

The concentration of COD in leachate has an impact on the composition of biogas, particularly the levels of H_2 and CO_2. Higher COD concentrations can lead to the inhibition of methanogenic bacteria and the increased activity of fermentative bacteria. In a study conducted by Guo et al. [47]. The performance of two different methods, one using graphite and the other using regular quartz sand, was compared in the treatment of low-concentration wastewater using UASB. The results demonstrated that the methane production rate in the graphite-enhanced reactor was 34 times higher than that in the regular quartz sand-enhanced reactor. The increase in methane content in biogas indicates that the addition of graphite significantly enhances the potential for biomass energy recovery in the AnMBR system.

AnMBR showed generally good organic matter removal performance and methane production efficiency when used to treat municipal wastewater. However, most of the available studies have an HRT of more than 6.0 h. There are few studies on the treatment performance at shorter HRTs, and the characteristics of organic matter composition of AnMBR effluent are rarely reported. loading, performance degradation, etc. Estrada-Arriaga et al. [35] investigated the organic removal and methane production performance during the shortening of HRT from 24.0 h to 3.2 h at 25 °C based on AnMBR system through experiments, and the experimental results found that the biogas production gradually increased with the shortening of HRT and reached 6.0 L/d at HRT 3.6 h. The biogas production per unit volume of the reactor reached 1.0 L/(L-d). The CH_4 content in biogas was maintained above 83% and CO_2 content was about 4% in the stable operation stage, which was significantly lower than the proportion of CO_2 in biogas of conventional anaerobic fermentation (40%), which was due to the biogas cycle promoting the dissolution of CO_2 in the mixture and the reduction of CO_2 to methane with lower content. The N_2 content in the biogas was around 13%, which mainly originated from dissolved N_2 in the feed water, indicating that AnMBR could achieve higher methanogenic capacity and efficient and stable COD removal than the conventional anaerobic biotechnology. 95% COD removal and 80.1–91.2% COD methanation efficiency were achieved, respectively.

In order to reduce solids accumulation and increase the conversion rate of biological methane, experiments were conducted in AnMBR systems by adding hydrolytic enzymes (a mixture of protease, lipase, and amylase). The study found that the addition of enzymes did not have a negative impact on the quality of the AnMBR permeate, and even slightly improved its quality. In addition, as particle concentration decreased, methane production increased, providing important guidance for optimizing AnMBR systems [102]. In AnMBRs treating wastewater rich in sulfates, the presence of sulfides can lead to the transformation of Fe^{2+} into a form that is less available for microorganisms, thus limiting its degradation of organic matter and methane production. Therefore, adding biodegradable chelating agents such as ethylenediamine-N,N′-disuccinic acid (EDDS) to the bioreactor can improve methane production. This is because the Fe-EDDS complex helps maintain the biological availability of Fe^{2+}. Additionally, after adding EDDS to ANMBRS, the COD concentration in the permeate was slightly lower than that without EDDS (12.7 mg/L and 17.2 mg/L, respectively), which may be due to the improvement of organic matter biodegradation in the presence of EDDS [103]. Therefore, the quality of the permeate in AnMBR with the addition of chemical substances can meet the requirements of wastewater reuse [103].

AnMBRs typically operate in a closed system, leading to a high concentration of methane in the effluent and resulting in energy wastage. When AnMBRs treat domestic wastewater with low COD concentrations, it is common for 40–60% of the total methane produced to be dissolved in the membrane permeate. To effectively recover dissolved methane, membrane contactors are in high demand, and AnMBR compact design and ease of operation make it feasible. Microporous hollow fiber membranes are often preferred as they offer higher gas desorption rates compared to non-porous membranes. In this case, concentration or partial pressure gradients are employed to promote the desorption and diffusion of dissolved gases through membrane pores. This is achieved by utilizing sweeping or vacuum conditions on one side of the membrane and countercurrent flow of the effluent feed on the other side. Commercial polypropylene membranes have been used for methane recovery due to their high hydrophobicity. However, pore wetting can diminish membrane performance. To enhance methane recovery, new membrane materials are being

Table 3.6 Summary of methane production and system performance of AnMBRs

Wastewater	Configuration	Temperature (°C)	HRT	Influent COD (mg/L)	COD removal (%)	Reference
Wastewater	Submerged	25	15–40	460	96	[23]
Municipal wastewater	Submerged	25	8	670 ± 100	–	[24]
Food	Submerged	35	20–120d	3000–2700	94 ± 8	[20]
Landfill leachate	Submerged	37	–	2.78–3.35	90	[42]
Domestic	External	37	15–60	685	55–90	[90]

developed, and polymer-fluorinated silica composite membranes serve as a promising example [114].

In conclusion, AnMBR reactors have enormous potential for producing high-quality biogas and efficiently treating wastewater. However, appropriate monitoring and management are required to achieve optimal results. Therefore, AnMBR reactors will become a more common biogas production and wastewater treatment technology in the future (Table 3.6).

3.8 Membrane Fouling

3.8.1 Membrane Fouling Mechanism

Membrane fouling is a significant challenge in the widespread implementation of AnMBR technology. It necessitates additional maintenance and cleaning, leading to increased operational costs, particularly during long-term treatment periods, which further affects operating expenses. The mode of operation, choice of membrane material, and membrane type employed all directly influence membrane fouling. These factors are key determinants of the size and quantity of foulants, including solutes, colloids, and microbial cells. Membrane fouling refers to the phenomenon where sludge flocs, colloidal particles, dissolved organic matter, inorganic salts, and various macromolecules adsorb, settle, and accumulate on the surface and within the membrane in the mixed liquor during reactor operation, resulting in an increase in transmembrane pressure or a decrease in membrane flux. Membrane fouling can lead to reduced membrane operating flux, directly impacting the operation and maintenance of the entire AnMBR system. It is one of the primary factors that limit the widespread adoption and application of AnMBR technology. The activities of microorganisms usually generate surplus sludge, and non-biodegradable pollutants in the wastewater also adsorb onto the sludge. The main substances causing AnMBR membrane fouling come from the sludge mixture. During the operation of the reactor, the membrane fibers are immersed in the sludge mixture, forming a dense filter cake layer. During AnMBR operation, microorganisms present in the mixed liquor can penetrate the membrane pores and proliferate on the inner surface of the membrane fibers, leading to the aggravation of membrane fouling. Consequently, the characteristics of the sludge mixture play a crucial role in determining the extent of membrane fouling in AnMBR systems. In the AnMBR process, the complex composition of the mixed liquor is the main cause of membrane fouling. The interaction between the membrane and the sludge mixture leads to membrane fouling. Microbial metabolites EPS and SMP have almost the same composition, which helps the aggregation and stability of sludge in reactor operation. However, a high EPS content can increase the viscosity of the mixed liquor, forming sludge flocs, adversely affecting microbial activity and causing membrane fouling. When the EPS content is too low, organic matter deposits and adsorbs on the membrane surface. In addition, Ca^{2+} combines with EPS to form flocs, but a high concentration of Ca^{2+} can cause inorganic membrane fouling. Nitrifying bacteria and filamentous bacteria are key microorganisms that produce high EPS and SMP and contribute to membrane biofouling. The operating conditions of the reactor also affect the production of SMP and EPS. For example, an appropriate SRT can reduce their production and alleviate membrane fouling. However, an SRT that is too long can lead to high sludge concentration, increased viscosity, and hindered mass transfer, resulting in severe membrane fouling. The SRT range varies for different types of wastewater, reflecting the influence of substrates. For example, the typical SRT range for synthetic wastewater is 20–40 days, while for leachate, it may exceed 125 days [91].

Potential foulants encompass inorganic precipitates, SMP, EPS, and cell. SMP refers to soluble proteins, polysaccharides, and humic-like substances, while EPS comprises insoluble carbohydrates, proteins, lipids, and nucleic acids embedded within highly hydrated gel-like matrices. These substances contribute to a reduction in permeate flux by diminishing the effective pore size of the membrane and augmenting frictional resistance to permeate flow [79].

In AnMBR systems, certain microorganisms, such as γ-proteobacteria, have the ability to colonize and form a biofouling layer on the membrane surface. This type of fouling occurs as a result of microbial metabolism, where metabolic by-products accumulate and adsorb onto the membrane surface, leading to the formation of a gel or cake layer and subsequently causing membrane fouling. Membrane fouling can take the form of inorganic fouling, characterized by the accumulation of crystals, particles, and colloids on the membrane surface or within the membrane pores, or organic fouling, which involves the build-up of organic matter and macromolecules (polymers) on the membrane surface. Through Fourier transform infrared spectroscopy (FTIR) and confocal laser scanning microscopy (CLSM) analysis of the cake layer, [66] found that organic fouling is mainly caused by SMPs and EPSs. In AnMBR systems, $MgNH_4PO_4 \cdot 6H_2O$ is a common inorganic fouling agent, while $K_2NH_4PO_4$ and $CaCO_3$ are also known inorganic fouling agents. The coexistence of biological and non-biological fouling further increases the filtration resistance. Martin Vincent et al. [72] found that compared to inorganic fouling, organic fouling is more prone to interception and deposition in the cake layer, leading to pore blocking. Therefore,

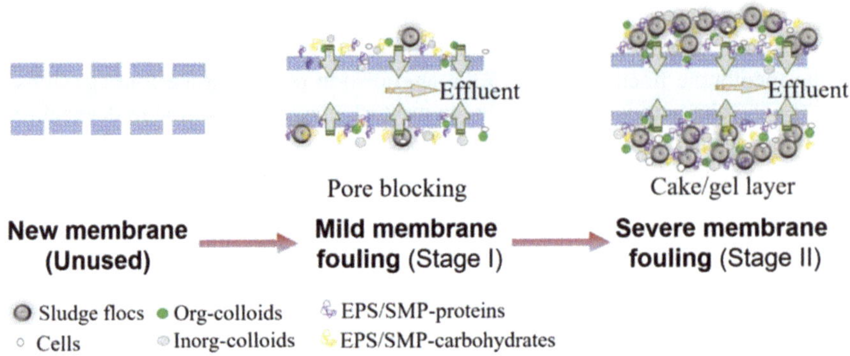

Fig. 3.12 Stage process of membrane contamination by EPS/SMP

membrane fouling typically starts with pore blocking, and then microorganisms begin to accumulate [78] (Fig. 3.12).

3.8.2 Microbial Community Structure and Membrane Fouling

While numerous factors influence the growth of biofilms, our understanding of how changes in community biodiversity reflect their growth status in response to environmental changes is limited. Recent research has highlighted the significant link between community evenness and functional stability, suggesting that evenness may be vital in governing biofilm growth. This study aimed to explore the influence of different evenness levels on biofilm growth rates by manipulating community evenness in a mesocosm experiment using selective stresses. The objective was to gain insights into the importance of community evenness in shaping the development of biofilms [40].

The transfer of microbial communities is the key to membrane fouling. In stable a AnMBR operation, when the microbial community is low in quantity and uniformity, membrane fouling can occur smoothly or slightly. Gao et al. [44] investigated the influence of temperature on microbial communities and membrane fouling in AnMBR treating wastewater. Experiments were conducted under different temperatures and concentrations. The results showed that membrane fouling became more severe at low temperatures, which was directly related to microbial activity and metabolic products. Changes in microbial community diversity were reflected by the size of the Shannon index. In addition, as the temperature increased, the diversity of microbial communities also increased. At 15 °C, 25 °C, and 35 °C, the Shannon index of microbial communities varied between 0.820–1.370, 0.963–1.846, and 2.218–2.995, respectively. Furthermore, the Gini coefficient can be used to evaluate the uniformity of microbial communities in the cake layer, which changes at different temperatures. The Pearson correlation analysis also showed a positive correlation between microbial community uniformity and TMP at a 95% confidence level [44].

EPS plays a crucial role in AnMBR systems, impacting the settling, flocculation, hydrophobicity, adhesion, and dewatering characteristics of sludge flocs, thereby influencing membrane fouling. Analysis of protein and polysaccharide content in loosely bound EPS (LB-EPS) and tightly bound EPS (TB-EPS) revealed interesting findings. In anaerobic sludge and cake layers, TB-EPS exhibited higher polysaccharide content compared to LB-EPS. However, the addition of graphite resulted in a decrease in polysaccharide content specifically in TB-EPS present in the filter cake layer, while the polysaccharide content of LB-EPS remained unaffected. Furthermore, there were no significant changes in the polysaccharide content of TB-EPS and LB-EPS in anaerobic sludge. As the proportion of filtrate in the influent increased, there was a slight reduction in polysaccharides in both anaerobic sludge and the filter cake layer. Conversely, the protein content in LB-EPS increased with the increasing proportion of filtrate, exhibiting a similar trend in anaerobic sludge and the filter cake layer. Prior to the addition of graphite, LB-EPS had higher protein content than TB-EPS in anaerobic sludge and the filter cake layer, but this decreased after the addition of graphite. LB-EPS demonstrated three times higher specific filtration resistance compared to TB-EPS and showed a higher ratio of protein to polysaccharides. The decrease in protein content in LB-EPS may be attributed to microbial consumption, stimulated by the addition of graphite, which is biochemically converted into biogas or utilized for microbial growth and maintenance of metabolic functions. Notably, there were no significant changes in the protein content of TB-EPS in the filter cake layer and sludge during the AnMBR operation. These results emphasize that the protein content in LB-EPS (sludge and cake layer) and the polysaccharide content in TB-EPS (cake layer) are the primary factors contributing to membrane fouling. Furthermore, graphite exerts a significant effect on EPS composition, effectively mitigating fouling. Particularly, graphite exhibits a more pronounced influence on the composition of LB-EPS (sludge and cake layer) in terms of protein content compared to TB-EPS (cake layer) polysaccharides [77].

The EPS components of cake layer were analyzed by EEM fluorescence spectra (Fig. 3.13). There were two peaks i.e., Ex/Em = 327/400 and Ex/Em = 312/448 (Fig. 3.8a). Peak 327/400 pertain humic acid like substances, which were significantly weakened after graphite addition (Fig. 3.13b). Moreover, peak Ex/Em = 312/448 in Fig. 3.8a represented protein like substances, which were less weakened after graphite addition [77] (Fig. 3.14).

EPS is a key biological substance, which mainly regulates properties of sludge flocs such as settling, flocculation, hydrophobicity, adhesion and dewatering ability, which greatly affects membrane fouling in AMBR system. The lose-bond EPS (LB-EPS) and tight bond EPS (TB-EPS) were analyzed for protein and polysaccharide. Results showed that polysaccharide content in LB-EPS was lower than TB-EPS in both anaerobic sludge and cake layer. Moreover, polysaccharide content in TB-EPS of the cake layer decreased after graphite addition, and there was no obvious effect on LB-EPS (polysaccharides). In addition, there were no significant changes in the polysaccharide content of both TB-EPS and LB-EPS in anaerobic sludge. With the

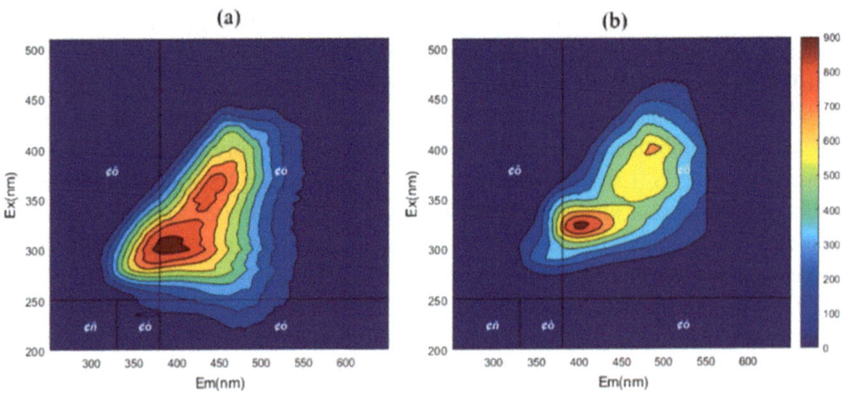

Fig. 3.13 EEM fluorescence spectra of EPS from cake layer, **a** before graphite addition, **b** after graphite addition

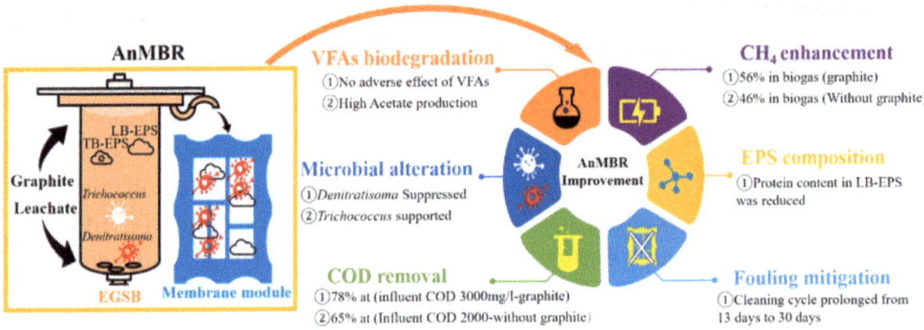

Fig. 3.14 Effect of graphite addition on overall performance of AnMBR

increase of leachate proportion in the influent, polysaccharide slightly decreased in both anaerobic sludge and cake layer [77].

Gao et al. [42] discovered a bacterial species, OP11, with a relatively high abundance in the fouling layers of two membranes. In contrast, species from the phyla *planctomycetes* and the *Lentisphaerae*, such as *Wesrepresent*, had low relative abundances in the fouling layer but high relative abundances in the suspended biomass. Similar differences were observed in other species. The results indicate that some minority species, such as OP11, attach to membrane surfaces, while others, including some that may play important roles in the metabolism of incoming organic matter, have less important or indirect roles in fouling.

In AnMBR, fouling is due to a fouling layer containing cells and the major protein EPS network. Specific populations preferentially attach to the membrane, leading to fouling. Except for one exception (the iron-reducing bacterium), the dominant cells within the fouling layer are a subset of the cells present in the suspended biomass of the bioreactor. Of particular note is the OP11 lineage within bacteria, which has a relatively high abundance in the fouling layer detected in the organism. Isolating and characterizing representatives of this species can provide useful information for biocorrosion control. Also noteworthy is the absence of *planctomycetes* and *lentisphaerae* bacteria in the fouling layer on crystalline PVDF (cPVDF) membranes, despite their abundance in the suspended biomass. Apparently, the mechanisms and rates of membrane attachment and detachment differ among different species. More research is needed to explain these differences. Finally, we note the significant differences in fouling rates and fouling layer composition between the two membranes used in this study, such as the presence of Planctomycetes on polyetherimide (PEI) membranes but not on cPVDF membranes. These observations support the value of combining membrane materials science with a better understanding of the potential for adhesion and biofilm formation by different microbial species [42].

3.8.3 Effect of Mixed Liquor Properties on Membrane Fouling

Membrane fouling is one of the main factors limiting the application of AnMBR, which can cause a decrease in membrane flux and directly affect the operation and maintenance of the entire AnMBR system. Currently, membrane fouling is generally divided into organic pollutants, inorganic pollutants, and microorganisms. Organic pollutants mainly include SMPs and EPS, of which polysaccharides, proteins, and humic substances are the main components. Due to the complexity of the mixed composition in the reactor, the formation of membrane fouling is difficult to describe using a single mechanism. However, among these pollutants, hydrophobic humic substances are easy to adsorb on the membrane surface or pores, leading to changes in the pore size distribution and increased surface hydrophobicity, which promotes membrane fouling and is considered the main cause of irreversible fouling. Polysaccharides have characteristics such as high molecular weight, gelation, and low biodegradation rate, which can easily complex with divalent or multivalent cations and form a gel layer that is easily attached by microorganisms on the membrane surface, making it a key substance for membrane fouling. At the same time, the non-covalent network structure formed by the combination of proteins and polysaccharides makes the structure and composition of membrane fouling more complex. The interaction between microbial attachment and other pollutants will promote the further development of membrane surface fouling towards the filter cake layer. It should be noted that in the initial stage of membrane fouling, SMPs play a major role, while in the later stage, EPS secreted by microorganisms become the main source of fouling. In addition, the "concentration polarization" phenomenon caused by the flow of pollutants to the membrane surface during continuous filtration is also a type of membrane fouling, which usually accumulates on the membrane surface in the form of a thin liquid layer.

The operation and maintenance of AnMBR are also affected by sludge concentration, mainly because sludge concentration directly affects factors related to membrane fouling such as the number of microorganisms in the reactor, circulation velocity, and sludge viscosity. Existing studies have shown that an increase in sludge concentration leads to an increase in the severity of AnMBR membrane fouling. In addition, the size and distribution of sludge particles also have an important influence on AnMBR membrane fouling. Generally, smaller sludge particle size is more easily deposited on the membrane surface, forming a denser cake layer, which further exacerbates membrane fouling. EPS and SMP in microbial metabolic products are mainly composed of proteins and polysaccharides. These metabolic products are easy to adsorb on the membrane surface and form a cake layer, which further exacerbates membrane fouling. It is worth noting that the impact of EPS and SMP on membrane fouling changes over time. In the early stages of membrane fouling, SMP has a more significant impact, while in the later stages, EPS becomes the main source of fouling.

Changes in TMP are closely related to membrane fouling. Deposition of biomass, microbial metabolites, and blockage by fine particles can lead to rapid TMP increase and shortened cleaning cycle time. Another reason for reducing the cleaning cycle may be related to the decrease in the abundance of functional microorganisms degrading EPS. Our study shows that the addition of graphite slows down the increase in TMP and extends the cleaning cycle to 30 days. Previous studies have shown that biomass (including EPS and microorganisms) is the main pollutant causing biological fouling. Before adding graphite, TMP increased rapidly, leading to accumulation of biomass and microbial metabolites on the membrane surface and a short cleaning cycle. These results suggest that the addition of graphite significantly slows down the rise rate of TMP and reduces membrane fouling. Membrane fouling is closely related to the chemical composition and concentration of substrates [77].

Anaerobic sludge plays a crucial role in the AnMBR system and exhibits characteristics such as high viscosity and a high concentration of MLSS. Compared to activated sludge, anaerobic sludge contains a higher content of biological polymers and inorganic matter, resulting in a higher pollution potential. The accumulation of solid particles, including small sludge particles, biological polymers, and inorganic matter, on the membrane surface is a common cause of membrane fouling in AnMBR systems. Membrane fouling can be classified into two types: cake layer formation on the membrane surface and pore blocking. Among these, cake layer formation is the primary contributor to membrane pollution, accounting for more than 80% of the total filtration resistance in most cases. Although cake layer formation leads to membrane fouling, the formed filter cake layer can act as an additional filter, providing rejection of various pollutants and pathogens. In the AnMBR system, the rejection performance of the filter cake layer primarily relies on its own characteristics rather than the underlying membrane. Therefore, inexpensive support materials such as meshes, woven fabrics, non-woven fabrics, or non-woven filter fabrics can be used as alternatives to microfiltration/ultrafiltration (MF/UF) membranes commonly employed in AnMBR systems. The establishment of a well-developed dynamic membrane layer, also known as a secondary membrane, can serve as a pre-filtering material, ensuring effective retention in aerobic dynamic membrane bioreactors (DMBRs) and anaerobic DMBRs, and offers distinct advantages such as low cost, high flux, and ease of cleaning [87].

3.8.4 Effect of Environmental and Operational Parameters on Membrane Fouling

In recent years, there have been significant advances in pollution reduction technologies for controlling the total amount of pollutants, including new AnMBR structures, optimized operating conditions, optimized gas–liquid ratios, as well as improved backwashing and chemical cleaning schemes. There are two cleaning methods based on the frequency and severity of fouling: the first is in-situ cleaning typically based on physical methods with a minimum time interval, and the second is non-in-situ cleaning usually conducted annually.

Backwashing is a common physical cleaning technique that uses the reflux principle to remove pollutants on the membrane surface. Zhen et al. studied the effectiveness of backwashing and considered various operating parameters such as backwash volume, flux, and duration. The study found that, for the same backwash volume, a higher backwash flux was more effective than a longer backwash period. Furthermore, the absolute permeability showed almost no change over time, indicating that backwashing can effectively restore the filtration performance of the membrane [127].

Anaerobic digestion gas sparging is a commonly used physical refinement technique. This technique injects gas molecules into the fluid flow passing over the membrane surface, generating shear stress that helps break up the cake layer and renew the pores. In AnMBR systems for treating domestic wastewater, gas sparging technology can enhance the system's treatment efficiency. Gas sparging is a common physical treatment technique that can improve the efficiency of AnMBR systems. When gas molecules are injected into the fluid, the resulting shear stress can disrupt the cake layer and renew the pores, thereby improving system performance. Researchers found that higher injection rates in gas-sparged AnMBR systems for treating domestic wastewater can enhance membrane permeability and flux. In addition, inducing vibration can enhance shear stress on the membrane surface, but the frequency and amplitude of vibration may be limited. To overcome these limitations, a new system called Magnetically Induced Membrane Vibration (MMV) has been developed and proven to be highly effective in overcoming amplitude, energy consumption, and resistance issues [60].

Chemical cleaning is a method for controlling membrane fouling. Different chemical agents, such as acids and alkalis, are used to reduce membrane decay and improve filtration performance, usually in combination with backwashing. However, optimizing the dosage of chemicals and evaluating their impact on the membrane surface remains a challenge that requires further research. The effects of physical and chemical cleaning on PVDF membranes have been determined, but for other types of membranes, the impact may be different.

Membrane fouling can result in a significant decrease in permeate flux, an increase in transmembrane pressure, and a loss of membrane processing efficiency. Depending on the type of contamination, fouling can be classified into several categories, including the location of fouling (internal/external), the nature of fouling (organic/inorganic), and fouling forms (reversible/irreversible, recoverable/residual). There are four fouling modes, including cake formation, pore blocking, intermediate blocking, and complete blocking. Recent studies have shown that cake formation is the most common type of fouling in anaerobic membrane bioreactors (AnMBRs), and microbial concentration is the main biological parameter of membrane processes. Microbes in the range of 0.1–15 μ can be completely removed by ultrafiltration (0.01–0.1 μ) and microfiltration ($\geq$ 0.1 μ) [21].

Microbes typically have hydrophobic surfaces, making them more prone to attach to the membrane and form biofilms. These microbes produce soluble EPS, also known as SMP, during their metabolic activities, which are considered the main fouling agents. The presence of EPS forces microbes to adhere to the membrane surface and form biofilms. In addition to biological factors, non-biological factors are also involved in the fouling mechanism, including high temperatures, as high temperatures can reduce particle size, making the aggregate structure tighter and smaller. Furthermore, TSS also affect fouling in AnMBR systems, with high TSS leading to sudden and rapid contamination. The properties of membrane materials also have a significant impact on fouling rates, with inorganic membranes being more resistant to blocking than organic membranes due to their smoother surface. These hydrophobic microbes typically produce soluble EPS, also known as SMP, which are considered the main fouling agents. He et al. [49] reported that the roughness of organic membranes increases with increasing molecular weight cut-off.

During the chemical cleaning of membranes, it is important to note that raising the pH of the mixed solution can lead to an increased presence of non-removable dirt. In a study conducted by Xue et al. [120] it was observed that sidestream AnMBR demonstrated greater suitability for filtrate treatment due to the reduced presence of experimental contaminants. Dereli et al. [27] conducted a study that demonstrated a direct relationship between suspended solids and membrane fouling when concentrations exceeded 20 g/L. However, when operated within a sludge age range of 30–50 days, scaling problems did not significantly alter. The performance of membranes is noticeably influenced by fouling parameters such as operating conditions, biomass, and membrane properties. Unlike other forms of fouling, biological fouling cannot always be significantly reduced through feed pretreatment due to the self-replication property of biofouling organisms. Nevertheless, recent strategies for controlling membrane fouling focus on in-situ chemical cleaning, mechanically assisted aeration flushing, enzymatic/bacterial degradation of fouling substances, nanomaterial-based membranes, and electrically assisted fouling reduction. These new strategies prioritize energy efficiency and the overall sustainability of pollution control rather than solely targeting membrane fouling control. Adding activated carbon has been found to be a simple and effective method for managing aerobic/anaerobic MBR membrane fouling. Similarly, the addition of millimeter-sized silica (1–2 μm) to granular activated carbon (GAC) in AnMBR has shown promising results in reducing scaling. Moreover, powdered activated carbon (PAC) has been found to be more effective than GAC in addressing membrane fouling issues.

According to a research conducted by Wang et al. [112] an increase in PAC dosage can lead to a flux enhancement of over 30%. This improvement can be attributed to the dynamic and permeable PAC layer formed on the membrane surface, which helps protect the membrane from fouling. Different types of fouling can be treated using various cleaning methods, categorized as in-situ/ex-situ, physical/chemical, and biological/biochemical. In terms of in-situ/ex-situ cleaning, it is determined by the location of the membrane module within the membrane tank. Backwashing and relaxation are common non-physical methods employed. Physical cleaning techniques in membrane bioreactor systems include the addition of particles/carriers, rotation, vibration, and the emerging ultrasonic energy technology, which has shown effectiveness. Membrane damage should always be considered during cleaning processes. Chemical cleaning of membranes typically involves the use of acids, alkalis, oxidants, and other chemicals such as surfactants, detergents, and metal chelating agents depending on the nature of the fouling. Preventive maintenance of AnMBR membranes is generally performed every 1–6 weeks, lasting approximately 1–3 h. During maintenance, chemicals like NaClO (300–2000 mg/L) and citric acid (0.2–1.5 wt%) are commonly utilized. Membrane recovery cleaning, on the other hand, is typically conducted once a year with a downtime of 24 h. Physical–chemical cleaning techniques, such as chemical enhanced backwashing (CEB) using NaClO (100–500 mg/L) and ultrasonic-assisted chemical cleaning, are commonly employed. Physical–chemical synergistic cleaning has proven effective in membrane cleaning for MBR systems. In terms of biochemical cleaning, enzymatic cleaning and population quenching are commonly employed methods. Enzymatic cleaning is utilized in AnMBR systems, while population control is still being developed at the laboratory scale [2, 112].

The application of three selective stressors [i.e., oxygen (O), cell retention (R), and pressure (P)] drove clear variations of species distribution in biofilms (Fig. 3.15). Intriguingly, the degree of such distribution was highly relevant to biofilm growth rate. For example, for the evenly distributed biofilms, i.e., the ones under the selective stressors of O1 and P1 (GC < 0.2), the pressure changing rates were as high as > 3.0 Pf d−1, while for the two most uneven communities (R1 and R2), the slowest pressure change rates (< 0.6 Pf d−1) were observed (Fig. 3.16). It suggests that an evenly distributed community favors the formation of biofilms. In other words, the biofilms with a more evenly distributed species assembly gain a stronger expanding capacity. It has been suggested that unevenness usually leads to lower functionality, and it is more significant when under environmental stresses. Since biofilm growth closely links to microbial functionality, it is understandable that biofilm formation is also governed by community evenness. In fact, the even distribution of species may introduce a guarantee that some will associate with each other for a rapid growth [40].

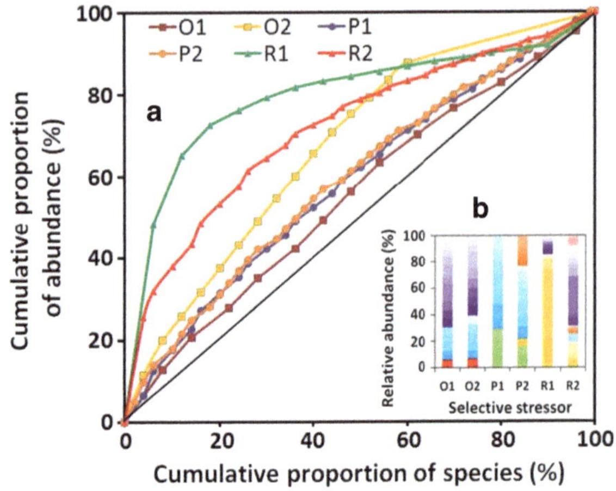

Fig. 3.15 Microbial abundance under changing selective stresses [40]

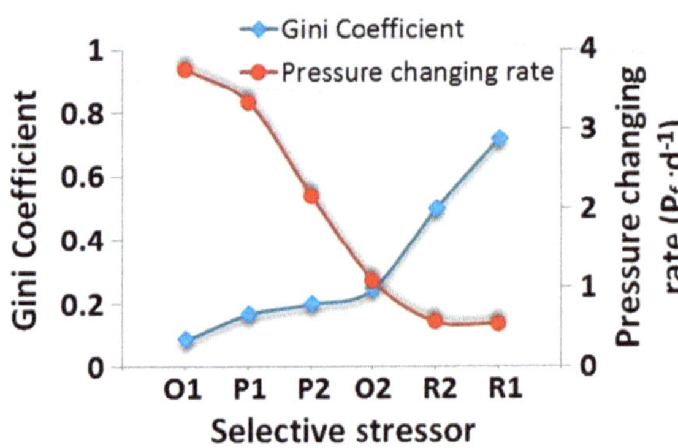

Fig. 3.16 The correlation between the evenness level and biofilm growing rate [40]

3.8.5 Role of Materials Addition on Membrane Fouling Control

Membrane fouling is inevitable, so developing low-cost membranes with good permeability and anti-fouling properties is beneficial to make AnMBR more competitive. In addition, slowing down the rate of membrane fouling by adding certain materials is a promising process. The two parts of AnMBR, anaerobic digestion unit and membrane component, have a crucial impact on the treatment of wastewater, so careful operation strategies are needed for both to achieve efficient treatment capacity. Membrane fouling reduces permeate flux and shortens membrane life, resulting in higher operating costs such as membrane replacement and energy demand. For example, Pretel et al. [82] reported that 60–75% of the energy consumption of AnMBR is related to membrane fouling.

Early AnMBR research mainly focused on enhancing digestion (i.e., anaerobic unit) or membrane fouling problems (i.e., membrane component). Some studies only focused on specific applications, such as biohydrogen production, pharmaceutical wastewater treatment, or single municipal wastewater treatment, ignoring the role of adding materials in AnMBR to simultaneously improve anaerobic digestion performance and reduce membrane fouling.

Adding materials to the AnMBR system, such as carbon-based coagulants or other carriers, can solve both problems at the same time: improving anaerobic biodegradation of organic matter and reducing in-situ membrane fouling. However, further research is needed to understand the impact of AnMBR challenges and the properties of added materials on pollution control mechanisms, such as quorum quenching (QQ) induction and the concentration and composition of microbial products (EPS and SMP) [78].

Common membrane materials include polymers such as PVDF, PES, PE, PSF, polyvinyl chloride and polytetrafluoroethylene (PTFE), as well as non-polymeric materials such as ceramics and stainless steel. Based on studies on membrane fouling in AnMBR, PVDF membranes have significantly lower fouling rates compared to PSF and cellulose membranes, and PVDF has good mechanical properties, such as high elongation at break and strong resistance, which improves its resistance to chemical cleaning. Modified PVDF membranes, such as PFPE-modified PVDF membranes, can effectively reduce membrane fouling and increase flux, membrane mass transfer coefficient, and methane recovery. Therefore, the efficient filtration of membrane materials and their response to fouling are crucial for the performance of MBR systems.

Rong et al. [87] found that using PVDF membranes under different temperature conditions (15–25 °C), the AnMBR system exhibited higher efficiency at high temperatures, but as the temperature decreased, microbial metabolic products accumulated on the membrane surface, resulting in severe organic pollution. In addition, the use of PEI membranes in AnMBR systems is more prone to scaling than PVDF membranes, demonstrating the importance and response of membrane materials to scaling. PVDF membranes, as a type of polymer membrane, have a strong chelation affinity for metal ions, such as Fe^{2+} and Al^{3+}. Due to the high deposition of polysaccharides on the surface of PVDF membranes, metal ions can be chelated and a loose gel layer is formed through intermolecular cohesive-chelating interactions in biopolymers.

In comparison to polymer membranes, inorganic membranes such as stainless steel and ceramics have longer lifespan, better filtration performance, and less membrane fouling possibility. Aslam et al. [12] found that the flux rate of hydrophilic ceramic membranes in AnMBR systems is 1.7 times higher than that of hydrophobic PVDF membranes. However, the high manufacturing cost of inorganic membranes limits their practical application. Nevertheless, with the development of global technology, the manufacturing technology of ceramic membranes is continuously improving, which may become more popular in the future. Currently, polymer flat membranes and ceramic flat-tube membranes are the most widely used membrane types in AnMBR systems. In ceramic flat membranes, metal ions competitively occupy limited binding sites of biopolymers, promoting the chelation of Ca^{2+} and

Mg^{2+} in dirt. Other studies have shown that the high content of uronic acids in loosely-bound EPS forms a complex structure with Ca^{2+}/Mg^{2+}, further strengthening the filter cake layer. Although ceramic and polymer membranes show different filtering resistance when processing similar mixed liquids, different membrane materials provide different growth environments for bacteria, thus resulting in different levels of pollution. Therefore, research on membrane materials needs to be strengthened. New and improved membrane materials can not only reduce membrane fouling, but also obtain better effluent quality. For example, Manni et al. used natural magnesite from Morocco to prepare a low-cost magnesite flat membrane and applied it to industrial wastewater treatment. Experimental results showed that the process had removal rates of 99% and 67% for turbidity and COD, respectively [71]. In addition, PES membranes blended with barium chloride showed excellent dye (methylene and blue-orange G) retention rates, optimal porosity, and the highest permeation flux. In another study, PVDF membranes were modified with Cu(II) and tannic acid, showing high hydrophilicity with antibacterial properties and requiring less chemical cleaning compared to unmodified membranes.

Research has shown that compared to biogas sparging, these suspended carriers can significantly reduce energy consumption in AnMBR systems while improving membrane performance, especially in highly polluted water bodies. In addition, these suspended carriers can provide a good bio-attachment surface, promoting microbial growth and activity on the membrane surface, and further reducing fouling. At the same time, some new membrane technologies, such as electrodialysis, electrosorption, ultrafiltration, etc., have also been introduced into the AnMBR system to reduce membrane fouling and improve water treatment efficiency. The application of these new membrane technologies can further improve the sustainability and economy of AnMBR systems, providing new options for the management and utilization of water resources [11].

It should be noted that adding suspended carriers (such as GAC, silica, and polyethylene terephthalate) to the AnMBR system can improve the removal efficiency of organic matter, regardless of whether these particles are in continuous or intermittent motion. This is because these particles have physical adsorption behavior, and the biofilm attached to their surface can further degrade organic matter through biological adsorption or biodegradation. However, when the upflow velocity of fluidized particles increases or the permeate flux decreases, the removal efficiency of organic matter will decrease. This relationship is because at high upflow velocity or low permeate flux, the filter cake effect on the membrane surface is small, and organic compounds cannot be effectively removed through the membrane [3].

It is worth noting that mechanical stress caused by moving particles related to the upflow velocity can damage the membrane surface. This is particularly likely to occur in the early stages of filtration when the cake layer is lacking or when organic membranes are used in AnMBRs. Loss of membrane integrity will have a negative impact on the permeate quality of AnMBRs. Although several studies have demonstrated that the membrane integrity is not compromised during short-term operation of AnMBRs with scouring particles, it was observed in a long-term pilot plant study lasting for two years that scouring particles damaged the membrane in the middle and bottom parts of the membrane module [95]. Therefore, the use of membranes with higher tear resistance in AFMBRs can achieve better permeate quality in the presence of scouring particles.

Adding low doses of adsorbent in AnMBRs can effectively reduce membrane fouling, as these systems possess the capability to adsorb small particle and degrade organic pollutants through attached growing bacteria and their scouring action. An example of this is the work conducted by the Stuckey group, who observed that the addition of PAC in AnMBRs significantly enhanced the quality of the permeate, irrespective of HRT and salt concentration [4, 36]. Baêta et al. [14] conducted a study and observed that the inclusion of PAC in AnMBRs resulted in higher COD removal rates (73–94% compared to 64–85%) and improved dye color removal rates (90–94% compared to 77–87%) when treating dyeing industrial wastewater, as compared to AnMBRs without PAC. Therefore, the addition of adsorbents in AnMBRs can ensure that treated urban wastewater meets the standards for reuse. It is important to note that excessive use of adsorbents may not enhance membrane filtration performance and can potentially act as pollutants, leading to the clogging of membrane pores or the formation of cake layers on the membrane. However, the inclusion of adsorbents can ultimately improve the quality of the permeate. The regeneration of adsorbents is another critical factor limiting their application. Careful control of adsorbent dosage and retention time is necessary to avoid sludge loss during the removal of adsorbents from AnMBRs [97].

The characteristics of membrane materials are crucial to the fouling rate, and inorganic membranes (especially ceramic membranes) have been proven to be more resistant to fouling than organic membranes due to their smooth surface characteristics. In addition to using methane-rich biogas, AnMBR is considered an ideal choice for removing COD, controlling sludge retention time (SRT), and hydraulic retention time (HRT) from landfill leachate. However, membrane fouling is a major problem associated with AnMBR application. Some researchers have studied the mechanism of membrane fouling and the composition of fouling to address this issue. Microorganisms, EPS, and SMP have been found to be the main components of membrane fouling. Accumulation, blocking, and adsorption of fouling agents on the membrane surface and in the membrane pores severely affect the filtration efficiency of the membrane, increasing operating costs [77].

The addition of coagulants (such as $FeCl_3$) can interact with soluble organic pollutants and colloidal organic pollutants to form larger particles, thus alleviating the problem of membrane fouling and is one of the effective ways in AnMBR. The addition of coagulants can reduce the concentration of organic compounds in the permeate. In theory, due to the larger size of the complex, the membrane can effectively retain the formed organic-coagulant complex, thus slowing down the rate of membrane fouling. In addition, since there is a low biomass and fewer metabolites (such as soluble extracellular polymer substances) in the bioreactor, researchers propose the use of phosphorylated polyvinyl alcohol as a polymer substrate to capture cells to slow down the rate of pollution in AnMBR. The research results showed that the AnMBR with embedded cells had equivalent treatment performance to

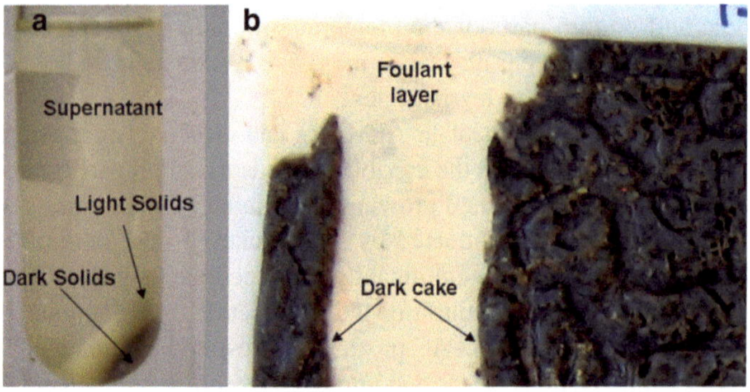

Fig. 3.17 **a** The mixed liquor separated into supernatant, light solids and dark solids after centrifugation at 12,000×g for 15 min, **b** plan view photograph of the fouled cPVDF membrane. A strip of loosely held dark cake has been removed to reveal a thin yellow foulant layer attached to the membrane

conventional AnMBR, and there was no significant difference in permeate quality [61].

The research conducted by Gao et al. [42] also demonstrated that fouling of polyetherimide occurred faster than that of cPVDF, further emphasizing the importance of membrane materials in fouling. Gao installed new cPVDF membranes in two membrane modules of an AnMBR system and ran the reactor for two months without membrane cleaning, as shown in Fig. 3.18. During the first five days of operation, the permeate flux rapidly decreased from 12 to 9 L/m^{2}h (LMH). Thereafter, the flux decreased at a slower rate, reaching a value of approximately 8 LMH on the 62nd day (Fig. 3.18). On days 40, 56, and 72, biomass suspension samples were taken from a sampling port located in the middle of the reactor vessel and centrifuged at 12,000×g for 15 min. The resulting precipitate consisted of a yellow layer (light solids) covered by a dark solid layer. On day 62, the cPVDF membranes were removed for analysis. The contaminated membrane surface was covered with a thick dark filter cake, which was removed by washing. Beneath it, a thin yellow layer of fouling was present, which was more difficult to remove (Fig. 3.17). On day 62, fresh PEI membranes (pore size ~ 30 kDa) were installed in the two membrane modules to examine the fouling differences between the two membrane types. The flux rapidly decreased, immediately dropping to < 6 LMH within a few hours and continuing to decrease to 4 LMH over 19 days, when the fouled PEI membranes were removed for analysis. PEI is a denser membrane than cPVDF, and thus, lower fluxes are expected compared to cPVDF membranes. This indicates that fouling of the PEI membrane occurred faster than that of the cPVDF membrane. Over an 81-day operational period, the COD concentration of the permeate remained stable (< 20 mg/L), indicating good overall performance (> 96% COD removal rate). The COD loading rate was about 5.0 kg/m^3-d, with an estimated COD removal rate of 1.05 gCOD/gVSS-d, which is comparable to high-rate anaerobic reactors such as UASB with granular sludge. Due to the ability to accumulate higher biomass concentrations,

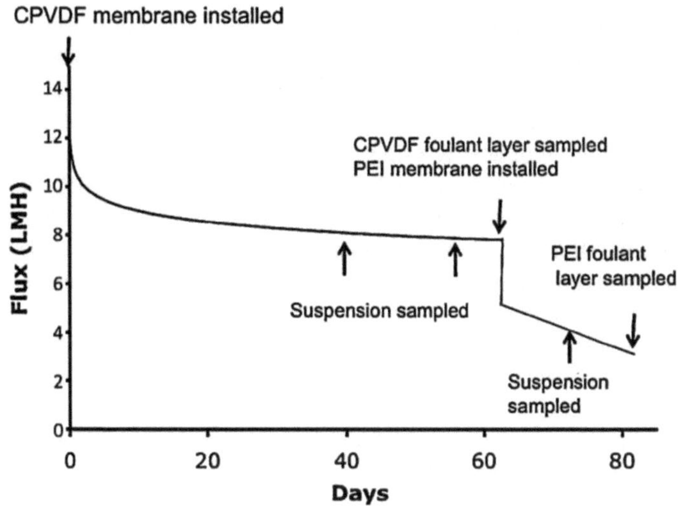

Fig. 3.18 Operational history of the anaerobic MBR. A new cPVDF membrane was installed on day 0, and a new PEI membrane on day 62

higher volumetric COD removal performance can be achieved. This suggests that the AnMBR system can operate at high rates with high-quality effluent.

Although various improvements have been made to AnMBR, efforts are still needed to achieve fine design and efficient utilization. Future research should consider the above theories to better understand the interactions between different phenomena and develop anti-pollution membranes. According to the literature, research on the gel layer involved in AnMBR is very limited. Finally, large-scale research is needed to ensure the applicability of AnMBR.

3.9 Prospects

AnMBR, as one of the latest advancements in wastewater treatment technologies, holds great promise in various applications such as wastewater treatment and energy recovery. A well-optimized AnMBR system has the capability to efficiently produce treated wastewater suitable for irrigation, as well as generate bioenergy and biofertilizers. Notably, AnMBR offers the advantage of lower capital expenditure compared to conventional treatment technologies. The decrease in temperature is typically the primary factor leading to the dissolution of methane. However, additional research is required to devise targeted approaches that can enhance methane production and maximize the recovery of dissolved methane. This is essential for improving energy yields and developing resilient systems capable of handling fluctuating temperatures in MWW treatment. Furthermore, it is crucial to address the challenges posed by varying organic loads and minimize the presence of greenhouse gases in the effluent. Co-digestion emerges as a potential solution to overcome these challenges and mitigate the impact of temperature and methane deficiency, thus ensuring a balanced energy supply in regions experiencing colder conditions. The pre-concentration of MWW using processes like FO has demonstrated significant potential in enhancing water and resource recovery in AnMBR. However, the cost-effectiveness of pre-enrichment remains a challenge, particularly considering the low concentration of MWW during winter. Further research is needed to assess the scalability and economic feasibility of pre-enrichment stages involving FO or other relevant technologies.

While membrane bioreactors offer the advantage of achieving high rates of COD removal, MWW can contain numerous emerging contaminants. Studying their fate under different operating conditions and investigating the effects of antibiotics on bacterial communities and membrane fouling are essential to determine the necessity of pretreatment or post-treatment.

Despite several advancements in AnMBR, membrane fouling remains a significant obstacle that reduces efficiency and hampers large-scale implementation. Therefore, various measures have been implemented to mitigate cake layer formation and membrane pore clogging. These techniques include biogas injection, liquid backwash and particle injection, the use of coagulants and fluidizing media, and chemical cleaning to remove pore clogging. Recently, the scientific community has observed the presence of a gel layer prior to the formation of the filter cake layer, highlighting the challenges posed by this gel layer and the need to understand its implications and develop effective mitigation strategies.

Integrating other technologies and approaches may be an effective means of controlling contamination in AnMBR systems. Combining electric fields or ultrasound with population bursting techniques can be successful in managing fouling. However, there is currently no reliable alternative to chemical cleaning for preventing pore clogging. Chemical cleaning in AnMBR systems may have detrimental effects on microorganisms, membrane integrity, and durability. Further studies are required to understand the impact of chemical by-products and to investigate the nature and evolution of interactions between fouling and commonly used chemical agents.

References

1. Abad V, Avila R, Vicent T, Font X (2019) Promoting circular economy in the surroundings of an organic fraction of municipal solid waste anaerobic digestion treatment plant: biogas production impact and economic factors. Bioresour Technol 283:10–17. https://doi.org/10.1016/j.biortech.2019.03.064
2. Abuabdou SMA, Ahmad W, Aun NC, Bashir MJK (2020) A review of anaerobic membrane bioreactors (AnMBR) for the treatment of highly contaminated landfill leachate and biogas production: effectiveness, limitations and future perspectives. J Clean Prod 255:120215. https://doi.org/10.1016/j.jclepro.2020.120215

3. Ahmad R, Aslam M, Park E, Chang S, Kwon D, Kim J (2018) Submerged low-cost pyrophyllite ceramic membrane filtration combined with GAC as fluidized particles for industrial wastewater treatment. Chemosphere 206:784–792. https://doi.org/10.1016/j.chemosphere.2018.05.045
4. Akram A, Stuckey DC (2008) Flux and performance improvement in a submerged anaerobic membrane bioreactor (SAMBR) using powdered activated carbon (PAC). Process Biochem 43(1):93–102. https://doi.org/10.1016/j.procbio.2007.10.020
5. Alibardi L, Bernava N, Cossu R, Spagni A (2016) Anaerobic dynamic membrane bioreactor for wastewater treatment at ambient temperature. Chem Eng J 284:130–138. https://doi.org/10.1016/j.cej.2015.08.111
6. Alibardi L, Cossu R, Saleem M, Spagni A (2014) Development and permeability of a dynamic membrane for anaerobic wastewater treatment. Bioresour Technol 161:236–244. https://doi.org/10.1016/j.biortech.2014.03.045
7. Amha YM, Corbett M, Smith AL (2019) Two-phase improves performance of anaerobic membrane bioreactor treatment of food waste at high organic loading rates. Environ Sci Technol 53(16):9572–9583
8. An Y, Wang Z, Wu Z, Yang D, Zhou Q (2009) Characterization of membrane foulants in an anaerobic non-woven fabric membrane bioreactor for municipal wastewater treatment. Chem Eng J 155(3):709–715. https://doi.org/10.1016/j.cej.2009.09.003
9. Aquino SF, Stuckey DC (2004) Soluble microbial products formation in anaerobic chemostats in the presence of toxic compounds. Water Res 38(2):255–266. https://doi.org/10.1016/j.watres.2003.09.031
10. Aslam A, Khan SJ, Shahzad HMA (2022) Anaerobic membrane bioreactors (AnMBRs) for municipal wastewater treatment-potential benefits, constraints, and future perspectives: an updated review. Sci Total Environ 802:149612. https://doi.org/10.1016/j.scitotenv.2021.149612
11. Aslam M, Charfi A, Lesage G, Heran M, Kim J (2017) Membrane bioreactors for wastewater treatment: a review of mechanical cleaning by scouring agents to control membrane fouling. Chem Eng J 307:897–913. https://doi.org/10.1016/j.cej.2016.08.144
12. Aslam M, McCarty PL, Bae J, Kim J (2014) The effect of fluidized media characteristics on membrane fouling and energy consumption in anaerobic fluidized membrane bioreactors. Sep Purif Technol 132:10–15. https://doi.org/10.1016/j.seppur.2014.04.049
13. Aydin S, Ince B, Ince O (2015) Application of real-time PCR to determination of combined effect of antibiotics on Bacteria, Methanogenic Archaea, Archaea in anaerobic sequencing batch reactors. Water Res 76:88–98. https://doi.org/10.1016/j.watres.2015.02.043
14. Baêta BEL, Luna HJ, Sanson AL, Silva SQ, Aquino SF (2013) Degradation of a model azo dye in submerged anaerobic membrane bioreactor (SAMBR) operated with powdered activated carbon (PAC). J Environ Manag 128:462–470. https://doi.org/10.1016/j.jenvman.2013.05.038
15. Barua S, Dhar BR (2017) Advances towards understanding and engineering direct interspecies electron transfer in anaerobic digestion. Bioresour Technol 244:698–707. https://doi.org/10.1016/j.biortech.2017.08.023
16. Becerra-Castro C, Lopes AR, Vaz-Moreira I, Silva EF, Manaia CM, Nunes OC (2015) Wastewater reuse in irrigation: A microbiological perspective on implications in soil fertility and human and environmental health. Environ Int 75:117–135. https://doi.org/10.1016/j.envint.2014.11.001
17. Berkessa YW, Yan B, Li T, Tan M, She Z, Jegatheesan V, Jiang H, Zhang Y (2018) Novel anaerobic membrane bioreactor (AnMBR) design for wastewater treatment at long HRT and high solid concentration. Biores Technol 250:281–289
18. Cai C, Hu S, Chen X, Ni B-J, Pu J, Yuan Z (2018) Effect of methane partial pressure on the performance of a membrane biofilm reactor coupling methane-dependent denitrification and anammox. Sci Total Environ 639:278–285. https://doi.org/10.1016/j.scitotenv.2018.05.164
19. Cao X, Chen H, Li Y, He Y (2011) Influence of alkalinity and load on the start-up process of anaerobic reactors. In: 2011 international symposium on water resource and environmental protection

20. Casu S, Crispino NA, Farina R, Mattioli D, Ferraris M, Spagni A (2012) Wastewater treatment in a submerged anaerobic membrane bioreactor. J Environ Sci Health Part A 47(2):204–209. https://doi.org/10.1080/10934529.2012.640562
21. Charfi A, Ben Amar N, Harmand J (2012) Analysis of fouling mechanisms in anaerobic membrane bioreactors. Water Res 46(8):2637–2650. https://doi.org/10.1016/j.watres.2012.02.021
22. Chen et al. (2022) A review on upgrading of the anammox-based nitrogen removal processes: Performance stability and control strategies. Bioresour. Technol. 364:127–992. https://doi.org/10.1016/j.biortech.2022.127992
23. Chen L, Gu Y, Cao C, Zhang J, Ng J-W, Tang C (2014) Performance of a submerged anaerobic membrane bioreactor with forward osmosis membrane for low-strength wastewater treatment. Water Res 50:114–123. https://doi.org/10.1016/j.watres.2013.12.009
24. Chen R, Nie Y, Kato H, Wu J, Utashiro T, Lu J, Yue S, Jiang H, Zhang L, Li Y-Y (2017) Methanogenic degradation of toilet-paper cellulose upon sewage treatment in an anaerobic membrane bioreactor at room temperature. Bioresour Technol 228:69–76. https://doi.org/10.1016/j.biortech.2016.12.089
25. Christenson L, Sims R (2011) Production and harvesting of microalgae for wastewater treatment, biofuels, and bioproducts. Biotechnol Adv 29(6):686–702. https://doi.org/10.1016/j.biotechadv.2011.05.015
26. Dai W, Xu X, Liu B, Yang F (2015) Toward energy-neutral wastewater treatment: a membrane combined process of anaerobic digestion and nitritation–anammox for biogas recovery and nitrogen removal. Chem Eng J 279:725–734
27. De Vela RJ (2021) A review of the factors affecting the performance of anaerobic membrane bioreactor and strategies to control membrane fouling. Rev Environ Sci Bio/Technol 20(3):607–644. https://doi.org/10.1007/s11157-021-09580-2
28. Dereli RK, Grelot A, Heffernan B, van der Zee FP, van Lier JB (2014) Implications of changes in solids retention time on long term evolution of sludge filterability in anaerobic membrane bioreactors treating high strength industrial wastewater. Water Res 59:11–22. https://doi.org/10.1016/j.watres.2014.03.073
29. Dereli RK, van der Zee FP, Ozturk I, van Lier JB (2019) Treatment of cheese whey by a cross-flow anaerobic membrane bioreactor: biological and filtration performance. Environ Res 168:109–117. https://doi.org/10.1016/j.envres.2018.09.021
30. Ding A, Fan Q, Cheng R, Sun G, Zhang M, Wu D (2018) Impacts of applied voltage on microbial electrolysis cell-anaerobic membrane bioreactor (MEC-AnMBR) and its membrane fouling mitigation mechanism. Chem Eng J 333:630–635. https://doi.org/10.1016/j.cej.2017.09.190
31. Elmoutez S, Abushaban A, Necibi MC, Sillanpää M, Liu J, Dhiba D, Chehbouni A, Taky M (2023) Design and operational aspects of anaerobic membrane bioreactor for efficient wastewater treatment and biogas production. Environ Challenges 10:100671. https://doi.org/10.1016/j.envc.2022.100671
32. Ersahin ME, Gimenez JB, Ozgun H, Tao Y, Spanjers H, van Lier JB (2016) Gas-lift anaerobic dynamic membrane bioreactors for high strength synthetic wastewater treatment: effect of biogas sparging velocity and HRT on treatment performance. Chem Eng J 305:46–53. https://doi.org/10.1016/j.cej.2016.02.003
33. Ersahin ME, Ozgun H, Tao Y, van Lier JB (2014) Applicability of dynamic membrane technology in anaerobic membrane bioreactors. Water Res 48:420–429. https://doi.org/10.1016/j.watres.2013.09.054
34. Ersahin ME, Tao Y, Ozgun H, Gimenez JB, Spanjers H, van Lier JB (2017) Impact of anaerobic dynamic membrane bioreactor configuration on treatment and filterability performance. J Membr Sci 526:387–394. https://doi.org/10.1016/j.memsci.2016.12.057
35. Estrada-Arriaga EB, Reynoso-Deloya MG, Guillen-Garces RA, Falcon-Rojas A, Garcia-Sanchez L (2021) Enhanced methane production and organic matter removal from tequila vinasses by anaerobic digestion assisted via bioelectrochemical power-to-gas. Bioresour Technol 320, Article 124344. https://doi.org/10.1016/j.biortech.2020.124344

36. Fang et al. (2018) Effect of high salinity in wastewater on surface properties of anammox granular sludge. Chemosphere 210:366–375. https://doi.org/10.1016/j.chemosphere.2018.07.038
37. Galib M, Elbeshbishy E, Reid R, Hussain A, Lee H-S (2016) Energy-positive food wastewater treatment using an anaerobic membrane bioreactor (AnMBR). J Environ Manage 182:477–485
38. Gao W, Han M, Xu CC, Liao B, Hong Y, Cumin J, Dagnew M (2016) Performance of submerged anaerobic membrane bioreactor for thermomechanical pulping wastewater treatment. J Water Process Eng 13:70–78
39. Gao D-W, Hu Q, Yao C, Ren N-Q, Wu W-M (2014) Integrated anaerobic fluidized-bed membrane bioreactor for domestic wastewater treatment. Chem Eng J 240:362–368
40. Gao DW, Tao Y, Fu Y, Liang H (2018) Egalitarianism in biofilms. Microb Ecol 76(1):49–51. https://doi.org/10.1007/s00248-018-1173-5
41. Gao D-W, Wen Z-D, Li B, Liang H (2013) Membrane fouling related to microbial community and extracellular polymeric substances at different temperatures. Bioresour Technol 143:172–177. https://doi.org/10.1016/j.biortech.2013.05.127
42. Gao D-W, Zhang T, Tang C-YY, Wu W-M, Wong C-Y, Lee YH, Yeh DH, Criddle CS (2010). Membrane fouling in an anaerobic membrane bioreactor: differences in relative abundance of bacterial species in the membrane foulant layer and in suspension. J Membr Sci 364(1):331–338. https://doi.org/10.1016/j.memsci.2010.08.031
43. Gao D, Li Y, Liang H (2022) Biofilm carriers for anaerobic ammonium oxidation: mechanisms, applications, and roles in mainstream systems. Bioresour Technol 353:127115. https://doi.org/10.1016/j.biortech.2022.127115
44. Gao D, Sui L, Liang H (2022) How microbial community and membrane biofouling respond to temperature changes in an anaerobic membrane bioreactor. Environ Technol Innov 28:102675. https://doi.org/10.1016/j.eti.2022.102675
45. Gouveia J, Plaza F, Garralon G, Fdz-Polanco F, Peña M (2015) Long-term operation of a pilot scale anaerobic membrane bioreactor (AnMBR) for the treatment of municipal wastewater under psychrophilic conditions. Biores Technol 185:225–233
46. Gu Y, Chen L, Ng J-W, Lee C, Chang VWC, Tang CY (2015) Development of anaerobic osmotic membrane bioreactor for low-strength wastewater treatment at mesophilic condition. J Membr Sci 490:197–208. https://doi.org/10.1016/j.memsci.2015.04.032
47. Guo Z, Gao L, Wang L, Liu W, Wang A (2018) Enhanced methane recovery and exoelectrogen-methanogen evolution from low-strength wastewater in an up-flow biofilm reactor with conductive granular graphite fillers. Front Environ Sci Eng 12:1–10
48. Gutwiński P, Cema G, Ziembińska-Buczyńska A, Surmacz-Górska J, Osadnik M (2016) Startup of the anammox process in a membrane bioreactor (AnMBR) from conventional activated sludge. Water Environ Res 88(12):2268–2274
49. He Y, Xu P, Li C, Zhang B (2005) High-concentration food wastewater treatment by an anaerobic membrane bioreactor. Water Res 39(17):4110–4118
50. Ho J, Sung S (2009) Anaerobic membrane bioreactor treatment of synthetic municipal wastewater at ambient temperature. Water Environ Res 81(9):922–928
51. Hou D, Lu L, Sun D, Ge Z, Huang X, Cath TY, Ren ZJ (2017) Microbial electrochemical nutrient recovery in anaerobic osmotic membrane bioreactors. Water Res 114:181–188. https://doi.org/10.1016/j.watres.2017.02.034
52. Hu Y, Cai X, Xue Y, Du R, Ji J, Chen R, Sano D, Li Y-Y (2022) Recent developments of anaerobic membrane bioreactors for municipal wastewater treatment and bioenergy recovery: focusing on novel configurations and energy balance analysis. J Clean Prod 356:131856
53. Hu YS, Cheng H, Ji JY, Li YY (2020) A review of anaerobic membrane bioreactors for municipal wastewater treatment with a focus on multicomponent biogas and membrane fouling control. Environ Sci Water Res Technol 6(10):2641–2663. https://doi.org/10.1039/d0ew00528b
54. Hu Y, Wang XC, Ngo HH, Sun Q, Yang Y (2018a) Anaerobic dynamic membrane bioreactor (AnDMBR) for wastewater treatment: a review. Bioresour Technol 247:1107–1118. https://doi.org/10.1016/j.biortech.2017.09.101

55. Hu Y, Yang Y, Yu S, Wang XC, Tang J (2018b) Psychrophilic anaerobic dynamic membrane bioreactor for domestic wastewater treatment: Effects of organic loading and sludge recycling. Bioresour Technol 270:62–69. https://doi.org/10.1016/j.biortech.2018.08.128
56. Huang Z, Ong SL, Ng HY (2011) Submerged anaerobic membrane bioreactor for low-strength wastewater treatment: effect of HRT and SRT on treatment performance and membrane fouling. Water Res 45(2):705–713. https://doi.org/10.1016/j.watres.2010.08.035
57. Hülsen T, Barry EM, Lu Y, Puyol D, Keller J, Batstone DJ (2016) Domestic wastewater treatment with purple phototrophic bacteria using a novel continuous photo anaerobic membrane bioreactor. Water Res 100:486–495. https://doi.org/10.1016/j.watres.2016.04.061
58. Jeong Y, Hermanowicz SW, Park C (2017) Treatment of food waste recycling wastewater using anaerobic ceramic membrane bioreactor for biogas production in mainstream treatment process of domestic wastewater. Water Res 123:86–95. https://doi.org/10.1016/j.watres.2017.06.049
59. Jeong Y, Kim Y, Jin Y, Hong S, Park C (2018) Comparison of filtration and treatment performance between polymeric and ceramic membranes in anaerobic membrane bioreactor treatment of domestic wastewater. Sep Purif Technol 199:182–188. https://doi.org/10.1016/j.seppur.2018.01.057
60. Jiménez-Benítez A, Ferrer J, Rogalla F, Vázquez JR, Seco A, Robles Á (2020) Energy and environmental impact of an anaerobic membrane bioreactor (AnMBR) demonstration plant treating urban wastewater. In: Current developments in biotechnology and bioengineering. Elsevier, pp 289–310
61. Juntawang C, Rongsayamanont C, Khan E (2017) Entrapped cells-based-anaerobic membrane bioreactor treating domestic wastewater: Performances, fouling, and bacterial community structure. Chemosphere 187:147–155. https://doi.org/10.1016/j.chemosphere.2017.08.113
62. Khan MA, Ngo HH, Guo WS, Liu YW, Nghiem LD, Chang SW, Nguyen DD, Zhang SC, Luo G, Jia H (2019) Optimization of hydraulic retention time and organic loading rate for volatile fatty acid production from low strength wastewater in an anaerobic membrane bioreactor. Biores Technol 271:100–108. https://doi.org/10.1016/j.biortech.2018.09.075
63. Li Y, Hu Q, Chen C-H, Wang X-L, Gao D-W (2017) Performance and microbial community structure in an integrated anaerobic fluidized-bed membrane bioreactor treating synthetic benzothiazole contaminated wastewater. Biores Technol 236:1–10
64. Lin H, Xie K, Mahendran B, Bagley D, Leung K, Liss S, Liao B (2009) Sludge properties and their effects on membrane fouling in submerged anaerobic membrane bioreactors (SAnMBRs). Water Res 43(15):3827–3837
65. Lin H, Zhang M, Wang F, Meng F, Liao B-Q, Hong H, Chen J, Gao W (2014) A critical review of extracellular polymeric substances (EPSs) in membrane bioreactors: characteristics, roles in membrane fouling and control strategies. J Membr Sci 460:110–125
66. Lin H, Peng W, Zhang M, Chen J, Hong H, Zhang Y (2013) A review on anaerobic membrane bioreactors: applications, membrane fouling and future perspectives. Desalination 314:169–188. https://doi.org/10.1016/j.desal.2013.01.019
67. Liu H, Wang Y, Yin B, Zhu Y, Fu B, Liu H (2016) Improving volatile fatty acid yield from sludge anaerobic fermentation through self-forming dynamic membrane separation. Bioresour Technol 218:92–100. https://doi.org/10.1016/j.biortech.2016.06.077
68. Luo H, Zeng Y, Cheng Y, He D, Pan X (2020) Recent advances in municipal landfill leachate: a review focusing on its characteristics, treatment, and toxicity assessment. Sci Total Environ 703:135468. https://doi.org/10.1016/j.scitotenv.2019.135468
69. Maaz M, Yasin M, Aslam M, Kumar G, Atabani AE, Idrees M, Anjum F, Jamil F, Ahmad R, Khan AL, Lesage G, Heran M, Kim J (2019) Anaerobic membrane bioreactors for wastewater treatment: novel configurations, fouling control and energy considerations. Bioresour Technol 283:358–372. https://doi.org/10.1016/j.biortech.2019.03.061
70. Maleki E, Catalan LJ, Liao B (2018) Effect of organic loading rate on the performance of a submerged anaerobic membrane bioreactor (SAnMBR) for malting wastewater treatment and biogas production. J Chem Technol Biotechnol 93(6):1636–1647

71. Manni A, Achiou B, Karim A, Harrati A, Sadik C, Ouammou M, Younssi SA, El Bouari A (2020) New low-cost ceramic microfiltration membrane made from natural magnesite for industrial wastewater treatment. J Environ Chem Eng 8(4):103906
72. Martin Vincent N, Tong J, Yu D, Zhang J, Wei Y (2018) Membrane fouling characteristics of a side-stream tubular anaerobic membrane bioreactor (AnMBR) treating domestic wastewater. Processes 6(5):50. https://www.mdpi.com/2227-9717/6/5/50
73. Martinez-Sosa D, Helmreich B, Netter T, Paris S, Bischof F, Horn H (2011) Anaerobic submerged membrane bioreactor (AnSMBR) for municipal wastewater treatment under mesophilic and psychrophilic temperature conditions. Biores Technol 102(22):10377–10385
74. Miao L, Yang G, Tao T, Peng Y (2019) Recent advances in nitrogen removal from landfill leachate using biological treatments—a review. J Environ Manag 235:178–185. https://doi.org/10.1016/j.jenvman.2019.01.057
75. Muratçobanoğlu H, Gökçek ÖB, Mert RA, Zan R, Demirel S (2020) Simultaneous synergistic effects of graphite addition and co-digestion of food waste and cow manure: biogas production and microbial community. Bioresour Technol 309:123365. https://doi.org/10.1016/j.biortech.2020.123365
76. Nabi M, Liang H, Cheng L, Yang W, Gao D (2022) A comprehensive review on the use of conductive materials to improve anaerobic digestion: focusing on landfill leachate treatment. J Environ Manage 309:114540
77. Nabi M, Gao D, Liang H, Cheng L, Yang W, Li Y (2022) Landfill leachate treatment by graphite engineered anaerobic membrane bioreactor: performance enhancement and membrane fouling mitigation. Environ Res 214:114010. https://doi.org/10.1016/j.envres.2022.114010
78. Nabi M, Liang H, Zhou Q, Cao J, Gao D (2023) In-situ membrane fouling control and performance improvement by adding materials in anaerobic membrane bioreactor: a review. Sci Total Environ 865:161262. https://doi.org/10.1016/j.scitotenv.2022.161262
79. Nagaoka H, Ueda S, Miya A (1996) Influence of bacterial extracellular polymers on the membrane separation activated sludge process. Water Sci Technol 34(9):165–172
80. Oberoi AS, Surendra KC, Wu D, Lu H, Wong JWC, Kumar Khanal S (2022) Anaerobic membrane bioreactors for pharmaceutical-laden wastewater treatment: a critical review. Bioresour Technol 361:127667. https://doi.org/10.1016/j.biortech.2022.127667
81. Peña M, do Nascimento T, Gouveia J, Escudero J, Gómez A, Letona A, Arrieta J, Fdz-Polanco F (2019) Anaerobic submerged membrane bioreactor (AnSMBR) treating municipal wastewater at ambient temperature: operation and potential use for agricultural irrigation. Bioresour Technol 282:285–293. https://doi.org/10.1016/j.biortech.2019.03.019
82. Pretel R, Robles A, Ruano MV, Seco A, Ferrer J (2014) The operating cost of an anaerobic membrane bioreactor (AnMBR) treating sulphate-rich urban wastewater. Sep Purif Technol 126:30–38. https://doi.org/10.1016/j.seppur.2014.02.013
83. Qi, Hu Liru, Fan Dawen, Gao (2017) Pilot-scale investigation on the treatment of cellulosic ethanol biorefinery waste water. J Chem Eng 309:409–416. https://doi.org/10.1016/j.cej.2016.10.066
84. Renou S, Givaudan JG, Poulain S, Dirassouyan F, Moulin P (2008) Landfill leachate treatment: review and opportunity. J Hazard Mater 150(3):468–493. https://doi.org/10.1016/j.jhazmat.2007.09.077
85. Rico C, Muñoz N, Fernández J, Rico JL (2015) High-load anaerobic co-digestion of cheese whey and liquid fraction of dairy manure in a one-stage UASB process: limits in co-substrates ratio and organic loading rate. Chem Eng J 262:794–802
86. Robles Á, Durán F, Giménez JB, Jiménez E, Ribes J, Serralta J, Seco A, Ferrer J, Rogalla F (2020) Anaerobic membrane bioreactors (AnMBR) treating urban wastewater in mild climates. Bioresour Technol 314:123763. https://doi.org/10.1016/j.biortech.2020.123763
87. Rong C, Wang T, Luo Z, Hu Y, Kong Z, Qin Y, Hanaoka T, Ito M, Kobayashi M, Li Y-Y (2022) Pilot plant demonstration of temperature impacts on the methanogenic performance and membrane fouling control of the anaerobic membrane bioreactor in treating real municipal wastewater. Bioresour Technol 354:127167. https://doi.org/10.1016/j.biortech.2022.127167

88. Rongwong W, Lee J, Goh K, Karahan HE, Bae T-H (2018) Membrane-based technologies for post-treatment of anaerobic effluents. NPJ Clean Water 1(1):21
89. Ruiz-Martínez A, Serralta J, Seco A, Ferrer J (2015) Effect of temperature on ammonium removal in Scenedesmus sp. Bioresour Technol 191:346–349. https://doi.org/10.1016/j.biortech.2015.05.070
90. Saddoud A, Ellouze M, Dhouib A, Sayadi S (2007) Anaerobic membrane bioreactor treatment of domestic wastewater in Tunisia. Desalination 207(1):205–215. https://doi.org/10.1016/j.desal.2006.08.005
91. Saha S, Hussain A, Lee J, Lee E, Lee HS (2023) An integrated leachate bed reactor-anaerobic membrane bioreactor system (LBR-AnMBR) for food waste stabilization and biogas recovery. Chemosphere 311, Article 137054. https://doi.org/10.1016/j.chemosphere.2022.137054
92. Sanguanpak S, Chiemchaisri C, Chiemchaisri W, Yamamoto K (2015) Influence of operating pH on biodegradation performance and fouling propensity in membrane bioreactors for landfill leachate treatment. Int Biodeterior Biodegradation 102:64–72. https://doi.org/10.1016/j.ibiod.2015.03.024
93. Shahid MK, Kashif A, Rout PR, Aslam M, Fuwad A, Choi Y, Park JH, Kumar G (2020) A brief review of anaerobic membrane bioreactors emphasizing recent advancements, fouling issues and future perspectives. J Environ Manage 270:110909
94. Shin C, Bae J (2018) Current status of the pilot-scale anaerobic membrane bioreactor treatments of domestic wastewaters: a critical review. Biores Technol 247:1038–1046
95. Shin C, Kim K, McCarty PL, Kim J, Bae J (2016) Integrity of hollow-fiber membranes in a pilot-scale anaerobic fluidized membrane bioreactor (AFMBR) after two-years of operation. Sep Purif Technol 162:101–105. https://doi.org/10.1016/j.seppur.2016.02.019
96. Skouteris G, Hermosilla D, López P, Negro C, Blanco Á (2012) Anaerobic membrane bioreactors for wastewater treatment: a review. Chem Eng J 198:138–148
97. Skouteris G, Saroj D, Melidis P, Hai FI, Ouki S (2015) The effect of activated carbon addition on membrane bioreactor processes for wastewater treatment and reclamation—a critical review. Bioresour Technol 185:399–410. https://doi.org/10.1016/j.biortech.2015.03.010
98. Smith AL, Stadler LB, Love NG, Skerlos SJ, Raskin L (2012) Perspectives on anaerobic membrane bioreactor treatment of domestic wastewater: a critical review. Bioresour Technol 122:149–159. https://doi.org/10.1016/j.biortech.2012.04.055
99. Song X, Luo W, Hai FI, Price WE, Guo W, Ngo HH, Nghiem LD (2018) Resource recovery from wastewater by anaerobic membrane bioreactors: opportunities and challenges. Bioresour Technol 270:669–677. https://doi.org/10.1016/j.biortech.2018.09.001
100. Svojitka J, Dvořák L, Studer M, Straub JO, Frömelt H, Wintgens T (2017) Performance of an anaerobic membrane bioreactor for pharmaceutical wastewater treatment. Biores Technol 229:180–189
101. Tang J, Wang XC, Hu Y, Ngo HH, Li Y (2017) Dynamic membrane-assisted fermentation of food wastes for enhancing lactic acid production. Bioresour Technol 234:40–47. https://doi.org/10.1016/j.biortech.2017.03.019
102. Teo CW, Wong PCY (2014) Enzyme augmentation of an anaerobic membrane bioreactor treating sewage containing organic particulates. Water Res 48:335–344. https://doi.org/10.1016/j.watres.2013.09.041
103. Thanh PM, Ketheesan B, Stuckey DC, Zhou Y (2017) Dosing of Ethylenediamine-N,N′-disuccinic acid (EDDS) to improve the bioavailability of Fe^{2+} in the presence of sulfide in a submerged anaerobic membrane bioreactor. Chem Eng J 330:175–182. https://doi.org/10.1016/j.cej.2017.07.158
104. Tian Y, Ji C, Wang K, Le-Clech P (2014) Assessment of an anaerobic membrane bio-electrochemical reactor (AnMBER) for wastewater treatment and energy recovery. J Membr Sci 450:242–248. https://doi.org/10.1016/j.memsci.2013.09.013
105. Trzcinski AP, Stuckey DC (2010) Treatment of municipal solid waste leachate using a submerged anaerobic membrane bioreactor at mesophilic and psychrophilic temperatures: analysis of recalcitrants in the permeate using GC-MS. Water Res 44(3):671–680. https://doi.org/10.1016/j.watres.2009.09.043

106. Umaiyakunjaram R, Shanmugam P (2016) Study on submerged anaerobic membrane bioreactor (SAMBR) treating high suspended solids raw tannery wastewater for biogas production. Biores Technol 216:785–792
107. Viruela A, Robles Á, Durán F, Ruano MV, Barat R, Ferrer J, Seco A (2018) Performance of an outdoor membrane photobioreactor for resource recovery from anaerobically treated sewage. J Cleaner Prod 178:665–674. https://doi.org/10.1016/j.jclepro.2017.12.223
108. Wang Z, Wu Z, Tang S (2009) Extracellular polymeric substances (EPS) properties and their effects on membrane fouling in a submerged membrane bioreactor. Water Res 43(9):2504–2512
109. Wang J, Wang S (2016) Removal of pharmaceuticals and personal care products (PPCPs) from wastewater: a review. J Environ Manag 182:620–640. https://doi.org/10.1016/j.jenvman.2016.07.049
110. Wang X, Chang VWC, Tang CY (2016) Osmotic membrane bioreactor (OMBR) technology for wastewater treatment and reclamation: advances, challenges, and prospects for the future. J Membr Sci 504:113–132. https://doi.org/10.1016/j.memsci.2016.01.010
111. Wang X, Wang C, Tang CY, Hu T, Li X, Ren Y (2017) Development of a novel anaerobic membrane bioreactor simultaneously integrating microfiltration and forward osmosis membranes for low-strength wastewater treatment. J Membr Sci 527:1–7. https://doi.org/10.1016/j.memsci.2016.12.062
112. Wang Z, Ma J, Tang CY, Kimura K, Wang Q, Han X (2014) Membrane cleaning in membrane bioreactors: a review. J Membr Sci 468:276–307. https://doi.org/10.1016/j.memsci.2014.05.060
113. Watanabe R, Nie Y, Wakahara S, Komori D, Li Y-Y (2017) Investigation on the response of anaerobic membrane bioreactor to temperature decrease from 25 °C to 10 °C in sewage treatment. Bioresour Technol 243:747–754. https://doi.org/10.1016/j.biortech.2017.07.001
114. Wongchitphimon S, Rongwong W, Chuah CY, Wang R, Bae T-H (2017) Polymer-fluorinated silica composite hollow fiber membranes for the recovery of biogas dissolved in anaerobic effluent. J Membr Sci 540:146–154. https://doi.org/10.1016/j.memsci.2017.06.050
115. Wu B, Kim J (2020) Anaerobic membrane bioreactors for nonpotable water reuse and energy recovery. J Environ Eng 146(2):03119002
116. Xiao X, Shi W, Huang Z, Ruan W, Miao H, Ren H, Zhao M (2017) Process stability and microbial response of anaerobic membrane bioreactor treating high-strength kitchen waste slurry under different organic loading rates. Int Biodeterior Biodegradation 121:35–43
117. Xie K, Lin HJ, Mahendran B, Bagley DM, Leung KT, Liss SN, Liao BQ (2010) Performance and fouling characteristics of a submerged anaerobic membrane bioreactor for kraft evaporator condensate treatment. Environ Technol 31(5):511–521. https://doi.org/10.1080/09593330903527898
118. Xie Z, Wang Z, Wang Q, Zhu C, Wu Z (2014) An anaerobic dynamic membrane bioreactor (AnDMBR) for landfill leachate treatment: performance and microbial community identification. Bioresour Technol 161:29–39. https://doi.org/10.1016/j.biortech.2014.03.014
119. Xing B-S, Zhang Y, Zhang R-Q, Ma J, Chang X-L, Tang X-F, Li Y-Y, Wang XC, Chen R (2023) An anaerobic dynamic membrane bioreactor (AnDMBR) system enhanced the biogas conversion efficiency and stability of mesophilic codigestion with waste activated sludge and food waste. Chem Eng J 471:144432. https://doi.org/10.1016/j.cej.2023.144432
120. Xue Y, Zhao H, Ge L, Chen Z, Dang Y, Sun D (2015) Comparison of the performance of waste leachate treatment in submerged and recirculated membrane bioreactors. Int Biodeterior Biodegradation 102:73–80. https://doi.org/10.1016/j.ibiod.2015.01.005
121. Yang J, Gou Y, Fang F, Guo J, Lu L, Zhou Y, Ma H (2018) Potential of wastewater treatment using a concentrated and suspended algal-bacterial consortium in a photo membrane bioreactor. Chem Eng J 335:154–160. https://doi.org/10.1016/j.cej.2017.10.149
122. Yee TL, Rathnayake T, Visvanathan C (2019) Performance evaluation of a thermophilic anaerobic membrane bioreactor for palm oil wastewater treatment. Membranes 9(4):55
123. Yue X, Koh YKK, Ng HY (2015) Effects of dissolved organic matters (DOMs) on membrane fouling in anaerobic ceramic membrane bioreactors (AnCMBRs) treating domestic wastewater. Water Res 86:96–107

124. Zayen A, Mnif S, Aloui F, Fki F, Loukil S, Bouaziz M, Sayadi S (2010) Anaerobic membrane bioreactor for the treatment of leachates from Jebel Chakir discharge in Tunisia. J Hazard Mater 177(1–3):918–923
125. Zhang X, Wang Z, Wu Z, Lu F, Tong J, Zang L (2010) Formation of dynamic membrane in an anaerobic membrane bioreactor for municipal wastewater treatment. Chem Eng J 165(1):175–183. https://doi.org/10.1016/j.cej.2010.09.013
126. Zhao T (2012) Study on the performance of anaerobic self-forming dynamic membrane (AnSFDM) for wastewater treatment and mechanism of AnSFDM formation and fouling. Dissertation
127. Zhen G, Pan Y, Lu X, Li Y-Y, Zhang Z, Niu C, Kumar G, Kobayashi T, Zhao Y, Xu K (2019) Anaerobic membrane bioreactor towards biowaste biorefinery and chemical energy harvest: recent progress, membrane fouling and future perspectives. Renew Sustain Energy Rev 115:109392. https://doi.org/10.1016/j.rser.2019.109392

Chapter 4
New Constructed Wetlands

4.1 Introduction

The advantages of cost-effectiveness, ecological friendliness and biological sources, constructed wetlands (CWs) also have some disadvantages, such as slow processing speed and huge land-use area [45]. It mainly depends on microorganisms and large plants to remove pollutants in wastewater. Therefore, unlike the traditional treatment system, it cannot directly treat any high concentration toxic pollutants in industrial wastewater. However, metals and other pollutants that are difficult to remove by traditional physical and chemical methods can be effectively removed by CWs. Wastewater with heavy load or rich nutrients may also have adverse effects, such as transfer of dominate species, leaching of nutrients in the system. Despite the biochemical properties could effectively remove organic pollutants, the total amount of pollutants cannot be completely removed.

In addition to biological processes, CWs are also treated by physicochemical processes. The types of substrates in wetland could determine the time span of continuous pollutant removal effect under its average loading rate. With the increase of age, the performance of CWs began to decline, because the physical and chemical processes gradually slowed down, such as the sediment adsorption, solid accretion, and particle accumulation. Therefore, the increase of concentration of pollutants in the regeneration water was detected over time. The removal of phosphorus is seriously affected by the removed of most phosphorus, and microbial membrane absorption and large-scale plant absorption only take a little portion [11].

Some simultaneous studies contradict the practice of CWs treated water for plant irrigation. For example, some researchers studied the pollution conditions while reusing the treated wastewater on the growth of tomato crops, and it was concluded that the fruits not in direct contact with the soil were not polluted. In addition, soil hydrology could also has adversely affected on the pollution condition. The study showed that the coal water slurry treated water could not be directly used to irrigate

G. Dawen and M. Nabi, *Novel Approaches Towards Wastewater Treatment*,
Springer Water, https://doi.org/10.1007/978-3-031-55189-5_4

tomato crops. It must applied the pretreatment with the necessary disinfection [1], which can be used under controlled conditions [29].

Several studies indicate that bullfrog tadpoles residing in a wetland supplemented with water from a CWs exhibit notable differences. The findings suggest that, in comparison to the control tadpoles, the tadpoles exposed directly to the effluent exhibit slower growth, a higher frequency of abnormalities, and less deformity [99]. Other studies showed that only use of CW as an independent treatment system is not enough to remove pollutants and pathogens to meet the requirement. The combination of CW with other treatment technologies, such as filtration, ultraviolet radiation, chlorination, ozonation, up flow anaerobic bioreactor (UASB), stabilization, anaerobic digestion, etc., is crucial for high-quality treatment [74]. Wetland is a long-term or short-term thin layer of water or intermittent water on the earth, and a soil over wet section with marshy or hygrophytic plants. It is the general name of various marshy and hygrophytic regions in land, flowing water, still water, estuarine and marine systems.

CWs are human-made environmental ecosystems designed for sewage treatment. They utilize a combination of physical, chemical, and biological processes to effectively remove pollutants, resulting in the purification of wastewater. By leveraging the joint action and organic integration of these three disciplines, CWs can effectively eliminate contaminants and achieve the desired sewage purification outcomes. At the same time, microorganisms in the CWs continuously promote the purification and treatment effect of sewage by absorbing, growing and reproducing the organic matter and nutrients of sewage, and the organic combination of green plants and microorganisms makes the entire CWs ecosystem more stable, so as to finally realize the resource utilization and harmless treatment of sewage.

CWs offer several advantages over conventional wastewater treatment methods, such as low cost, low energy consumption, low maintenance, high aesthetic value, and ecological benefits. CWs can remove a wide range of contaminants from wastewater, such as organic matter, nutrients, pathogens, metals, and emerging pollutants. However, CWs also face some challenges and limitations that need to be addressed in order to improve their performance and sustainability. Some of these challenges are: the lack of standardized design criteria and guidelines for different types of CWs and wastewater; the variability and uncertainty of the removal efficiency and long-term stability of CWs; the potential clogging of the wetland media due to the accumulation of solids and biomass; the greenhouse gas emissions and odor production from CWs; the disposal or reuse of the harvested plants and accumulated sludge from CWs. Therefore, this chapter focus on: developing novel and innovative wetland materials, configurations, and hybrid system; optimizing the design and operation parameters of CWs for different wastewater types and climatic conditions; evaluating the performance and environmental impacts of CWs using advanced monitoring and modelling tools; enhancing the removal of emerging pollutants and pathogens from wastewater using CWs, and exploring the potential of CWs for water reuse, nutrient recovery, energy production, and carbon sequestration.

4.2 New Types of Wetlands Substrates

As the largest volume construction part of CWs, the substrate is known as the media that could support the growth of organisms and microorganisms' communities, and it also makes a valuable contribution to the water treatment. Various water contaminants, such as non-biodegradable pollutants and toxic metals, can be primarily removed through traditional chemical, physical, and biological mechanisms. These mechanisms include processes such as filtration, adsorption, interception, and ion exchange. These techniques play a vital role in the removal of pollutants from water sources, ensuring improved water quality and minimizing potential harm to the environment and human health [55]. After many studies in the development of CWs, different types of substrates have been used for wetland construction, such as industrial, natural, and artificial substrates classified based on their sources and characteristics [160]. Traditional substrates have some application problems, like clogging, low efficiency of removal, high cost, second pollutions, and other issues [137]. Therefore, it is highly desirable to explore some novel substrates that have great performance, low cost, and high safety to tackle the current issues. The selection of substrates mainly determines the enhancements of the CW efficiencies. In more specific, adaptability and cost are the two fundamental factors that should be primarily considered during the selection. In this part, the conventional substrates were briefly reviewed, then it would mainly focus on the newly emerged substrates with the new classification based on their main characteristics to remove pollutants in CWs.

4.2.1 Conventional Substrates

Substrates in CWs, also known as fillers, are generally composed of several media such as clay, coarse materials, rocks, broken tiles, industrial waste, fly ash, organic materials, shale, zeolite, etc. Therefore, various materials including soil, sand, minerals, organic materials and industrial wastes such as fly ash, steel slag and pulverized coal can be used as substrates of CWs. CW substrate could provide nutrition for the growth of organisms, it also has ability to remove pollutants by adsorption, precipitation and filtration. Natural substrates, like Zeolite, gravel, diatomite, limestone, and volcanic rock, are thought very suitable for the CWs construction with excellent properties of adsorption and ion exchange capacity. Industrial waste including fly ash, coal cinder and construction waste refers to solid, semi-solid, liquid or waste materials that were produced from industrial processes. These industrial wastes are good alternatives to adsorbent materials, which reflects the principles of waste utilization. The artificially modified matrix is usually made of natural substances or wastes, including ceramsite, activated carbon, synthetic fiber and modified swelling.

4.2.2 *Emerged Substrates*

As the increase of emerging pollutants, the construction of CWs should be upgraded to enhance treatment efficiency. Thus, each part of CWs should be redesigned, particularly the substrates. Determining the properties and advantages of various substrates should be helpful for the reconstruction of CWs [55]. Emerged substrates can be divided into three groups: organic substrates, inorganic substrates and mixed substrates by their characteristics and sources. In general, emerged substrates have resulted in great gains in the physical factors, like electrical conductivities, hydraulic, particle size, porosity, mechanical strength, and in chemical factors (e.g., chemical stability, surface charge, and biological characteristics) [160]. A wide variety and the main information of emerged substrates are showed in Table 4.1. Organic materials are well known for its abilities to provide carbon sources and serve as microbial carriers and metal adsorbents [21]. They are cheap and promising alternative fillers for wastewater remediation. More recently, organic substrates like alum sludge, PHBV and PLA blend, bio-ceramic, biochar became established as CWs substrates. Inorganic substrates mainly come from the aluminum, calcium, and iron, which have higher ion-exchange capacities.

4.2.2.1 Organic Substrates

(a) **Alum sludge materials**

Alum sludge is an aluminum-containing sewage treatment residual in sludge form, and its high P adsorption capacity has been applied for phosphorus (P) removal [7]. Zhao et al. [175] used the wastewater treatment cake-like sludge in Ireland to demonstrate excellent nitrification and removal performances of organics, and the enhancement of phosphorus removal rate was also related to the amount of Al. Alum sludge could have agronomic value for P deposition. Babatunde and Zhao [7] conducted an experiment on phosphorus fractionation. Their findings revealed that the diffusion of adsorbed P in pores was influenced by the presence of high-concentration soluble P. Additionally, a significant portion of labile inorganic P was found to be stored within the interior of the alum sludge over the long term. Consequently, exchangeable organic P is more readily decomposed compared to other forms of P, although its overall quantity is relatively small.

Although the inorganic form were most of P stored in, Qualls and Richardson [94] found out four ca-bound P deposit processes: (a) precipitation by calcareous algae, (b) coprecipitation with carbonate, (c) chemical precipitation of calcium phosphates, or (d) adsorption/precipitation of PO_4 on the calcium carbonate surfaces. And they also found three reasons that contribute to deposit insoluble organic P: (a) plants largely uptake P under the ground with plant residues in partial decay, high solution concentrations, humification and death part.; (b) in the process of decomposition and humification of plant residues, microorganisms immobilize phosphorus to produce detritus complexes rich in phosphorus, and phosphorus comes from dead microbial

Table 4.1 Summary of the main characteristic information about substrates used in CWs

Substrate categories	Substrates	Removal pollutants (removal efficient %)	Properties	Advantages	Disadvantages
Natural substrates	Zeolite	BOD_5 (69–95%) COD (22–76%) NH_4^+-N (28 ± 40%) PO_4^{3-}-P/TP (20 ± 33%)	Porous silicate framework, strong ability to exchange other metal elements	Higher removal Efficiencies of COD and N	Mining Zeolite is not environmentally friendly
	Soil	TN (75%) TP (80%)	High moisture	Cheap to get	Weathering and erosion could break down it easily
	Gravel	COD (60%) TN (68%) TP (20–40%)		High hydraulic permeability coefficient	Low removal rates of P and N
	Shale		A by-product of maricultural	Benefit reuse of waste	The particles are fragile
	Limestone	COD (22–58%) TN (42–52%) TP (54–68%)	Good processability and polishing property	Can be used as adsorbent	Not good for the survival of microorganisms
	Volcanic		Good water transmissibility	Can get raw materials easily	
	Vermiculite		Strong ability of ion exchange and adsorption	No side effects	
Industrial and agricultural substrates	Quartz sand	NH_4^+-N (83%)	Compression and wear resistance, high mechanical strength	Cheap to obtain	
	Oyster shell		Maricultural wastes	Benefit reuse of waste	The varies adsorption capacity of pollutants
	Fly ash	TP (70%)	Porous honeycomb surface	The adsorption capacity is high	The particles are too fine
	Steel slag	PO_4^{3-}P (94.2%)	Crystalline structure	P removal performance	The effluent is usually alkaline

(continued)

Table 4.1 (continued)

Substrate categories	Substrates	Removal pollutants (removal efficient %)	Properties	Advantages	Disadvantages
Synthetic substrates	Bio-ceramic	TP (43%)	Developed micropore structure	Good adsorption capacity	
	Activated carbon		Developed pore structure	High adsorption	Low removal rates of pollutants
	Biochar	COD (93%) TN (42–52%) NH_4^+-N (86%) TP (87%)	Feedstock and pyrolysis conditions depended the properties	Reusing the waste biomass and effective adsorbent	High energy requirement of pyrolysis

biomass; and (c) enzymatic and physicochemical incorporation of phosphorus in humid substances during formation. The mechanism for P removal is mainly depend on adsorption and precipitation with alum sludge, which appears as –OH was replaced by phosphate ions. Therefore, ligand exchange could be the dominating pathway for adsorption and precipitation. Being that decomposed phosphate by alum sludge may not that easy due to the Al and Fe fraction associated with the inorganic [141] provide that alum sludge may generate secondary pollution, because of the releasing of Al^{3+} during the utilization. Beside the alum sludge, other Al-rich materials, like red/fly-ash brick, bauxite, and raw/acid activated red mud, have also been used as substrates in CWs. P-sorption substrates are strongly relied on the fraction of Ca, Fe or their oxides or hydroxides. Ca-rich materials and Fe-rich materials could be successfully applied for P removal in CWs efficiently. For example, Yang et al. [160] summarized Ca-rich and Fe-rich materials (both in natural/artificial) that applied in CWs, like Ca-rich materials (limestone, calcite, marble, wollastonite, apatite, steel slag, ceramite), Fe-rich materials (ochre, peat, spongy iron, micro-electrolysis of iron scraps and active carbon).

The above researches show that alum sludge has good abilities in phosphorus adsorption. However, problems with easy blocking and high energy consumption in complex process still remain unresolved and may even get worse. Thus, it needs develop an economical and efficient way to modify alum sludge, which should greatly improve the adsorption performance of adsorbents. And the advantages of cost-effective and environmental protection should be considered to not only prevent blockage, but also reduce energy consumption. Also, alum sludge contains a large number of metal aluminum ions, which can be dissolved by acidification. Therefore, production and release of secondary pollutants to environment should try to avoided, to achieve the sustainable development.

(b) **Poly-3-hydroxybutyrate-co-3-hydroxyvalerate (PHBV) and Polyacetic acid (PLA) blend**

As a substitute for organic carbon sources, Poly-3-hydroxybutyrate-co-3-hydroxyvalerate (PHBV) are biodegradable polymers that are artificially made and have attracted a lot of interest lately. This is primarily due to their superior efficiency in denitrification and their increased capacity to stably limit the release of dissolved organic carbons. These polymers offer an advantageous solution for promoting effective denitrification processes while ensuring a controlled and stable supply of organic carbon [111]. Organic carbon could be a good electron donor to stimulate nitrogen removal performance, especially for low C/N ratio situation in microbial denitrification process. The soluble carbon sources have been used in the construction of CWs to remove nitrate, such as methanol, fructose, and hydrolyzed sludge [161]. The process of denitrification depends on the availability of soluble carbon sources, which can vary depending on the quality of the effluent water. This makes the process more challenging. Natural material is easily obtained and treated which could be used as substrates in CWs, such as maize cobs, cattail litter, wheat straw, and rice husk. But those have problems on denitrification efficiency and the color stain. PHBV, a highly biodegradable polymer produced by microorganisms, is often underutilized due to its high cost. However, blending PHBV with PLA offers a cost-effective solution. This blend serves as both a carbon source and a biofilm carrier in denitrification processes, reducing expenses and enhancing efficiency. The releasing of amount of carbon source was measured by the effluent COD concentration. In general, the effluent COD depends on the carbon sources biodegradability and the enzymatic hydrolysis in biodegradation process. PHBV/PLA provided the large amount of organic matters, which could increase microorganisms population. Some researchers found that the effluent COD will remain in stable with PHBV/PLA added, because the growth of microbial could achieve balance within the denitrification process. The solid carbon source, PHBV/PLA, could serve as a biofilm, which provide more stable environment for microbial growth [161]. And it was added as both external carbon source and energy source to improve ammonia removal efficiency. Compared with PHBV/PLA, many studies found that the certain accumulation of NH_4^+-N was appeared after other treatments. The free NH_4^+-N may be harmful to aquatic plants and microorganisms by consuming the dissolved oxygen (DO) [161]. Nitrification mainly occurs in the chemo-autotrophic process with aerobic condition, but lower oxygen conditions may limit the reaction. Many organisms consumed the oxygen to utilize carbon sources. In contract, it could limit NH_4^+-N removal efficiency due to the shortage of oxygen in CWs. To supplement oxygen under limited aeration, PHBV/PLA was used.

Higher NH_4^+-N removal efficiency related to the aerated condition in CWs. Thus, artificial aeration has been used to improve nitrification. Treatments like limited aeration intermittent aeration, and continuous aeration were investigated on domestic wastewater treatment in CWs. Limited aeration has distinct advantages in denitrification processes as it facilitates the degradation of nitrified nitrogen through heterotrophic microbial and anaerobic pathways. Moreover, the addition of PHBV/

PLA enhances aeration conditions, resulting in higher rates of NO_3^--N and NH_4^+-N removal. Studies on substrate utilization in VSSF CWs have demonstrated impressive removal rates, reaching approximately 91.00% for NH_4^+-N and 97.03% for total nitrogen [109]. PHBV is an intracellular microbial storage substance, so it can be metabolized by many microorganisms. PLA is a chemical polymer that is more difficult to degrade than PHBV. Due to their distinct biodegradability, they might generate small size pores or channels in PHBV/PLA carrier during cultivation. These structures offer habitats for the process of anoxic and anaerobic denitrification. Therefore, the composition and configuration of carbon sources have good effect on the rapid colonization.

In order to identified the microbial abundance, the results of pyrosequencing have been used to show that denitrifying bacteria were dominant in both CWs. In addition, microbial community structure could be changed by limited aeration. But it had no obvious inhibition on the activity of denitrifying bacteria. PHBV/PLA is kind of the porous ceramsite that could provide an aerobic environment for the microbial growth, and it also provide the fermentation process for anaerobic digestion. In conclusion, limited aeration is a viable option to remove nitrogen via CWs supported by PHBV/PLA because it can provide sufficient oxygen for nitrification and denitrification processes while minimizing the energy consumption and cost of the system [161].

(c) **Bio-ceramic**

Bio-ceramics are elliptical, spherical or irregular granular that are made by artificial matrices with clay (including bentonite, kaolin, bleach, etc.). A series of processes such as drying, grinding, and pelletizing were included in bio-ceramics production. Despite biofilm growth being slower on the surface area of bio-ceramic compared to traditional substrates, bio-ceramic excels in preventing biofilm consolidation, internal blockage of fillers, and the accumulation of floating mud on the water surface. As a result, it demonstrates a high potential for efficient nitrogen and phosphorus removal relay on the high adsorption stability, great capacity, and excellent affinity. With further research on the use of bio-ceramics in CWs, some studies have found that when the particle size belong in range of 5–25 mm, have low COD removal efficiency. Jin et al. [57] stated that when the concentration is about 480–500 mg/l, the COD removal rate by bio-ceramics is only about 27.4%. However, Min et al. [87] indicated that the bio-ceramics saturated adsorption capacity of phosphorus could reached 238.10 mg/kg with the low pollution removal rate of 31.2%. The possible explanation should be that the low fraction of calcium oxide is not good for the chemisorption of phosphorus. Zhao et al. [175] compared six substrates (limestone, ceramic medium, zeolite, bio-ceramics, shale and coarse sand) with their abilities of the NH_4^+-N adsorption efficiency. The highest saturated adsorption was about 666.68 mg/kg, which has greater capacity than the zeolite (1250.00 mg/kg). Jin et al. [58] used a lab-scale simulated CWs to treat sewage water. Four kinds of substrates, quartz sand, bio-ceramsite, anthracite and shale, were used to determine the best treatment effect. The results showed that: (1) dissolved organic phosphorus could be removed by Bio-ceramsite efficiently; and anthracite had the best total phosphorus removal rate i.e., 85.87%. (2) Precipitation, and biological and chemical adsorption could be the primary processes for bio-ceramsite to remove COD, TP and chromaticity.

Furthermore, in addition to its excellent performance in eliminating organic pollutants, bio-ceramic also serves as a cost benefit and efficient material for removing heavy metals in CWs. Mu et al. [91] conducted research on treating Cr(VI)-contaminated wastewater. They combined CWs with microbial fuel cells to address the high mobilized and toxic Cr(VI), which demonstrated the potential for Cr(VI) removal. In that study, calcite, zeolite, bio-ceramic, and volcanic rock were used with up-flow CWs-microbial fuel cell systems to examine the absorption capacity of fillers during the extended operation period. They found that although bio-ceramics only had a 72.2% removal rate of Cr(VI), It is simple to immobilize microorganisms.

(d) **Biochar**

Biochar is an insoluble solid material rich in carbon obtained by pyrolysis of biomass at a certain temperature under anaerobic conditions. It has stable physical and chemical properties, reasonable pore distribution, large surface area and strong ability of adsorption characteristics, and is widely utilized in soil treatment, climate regulation and wastewater treatment. Generally speaking, raw materials such as shells, wood, straw, sludge and manure can all be sources of biochar and could be reused as shown in Fig. 4.1. Its phytochemical properties are highly similar to those of biomass materials in terms of type and degree of pyrolysis temperature. Specifically, the amount of surficial functional groups, the degree of aromaticity and the size of biochar are mainly determined by temperature, which are key characteristics for treating polluted water [154]. There are lots of researches about wetlands constructed on biochar substrates, which found that biochar has significant effects on the treatment of TP, COD and TN in sewage. For example, the main way to remove nitrogen depends on nitrification–denitrification, and biochar can enhance the nitrogen removal performance of wetlands. Wang et al. [144] explicate that under the optimal hydraulic retention time (2.5 d at 8–12 °C) in a baffled wetland, the removal rates of NH_4^+-N, COD, phosphate (PO_3^{4}-P) and suspended solids (SS) were 72.38, 90.51, 90.73 and 94.57%. Lu et al. [79] studied on the adsorption capacity of biochar saturated with NH_4^+-N and found it to be extremely high at 1.35 mg/g. Furthermore, the biochar contribute low desorption rate of NH_4^+-N, indicating that biochar has stability adsorbed ability. Wan et al. [133] discovered that phosphorus was highly absorbed by biochar with an absorption rate of 0.7146 mg/g. However, its adsorption capacity for NH_4^+-N is not as good as that for phosphorus due to the biochar surface's negative charge with low anion exchange ability and functional groups, which limits the adsorption efficiency of phosphate [78]. Nitrogen and phosphorus are mainly adsorbed by biochar through physical and chemical adsorption processes. The biochar's specific surface area can affect physical adsorption by chemisorption that highly correlated with the type of biochar and its number of functional groups. During nitrogen adsorption, the functional groups could react with nitrogen through electrostatic adsorption and ion exchange. During phosphate adsorption, the morphology of metal compounds have

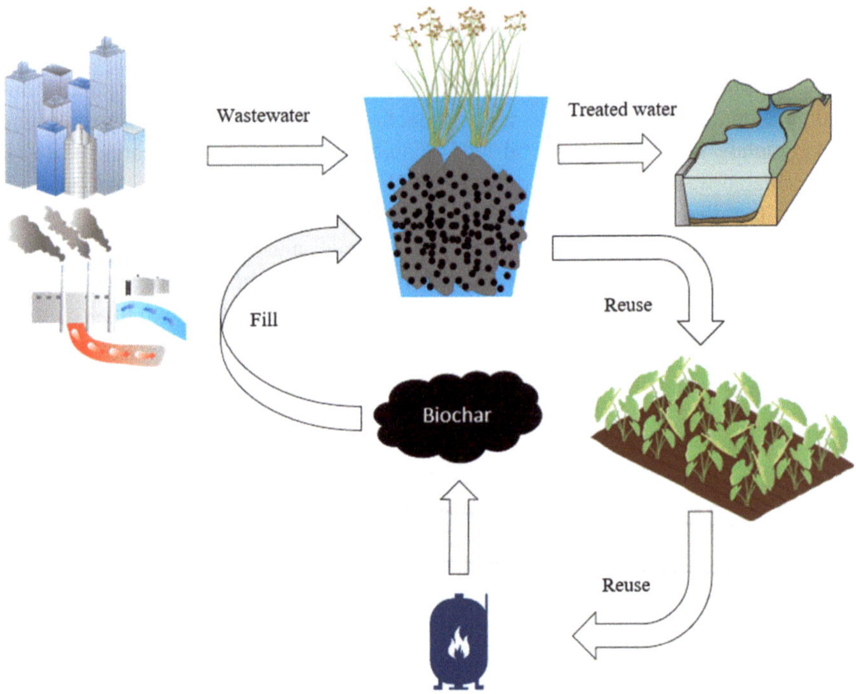

Fig. 4.1 The use of biochar substrate for wastewater treatment in CWs and its recycling circle [36]

a great influence on the adsorption strength because surface deposition through the reaction of metal compounds is the main process of phosphate removal [163].

Biochar is solid carbon that become a kind of adsorption material with good performance because of containing amount of oxygen with rich porosity, huge surface area and functional groups. However, the adsorption performance is mainly influenced by the selection of raw materials, preparation conditions and pretreatment methods, which leads to significant differences in adsorption capacity for various pollutants. In addition, biochar itself has some defects greatly limits its phosphorus and other substances adsorption, such as surface negative charge, low capacity for anion exchange and less effective functional groups and so on. Therefore, the modification of biochar to improve its adsorption capacity is a research hotspot at present. Hu et al. [52] focused on the effect of wastewater herbicide atrazine removal by adding biochar in the laboratory scale reed bed CWs. They found that the removal rate of atrazine increased from 50 to 70% and COD was from 66.7% to 86.7%, respectively, by adding sulfuric acid modified biochar. Generally speaking, adding biochar could improve the removal performance of nitrogen and phosphorus in reed bed CWs, and the effluent could reach China's surface water quality standard with Class III, which is better than the inferior Class V. Modified biochar, as a synergist, has a certain application potential to treat herbicide wastewater and has a wide application prospect.

At present, the modification methods of biochar mostly adopt coprecipitation, impregnation, chemical precipitation and other ways to load different forms of metal elements on biochar to provide more adsorption sites, therefore it improves the ability to absorb P. This process uses biochar as carrier, which solves the problems that metal elements are difficult to recover and easy to form secondary pollution after entering water. Furthermore, the incorporation of Layered double hydroxides (LDHs) within biochar presents a synergistic approach to address the issue of limited adsorption capacity for phosphate due to the tight packing of LDHs. This synergistic strategy is characterized by the following aspects: Firstly, utilizing biochar as a matrix carrier facilitates the dispersion of LDHs. Secondly, the presence of LDHs enhances the number of surface active sites, thereby augmenting its adsorption capacity. The interaction between biochar and LDHs further amplifies their combined adsorption capabilities.

Similarly, due to the intrinsic magnetic interaction between γ-Fe_2O_3 and Fe_3O_4, it is easy to agglomerate, which leads to the decrease of specific surface area and catalytic activity for pollutants. Therefore, the magnetic modification of biochar is also very important. In general, the modification of biochar not only enhances its adsorption performance through synergistic effects but also facilitates the separation of biochar from water [15].

(e) **Peat**

Peat is kind of a spongy material formed with the decomposition of organic matter, which primarily includes plant material, such as fens, muskegs, swamps, moors and bogs. Cellulose and lignin are the main components in peat [95]. Lignin also has the polar functional groups, including ethers aldehydes, alcohols, acids, ketones, phenolic and hydroxides which contribute to form chemical bonds. The p-hydroxy group is the main component of lignin and Humus. Therefore, polar organic molecules and transition metals could be adsorbed by peat with its high adsorption capacity. In addition, its highly porous structure play an important role in microscopic studies. The porosity of partially decomposed peat is about 95%, and the specific surface area is 200 m^2/g [34].

Exchange cations is one of the characteristic of peat. McKay and Poots [85] calculated the surface area of sphagnum by using acid alkaline dyes. In the range of 150–250/m particle size, the adsorption capacity of basic dyes is 122.2 m^2/g, that of acid dyes is 11.8 m^2/g, and that of nitrogen is 27.3 m^2/g. Negatively charged groups in humus and rot soil has a strong attraction to basic cations. On the contrary, peat has high cation exchange capacity and low anion exchange capacity, especially in the case of acidic dyes, where negatively charged groups repel anions, resulting in a low specific surface area [125]. There are several factors could affect the exchangeable cation properties of peat, which includes the degree of peat decomposition present during peat formation, environmental conditions and collection methods. Indeed, when peat is extracted from swamps and subjected to processes such as drainage, air-drying, drying, milling, or crushing, certain properties of peat undergo significant changes, despite the composition of peat largely remaining the same. Studies have revealed that various types of water-holding capacities of peat undergo a reduction to varying degrees following processes such as drainage, air-drying, drying, milling, or crushing. Additionally, the porosity and permeability of peat also undergo changes during these processes. As a result, the utilization of air-dried peatland in the treatment of peatland wastewater encounters limitations in several research studies.

It has been found that peat is good for removing some organic matter, suspended solids, and even some heavy metals from CWs [34]. Gunes [43] used this unconventional medium in CWs. The COD removal rate was significantly improved, and it was concluded that most of the influent COD was captured by plant roots and peat matrix [100]. Because peat is a polar material, as a natural adsorbent, it can also adsorb most metals [34]. Therefore, peat is good for withdrawing non-ferrous metals and polyvalent metal cations from wastewater. One of the benefits of having a high chelating ability is that it can be used to develop methods for removing various metals and organic compounds from the environment. Humic acid, for example, can help eliminate metal ions such as chromium, copper, antimony, cadmium, iron, phosphates, nickel, lead, silver and mercury. Dissanayake and Weerasooriya [37] found that the removal of humic acid led to a decrease in the adsorption ability of metals. Humic acid acts as a chelating agent, forming bonds with metal ions through hydroxyl groups in aromatic carboxylic acids. Additionally, humic acid's cation exchange and surface adsorption properties contribute to its effective metal adsorption.

4.2.2.2 Inorganic Substrates

(a) **Construction Wastes**

Construction waste is a by-product of urbanization, which has a large amount and low use rate, which results in extensive pollution of solid waste. Based on the research of Wang et al. [135], the municipal solid waste produced in China accounts for about 29% of the world every year, of which the amount of construction waste is about 40%. Red brick, fly ash brick and other broken bricks are typical construction waste. However, it is porous and rich in iron, calcium, aluminum and other elements, so it is expected to become an ideal substrate for phosphorus removal from coal water slurry in resource recovery and utilization. Some researchers have studied the recycling of construction waste. For example, Wang et al. [145] provided that broken bricks could improve the growth of aquatic plants and some microorganisms in coal water slurry, and are effective materials for treating pig wastewater by coal slurry. It can be seen that some achievements have been made in the recycling of construction waste, but the research in CWs has yet to be developed. Figure 4.2 illustrates a microscopic system diagram of a CW substrate designed for treating construction waste. The substrate incorporates a multi-dimensional graded matrix, consisting of various layers. The bottom support layer is filled with coarse gravel, while the middle section utilizes red brick and fly ash brick as filling materials. The surface layer can consist of gravel or similar materials, promoting plant growth and facilitating the dispersion of wastewater within the system.

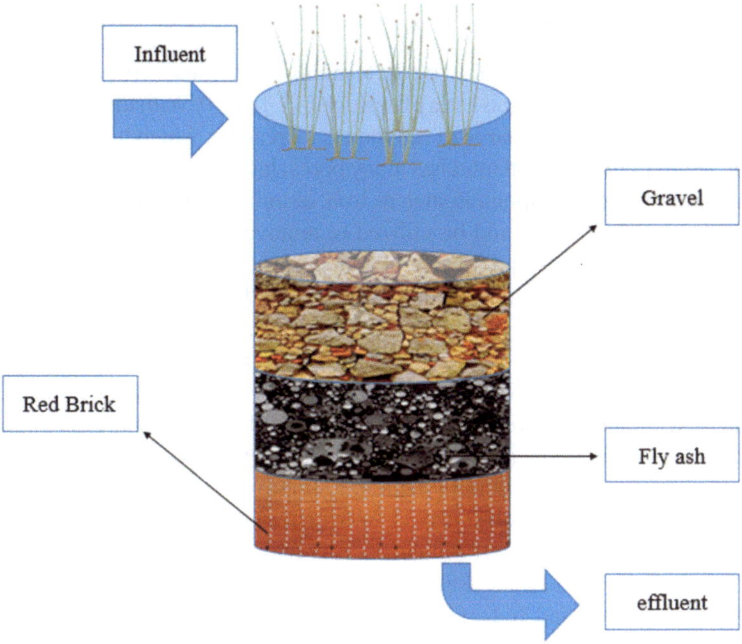

Fig. 4.2 VFCW systems schematic diagram (substrates of fly-ash brick, red brick, and gravel) in laboratory-scale [55]

Typical construction waste (red brick, gravel, fly ash, etc.) can be used as the phosphorus removal substrate of intermittent vertical flow circulation CWs, which has good phosphorus removal capacity and promotes the growth of microorganisms. The transformation and retention of phosphorus in VFCW with different substrate combinations still need to be studied. One thing that is certain due to the rising costs and limited availability of natural materials, the utilization of waste materials as substrates has enabled the development of low-cost alternatives. Due to the local geographical conditions, availability of materials, and cost of material processing, the use cost of substrates may achieve 41% of the total construction cost [25]. Therefore, the recycling of materials has attracted attention. For example, synthesized ceramsite, oyster shell, waterworks sludge and alum sludge [156], a by-product of waterworks, and Wollastonite can be reused as the substrate of commercially valuable CWs. Hence, the reuse of construction waste as the primary material for CWs substrates offers significant cost savings. Firstly, it reduces the construction costs of the wetland system by utilizing readily available and renewable materials. Secondly, it provides a cost-effective solution for waste reduction. Lastly, it enhances the economic value by repurposing and deriving benefits from the construction waste [112].

(b) **Red mud (activated)**

Red mud is a solid residue produced by the leaching of alumina from bauxite by strong alkali. It is strongly alkaline and has a pH value of 12.1–13.0. With the development of aluminum industry, the discharge of red mud is increasing, which has caused serious environmental pollution. Red mud has many pores, large surface area, and contains a large amount of metal compounds such as iron, aluminum and calcium. It is an ideal recycling material, which could be utilized to remove phosphorus, nitrogen, heavy metals, organic pollutants and other pollutants in water. It is a good adsorbent. Liu et al. [76] studied the phosphorus removal effect of red mud on reclaimed water, and found that the mixing ratio of red mud was 2.5%. The removal rate of total phosphorus could reach 82%. Zhang et al. [167] found that adding appropriate amount of red mud is conducive to the growth of ryegrass, which is very suitable for growing in the red mud mixed matrix. However, the high alkalinity of red mud is the main problem, because it limits the application of adsorption performance, so acid modification treatment has become an important method to improve its adsorption performance. The results show that hydrochloric acid soaking treatment can reduce its alkalinity and enhance the adsorption performance of red mud, so as to reduce environmental pollution. Red mud can be used as the substrate filler to remove water pollutants, but adding acid modified red mud will change the physiochemical properties of wetland substrate, and could affect the growth of plants and the living environment of microorganisms, and then affect the decontamination ability and long-term operation of wetland. Limited research exists on the comprehensive effects of acid-modified red mud in wetland systems, particularly regarding nitrogen and phosphorus removal, plant growth, and microbial community changes. Further investigation is necessary to understand the potential benefits of acid-modified red mud in these areas.

Adding of acid modified red mud can accelerate the remove rate about nitrogen and phosphorus that present in CWs ecosystem, and the increase of adding concentration in the early stage could also improve the treatment efficiency. And the removal rate of phosphorus increases with adding concentration. Adding acid-modified red mud can promote the utilization of total nitrogen and total phosphorus in sediment by plants, and adding acid-modified red mud is more conducive to the long-term growth of wetland aquatic plants. The introduction of acid-modified red mud causes notable alterations in the microbial community structure within the sediment. As the concentration of acid-modified red mud increases, the dominant species of sediment microorganisms change, favoring microorganisms better adapted to the acid-modified red mud environment. These acid-tolerant microorganisms progressively dominate the community as their abundance increases. Appropriate addition of acid modified red mud is beneficial to the treatment efficiency of CWs ecosystem [49].

(c) **Calcium silicate hydrate**

Hydrated calcium silicate is a kind of inorganic material synthesized by hydrothermal synthesis of natural materials such as quicklime and quartz powder, which are cheap, easily available, non-toxic and harmless. Its phosphorus removal is based on the principle of crystallization. The formation of crystal nucleus plays a decisive role in the crystallization process. Hydrated calcium silicate contains soluble OH^- and Ca^{2+}, so the relative concentration of $PO_4{}^{3-}$, Ca^{2+} and OH^- in water is low and becomes metastable. When it contacts with seed crystal, it will also form concentrated adsorption layer on the surface of seed crystal, and local calcium phosphate precipitation will promote the crystallization reaction process, thus improving phosphorus removal efficiency. At present, the research on phosphorus removal materials is very extensive, and the common phosphorus removal materials include fly ash, zeolite, montmorillonite, lime, various iron salts, aluminum salts, inorganic–organic composite cationic flocculants and so on. Hydrated calcium silicate has many phosphorus removal characteristics due to its different material characteristics and phosphorus removal principles. Firstly, hydrated calcium silicate has strong phosphorus removal ability. For common adsorbents, the phosphorus adsorption capacity of fly ash is one part of zeolite and the other part of montmorillonite. The adsorption capacity of most natural adsorbents is only between. Chen et al. [23] found that when xonotlite seed reaction was added to simulated wastewater, the phosphorus removal ability of hydrated calcium silicate could reach the phosphorus removal rate in water, which was much higher than other phosphorus removal materials. Because of its good phosphorus removal performance, it is considered that this material can replace the traditional chemical phosphorus removal agent for advanced treatment of sewage, and has been widely valued.

Secondly, the product recovery rate of hydrated calcium silicate is high. The phosphorus content of the product after the use of xonotlite is higher than that of the high-quality phosphate rock in Morocco. And hydrated calcium silicate can be reused. The reason is that crystalline active regions can be formed on the surface after initial use, which is beneficial to crystallization.

The biggest characteristic of calcium silicate hydrate is that it is non-toxic and harmless. Its raw materials are quicklime and quartz, which are green inorganic materials. It is basically free of phosphorus and heavy metals. The final product of its reaction with phosphorus is hydroxyapatite, which has no secondary pollution and is widely used. It can be reused after recycling [68].

Mostly, calcium silicate hydrate can enhance the phosphorus removal capacity of CWs significantly. The phosphorus removal rate can be increased by 20–40% after adding calcium silicate hydrate. Particle size, dosage and hydraulic retention time of hydrated calcium silicate have significant effects on phosphorus removal rate. Decreasing particle size, increasing dosage, and extending hydraulic retention time in the system can improve phosphorus removal efficiency. The presence of wetland plants enhances phosphorus removal, as better plant growth correlates with improved system performance. Adding calcium silicate hydrate alone can elevate the pH of the system, but the buffering effect of plants helps stabilize the effluent pH at approximately 8.2 [173].

(d) **Zeolite/Mg-Fe-LDHs**

Among many new functional materials, LDHs are anionic layered compounds similar to montmorillonite, also known as anionic clay and hydrotalcite. Their composition and structure are special. LDHs consist of positively charged sheets comprising divalent and trivalent metal ions on both sides, with anions and water molecules situated in between. Additionally, LDHs possess a microporous structure on their surface. These materials exhibit both acidic and alkaline characteristics, allowing for interlayer anion exchangeability and possessing a microporous structure. Moreover, the divalent and trivalent metal ions in its composition structure can make LDHs have different chemical properties, and different kinds of metal ions will make their physical configuration and surface properties change differently. Therefore, the special configuration of LDHs can be used to the VFCWs substrate by bimetallic compound synthesis [19].

MgFe-LDHs are generally prepared by coprecipitation of $MgCl_2$ solution and $FeCl_3$ solution under alkaline conditions, and the ratio is generally 2:1 showed in Fig. 4.3. The results show that it is feasible to use MgFe-LDHs membrane to cover the typical substrate of VFCWs. The purification effects of CODCr, ammonia nitrogen and total phosphorus were improved to varying degrees by the modified matrix. The average removal rates of these pollutants were high, with CODCr being more than 80%, total phosphorus being more than 60%, and ammonia nitrogen being more than 90%.

The modification of MgFe-LDHs coated matrix mainly affects the physiochemical properties of the matrix, and enhances the physical and chemical interception and chemical adsorption of pollutants by the matrix; It has little effect on the biological purification effect of the substrate, which is mainly reflected in the ability of the substrate to attach and grow biofilm. MgFe-LDHs coating modified substrate has good purification effect on VFCWs [169].

(e) **Filtrate/LECA**

LECA is widely utilized as a substrate in CWs due to several beneficial attributes. These include its capacity for phosphate removal, excellent mechanical strength, high

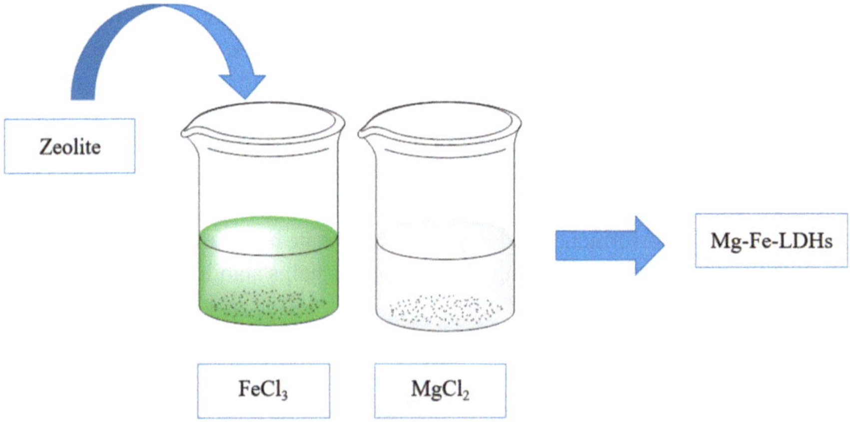

Fig. 4.3 Simplified schematic diagram of synthetic method of Mg-Fe-LDHs

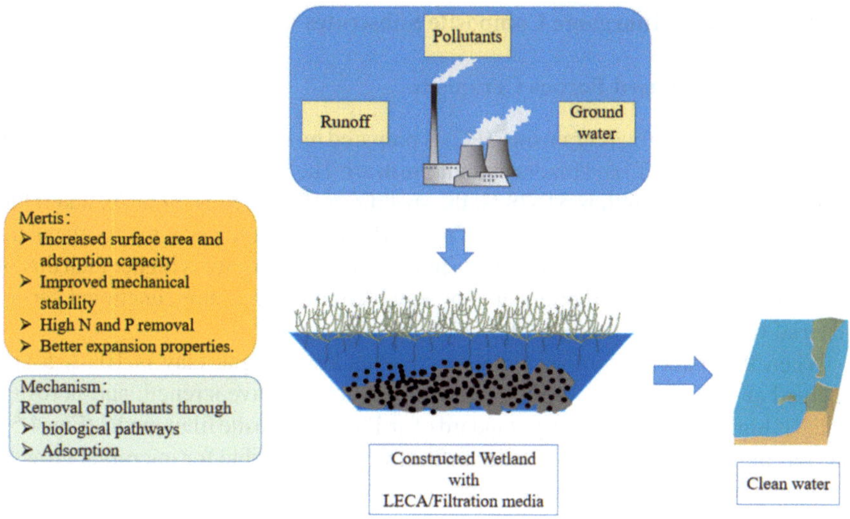

Fig. 4.4 Schematic diagram of LECA properties that improved performance of CWs

hydraulic conductivity, and support for plant and biofilm growth, as shown in Fig. 4.4. These properties make LECA an ideal choice for constructing efficient and effective CW systems. Compared with other substrates, LECA has a higher capacity to remove P. The literature reports LECA could treat many types of wastewater contaminants, such as organic matter, nitrogenous compounds, and heavy metals [86].

LECA is a lightweight aggregate with high strength. LECA, as a mature material, has been widely used in the construction of urban green infrastructure, such as the construction of materials for roofs and walls, the utilization of permeable pavement, and insulation concrete. In the early 1990s, LECA was first used as the substrate of CWs, and its cation exchange capacity was estimated to be 9.5 c mol/kg [90]. In general, clay minerals are used as raw materials to produce LECA, such as Montmorillonite or Illite. And their chemical composition generally changes with the chemical properties of the raw materials. The composition of LECA varies with the properties of the raw clay materials based on the hydrous aluminosilicates containing different amounts of iron, magnesium, and other alkaline metals. Recently, the production of modified LECA with artificial, natural, and some recycled materials such as shale, granite, and mining residues, as well as industrial by-products such as fly ash and wastewater sludge, have been investigated. During the combustion process, clay rapidly expands through the evaporation of pore water, the addition of carbonate and iron oxide decomposition. Due to chemical changes during the combustion process, the composition of LECA's final product is slightly different from that of the raw materials; it lacks hydrated minerals and carbon. LECA is generally composed of 5–6 main components: 1–4% of MgO, CaO, Na_2O, 15–18% of Al_2O_3, 60–70% of SiO_2, and 4–7% of Fe_2O_3, while the content of other components is less than 1% [90].

4.2.2.3 Organic/Inorganic Composite Substrates

(a) **Iron Oxide-Based Porous Ceramsite**

Iron oxide-based porous ceramsite (IPC) is prepared by calcining pore-forming agent sawdust, goethite (GT) and binder attapulgite in air. The pore size of IPC is 10–20 μm, the compressive strength is 51–78 N, the specific surface area is 79 m^2/g, and it has good pore connectivity. IPC is a kind of porous biological ceramsite, microorganisms can grow in the deep pore of IPC, and its biological load capacity is higher than that of commercial ceramsite and granular zeolite, which provides a new functional carrier material for biological aerated filter [170].

When the mass ratio of goethite, sawdust and attapulgite clay is 10: 2: 5 and calcined at 700 °C for 3h, the IPC has the best compressive strength and porosity, which is higher than the industry standard of artificial ceramsite filter media (Fig. 4.5). IPC has high porosity and non-toxic surface, which is suitable for microbial growth. It not only overcomes the shortcomings of common ceramsite and granular zeolite in surface glaze formation and low opening rate, but also has low calcination temperature, energy saving and consumption reduction, reduces the production cost of CWs matrix and broadens the application field of goethite [10]. The results showed that IPC formed a d-size macroporous structure, which was beneficial to the adsorption of microorganisms in biological aerated filter. IPC has good cell affinity and compatibility with attapulgite (binder) and sawdust. Through the experiment of microbial loading in CWs, it is proved that environmental microorganisms can be loaded into the inner surface pores of substrate through the opening pores of IPC [10].

(b) **Nanomaterials**

Nano-iron (nZVI), nano-iron/copper (Fe–Cu) bimetallic materials and nano-iron sulfide (S-nZVI) have attracted wide attention. nZVI has been widely used in the degradation of NO_3^- because of its excellent electron donor ability, high in-situ

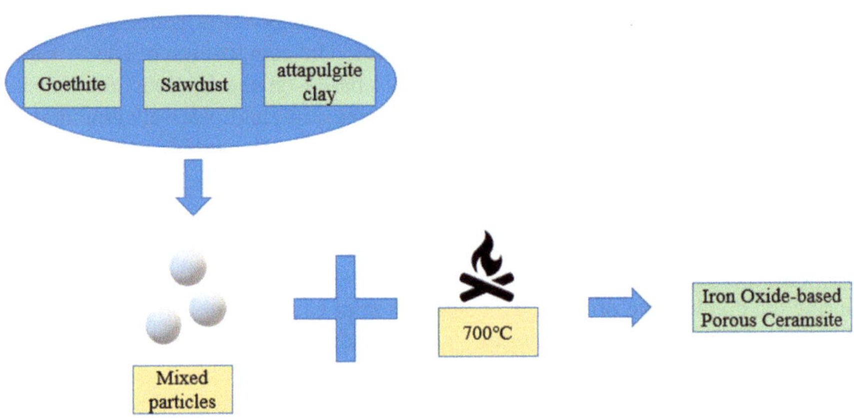

Fig. 4.5 Formation diagram of iron oxide-based porous ceramsite

reactivity, environmental friendliness, wide sources and low price. However, pure nZVI produces little N_2 when reducing NO_3^-, and instead it produces a lot of ammonia. Therefore, many studies have tried to improve the removal of NO_3^- to produce N_2 by optimizing nZVI. For example, Cu-nFe0 and Fe0/(Fe/Pd) composite metal materials have been successfully used to increase the amount of N_2 produced. Because Cu has the advantages of good selectivity, low cost and high activity, Fe-Cu bimetallic materials have become a good choice to improve the selective removal ability of NO NO_3^-. S-nZVI, as a kind of excellent material, has been widely used in water treatment [120]. When S-nZVI comes into contact with sewage, Fe^{2+} will be produced, whichis treated as electron donor in the complete denitrification process. In addition, Fe^{3+} and sulfate produced by S-nZVI is good as electron acceptors for anaerobic CH_4-oxidizing bacteria, and promote anaerobic oxidation of CH_4, thus reducing CH_4 emission from CWs. Dong et al. [38] study shows that nZVI can affect (promote or inhibit) the biological CH_4 conversion of CO_2 in both directions. Therefore, adding Fe-Cu and S-nZVI to CWs could effectively enhance the efficiency of NO_3^- removal and reduce greenhouse gas emissions. However, the above materials have some limitations, including easy oxidation by non-target chemicals and particle aggregation. nZVI is extensively employed in environmental remediation due to its large specific surface area, high reducibility, and reactivity. However, when used alone, nZVI is prone to oxidation or even spontaneous ignition due to the precipitation of metal hydroxides and carbonates. It may also agglomerate, forming chain-like aggregates, thereby reducing its dispersion and stability. In order to solve the above problems, the load method is often used in the practical application of nZVI. At present, montmorillonite, activated carbon and bentonite are commonly used to fix nZVI, and there are also reports that slag powder is used as carrier to load nZVI. Slag powder material is a by-product of blast furnace ironmaking [22]. Using it in CWs can effectively remove phosphorus from sewage, and the proportion of phosphorus released after adsorption saturation is small, which generally does not cause secondary pollution to water environment. Therefore, the combination of various materials and nZVI can effectively improve TN and TP removal in sewage, and provide a new type of environmental protection material which can be prepared in large quantities, and is cheap, safe and green. It is not only waste utilization, but also provide new adsorption materials for water treatment process [22].

(c) **Zeolite and biochar**

Composite matrix is a composite material rich in organic and inorganic components. The organic components of the composite matrix are mainly carbon sources, and the inorganic components are mainly metal salts or metal oxides (Fig. 4.6). The composite matrix has the characteristics of autonomous controllable and efficient enhanced simultaneous nitrogen and phosphorus removal. In a study conducted by Cheng et al. [27] on the nitrogen removal efficiency of wastewater treatment plant tail water using a tidal flow CWs, four reactors were established. Reactors A and C were layered filled with zeolite and biochar substrate, while reactors B and D were mixed filled with the same substrate. The study revealed that the zeolite and biochar substrate exhibited superior performance compared to ceramsite.

Fig. 4.6 The carrier membrane hanging condition with zeolite and biochar substrate and ceramsite

After adding Acorus calamus to the CWs system, there are four wetland systems with different nitrate concentration (0.13, 0.99, 1.88 and 2.34 mg/l) in the CWs effluent. As shown in Fig. 4.7, wetland A had the best removal effect. It is because the metabolism and utilization of nitrate are promoted by the absorption of plant roots and the secretion of rhizosphere organic matter; Beside that, the system has adapted to the high influent load, and the main functional microorganisms have been domesticated and enriched to promote the dual synergy between microorganisms and plants, thus strengthening the removal of nitrate.

The four reactors achieved relatively stable nitrogen removal efficiency. The ammonia nitrogen removal rate fluctuated between 66.74% and 77.87% as shown in Fig. 4.8. Reactor A had a total nitrogen removal rate of over 90%, and the effluent total nitrogen concentration was kept below 1 mg/L, meeting the Class III standard of surface water environment. In the effluent of reactors B, C and D, total nitrogen concentrations was about 1.75, 2.92 and 2.97 mg/l. Although they did not meet the requirements of Class III surface water, most of the nitrogen was removed. When the solid carbon source studied by Gong et al. [42] who used to treat the tail water of sewage plant in CWs, the total nitrogen in the effluent is as high as 5 mg/L, which shows that the new modified carrier used in this study can improve the nitrogen removal efficiency of CWs to a certain extent.

The results show that the functional carrier modified by biochar and zeolite works better in the incomplete drainage system than in the complete drainage system. Reactor A had the highest removal rate, followed by reactor B, reactor C and reactor D. The matrix filling mode with layered accumulation was also more resilient to disturbance. This may be due to biochar and zeolite could improve more favorable environment to the growth of microbial and microbial attachment by mixed

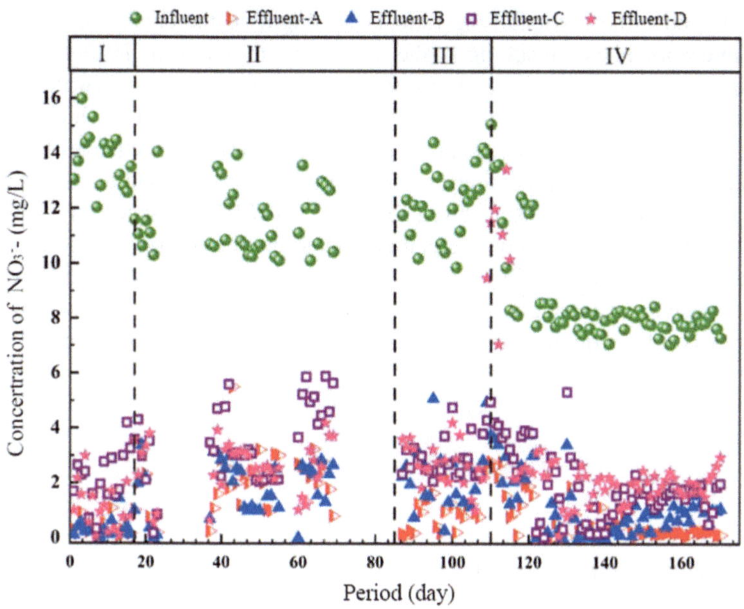

Fig. 4.7 Changes of nitrate concentration in influent and effluent in different types of CWs

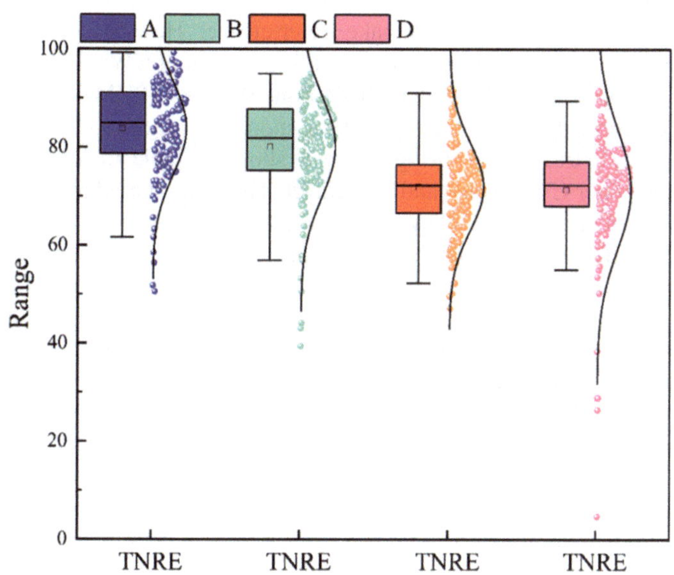

Fig. 4.8 T Box diagram distribution of total nitrogen removal efficiency in four groups of CWs

filling. Because the carrier filling ratio is not high, so the ceramsite cannot well retain microorganisms, resulting in the loss of microorganisms in the system and the impact resistance of wetland system. The porous ceramsite matrix based on iron oxide was synthesized from attapulgite, goethite and sawdust, which is rich in carbon, aluminum, iron and magnesium, and is beneficial to ion exchange of ammonia nitrogen and phosphate precipitation. The ammonia nitrogen and total phosphorus removal in simulated domestic sewage are 94.2 and 72.3%, respectively, which are much higher than those of commercial ceramsite matrix. Therefore, the composite matrix has the advantages of both organic matrix and inorganic matrix, and has strong comprehensive treatment ability for nitrogen and phosphorus pollutants in wastewater, so it is a high-quality simultaneous nitrogen and phosphorus removal matrix among new substrates.

4.3 New CWs

4.3.1 Development of CWs

People started to study CWs in the 1950s. A German scientist Seidel discovered that reeds can remove many things from the water that make it dirty or unhealthy [107]. He realized that wetlands can help make water cleaner. He made some pools with plants like reeds in them. These were like early versions of CWs. In the 1970s, Dr. Kickuth came up with an idea called "Root Zone Method" (RZM) [105, 106]. This idea helped people do more research and use more CWs. It also showed that CWs were a new way to treat sewage using nature, as shown in Table 4.2. In the late 1980s, people made better CWs using sand and gravel. They also started to use them more widely in different countries. They learned more about how they work and how to design them better [8]. By the late 1990s, people used CWs to treat different kinds of sewage water, such as water from farms or factories. They also made different kinds of CWs with different plants, materials, and ways to operate them. China started to use CWs later than other countries. The first one was built in China in the early 1990s. But in the past 30 years, China has done a lot of research and made a lot of progress in using CWs. By 2006, there were about 200 CWs in China [71], and by 2011, there were 425 [168]. By the end of 2020, there were 1171 CWs in China. This number was found from different sources, such as books, websites, news, and project materials. Compared to the past data, in the last 10 years, there were 75 more CWs built every year. The data also show that CWs are in all parts of China, but more in the places that are richer and less in the places that are poorer. They are also more in the southeast and less in the northwest.

Table 4.2 Global status and the historical development of CWs

Years	Events
1953	Dr. Kathe Seidel of Germany found that reed can remove a large amount of organic and inorganic substances
1960s	NASA National Space Technology Laboratory developed a composite system combining anaerobic microorganisms and reed
1970s	Dr. Kiekuth put forward the famous root zone theory, laying the foundation for the application of CWs
1980s	CWs started to spread in the world
1990s	Establishment of China's first surface flow CWs for domestic sewage treatment
2010s	More than 400 CWs established in China

4.3.2 New Concept in CWs

CW is an outstanding technology in ecological and economic aspects, but sometimes it cannot meet the emission standards and requirements. A novel concept has emerged, which centers around strengthening the CW as the core element of a combined ecological restoration and water purification process. This innovative approach aims to enhance the treatment performance of the CW significantly [74]. The new CWs is a new type of wetland that uses its advantages and overcomes technical limitations by coupling individual technologies. For example, through aeration and segmented influent technology, the system's dissolved oxygen level and carbon source utilization efficiency are improved, microbial nitrification and denitrification are strengthened, and removal effects of organic matter, NH_4^+-N improved. By utilizing the artificial bow shed and seasonal plant configuration technology, the efficiency of the CWs during winter can be significantly improved. This is achieved by harnessing the combined benefits of internal thermal insulation within the wetland and the growth of cold-resistant plants. As a result, the removal rate of organic matter and NH_4^+-N can be increased by over 40% during the winter season. Throughout the year, the treated water from the purification process consistently maintains stable levels of TIN, TP, and COD. These levels surpass the requirements set by Class VI standards for surface water quality. Consequently, the purified water is suitable for multiple forms of recycling, facilitating various applications for recycled water. Project construction investment $\leq$ 1000 CNY/m^3/d, operation cost $\leq$ 0.2 CNY/m^3/d [74].

Membrane bioreactor (MBR) is a treatment method that combines the activated sludge process and membrane separation technology, resulting in superior treatment efficiency and enhanced the quality of effluent regards to TSS and COD. In many cases, if the film is replaced with low-cost materials to minimize the cost investment of the membrane, the treatment efficiency will be affected. As an alternative, the combination of MBR and CWs can be utilized in this instance. There is synergy between joint governance systems. By decreasing TSS and organic matter, MBR lowers clogging occurrences and extends CWs' service life. Simultaneously, CWs eliminate nutrients retained in MBR effluent and recycle top-notch water for reuse. Combined dynamic membrane bioreactor and vertical continuous flow water (DMBR-IVCW) are employed to manage high concentration synthetic urban wastewater. The removal rates of COD, TN, TP and TSS were achieved 93.3%, 93%, 90.4% and 99.1% respectively. However, DMBR contributes little to the removal of 10–20% of TN and TP and 31.2% of COD [74].

In the recent, clogging was a significant operational challenge for CWs. This issue is influenced by two key factors: the concentration of TSS and organic matters. Organic matter could benefit for biofilm growing, exacerbating the problem over time. As a result, substances accumulate within the matrix pores of the wetland riverbed, leading to ponding near the inlet area. The ponding causes anaerobic conditions in the riverbed, which reduces the efficiency of subsurface flow wetlands [174]. The wastewater undergoes an anaerobic treatment process before being discharged into the wetland, providing a cost-effective and affordable solution. This method is sustainable and self-controlled when proper conditions are maintained, and it results in low sludge production. In comparison to the coal water machine, it requires much less operating space, specifically < 0.1 m^2 per population equivalent (PE). Although anaerobic treatment alone may not be as effective as CWs in removing organic matter and pollutants, it serves as an essential component in strengthening sewage treatment. To enhance the treatment efficiency of CWs further, other technologies can be introduced. Among these, coupling anaerobic treatment with CWs emerges as an excellent choice, allowing for improved treatment results. Anaerobic treatment has the advantage of less sludge, which significantly reduces the blockage of CWs [84]. At the same time, the biodegradability of pollutants is improved after wastewater pretreatment, which makes the CWs more completely remove pollutants. Now, one of anaerobic treatment process is UASB. It is reported that the UASB treatment has achieved 66% removal rate of TSS (301 mg/l for influent TSS and 102 mg/l for effluent TSS), 62% removal rate of COD and 68% for BOD removal. The removal rate of organics, SS, nutrients and pathogens by CWs was improved, reaching an average removal rate of 74 (± 12%) of COD, 83 (± 9%) of BOD, 68 (± 17%) of TSS, and 49 (± 22%) of TN. 51 (± 26)% of TP and 94 (± 13)% of fecal coliform [3]. The combination of UASB and CWs can be used for the treatment of drug wastes. The integrated system will increase to 95% of the average COD removal rate, while the COD removal rate of UASB alone will only be 75% [128].

Although slow sand filter (SSF) has been explored as an important part of domestic water treatment, its usage is limited due to the clogging of the sand bed [67]. Therefore, the study on disinfection and polishing of secondary effluent by using sand filter in the third stage has proved that this is a cost–benefit and promising treatment method [45]. Total coliform could decrease by 0.3–3.5 logarithmic units, fecal coliform decreased by 2 logarithmic units, and coliform decreased by 2.3–3.7 logarithmic units. It was reported that the number of enterococci in secondary treatment wastewater was 0.7–2.6 log units [65]. A study reported the effectiveness of two systems, namely the granular activated carbon (GAC) sandwich slow sand filter and the large duckweed CWs stabilization tank (CWs ST), for removing drugs and personal care products such as DEET, caffeine, and paracetamol [70]. The authors found that the removal rates of DEET, caffeine, paracetamol and by CW-ST-SSF were 95.9%, 98.1% and 99.1%, respectively, while the removal rates of DEET, CAF and PAR by CWs ST-SSF without SSF were 27.1%, 65.8%, and 92.2%, respectively. Combining SSF and CWs improves water treatment efficiency, allowing safe reuse and protecting drinking water sources from pathogens.

Ozonation and ultraviolet treatment are widely acknowledged disinfection methods. Ozone acts as a powerful oxidizing agent, degrading both organic and inorganic substances by generating hydroxyl and other free radicals directly. It is highly effective in neutralizing a diverse array of microorganisms, making it a suitable option for treating disinfected water [134]. In multiple European sewage treatment plants, UV rays are utilized as a disinfection method for drinking water. UV radiation within the 240–280 nm range causes permanent damage to the DNA of bacteria and viruses, ensuring complete sterilization of the water [151]. It is suggested to integrate advanced oxidation methods with currently available cost–benefit technologies to reduce capital investment and maintain high quality wastewater for reuse. Raw sewage undergoes a series of treatment processes, with ozone oxidation and an anaerobic reactor integrated in tandem. This combination effectively treats the sewage, and the resulting effluent is further polished before discharge. A study showed that the method reduced 84.4% of BOD_5, 81.6% of COD, 78.9% of TOC and 86.9% of TC, and effectively removed 97.6% of dyes, 99% of turbidity, 99.9% of total coliform and 100% of coliform [30]. Researchers reported that the concentration of micro pollutants naproxen and ibuprofen decreased about 97.3% and 97.2% respectively when the ozone dose was 15 mg/l through ozone oxidation that were used to treat the influent water of sewage treatment plants [64]. The use of an independent continuous disinfection system could only reduce fecal coliform by 1.5 Log. However, by integrating CWs with a UV system, it is possible to lower it to 9.1 Log [6].

A stabilization tank is a derivative of CWs, which consists of several facultative, anaerobic, and mature tanks [28]. It can effectively remove organics and pathogens. As a result, it is typically paired with other treatment processes to treat secondary effluent. Its effectiveness is heavily influenced by temperature and sunlight, which are deemed essential to its operation. Consequently, this technique is primarily employed in tropical and temperate climate regions. Researchers studied the removal effect of stabilization tank combined with CWs on BOD and TSS of pretreatment raw material separation liquid (Fig. 4.9). They found that the better effluent quality water could be achieve through the combined system [124]. In certain scenarios, the quality of the effluent is adequate for reusing such as in agricultural irrigation. However, this is not always the case that has great performance. The pretreated pig plasma was treated with HSSF CWs and stabilization tank within 7 days of hydraulic retention time, and the study revealed that the combined system outperformed the single HSSF CWs. The following pollutants were effectively eliminated around 148.1% of Mn, 102.5% of COD, 113.4% of TP, 87.9% of TC, 94% of NH_4^+-N, 64.2% of Cu, 36.6% of NO_3^-, and 32.1% of Ca^{2+}. However, the levels of Fe, SO_4^{2-}, SS, Zn, and NO_2^- increased. Apart from this, the combined system was efficient in eliminating fecal contaminants [92].

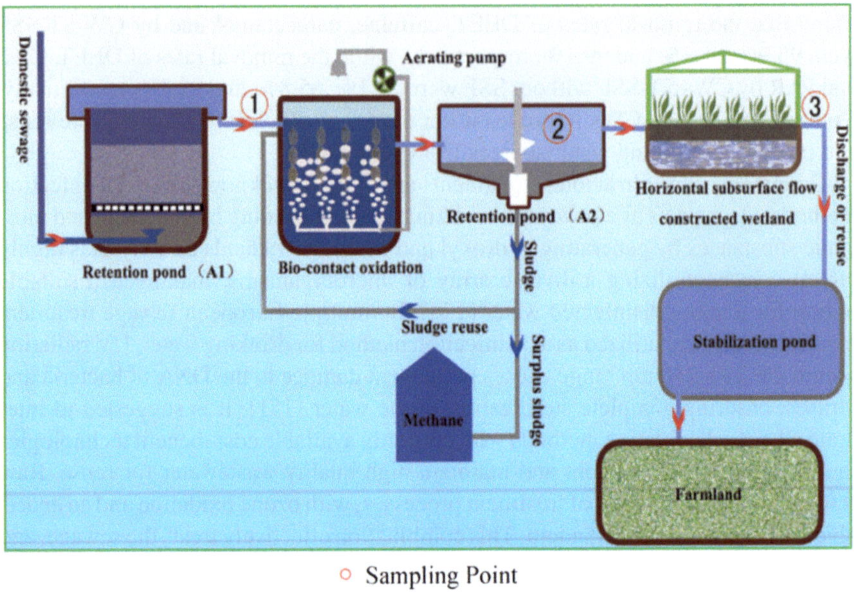

Fig. 4.9 The flow chart of the combined system with Stabilization pond and Horizontal subsurface flow CWs [40]

The research on CWs technology started late in China, and it was not emphasized and promoted until the end of the twentieth century. The first study about CWs application in China was the construction of the reed wetland project, which initiated in Tianjin in 1987. In the 1990s, the first surface flow CWs in China employed for treating domestic sewage was established in Changping, Beijing. Subsequently, China has successively established Chengdu Living Water Park and Yaonigou Artificial Wetland on the north bank of Fuxian Lake in Yunnan. The practical application of artificial wetlands has gradually increased, realizing the purposes of landscape viewing and water purification. Since the twenty-first century, CWs as a new sewage treatment technology have been widely used in China. According to Zhang et al. [168], the number of CWs projects established in the south and north of China has increased significantly, especially in small and medium-sized cities since 2000. And the study showed that the development trend of CWs to control water pollution in China and the distribution of CWs types in China through the analysis of collected data. The results showed that, as of 2010, sewage treatment plants were still the main sewage treatment plants in China, and the sewage treated by CWs accounted for only 0.82%, However, an assessment of the adoption rate of different types of CWs across various provincial administrative region and coastal areas along the river in southern China has reached more than 80%, and the application scope is very wide.

4.3.3 *New Composite Constructures*

4.3.3.1 New Spiral Composite CWs

More attention has been paid to the construction of CWs on a small area. In order to save space, a tower composite CWs with vertical construction and good nitrogen removal effect can meet the requirements.

As shown in Fig. 4.10, spiral composite CWs are spiral steps, and each step is a wetland reactor. Using the structural characteristics of spiral composite CWs to enhance the COD removal efficiency. Compared with common compound wetland, spiral stepped compound CWs can make sewage fall from a high place, solve the problem of sewage migration, and omit pumping link. Building artificial wetlands on limited areas has attracted more and more attention. In order to save space, Ye and Li [162] designed a tower-type composite artificial wetland with good nitrogen removal effect, which is constructed in a vertical direction. Moreover, the spiral CWs can be used for high-head sewage treatment, and the falling water aeration can also increase the concentration of DO in sewage, which is beneficial to the removal of pollutants in water. The control experiment was conducted for the removal efficiency of COD, ammonium, and TP in such CWs. Constructing CWs in vertical direction instead of horizontal direction can greatly reduce the land area, keep good purification effect with traditional wetlands, and reasonably meet the needs of urban development. This provides a valuable research direction for the design and development of CWs in the future. When considering the development of spiral CWs for various purposes like spiral balconies, stairs, or buildings, factors such as hills, convex terrain, and artificial convex structures should be considered during the planning and design phase. It is important to consider different combinations of composite CWs, excluding the alternate combination of subsurface flow CWs (SSFCWs) and surface flow CWs (SFCWs) [177]. Various combination modes can coordinate the advantages and shortcomings of different types of wetlands to the maximum extent, so as to achieve the best treatment effect. Seasonal changes should be taken into consideration when assessing the treatment efficacy of a CWs, which serves as an outdoor sewage treatment system. Considering the varying degrees of siltation that exist in different CWs, taking SFCWs as an example, because of its water level, it is easier to lead to the loss of upper substrate, and the siltation situation is generally more serious than that of SSFCWs. The same spiral CWs can't solve this problem, so it still needs to be research on the wetland in the future [177].

4.3.3.2 Prefabricated Stepped Waterfront Wetland

The prefabricated stepped waterfront wetland mainly includes: pretreatment system, stepped wetland system and sedimentation tank. The pretreatment system comprises an aeration room, a grid area, and a low-power water pump. The water flow undergoes a sequential passage through the aeration room and grid. The grid effectively captures

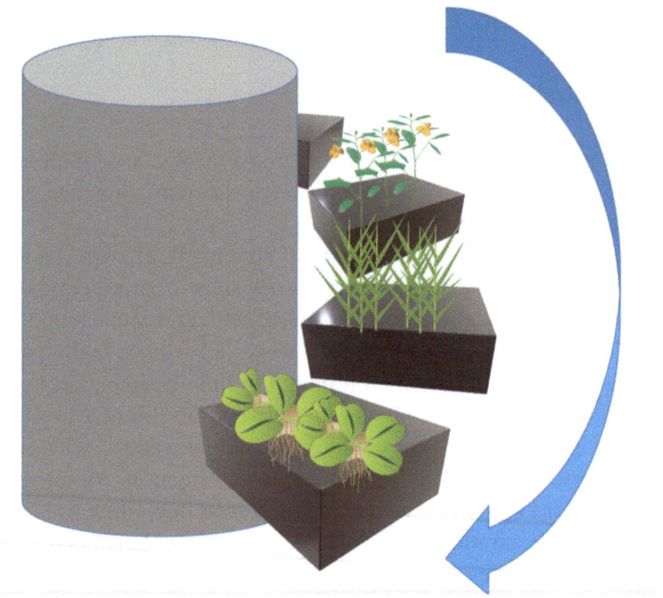

Fig. 4.10 Schematic diagram of spiral composite CWs with multiple steps and various grow plants

large suspended particles present in the initial water, while the aeration process introduces oxygen to enrich the wetland water [164]. Once treated, the water proceeds to the next phase, which involves a stepped wetland system typically composed of multiple levels. The primary focus of the multi-level wetlands is presently geared towards the removal of nitrogen and phosphorus from the water. Wetlands use efficient fillers like hollow bricks, cinders, and zeolites to intercept nitrogen and phosphorus effectively. Aquatic plants such as reeds and calamus with good pollutant absorption capacity can be selected as wetland plants in multi-stage wetlands [80]. Sewage can pass through the three-level wetland system by gravity, and then enter the sedimentation tank for precipitation following the treatment process within the CWs system, and the overlying water flows into the river along the outlet pipeline to realize the purification of river water quality. The soil layer of wetland is taken from the soil layer of waterfront area along the river. The experimental area ensures ventilation and simulates sunlight to ensure the normal growth of plants.

The screening and configuration of fillers in stepped wetlands should not only possess strong adsorption properties that CWs effectively remove nitrogen and phosphorus pollutants from water, but also cooperate with plants and microorganisms to form a plant-filler-microorganism collaborative purification system, so as to further improve the interception ability of pollutants in stepped wetlands. At present, there are many kinds of adsorption fillers used for water pollution purification. Among many adsorption fillers, volcanic rocks, zeolites and cinders have good adsorption performance for nitrogen and phosphorus pollutants, and their prices are low.

Fig. 4.11 Schematic diagram of prefabricated waterfront wetland design

It has the advantages of stable performance, simple operation and strong manual controllability, and has a wide application foundation [80].

The removal efficacy of ammonium in CWs by seven kinds of fillers was investigated. The results show that volcanic rocks, zeolites and cinders have good adsorption capacity for ammonia nitrogen in water, and good adsorption and interception for phosphorus in water. The construction of stepped wetlands involves an industrial design approach. The traditional CWs module was enhanced by considering construction and installation issues beforehand. The wetland packing layout was transformed into separate packing blocks fixed with ecological cloth belts. This design allows for easy removal and replacement of individual packing blocks without the need to clean a large area. This approach facilitates timely renewal and regeneration of the packing, effectively addressing the problem of packing blockage [164] (Fig. 4.11).

The design advantages of the prefabricated stepped wetland system are as follows [164]:

1. Prefabricated Design: The system employs prefabricated packing blocks, inspired by the concept of prefabricated buildings. This allows for easy replacement of blocked wetland fillers, solving the issue of wetland blockage efficiently.
2. Effective Phosphorus Interception: The hollow bricks in the prefabricated ecological filler blocks, made of concrete and diatomite, effectively intercept phosphorus in water.
3. Strong Interception Capabilities: Aquatic plants like *Phragmites australis, Iris Siberia,* and *Acorus gramineus*, selected for the wetland, demonstrate robust tolerance to the environment, making them suitable for year-round planting. These plants effectively intercept nitrogen and phosphorus in rivers, reducing pollutant concentrations in effluent.
4. Synergistic Action: The combination of ecological fibers, green zeolite, coal cinder, and volcanic rocks within the prefabricated packing blocks promotes biofilm formation, facilitating microbial and plant interception of nitrogen and phosphorus in the water. Additionally, green zeolite and volcanic rocks efficiently intercept ammonia nitrogen, while coal cinder excels at phosphorus removal.
5. Sedimentation Tank: The purified water passes through a sedimentation tank, where sand and gravel particles carried through the wetland system are allowed to settle. This process prolongs the Hydraulic Retention Time (HRT) and ensures clear effluent with pollutant concentrations meeting standards.

4.3.3.3 CWs Combined with Microbial Fuel Cell (MFC)

The utilization of MFC technology is good for energy conservation in wastewater treatment. As one of the earliest forms of bio-electrochemical systems (BESs), MFCs employ microorganisms as biocatalysts to facilitate the anaerobic oxidation of organic matter and generate electrons. CWs have excellent effects in economy, energy and environment. With the continuous recognition of CWs, researchers realized that the treatment efficiency of CWs can be improved by strengthening. Compared with traditional CWs, new CWs achieve various goals by combining with various emerging technologies. The effectiveness of CWs in treating wastewater is well established, but their anaerobic nature often results in slow treatment performance. The integration of MFCs into CWs provides a solution to these challenges, significantly enhancing the processing performance of the system [157]. MFC uses proton exchange membrane to separate anode (anaerobic) and cathode (aerobic) chambers. The two chambers are connected by wires. The transfer of protons between two chambers is facilitated by a membrane, which generates an electric current through the potential difference created between the two chambers [115]. Since there is an aerobic area on the surface and an anaerobic part in the deep layer of CWs, it is very feasible to combine CWs with MFC. MFC can provide conductive materials in the deep or bottom layers and surfaces to simulate the reaction in CWs. The whole process improves the processing performance of CW-MFC, thereby reducing its area [46]. This also shows the practical application of MFC technology in the field. In the past ten years, the integration of CWs with MFC technology has shown significant improvements in processing performance, energy generation, bioenergy storage, design and operation, as well as successful field application, thus garnering widespread attention [115, 116]. Recently, double cavity soil film based CW-MFC has been innovated, which has proved some advantages compared with classical single cavity CW-MFC, such as (i) By reducing resistance, the technology is able to improve both power generation performance; (ii) The separation of the cathode and anode prevents mixing, facilitating easy recovery of synthetic products at each electrode; (iii) Wastewater is able to flow sequentially under gravity, resulting in significant energy savings [47, 88].

CW-MFC stands out for its capability to generate electricity from microorganisms' life activities using various biological energy in sewage as electronic donors. Additionally, the process enhances biological life activities, resulting in a more efficient sewage removal rate. This reciprocal relationship makes CW-MFC a promising technology for sustainable wastewater management (Fig. 4.12).

With the depletion of fossil energy and the intensification of global warming, new biotechnology based on biomass resources is highly expected. In this case, the discovery of electrochemical active bacteria (EAB) has attracted the attention of academia and industry. EAB is the general name of microorganisms that have electrical interaction with the outside world. Microorganisms with electrochemical activity can generate electricity by oxidizing and decomposing organic substances and releasing electrons to the electrode outside the cell. These bacteria are called electricity generating bacteria. On the other hand, EAB can conduct reductive metabolism

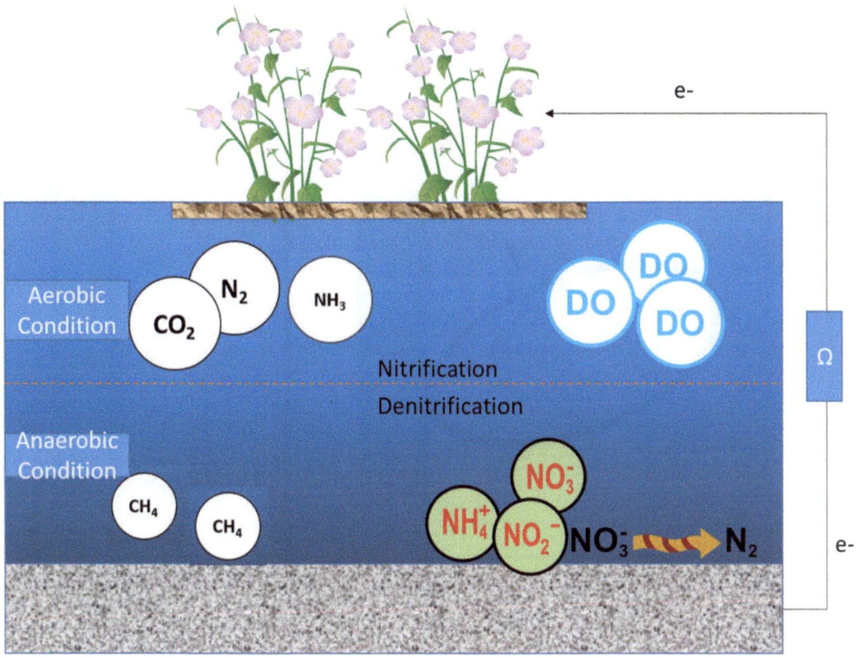

Fig. 4.12 Schematic representation depicting the microbially-catalyzed processes

by accepting electrons from external electrodes, and is expected to use carbon dioxide and cheap low-molecular compounds to produce useful substances. Based on the above characteristics of EAB, it is expected to develop microbial electrochemical systems (MES) such as MFC and MES with less load on the earth's environment (Fig. 4.13).

(a) **Composition of CW-MFC**

MFC usually consists of anode and cathode. The anode uses graphite with high conductivity and easy to be attached by microorganisms, while the cathode mostly uses air cathode. The air cathode is composed of conductive substrate, catalyst and air diffusion layer. The anode is buried in the anaerobic layer of CW. As the microorganism decomposes the organic matter in the sewage, the microorganism attached to the anode will transfer the electron to the electrode. The electrons move to the cathode through the conductive substrate. At the cathode, oxygen in the air reacts with electrons and bromine to form water through the action of oxygen reduction catalyst. When the above reactions occur continuously, the external circuit generates current and obtains power equivalent to the product of potential difference between anode and cathode and current.

Compared with CW, CW-MFC not only promotes the progress of energy-saving wastewater treatment, but also improves the removal efficiency of pollutants. Several studies have demonstrated that microorganisms such as iron-reducing bacteria

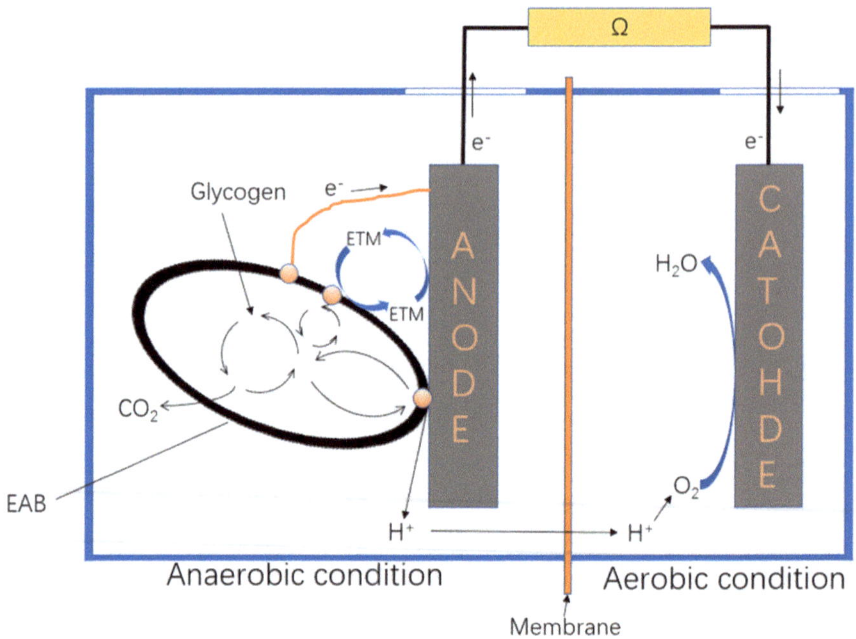

Fig. 4.13 Work mechanism diagram of MFC that shows the transaction of electrons and protons

(IRB), sulfur-oxidizing bacteria (SOB), and sulfate-reducing bacteria (SRB) work in synergy to convert iron and sulfur in FeS between different valence states. This promotes electron transfer and enhances the decolorization, degradation, and electricity production performance of azo dyes within the CW-MFC system. Furthermore, the addition of FeS can provide electrons for azo dye wastewater treatment, reducing the need for organic carbon sources [172]. At the same time, research shows that CW-MFC can effectively strengthen the sewage treatment capacity. Research has shown that the A4 unit configuration (CW-MFC-with plants) has the highest capacity for pollutant removal per unit area. This considerably enhances the area efficiency of wetland systems, effectively mitigating the challenge of requiring large amounts of land space for wetland processes [171].

(b) **Application of CW-MFC**

Currently, CW-MFC has garnered significant attention due to its numerous advantages, leading to an increase in various research studies focused on this technology. The research primarily revolves around two aspects: the influence of new substrates on CW-MFC and the enhanced treatment of pollutants using CW-MFC. Moreover, there is high anticipation that CW-MFC can address the issues of blockages and pollution buildup commonly encountered in traditional CWs.

Industrial sewage contains a significant amount of heavy metals, which can severely impact the water environment and pose health risks through the biological chain. To address this issue, researchers conducted a recent study and developed a downstream Leersia hexandra Swartz CW-MFC (DLCW-MFC) for treating Cr (vi) and 4-chlorophenol (4-CP) contaminated wastewater. Impressive removal rates of 99.0% for Cr (vi) and 78.6% for 4-CP were achieved with the DLCW-MFC. This technology shows great promise for simultaneously detoxifying compound mixtures found in polluted wastewater while generating electricity [142]. Wine, being one of the most popular beverages globally, generates a significant amount of wastewater during its production, making wine wastewater a prominent environmental concern. This type of wastewater is especially noteworthy due to its high concentrations of ethanol and phenolic compounds, which present challenges for microbial growth and activity. These characteristics make the treatment of wine wastewater a complex process. A team of researchers has innovatively integrated ethanol-resistant yeast with laccase activity into a CW-MFC system. Their aim was to mitigate the negative impact of ethanol-polluted wastewater on microorganisms, while also generating electricity and reducing costs. The results show that the maximum COD and phenol removal rate could achieve 79% and 85% were observed [61]. Carbamazepine (CBZ), a common pollutant found in wastewater from drug manufacturing and personal care products, presents a major removal challenge due to its persistence. Conventional treatment methods may even generate toxic byproducts. However, a CW-MFC system has shown impressive results, operating continuously at room temperature for nearly 80 days with an average removal rate of 97%. These findings highlight the promising potential of CW-MFC as an effective technology for processing PPCPs in wastewater [153]. Coincidentally, the removal efficiency of tetracycline (TC) and sulfamethoxazole (SMX) increased by 6.0–8.7% and 0.8–1.3% respectively in the sponge iron-enhanced CW-MFC, in which s-Fe-0 also significantly reduced the total *sui* and *tet* genes by 84.10–94.11% and 49.61–60.63%. Improvements made to the electrode layer of MFC-CWs can significantly influence the composition of microbial communities, facilitating the transformation of antibiotic removal, bioelectricity generation, and accumulation of antibiotic-resistant genes[146].

In light of the benefits of CW-MFC systems for treating emerging pollutants, researchers have endeavored to address the limitations of conventional CWs. Several studies have demonstrated that CW-MFC systems can effectively treat wastewater while mitigating packing pollution in reactors of varying scales (including laboratory and pilot scales), using different fillers, such as activated carbon and gravel. Additionally, MFCs have a substantial impact on reducing filler blockage in the pilot-scale CWs system by minimizing the accumulation of pollutants within the filler material [158].

CW-MFC has a great impact on the community structure and diversity of microorganisms through the strengthening of its own electrode layer, so its treatment performance is far better than that of traditional CWs, and some nitrogen removal rates even reach 2 times [143]. At the same time, CW-MFC can also generate electric energy, which has far-reaching significance for the realization of the carbon cycle in the future.

CW-MFC technology is a pioneering method for improving wastewater treatment efficiency while simultaneously generating electricity, making it the most feasible

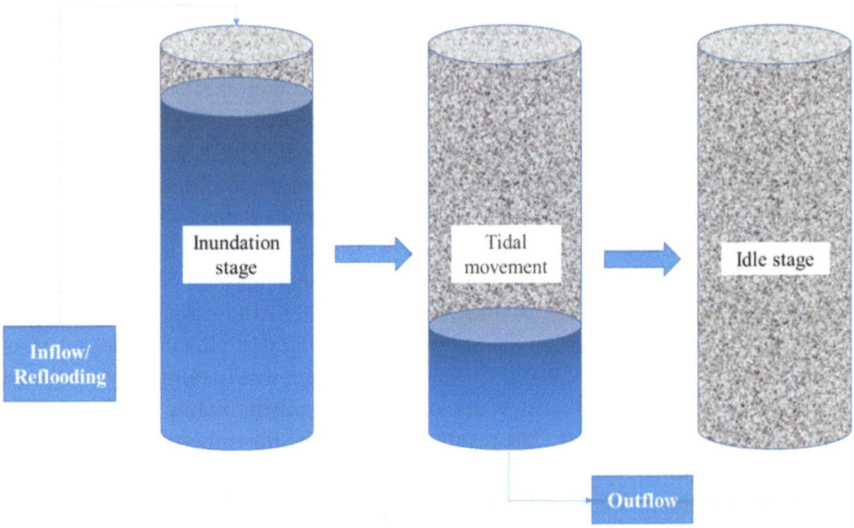

Fig. 4.14 Schematic representation of Z-TFCW function in different stages during the flow event

and practical option for intensifying desired processes among other integrated bio-electrochemical systems. The technology has yet to be explored at the field scale, while process-related aspects and numerous designs and of CW-MFC have been studied at laboratory, apart from electroactive wetlands. To upscale the technology, current problems need to be addressed, and further optimization of the system is necessary.

4.3.3.4 Tidal Flow CWs (TFCWs)

See Fig. 4.14.

4.3.3.5 Composition of TFCWs

CWs have undergone significant development in the past 50 years, These wetlands can be enhanced through the aeration machinery and integration of hydraulic to boost treatment efficiency. The increased understanding of innovative CWs developed new types with multifunction, such as TFCWs, that demonstrate superior treatment performance and compete effectively with conventional technologies. Numerous types of wetlands have been implemented globally, and SSF garnered the attention due to their superior treatment performance. In SSF wetlands, wastewater is routed from pipelines into substrates at the bottom of the wetland, which support a large surface area for bacterial biofilms that improve the interaction with the wastewater. However, there is less DO in SSF and SS in sewage is easily blocked, the treatment efficiency of SSF is not as good as that of conventional sewage treatment device. In autumn and winter, the plants in the CWs wither and decay, leading to a substantial accumulation of decaying substrate. This hinders the entry of oxygen into the wetland substrate, thereby reducing the treatment efficiency. To address this issue, manual cleaning of the wetland is often required to restore its original effectiveness. Currently, aerated ducts are commonly employed within the wetland to introduce oxygen and raise dissolved DO. This approach helps counteract the decline in sewage treatment efficiency during the colder months. In summary, the need for manual cleaning and the added cost of implementing aerated ducts to maintain the CWs pose significant challenges to its widespread adoption. These additional expenses make it challenging to promote and popularize the use of CWs for wastewater treatment.

Through observing the phenomena in nature, people have developed TFCWs. TFCWs is a cyclic phenomenon of repeated aerobic and anoxic by simulating tidal phenomena in nature. In the TFCWs, the sewage is putted into with the water level increases, making the sewage fill the CWs to create an anoxic environment. In an anoxic environment, anaerobic bacteria thrive, leading to an increase in denitrifying bacteria. Simultaneously, incoming sewage introduces BOD, providing a rich carbon source that enhances the activity of denitrifying bacteria. As a result, denitrification becomes evident and pronounced in this environment. While inhibiting denitrifying bacteria, aerobic bacteria such as nitrifying bacteria get a good growth environment. At the same time, the excessive proliferation of denitrifying bacteria caused by long-term sewage circulation should be avoided, which will further aggravate the blockage of CWs. In conclusion, the cyclic rising and falling tide processes create an environmental system that facilitates the coexistence of aerobic and anaerobic microorganisms.

TFCWs utilize anoxic and anaerobic cycles, occurring at least twice daily within the wetland substrates. The design of these wetlands heavily relies on the cation exchange capacity of the wetland media. As water overflows the wetland, NH_4^+ ions are adsorbed onto the negatively charged surface present in the aggregate/root matrix. During discharge, the previously adsorbed ammonium ions remain in the thin biofilm exposed to atmospheric oxygen, leading to nitrification within the biofilm. In the subsequent flood cycle, NO_3^- is released into the bulk water through desorption and then denitrified by bacteria. It's worth noting that there is no inhibition of BOD on nitrification during the drainage stage; however, high BOD and nitrate levels can exist during the flooding stage, creating optimal conditions for denitrification.

4.3.3.6 Application of TFCWs

As the structure of tidal flow wetland has no major difference from that of traditional CWs, the difference lies in the way of water inflow and outflow. Therefore, tidal flow wetland is easy to construct and operate, and is adopted in the laboratory. At the same time, tidal flow wetlands have combined with other technologies in quite a number of cases, giving play to excellent performance.

The electrolytic integrated TFCWs (EC-TFCW) system was utilized by a research team. The combination of electrolytic integration, microbial degradation, and flocculation led to a synergistic effect, resulting in the exceptional performance of EC-TFCWs, and the average removal rate of SMX reaches 81.86% [77]. Coincidentally, some studies have shown that the combination of TFCWs and MFC system was utilized, which can simultaneously enhance nitrogen and antibiotic removal and power generation [178].

Biochar is widely used in various sewage treatment due of its many advantages. A research team combined cellulose carbon source and TFCWs with MFC, and the results verified that the treatment effect of wastewater can be improved at low C/N ratio [167]. Biochar was utilized by another research team to stabilize a salt-tolerant aerobic denitrifying bacterium (Zobellella sp. A63), which was used as a substrate to the TFCWs. The removal efficiency of TN, NO_3^--N, and COD were 72.6%, 68.2%, and 82.5% respectively [173].

Our team recently introduced 4 different TFCWs, which CWs-A and B is full discharge of water, and CWs-C and D is part discharge of water. Moreover, CWs-A and C is uniform distribution of biofilm carrier, and CWs-C and D is layered distribution of biofilm carrier (Fig. 4.15).

This TFCWs system is to remove the nitrogen compounds of eutrophic water. In the first stage, with a stable influent concentration of ammonium of 5–6 mg/l, the initial ammonia nitrogen removal rate is less than 50%. This is because

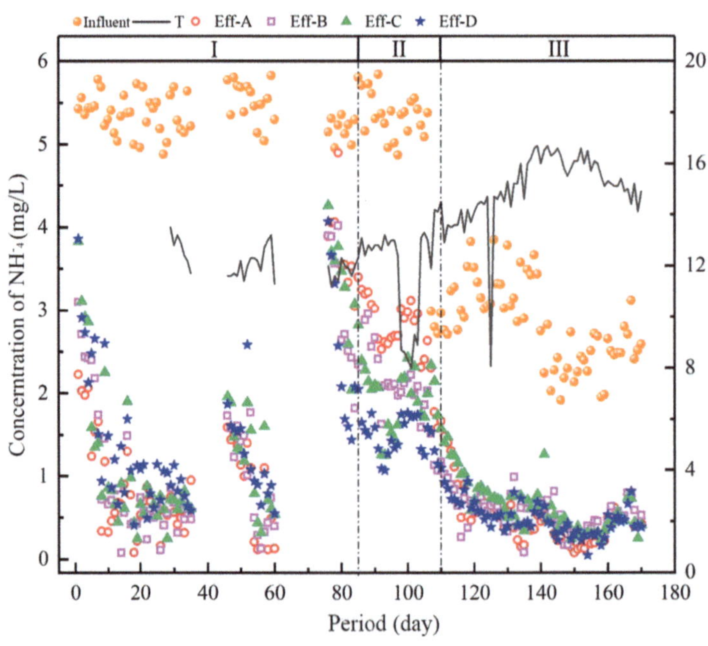

Fig. 4.15 Various in ammonia concentration in effluent and influent of different types of tidal flow CWs

the microorganisms have not fully adapted to environmental changes, and some of the sludge in the inoculated sludge dies and cracks, which can also lead to an increase in the effluent ammonium concentration. After 10 days of operation, the four wetland devices achieved relatively stable nitrogen removal efficiency, with ammonia nitrogen removal rates fluctuating between 82.29 and 89.28%. Among them, the ammonia nitrogen removal efficiency of CWs-A and B was slightly higher than that of CWs-C and D. Due to the pandemic of COVID-19, the device was temporarily suspended for 10 days. When restarted, the ammonium concentration in the effluent rapidly increased to around 1.8 mg/l, with a removal rate of less than 70%. However, in the subsequent period, the ammonium concentration in the effluent continuously decreased, and finally reached a stable ammonia nitrogen removal effect again at the 55th hour. The mean rate of elimination of ammonium was raised to 85.05–93.41%. The maximum removal rate reached as high as 97.84%, resulting in an effluent ammonia nitrogen concentration of only 0.12 mg/l. The ammonia nitrogen removal performance of CWs-A and B is generally better than that of CWs-C and D. In each operating mode, CWs-A has a better ammonium removal effect than B. Similarly, CWs-C has a better ammonium removal efficiency than D. This indicates that the fully emptied operating mode provides a larger oxygen contact area, thereby enhancing the ammonia nitrogen removal effect. And the layered packing discharge has a slightly inferior effect on the oxidation enhancement of ammonia nitrogen compared to the mixed arrangement. Subsequently, the device was suspended for half a month and continued to operate. The concentration of ammonia nitrogen in the effluent go up again, but showed strong recovery performance in subsequent operation. When the concentration rapidly decreased, the removal rate gradually increased. On the 85th day, when the removal rate of ammonia nitrogen returns to 50%, the hydraulic retention time was reduced to 12 h (inundation time is 11 h, evacuation time is 1 h). The sudden reduction of hydraulic retention time did not cause substantial disturbance to the stability of the system, and the effluent still showed a continuous decreasing trend. At the 91st cycle, the concentration of the effluent reached its lowest, and the ammonia nitrogen removal rates of the four devices were 54.62%, 58.73%, 64.55%, and 77.23%, respectively. It is worth noting that after resuming operation from the 75th cycle, the ammonia nitrogen removal performance of CWs-C and D is due to CWs-A and B. Moreover, under the same operating mode, the ammonia nitrogen removal efficiency of CWs-B and D began to be better than that of CWs-A and C. The sudden increase in ammonia nitrogen concentration during the 92–106 cycle is due to temperature fluctuations, with the inlet temperature only below 10 °C, which has a significant inhibitory effect on microbial activity and weakens the metabolic effect of microorganisms on ammonia nitrogen. As the temperature recovers (to around 13 °C), the removal effect of ammonia nitrogen gradually recovers. Finally, in the 110th cycle, the removal rates reached 43.78%, 60.53%, 46.81%, and 63.03%, respectively, showing significant differences. After plant transplantation, the effluent of the four devices maintained a gradually decreasing trend, and a stable ammonia nitrogen removal rate was maintained after the 125th cycle. At this time, microbial plants have formed a stable and efficient collaborative relationship, jointly achieving nitrogen removal in eutrophic water through metabolism.

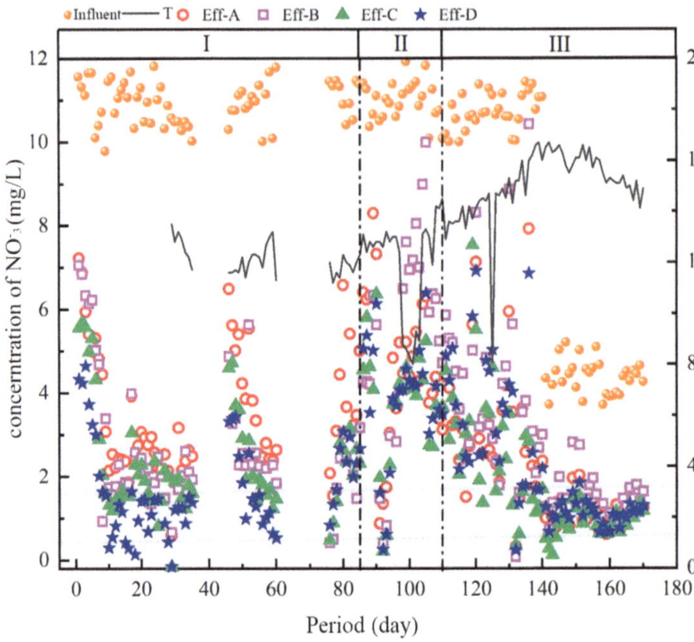

Fig. 4.16 Various in nitrate concentration in effluent and influent of different types of tidal flow CWs

During the 125th to 170th cycle, the highest ammonia nitrogen removal rates of the four devices can reach 96.34%, 97.21%, 90.79%, and 97.83%, respectively. The increase in ammonia nitrogen concentration in the effluent after the 136th and 160th cycles is closely related to temperature changes (Fig. 4.16).

Similarly, while maintaining the same nitrate concentration in the influent, the initial nitrate removal rate of CWs-A and B in Stage 1 is less than 40%, while the nitrate removal rate of CWs-C and D is nearly 60%. The organics produced after the death and decomposition of microorganisms can be used as the carbon source of denitrification to promote the reduction of nitrate. The presence of the bottom submerged zone in CWs-C and D becomes the anaerobic fermentation site after sludge death, thereby improving the removal of nitrate. After 10 days of operation, the nitrate removal performance also reached a stable state. The initial nitrate content fluctuates around 2 mg/l. The nitrate removal rates of the four wetland devices in this stage were 75.76 ± 5.17%, 79.33 ± 4.87%, 82.63 ± 5.37%, and 88.48 ± 3.79%, respectively. The average nitrate concentration in the effluent of CWs B and D arranged in layers is lower than that of CWs A and C. This may be because the concentrated distribution of the download body in the layered situation is conducive to the transfer of molecular information between various functional bacteria, promoting the improvement of microbial activity, and the cooperation between microorganisms to ensure that the overall environment of the area is more suitable for the growth of functional bacteria. After a 10 day suspension of operation, the nitrate concentration in the effluent also rapidly increased to the initial level, and then reached a stable nitrate concentration again within 10 days. The difference is that after experiencing a 15 day pause again, the nitrate concentration in the inlet water is 11 mg/l. At a concentration of 45 mg/l, the average concentration of nitrate in the effluent is only 1.14 mg/l. This is because in the case of long-term lack of substrate, some microorganisms lyse and die, and release intracellular proteins to be used as electron donor in the denitrification process, thus promoting the reduction of nitrate. As the device continues to operate, the available organic matter gradually decreases, and the nitrate concentration in the effluent continuously increases, reaching stability on the 85th day. The average nitrate concentration in the effluent is 3.44 mg/l, which is slightly higher than the nitrate concentration in the effluent on the 35th day. This indicates that the recovery of microbial activity at this time is in balance with the autolysis of dead microorganisms. Starting from the 86th day, after the hydraulic retention time was reduced to 12 h, the nitrate nitrogen concentration in the second stage effluent rapidly increased and decreased. This is mainly due to the sudden increase in nitrate load, which prevents microorganisms from immediately adjusting the activity of metabolic proteins within the bacteria for timely utilization. However, the system has a higher ability to resist shock loading of nitrate, resulting in a rapid decrease in nitrate concentration. At the 92nd cycle, the nitrate concentration drops to the lowest, with an average level of only 0.73 mg/L. Unlike the first stage, the nitrate removal performance of CWs A, B, C, and D did not differ significantly in the second stage, with average levels of 45.35 $\pm$ 6.72%, 58.91 $\pm$ 6%, 59.11 $\pm$ 13.93%, and 69.40 $\pm$ 7.39%, respectively.

In addition, the box plot distribution of total nitrogen removal rate was used to comprehensively evaluate the performance differences of different wetland devices in efficient biological nitrogen removal from eutrophic water bodies. From Fig. 4.17, it can be seen that the overall nitrogen removal performance of CWs-C and D is better than that of CWs-A and B. Therefore, it can be considered that the anaerobic environment in the substrate submerged area under quantitative drainage meets the environmental requirements of the two anaerobic functional bacteria, Femmox and NDFO, ensuring their high expression of activity and promoting their synergistic effect to achieve synchronous removal of ammonia nitrogen and nitrate from eutrophic water bodies. In addition, although CWs-C does not perform as well as CWs-B in ammonia nitrogen removal, nitrate nitrogen accounts for the majority of eutrophic water bodies, and the presence of plants further narrows the difference in the performance of functional carriers in driving microbial autotrophic nitrogen removal. In addition, from the perspective of carrier arrangement, the wetland device under mixed filling mode has a higher total nitrogen removal performance than the layered arrangement. On the one hand, under the mixed arrangement, nitrate and ammonia nitrogen at various locations in the wetland can be absorbed by the biochar on the carrier surface, which improves the concentration of ammonia nitrogen and nitrate nitrogen in the microenvironment near the biofilm, promotes the activity of microorganisms, and avoids the substrate environment with relatively low average of each carrier surface in the same area under layered filling. On the other hand, the uniform distribution of the mixed filling method enhances the retention capacity

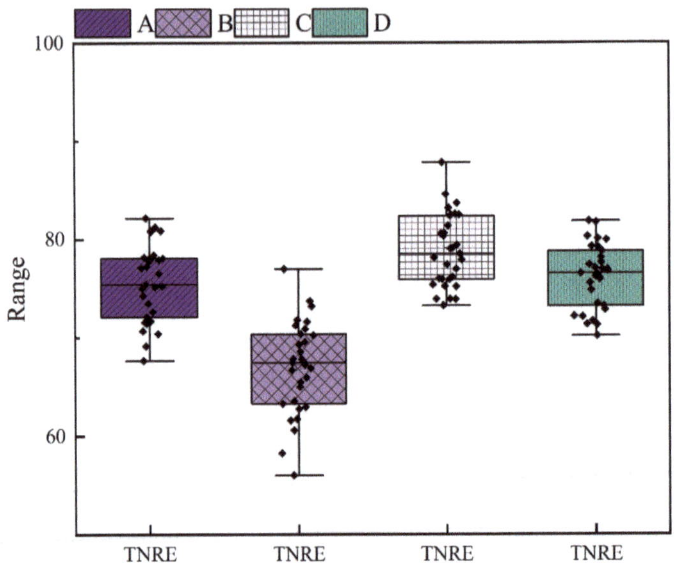

Fig. 4.17 Box diagram distribution of total nitrogen removal efficiency in four groups of CWs

of microorganisms, and the interaction between ceramic particles and functional carriers achieves synergistic effects between different substrates, thereby promoting the synchronous removal of ammonia nitrogen and nitrate.

In CWs-A and B, 81.13% of ammonia nitrogen was rapidly removed within the first 6 h, while only 63.46% ammonia nitrogen was removed in CWs-C and D. In comparison, the removal efficiency of CWs-A and B is better than that of CWs-C and D, and the final remaining ammonia nitrogen concentration is only 0.334 mg/l, which is less than 0.543 mg/l in CWs-C and D, indicating that the full emission operation mode is more conducive to ammonia nitrogen removal. This may be because after more liquid is discharged, there is relatively more substrate in effective contact with air, which is beneficial for the biofilm to absorb oxygen and promote the oxidation of ammonia nitrogen. Figure 4.18 also show the corresponding results. On the contrary, although C and D have poorer ammonia nitrogen removal performance compared to CWs A and B under full emission mode, their corresponding nitrate removal efficiency is significantly better. 58.43% of the nitrate in the influent was quickly removed in the first eight hours, and still showed a fast nitrate removal rate, but only 33.86% of the nitrate was removed in the first eight hours in CWs-A and B, which further proved that the presence of submerged areas under the quantitative discharge mode is conducive to the growth of anaerobic autotrophic/heterotrophic denitrification bacteria, thus improving nitrate reduction. During 12 consecutive hours of cultivation, the concentration in the final two bottles decreased to 0.334 mg/l and 0.543 mg/l, and the ammonia nitrogen removal rate approached zero. The final remaining nitrate concentrations were 1.364 mg/l and 0.723 mg/l, and still maintained a rapid decline

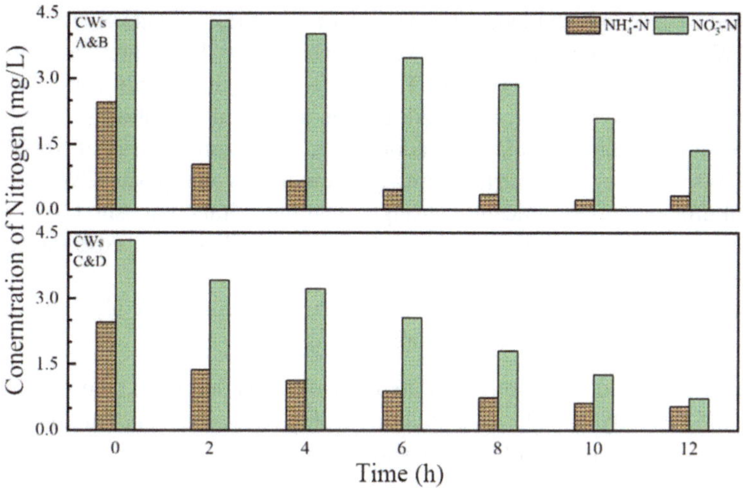

Fig. 4.18 Changes of nitrogen concentration in two types of TFCWs at different time points in the typical cycle

pattern. This indicates that extending the hydraulic retention time appropriately is beneficial for further nitrate removal, thereby improving the total nitrogen removal rate. However, the experimental results of a typical cycle show that when the total nitrogen concentration in the influent water is 6.875 mg/l, the total nitrogen concentration in the effluent of CWs-C and D is only 1.266 mg/l, which meets the nitrogen concentration requirements for surface Class IV water. This indicates that the wetland has high application potential in treating low concentration eutrophic water. Especially in temperate regions, higher water temperatures can better promote the activity of functional bacteria, thereby reducing the total nitrogen concentration of the final effluent and facilitating efficient autotrophic denitrification.

4.3.4 Comparison of Different CWs

In order to show more intuitively the advantages of the new types of CWs in saving land area, comparing the land area of different types of CWs with different structures is necessary.

It can be observed from Table 4.3 that when it can be seen that although the influent concentration of the CWs listed is different, the removal rate is basically 70–90%, and the new spiral composite CWs also meet this standard. Except for the tidal-subsurface flow combined CWs studied by Gao et al. [41], the removal rate of NH_4^+-N in other CWs in the table is relatively low compared with COD and TP. The removal of NH_4^+-N relies on oxygen, but SSFCWs, especially VSSFCWs, have low dissolved oxygen, which hinders the degradation of NH_4^+-N. Gao et al. [41] used a tidal CWs that can draw air into the wetland bed due to the air pressure difference caused by the tidal operation mode. This can enhance the oxygen enrichment and oxygenation of the wetland, and facilitate the aerobic degradation of NH_4^+-N by microorganisms. Xiong et al. [155] adopted the same combination of SFCWs and HSSFCWs. Although the surface flow-horizontal flow composite CWs for COD removal efficiency is significantly stronger than the spiral composite CWs, we can use the structural characteristics of spiral composite CWs to improve the removal rate and solve this shortcoming.

In the actual construction process of the new spiral composite CWs, it is crucial to pay attention to the following considerations:

1. When addressing rainwater treatment on viaducts and roofs, the design stage should carefully consider construction height and location, particularly focusing on the safety of road traffic under viaducts.
2. If the bed body of this system can be made using transparent materials, incorporating landscape design would be highly beneficial.
3. The construction of the spiral composite CWs should prioritize both aesthetics and building material safety. For instance, opting for a combination of 3.3 cm thick double laminated glass and a stainless-steel keel frame proves to be an excellent choice. This material can handle a maximum bearing capacity of 1000 kg per square meter, making it particularly suitable for high-rise buildings.

Gao and Hu [40] studied marsh meadow Calamagrostis angustifolia wetland (MMCW), Ecotone wetland (EW) and Carex Schmidtii Meinsh. marsh (CSMM) in Sanjiang Plain, and they found that t the differences of N_2O fluxes between MMCW and EW and between MMCW and CSMM were due to the significant differences of the soil volumetric moisture content (Fig. 4.19b). However, the differences of N_2O fluxes between CSMM and EW were due to the significant difference from soil NH_4^+-N content (Fig. 4.19c).

For the horizontal subsurface flow CWs (HSSFCWs) with bio-contact oxidation (BCO) pre-treatment, Song et al. [114] found that as shown in Fig. 4.20, the average COD concentrate in the raw sewage was 160 ($\pm$ 8.34) mg/L, and ranged from 100 to 250 mg/L. The removal efficiency was about 70% with an average COD concentration in effluent of 57.30 ($\pm$8.25) mg/L. The temperature of the influent wastewater during the monitoring period varied between 8 and16 °C, NH_3-N was partially removed by the bio-contact oxidation (60.47%) for sewage treatment. The removal efficiency of TP by bio-contact oxidation was about 16%. The concentration of TP in the effluent was basically decided by the influent, and the mean concentration of influent and effluent was 3.43 ($\pm$0.67) mg/L and 2.88 ($\pm$0.64) mg/L, respectively.

Table 4.3 Purification effect of COD, NH_4^+-N and TP by different composite CWs [177]

Wetland categories	COD			NH_4^+-N				TP	
	Inlet concentration	Effluent concentration	Removal rate	Inlet concentration	Effluent concentration	Removal rate	Inlet concentration	Effluent concentration	Removal rate
Composite vertical flow CWs	100.0	17.5	82.5	10.0	2.7	73.0	1.0	0.26	74.0
Surface-horizontal flow composite CWs	273.2	38.8	85.8	25.3	8.4	66.9	3.5	0.9	91.76
Tidal-subsurface flow combined CWs	214.2	89.4	58.3	10.2	2.9	71.0	5.4	2.9	46.9
Stepped CWs	213–381	14.1–30.8	90.6	48–112	9.9–19.6	66.7	4.8–12.1	Less 1.0	–
Spiral composite CWs	198.5	39.2	80.3	28.5	7.8	72.6	2.4	0.4	81.5

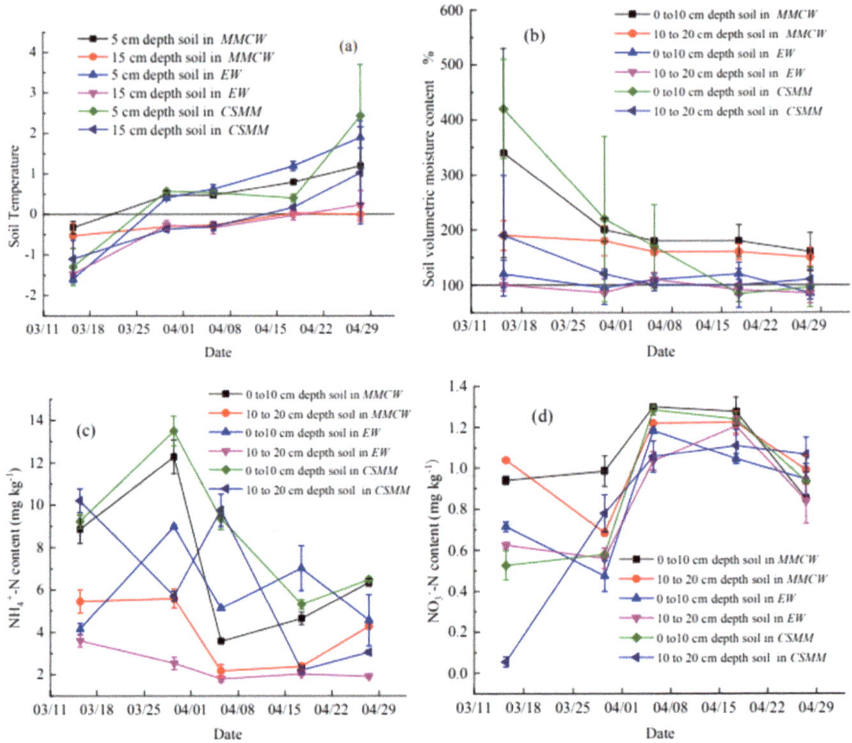

Fig. 4.19 Spatial variations and distribution in soil temperature (**a**), soil volumetric moisture content (**b**), soil NH_4^+-N (**c**) and NO_3^--N (**d**) content

4.4 Microbiology of New Constructed Wetlands

4.4.1 Microbial Community Structure

In nature, various microorganisms coexist with healthy and asymptomatic plants, which have been considered as an important part on effecting the growth of plants. Plant microbiome, such as bacteria, fungi, archaea and protozoa, participate in plant competition under various environmental pressures. In polluted environment, heavy metals are transformed between root microorganisms and root exudates through positive correlation interaction, including bacterial iron carrier, enzyme detoxification, biosorption and active efflux transport, so as to relieve environmental pressure [32]. The interaction between plants and microorganisms is the key to plant colonization. Some studies have reported of root microbial communities have a wide range of beneficial effects on plant health, which could induce plant immune system resistance for disease inhibition and increase nutrient acquisition. The interaction between plant roots and their associated microbial communities is not unidirectional. The diversity

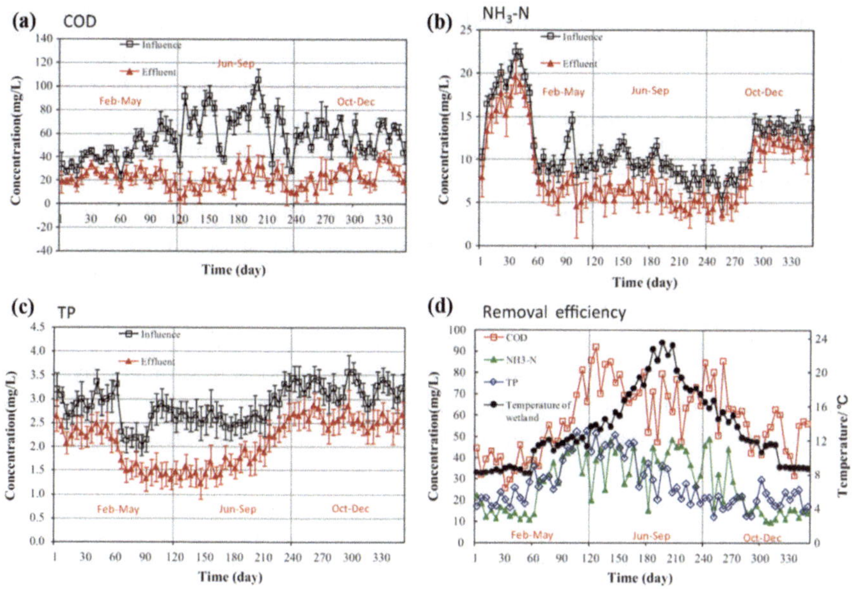

Fig. 4.20 Pollutants removal efficiencies in greenhouse-structured wetland (**a** chemical oxygen demand (COD), **b** ammonia–nitrogen (NH_3-N), **c** total phosphorus (TP), and **d** removal efficiencies in greenhouse-structured wetland)

of microbial communities can be used to treat the different pollutants in wastewater systems, compost and soil polluted environments in terms of form, quantity, substrate utilization and activity [53]. Recent studies have shown that bacterial colonization communities in plants significantly increased the ability of phytoremediation. For example, in the whole biological treatment process of ordinary sewage treatment plant [165], a microbial community play an important role for degrading the organic industrial waste through the oxidative metabolism. The bacterial population formed based on the basic structure of plants. Biological treatment relies on a large number of bacterial species and protozoa, and finally converts to carbon dioxide, biomass, and some small compounds [165]. Previous studies have focused on using molecular techniques to identify bacterial species and further to understand those cumulative functional capabilities.

However, the emergence of molecular technology cannot link microbial function with gene characteristics. But in analyzing the structure of microbial communities in the environment, it helps to elucidate the contribution of microbial communities to pollutant degradation [96]. Many studies have verified the effectiveness of rhizosphere microorganisms in treating organic pollutants. For example, some studies have found that plant root microorganisms are helpful to the degradation of organic pollution. In a research, p-chlorobenzene including p-chloroaniline, p-chlorophenol and p-chlorotoluene with concentration below 120 mg/l was degraded by reed wetland root microorganisms under simulated natural conditions. After 12 days of degradation, the degradation rates basically reached the maximum, which were 79.8%, 80.3% and 78.75% respectively [140] In other study, the root system of *Fescue arundinacea* and *Festuca arundinacea* was degraded by treating soil with pyrene labeled with ^{14}C, and finally it was found that 8 and 30% of pyrene mineralized into CO_2 and H_2O respectively [24]. In order to strengthen the effect of phytoremediation and pollution control, the dominant strains were inoculated to strengthen the degradation and purification of microbial pollutants in roots. In some studies, three *Arbuscular mycorrhizal* (VA) mycorrhizal fungi were inoculated into PAHs contamination to study the degradation of polycyclic aromatic hydrocarbons. In this investigation, three VA mycorrhizal fungi were inoculated into Phillionic onion. After comparing the results for 16 days, it was found that 75–95% of monocyclic aromatic hydrocarbons remained in the control group, while the experimental group of VA fungi in inoculated plants decreased to 2–40% [4]. Arbuscular mycorrhizal fungi can promote plant growth and nutrient absorption by reducing rhizosphere toxicity. Large microbial communities such as fungi and Rhizobia can release various volatile organic compound, which can promote plant growth and play a vital role in the expression of phytoremediation. Although phytoremediation offers great promise, it has not yet formed a viable treatment system, because its removal efficiency of organic pollutants varies with its concentration and type, and the viability of plants varies with seasons and the type of planting medium.

4.4.1.1 Bacterial Biofilm

Biofilm is recognized for its ability to adsorb organic pollutants, playing a significant role in the transport and fate of pollutants in aquatic environments. Previous studies have demonstrated the adsorption capabilities of biofilms in various contexts. For instance, biofilms grown in rotating annular bioreactors using river water as inoculum were found to absorb herbicides [66]. Additionally, river biofilms have been shown to adsorb pollutants such as tetrabutyltin, p,p-DDT, diclofop-methyl, triallate, lindane, atrazine, and parathionmethyl [50, 104, 139]. Researchers have suggested that the adsorption of organic pollutants by biofilm follows a similar competitive mechanism observed in metal ion adsorption [119].

However, despite the reports on biofilm-mediated removal of organic pollutants in wastewater treatment processes [9, 14, 18], there is a lack of research on di(2-ethylhexyl) phthalate (DEHP) removal via adsorption by biofilms formed with pure cultures. According to Wen et al. [147], they confirmed that DEHP was removed by both cell suspension and biofilm grown on vermiculite. Although biofilms formed by pure cultures may not exist in natural environments, the findings suggest that biofilms containing bacterial cells like strain C21 are capable of adsorbing DEHP. In summary, the removal of DEHP in the system with biofilm can be attributed to the combined effects of vermiculite adsorption, biodegradation by biofilm biomass and suspended bacterial cells, and adsorption by the biofilm biomass. These three activities synergistically contribute to the removal of DEHP.

The mechanism of DEHP removal by the biofilm formed by Arthrobacter sp. strain C21 on vermiculite as illustrated in Fig. 4.21, (a) the cells with EPS attach and stick on the surface of vermiculite and form basic skeleton of biofilm via EPS; (b) the cells grow progressively and gradually form the compact biofilm; (c) at the beginning of the biofilm formation, PEs removal is mainly contributed by biodegradation and adsorption by vermiculite; (d) the adsorption enhances the mass transfer and thus biodegradation rate; (e) as biofilm becomes mature (5–36 h), DEHP removal mainly relies on biodegradation; and (f) at the end, mature, stable biofilm forms (36–54 h), adsorption of DEHP by biofilm, especially hydrophobic components protein rather the adsorption by vermiculite becomes main adsorption force for DEHP removal.

The formation of biofilm accompanied with the increase in bacterial biomass with much more cells in the presence of vermiculite than those grew only in suspension. Table 4.4 summarizes the change in total biomass over a 54 h period in the absence (marked as Suspension) and presence of vermiculite (marked as Biofilm). The initial biomass was similar (0.032 versus 0.034 mg/ml). At 36 h, suspension had 0.51 mg/ml while biofilm had 0.787 mg/ml and then both stabilized at these levels (Table 4.4). The amount of biomass in the presence of vermiculite was much more than that without vermiculite by at least 50%.

EPS generally exist in biofilm and is formed under different environmental conditions; extracellular proteins and polysaccharides are major components [150]. The change in the contents of extracellular proteins and polysaccharides was

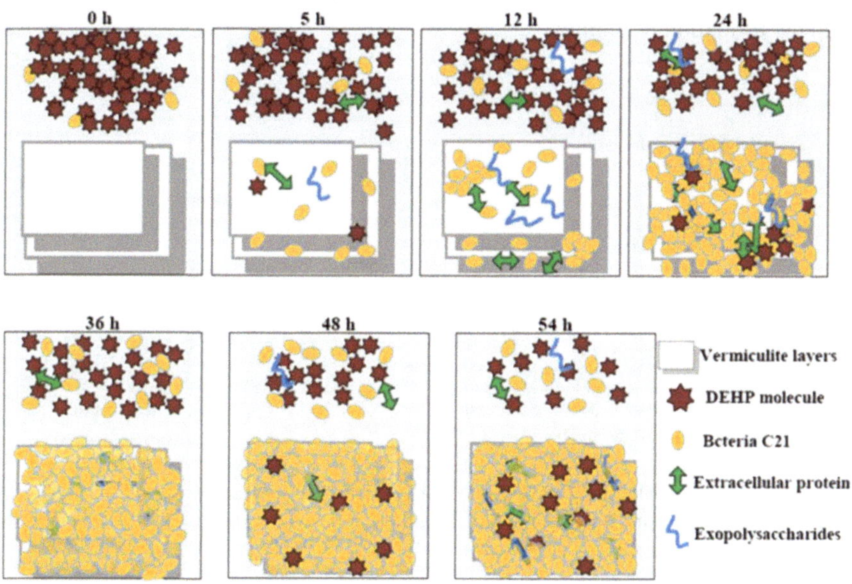

Fig. 4.21 Conceptual illustration of the reposed of the DEHP removal during biofilm development on vermiculite [147]

Table 4.4 Changes of biomass content and DEHP consumption with time in biofilm and suspension cultivation systems [147]

	Total biomass (mg/mL)		Biomass increased (mg/mL)		DEHP removed (mg/mL)		DEHP decreased (mg/mL)	
	Suspension	Biofilm	Suspension	Biofilm	Suspension	Biofilm	Suspension	Biofilm
5h	0.032	0.0340	0.032	0.034	0.513	0.979	0.513	0.979
12h	0.093	0.286	0.061	0.252	1.494	2.857	0.981	1.878
24h	0.304	0.574	0.211	0.288	3.666	6.027	2.172	3.170
36h	0.510	0.787	0.206	0.213	5.649	6.817	1.983	0.790
48h	0.531	0.793	0.021	0.006	6.251	8.643	0.872	1.826
54h	0.540	0.754	0.009	0	7.161	9.099	0.640	0.456

analyzed during the biofilm formation process (Fig. 4.22a). During the incubation period, the content of proteins was always higher than that of polysaccharide. This is similar as previous observation that in wastewater treatment system, the amount of extracellular protein in biofilm is always much higher than polysaccharide [33]. In this study, some extracellular proteins could be extracellular enzyme with the ability of degrading extracellular polysaccharides, and this would lead to the decomposition of these biomacromolecules. Compared to the cell suspension cultivation systems, EPS production in biofilm was higher in the early stage of biofilm cultivation (0–12 h) (Fig. 4.22b). However, it showed the opposite tendency with the formation of biofilm.

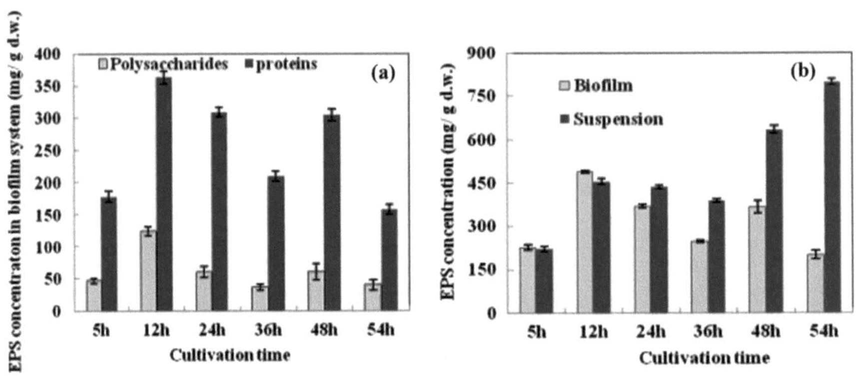

Fig. 4.22 **a** Changes of the contents of extracellular proteins and polysaccharides during incubation period in biofilm system; **b** comparison of EPS concentration between biofilm and suspension systems [147]

4.4.1.2 Electroactive Bacteria (EABs)

Microbes produce EPS as a major structural material of biofilm. EPS perform a variety of functions in the formation of biofilm (adhesion of bacteria, adsorption of pollutants, nutrients supply etc.). With the formation of biofilm, more and more bacteria adhered to each other with EPS. Studies have shown that seventy to eighty percent of EPS was exopolysaccharides and extracellular proteins with small fraction of humic acid, nucleic acids and lipids [59, 72]. Due to the co-presence of the other substances, EPS contained a variety of polar and nonpolar functional groups, such as carboxyl, hydroxyl, mercapto, phenolic hydroxyl and phosphate group, which would determine the chemical and biological properties of biofilm, and give biofilm the ability to sorb organic compounds and ions.

Wen et al. [147] also used Pearson correlation analysis to reveal a significant positive correlation between EPS production and the removal of Di-(2-ethylhexyl) phthalate (DEHP) in the biofilm system (rp > 0.86). This indicates that EPS contributes to DEHP removal through adsorption. EPS molecules contain both hydrophilic components (primarily polysaccharides) and hydrophobic components (mostly proteins), which influence their ability to adsorb organic pollutants. The presence of hydrophobic functional groups in EPS enhances the adsorption capacity for organic compounds. In this study, the protein content in the EPS biomass was significantly higher than that of polysaccharides, suggesting that DEHP adsorption by the biofilm was primarily facilitated by EPS.

In the microbial community structure of CW-MFC, the relative abundance of EAB is higher compared to other microorganisms. They belong to different phyla and genera. In terms of phyla, Proteobacteria, Firmicutes, Actinobacteria, and Acidobacteria are commonly found and responsible for transferring electrons to the anode electrode. Proteobacteria is a well-known group of EABs that play a key role in biodegradation of nitrogen, phosphorus, and other pollutants. Previous research has shown that the high pollutant degradation rate of CW-MFC is due to the high growth rate of Proteobacteria [78]. Other EAB phyla such as Planctomycetes and Actinobacteriota are also enriched in CW-MFC compared to CWs. Therefore, an enrichment of many microorganisms related to nitrate-degrading-bacteria (Proteobacteria, Actinobacteriota, Patescibacteria, Firmicutes, and Bacteroidota) is observed in CW-MFC systems. This leads to higher efficiency of electricity generation, nitrogen and phosphorus removal compared to CWs [121].

In terms of genus, some commonly found EABs include *Shewanella, Geobacter, Rhodoferax*, and *Pseudomonas*. Among them, *Shewanella* and *Geobacter* are the most typical and well-studied EAB genera. *Shewanella* mainly exists in marine and freshwater environments and has a broad range of metabolic pathways that allows it to grow under different environmental conditions. *Rhodoferax* and Pseudomonas are also widely distributed EABs with important applications in environmental remediation and wastewater treatment. *Geobacter* is mainly distributed in soil and sediment and has excellent electrochemical and degradation properties, making it widely used in MFC and other electrochemical technologies, which is associated with metal mineral reduction and denitrification [69], while Nitrosomonas and Nitrobacter are core genera of nitrifying bacteria. Previous studies have reported that Nitrosomonas can be used for electro-annamox, where ammonia is used as the electron donor and electrons are transferred to an anode electrode [143].

4.4.2 Responses of Microbial Communities Towards Plants in Constructed Wetlands

4.4.2.1 Regulation of Root Microbial Community by Plants

Root exudates are small molecular organic acids, amino acids, fatty acids, sugars and other substances secreted by plants to the surrounding environment through roots in order to increase the uptake of nutrients or adapt to the stress of the external environment. They can change the pH value and redox potential (Eh) of rhizosphere soil, and have chemical reactions such as chelation or complexation precipitation with heavy metals, which can affect the number and activity of soil microorganisms [103]. There are more than 200 kinds of them, which are usually classified according to chemical components, mainly including amino acids, organic acids, sugars, inorganic ions, enzymes and other components. Among them, low molecular weight phenolic compounds, amino acids, sugars and organic acids can determine the types and quantities of rhizosphere microorganisms and soil enzymes. Low molecular weight soluble organic matter, although its content is low, can change the physical and chemical process of rhizosphere soil and promote the dissolution of insoluble nutrients. By utilizing the synergistic effect between plants and rhizosphere organisms that promote growth, root exudates can provide nutrients, substrates and signal molecules to rhizosphere microorganisms, thus increasing the population and activity of rhizosphere microorganisms, and playing an important role in plant nutrition, physiological traits and rhizosphere microenvironment [96]. Studies have shown that compared with non-rhizosphere soil, the specific carbon source and energy source of rhizosphere exudates can promote the growth and reproduction of rhizosphere microorganisms, and significantly increase the number and activity of rhizosphere microorganisms to 5–20 times [56].

However, some studies have found that there are a large number of microbial taxa in and around plant roots, including a relatively high proportion of flora that adapt to special metabolic niche and grow slowly. Plants load primary metabolites such as sugar, amino acids and organic acids into phloem, then transport them to roots and secrete them into rhizosphere, which cannot be fully utilized by microbial flora [54]. In recent years, with the progress of microbial assembly, metabolic footprint technology, natural variation and genome-wide association, the role of secondary metabolites secreted by plants in shaping the composition and function of microbial communities has been discovered step by step. In most cases, secondary metabolites affect microbiome mainly by changing the taxonomic composition of microbiome or the characterization of resistance to pathogenic bacteria. For example, benzoxazolinone, the main defense compound secreted by grass to rhizosphere, changes the composition of microbial community after exudation, and then activates plant defense [62].

Plant-related microorganisms can alter the evolutionary response of plants to environmental stress in at least three non-mutually exclusive ways: by changing the adaptability of individual plant genotypes, the expression of plant traits related to adaptability, and the direction of natural selection in populations under environmental pressure through the influence of microorganisms on reproductive adaptability. Plants usually select microbial groups that can promote stress resistance under abiotic or biotic stress conditions [123]. For example, plant root exudates can enhance the biodegradation effect of PAHs in soil. Compounds released from ryegrass roots can up-regulate the expression level of polycyclic aromatic hydrocarbon cyclohydroxylated dioxygenase dioxygenase alpha subunit (PAH-RHD alpha) gene in rhizosphere bacteria and mediate the specific functional bacterial community of PAHs degradation in soil [63]. Soil microorganisms such as Pseudomonas and Nocardioides encode related genes of PAHs degradation [44], and gene abundance and number of functional bacteria are important characterizations of PAHs degradation potential. Therefore, the structural characteristics of bacterial community are key factors affecting the remediation effect of rhizosphere microorganisms, and the abundance of PAHs degradation genes can be used as an important index to measure the degradation ability of rhizosphere microorganisms. The detailed mechanism procedure is shown in Fig. 4.23.

In recent years, the ability of root exudates to degrade soil pollutants and its influence on related degradation genes have become a new research hotspot in soil

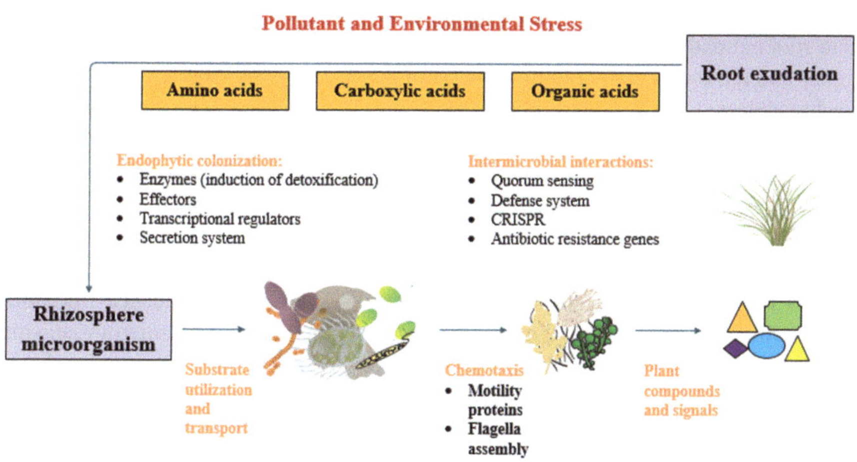

Fig. 4.23 Schemic diagram of the relationship between plants and their root microbial community structure

remediation. However, the degradation efficiency of PAHs and the response characteristics of microbial community to root exudates are still unclear. Some recent studies have identified a variety of secondary metabolites that affect the composition and function of microbial groups, but they belong to different chemical functional classifications and have different mechanisms of action. At the same time, it has been proved that secondary metabolites can fine-tune the composition and function of plant microbiome. However, the effects of secondary metabolites secreted by plants on microbiome by regulating rhizosphere exudates by acting as signal molecules, nutrient elements or toxins still need further discussion.

4.4.2.2 Effect of Microbial Interaction on Root Microbial Community

Microbial communities play a vital role in nitrogen removal in CWs, but they can also be affected by various factors such as antibiotics, plants, and environmental conditions. One of the factors that can affect microbial interaction and root microbial community is the presence of antibiotics in wastewater. Antibiotics are considered to pose potential risks to microbial communities in CWs, as they can inhibit or select for certain microorganisms, alter microbial diversity and composition, and induce antibiotic resistance genes. In the rhizosphere of CWs, root exudates are proposed to play a primary role in regulating microbial community dynamics. Root exudates are organic compounds released by plant roots that can provide nutrients, signals, and substrates for microorganisms. However, root exudates can also interact with antibiotics and affect their bioavailability, degradation, and toxicity to microorganisms. Tong et al. [122] found that root exudates from different plant species (Phragmites australis, Typha angustifolia, and Iris pseudacorus) had different effects on the removal and degradation of tetracycline (an antibiotic) in CWs. They also found that root exudates influenced the abundance and diversity of tetracycline resistance genes and tetracycline-degrading bacteria in the rhizosphere. Another study by Shirdashtzadeh et al. [113] showed that root exudates from Phragmites australis enhanced the removal of sulfamethoxazole (another antibiotic) in CWs by stimulating the growth and activity of sulfamethoxazole-degrading bacteria. These studies suggest that root exudates can modulate the effect of antibiotics on microbial interaction and root microbial community in CWs.

Another factor that can affect microbial interaction and root microbial community is the type and diversity of plants in CWs. Plants can influence microbial communities by providing oxygen, organic matter, and habitat for microorganisms. Different plant species may have different effects on microbial communities due to their different physiological and morphological characteristics. For instance, a study by Shirdashtzadeh et al. [113] compared the performance and microbial communities of three types of CWs (subsurface flow, surface flow, and floating) planted with different plant species (*Phragmites australis*, *Typha latifolia*, *Carex appressa*, *Juncus ingens*, and *Lemna minor*) for treating stormwater runoff. They found that plant type and diversity significantly affected the nitrogen removal efficiency and the structure and composition of microbial communities in CWs. They also found that plant type and diversity influenced the abundance and diversity of nitrogen-cycling genes (such as nitrification, denitrification, anammox, comammox, and DNRA genes) in the rhizosphere. These results indicate that plant type and diversity can shape microbial interaction and root microbial community in CWs.

A third factor that can affect microbial interaction and root microbial community is the environmental conditions in CWs. Environmental conditions such as temperature, pH, dissolved oxygen, nitrogen concentration, and hydraulic loading rate can affect the microenvironment and the metabolic activity of microorganisms in CWs. Different environmental conditions may favor or inhibit different microbial processes such as nitrification, denitrification, anammox, comammox, organic matter degradation, and phosphate removal. For example, a review by Verduzo Garibay et al. [126] discussed the effects of environmental factors on the structure, diversity, and activity of microbial communities in CWs. They reported that temperature affects the growth rate and enzyme activity of microorganisms, pH affects the availability of substrates and electron acceptors; depth affects the oxygen diffusion and redox potential; substrate affects the surface area and porosity; availability of organic carbon affects the competition between heterotrophic and autotrophic microorganisms; and presence of plants affects the oxygen supply and root exudation. They also reported that different environmental factors affect different pollutant removal pathways (such as nitrogen, phosphorus, and organic matter removal) and performance within different CWs configurations. These findings imply that environmental conditions can alter microbial interaction and root microbial community in CWs.

Wetland ecosystems are natural systems that use water, soil, and microorganisms to work together with wetland plants to provide many benefits such as preventing floods, regulating climate, reducing soil erosion, and degrading environmental pollutants. Plants have roots that are very important for their functions, because they help them get nutrients from the soil and water, and they also make a lot of substances that attract and feed microorganisms. The type of plant, the environment where it grows, and the part of the plant where the microorganisms live can affect what kinds of bacteria and fungi are found in the soil around the roots and inside the plant tissues. Some studies have shown that the substances that the roots make can select for certain microorganisms, which affects how diverse and different they are, and that some microorganisms are more common in the soil around certain plants than others. Wetland plants are influenced by both the soil and the water around them, but it is not clear yet if different wetland plants have different microorganisms in their soil and tissues.

Wetland plants offer important places to live, food, and energy for microorganisms, but the water and soil in wetland ecosystems are not the same everywhere. The number, type, and arrangement of microorganisms can change depending on the environment [136]. One of the main goals of microbial ecology is to understand how microorganisms adapt to different environments [13]. Some studies have found that in wetland ecosystems, the microorganisms in the water and soil change depending on the environment [75].

Root microbial communities are beneficial for plants to cope with abiotic stress, such as drought, salinity, and heavy metals. They can either directly protect plants from stress by producing certain substances or enzymes, or indirectly enhance plant stress tolerance by interacting with other microorganisms [48, 73]. Root microbial communities also help plants adapt to environmental stress conditions by changing their structure and composition. For example, a recent study found that rice root microbial community structure was significantly altered by short-term drought stress, but it could recover to the pre-stress level after the stress was relieved [101]. However, long-term drought stress had a lasting and severe impact on the endophytic microbial community in rice roots, which could not be fully restored after the stress was relieved. The study also found that actinomycetes, a group of bacteria, were greatly enriched in rice roots after long-term drought stress, accounting for more than 80% of the bacterial community. One of the actinomycetes, Streptomyces sp. SLBN-177, was isolated from the most abundant OTU (operational taxonomic unit) in the root microbial community. This bacterium could promote root growth under drought conditions when inoculated to rice plants [Prolonged drought imparts lasting]. Similarly, another study showed that sorghum root microbial communities were significantly affected by drought stress, and actinomycetes were moderately enriched. The enrichment of actinomycetes was associated with increased root growth in sorghum [101]. In addition, root microorganisms can also help plants deal with salinity and heavy metal stress by producing enzymes that can remove reactive oxygen species (ROS) from plant cells. ROS are harmful molecules that can damage plant cells under stress conditions. When plants are exposed to salinity and heavy metal stress, their chlorophyll a and b content and productivity decrease significantly, and their endogenous hydrogen peroxide level increases rapidly. However, root microorganisms can produce enzymes such as alkyl hydroperoxide reductase and glutathione-S-transferase that can help plants scavenge ROS and protect plant cells from oxidative damage [97]. These studies demonstrate that root microbial communities play an important role in helping plants resist abiotic stress and survive in harsh environments.

4.4.3 Responses of Microbial Communities Towards Environmental Factors

4.4.3.1 Seasonality

The microbial communities and their nitrogen transformation activities is seasonality. Seasonality refers to the periodic changes in environmental conditions that occur due to the Earth's rotation around the sun, such as temperature, light, precipitation, and vegetation. Seasonality can influence the microenvironment and the metabolic activity of microorganisms in CWs by affecting various parameters such as oxygen diffusion, redox potential, substrate availability, pH, and plant growth. Different seasons may favor or inhibit different microbial processes such as nitrification, denitrification, anammox, comammox, organic matter degradation, and phosphate removal.

Shirdashtzadeh et al. [113] analyzed the seasonal variation of microbial communities and their nitrogen transformation activities in a HSSFCWs treating domestic wastewater. They used molecular techniques such as quantitative PCR (qPCR) and high-throughput sequencing to quantify and identify the microorganisms involved in different nitrogen transformation pathways (such as nitrification, denitrification, anammox, comammox, DNRA). They found that seasonal variation had a significant impact on the abundance and diversity of nitrogen-cycling microorganisms. The abundance of nitrifying bacteria, denitrifying bacteri, anammox bacteria, comammox bacteria, and DNRA bacteria was higher in summer than in winter. The diversity of nitrifying bacteria, denitrifying bacteria (nirS), anammox bacteria, comammox bacteria, and DNRA bacteria was also higher in summer than in winter. The seasonal variation of microbial communities was correlated with environmental factors such as temperature, DO, NO_3^-, NH_4^+, TN, TP, COD, pH, EC, oxidation–reduction potential (ORP), hydraulic retention time (HRT), hydraulic loading rate (HLR), and plant biomass [113].

For instance, Matamoros and Salvadó [83] conducted a study on the removal efficiency of 18 pharmaceuticals in CWs planted with Daphnia sp. They observed that during the summer season, the removal efficiencies of cashmeran, triclosan, diclofenac, galaxolide, ketoprofen, naproxen, and tonalide were significantly higher compared to the winter season ($p = 0.002$), indicating a significant seasonal effect. However, some pharmaceuticals like diclofenac, ketoprofen, ibuprofen, and carbamazepine showed no significant differences between the warm and cold seasons. This could be attributed to higher concentrations of multiple pharmaceuticals during the cold season, which promote their elimination since the removal of pharmaceuticals in CWs follows first-order degradation kinetics.

4.4.3.2 pH

The efficiency of CWs systems can be influenced by the pH levels of the water they contain. Sanusi et al. [102] found that acidic rainfalls, with pH levels ranging from 4.4 to 5.7 in eastern France and as low as 2.9 in certain areas of China, could potentially impact CWs. Furthermore, a separate study investigating the influence of nutrients and pH on the effectiveness of VFCWs indicated that the effects of nutrient deficiency and low pH on the operation of such systems have been minimally explored. However, initial field studies suggest that operational issues may arise as a result of these factors.

In general, it is important to maintain an optimal pH range between 6.5 and 8.5 for CWs to operate efficiently. If the pH is too low or too high, it can affect microbial activity in the system and reduce its efficiency. For example, if the pH is too low, it can inhibit the growth of plants and microorganisms that are responsible for removing pollutants from the water [166]. On the other hand, if the pH is too high, it can cause precipitation of minerals such as calcium carbonate which can clog the pores in the soil and reduce its permeability. The pH level in CWs plays a crucial role in the removal of pharmaceuticals. It impacts the behavior of pharmaceutical compounds within the wetland environment by disrupting microbiologically mediated removal processes and influencing the sorption of pharmaceuticals onto the substrate of the wetland. The composition of the bacterial community in the wetland is often associated with the aquatic chemistry, particularly the pH level. Methanogenic bacteria have an optimal pH range of 6.5–7.5, while nitrifying bacteria thrive best at pH levels of 7.0–8.0. The pH indirectly affects the removal efficiency of pharmaceuticals in CWs by influencing microbial metabolism. The fate and distribution of pharmaceutical compounds in the wetland are highly influenced by the pH level since it determines the proportion and form (neutral or charged species) in which antibiotics exist. This, in turn, affects the sorption process of antibiotics onto the substrate of the wetland through electrostatic interactions [166].

Ammonia-oxidizing archaea (AOA) has been proved to be ubiquitous in aquatic and Terrestrial ecosystem. Chen et al. [20] studied the community diversity and distribution of AOA in wetland soils of different vegetation and depths. The results showed that the AOA amoA gene in the two wetland soils belonged to three lineages: Nitroso algae, Nitroso algae and Nitroso algal clusters. The different dominant positions of these AOA lineages indicate differences in their adaptation to acidic habitats, anoxic/anoxic alternation, organic matter, and other environmental factors, indicating a high diversity of AOA in swampy soils. The main driving factors of AOA community are pH, as well as organic carbon and ammonium nitrogen, which, along with many other environmental factors, also play an important role. The results of canonical correspondence analysis (CCA) indicated that there were positive correlations between pH and total organic carbon (TOC) content with the abundance of AOA communities represented by *Deyeuxia* (Qde) and Zhalong wetland (Zde), while negative correlations were observed with the Honghe and Sde AOA communities. The contents of NH_4^+-N and NO_3^--N showed positive correlations with the Honghe and Sanjiang wetland (Sde) AOA communities but negative correlations with the Qde and Zde AOA communities. The significant differences among the Honghe and Qixinghe AOA communities were consistent with the results of cluster analysis. The AOA communities in the Honghe wetland were distinct from the other AOA communities. Mantel regression analysis further supported the CCA results, highlighting that pH and TOC content were the most influential environmental parameters impacting the AOA communities ($P \leq 0.01$), as shown in Table 4.5. Additionally, NH_4^+-N content showed a strong correlation with the AOA communities ($P = 0.04$), while NO_3^--N content did not show a significant correlation.

In conclusion, maintaining an optimal pH range is important for ensuring that CWs operate efficiently. The optimal pH range for CWs is between 6.5 and 8.5. If the pH is too low or too high, it can affect microbial activity in the system and reduce its efficiency.

Table 4.5 Mantel regression analysis of the correlation between AOA communities and environmental parameters [20]

Environmental parameters	Forward sequences community		Reverse sequences community	
	R	*P*	*R*	*P*
pH	0.7595	0.006	0.7616	0.011
TOC	0.6493	0.013	0.6614	0.010
NH_4^+-N	0.4235	0.040	0.3914	0.044
NO_3^--N	0.1120	0.226	0.1047	0.218
All	0.4274	0.024	0.4001	0.029

4.4.3.3 Dissolved Oxygen Concentration

Intermittent aeration will change the distribution of dissolved oxygen in the whole reaction. In previous study, researchers found that TFCWs can achieve better efficiency than conventional CWs in secondary wastewater treatment [176]. Different dissolved oxygen concentration was distributed in different height along the reactor, which indicated an aerated or anoxic condition occurred in one sewage treatment cycle. Meanwhile falling water bring the air into CWs during descent process, which also change dissolved oxygen concentration in the surface and top of CWs. However, O_2 supply system shows that tidal aeration is the dominant position rather than falling water aeration [176]. Different dissolved oxygen cause the efficiency of nitrification, which change the microbial community of TFCWs.

4.5 Removal of Emerging Pollutants

Emergent macrophytes are a unique group of aquatic plants that grow in shallow water environments where their basal portion remains submerged below the water surface while their leaves, stems, and reproductive parts rise above it. These plants typically thrive in water depths ranging from 0.5 to 1.5 m and are commonly found in wetlands, lakes, rivers, and estuaries. The root and rhizomes of emergent macrophytes are adapted to subsist in sediment with anaerobic conditions, which means they lack oxygen. These parts of the plant receive the necessary oxygen supply from the aerial portions of the plant, which are exposed to atmospheric air. Moreover, young foliage can perform anaerobic respiration, allowing them to survive in low-oxygen environments. One of the significant roles of emergent macrophytes is their ability to act as nutrient pumps. They assimilate nutrients such as potassium, nitrogen, and phosphorus from the sediment and transfer them to the water column, where they contribute to the seasonal changes in nutrient availability. The process of nutrient transfer occurs through leaching and plant litters, which are the fallen parts of the plant. In addition to the nutrient cycle, emergent macrophytes also require carbon dioxide for their growth and development. They obtain CO_2 from the atmosphere and use it for inorganic carbon fixation. Carbon fixation is a vital process by which the plant converts carbon dioxide into organic compounds that are used for energy production and growth. Finally, the aerenchyma tissues of emergent macrophytes play a crucial role in their survival. These tissues are specialized air channels within the plant that store and transfer oxygen to other parts of the plant, especially the roots, which lack access to atmospheric oxygen. These adaptations allow emergent macrophytes to survive in conditions that would be hostile to other plants [130, 132].

Submerged plants are a type of rooted plant that grows entirely underwater in the illuminated region of an aquatic ecosystem. They contribute significantly to maintaining water quality and creating habitats for aquatic organisms. In the context of CWs, submerged plant species rely on the water column to obtain essential resources such as water, nutrients, and dissolved gases. CWs are purposely designed engineered systems that replicate the functions of natural wetlands. Their primary objective is to treat wastewater or stormwater runoff prior to its release into the surrounding environment. Moreover, the root system of submerged plants provides anchorage and stability to the plant, while also absorbing nutrients from the substrate of the wetland. This helps to maintain sediment stability and prevents erosion, which is important for the overall health of the aquatic ecosystem. The dissected leaves of submerged plants have a large surface area that allows for efficient absorption of nutrients and photosynthesis. This is crucial for the survival of submerged plants, as they must compete with other plants and algae for resources within their environment. Submerged plants can compete with phytoplankton for nutrients, such as nitrogen and phosphorus. This competition can help prevent excessive algal growth and maintain a balance between different types of aquatic vegetation. Submerged plants are commonly used in multistage systems for secondary treatment of wastewater. In these systems, the plants help to remove excess nutrients, organic matter, and pollutants from the water before it is discharged back into the environment. This process is critical for maintaining water quality and protecting the health of downstream ecosystems [12].

Wetland plants or macrophytes (aquatic plants) possess intrinsic characteristics that make them essential components of CWs. However, not all macrophyte species are suitable for wastewater treatment [39]. Only those species capable of withstanding inundation and exposure to varying pollutant loads in wetlands are suitable for treatment purposes [39]. While physical and microbial processes play a significant role in contaminant removal in CWs, macrophytes offer additional functions that enhance wastewater treatment [17, 93]. For example, photoautotrophic plants assimilate inorganic carbon from the atmosphere, producing organic matter that serves as an energy source for heterotrophs such as bacteria, fungi, and other animals. Additionally, optimal light and nutrient availability promote high primary productivity in wetland-dominated ecosystems [149]. As a result, the associated heterotrophic activity accelerates the decomposition and transformation of organic matter and related substances in the aquatic environment [16, 17].

Wetland plants/macrophytes have unique properties that allow them to grow and thrive in aquatic environments. These properties, such as their ability to photosynthesize, uptake nutrients from the water, and tolerate high levels of moisture, make them ideal for use in CWs [26]. While macrophytes are important for treating wastewater in CWs, not all species are suitable for this purpose. Certain species may not be able to tolerate the conditions present in a treatment wetland, such as fluctuating water levels or high pollutant loads. The most appropriate macrophyte species for treatment wetlands are those that can tolerate the conditions present in these environments. This includes species that can withstand inundation, exposure to stormwater, and varying pollutant loads. While wetland plants/macrophytes are important for wastewater treatment. Physical and microbial processes are also crucial for removing contaminants from the water. These processes include sedimentation, filtration, adsorption, and microbial activity. Although physical and microbial processes play a significant role in contaminant removal, macrophytes also contribute to the treatment process. For example, they can provide a surface area for microbial growth and support the development of microbial communities that break down pollutants [26]. Photoautotrophic plants are able to generate organic matter by taking in carbon dioxide from the atmosphere and converting it into organic carbon through photosynthesis. This organic matter can then serve as an energy source for heterotrophic organisms, such as bacteria and fungi, which break down pollutants. The growth and productivity of wetland plants/macrophytes is influenced by several factors, including light availability and nutrient supply. When these factors are present in sufficient amounts, the primary productivity of the ecosystem can increase, leading to greater pollutant removal. The activity of heterotrophic organisms, such as bacteria and fungi, can enhance the breakdown and decomposition of organic matter and related substances in an aquatic environment. This can lead to more efficient wastewater treatment in CWs [26].

Microplastics (MPs) are small plastic particles with sizes ranging from 1 to 5 mm. They are characterized by their strong hydrophobicity, which makes them highly effective in adsorbing a wide range of organic pollutants (OPs), including polychlorinated biphenyls (PCBs), polycyclic aromatic hydrocarbons (PAHs), phthalic acid esters (PAEs), and organochlorine pesticides (OCPs). The ability of MPs to adsorb OPs has been demonstrated in studies such as the one conducted by Alimi et al. [2]. The co-occurrence of MPs and OPs has been reported in many countries with coastal wetlands, including the United States, the United Kingdom, Japan, South Africa, and others. Researchers such as Mizukawa et al. [89] have found that MPs can act as carriers of OPs, possibly contributing to their widespread distribution in aquatic environments. And the results have shown that various types of OPs can be detected on the surface of MPs, indicating their potential role in transporting these contaminants. Examples of OPs found on MPs include PCBs, PAHs, PAEs, and OCPs like dichlorodiphenyltrichloroethane (DDT), hexachlorocyclohexane, and chlordane. The presence of these contaminants on MPs highlights the importance of understanding the potential ecological risks associated with the co-occurrence of MPs and OPs. Aquatic organisms readily consume MPs, resulting in the biological uptake of both MPs and the OPs they contain. This process, which has been studied by researchers like Rochman et al. [98] can result in damage to human health through the food chain. The potential effects of this phenomenon on ecosystems and the food web that humans rely on are still not fully understood. MPs possess the capacity to engage with OPs via adsorption and desorption mechanisms, thereby influencing their environmental dynamics including bioaccumulation, degradation, and migration. This interaction can also lead to the transfer of these pollutants across different environmental compartments. Researchers like Wang et al. [138] have highlighted the importance of studying the interactions between MPs and OPs to better understand the potential risks associated with their co-occurrence. The co-occurrence of MPs and OPs poses a potential harm to the environment. The growing identification of MPs in various environmental media and their role in carrying OPs has raised concerns about the long-term impacts of this phenomenon on ecosystems. The potential health risks associated with the ingestion of MPs and their carried OPs by humans is also a significant concern. While much attention has been given to the impact of MPs, there is still a lack of comprehensive understanding regarding the intricate effects that arise from the interaction between MPs and OPs) in wetlands, both in aquatic and terrestrial systems. Researchers such as Su et al. [117] have called for more research to be conducted in this area to better explore the potential risks and implications of this phenomenon on wetland ecosystems.

The application of area can increase the content of soil NH_4^+-N, which serves as an electron donor, thereby accelerating the reductive dechlorination of HCB. Additionally, moderate supplementation of nitrate can shift more available electrons towards dechlorination and regulate the growth of core microbial groups, enhancing the anaerobic removal of PCP. Overall, these findings highlight the promising potential of anaerobic reductive dechlorination mediated by microbial activities for the removal of chlorinated organic pollutants from the environment. The interaction between different reduction processes and the importance of optimizing substrate and product concentrations warrant further investigation.

4.5.1 Adsorption and Microbial Processes

Microorganisms play a crucial role in pollutant removal in CWs. They contribute to the reduction of COD, BOD, organic pollutants, and inorganic nitrogen found in textile wastewater [81]. The roots of plants provide a large surface area for bacterial attachment, allowing them to take up nutrients from the surrounding soil solution. In addition, plant roots release oxygen into the rhizosphere, creating an aerobic environment that facilitates the degradation of pollutants. Heterotrophic microorganisms utilize organic pollutants as a carbon source and under aerobic conditions produce CO_2 or methane under anaerobic conditions [131].

Oxygen required for the aerobic degradation of pollutants is also supplied by the release of oxygen from the plant roots into the surrounding substrate. Furthermore, microorganisms form a biofilm on the surface of the bed material, which aids in filtration. These microorganisms are capable of accumulating small amounts of heavy metals and reducing them into non-mobile forms [109].

When wastewater passes through the plants and filter media in CWs, treatment occur. Rhizofiltration and adsorption occur as a result of the interaction between wastewater and plant roots within the filter media. The root hairs of plants provide an aerobic condition that supports microbial activities, leading to the biological degradation of organic pollutants in wastewater [5, 16, 108]. The presence of dense macrophytes in CWs reduces light penetration into the water column, inhibiting the photosynthetic activities of phytoplankton. This reduction in light limits algal growth, which is particularly important in FSFCW where excessive phytoplankton growth can increase suspended solids in the outflow [129, 132].

The microclimatic effect of wetland plants is especially significant in CWs. In SSFCW, plant cover and litter provide insulation for the soil during winter frost and high temperatures in spring. Stands of emergent vegetation in CWs reduce wind velocity near the water surface, aiding in suspended solids removal through improved sedimentation and reduced resuspension [5, 110, 129, 132]. Additionally, the physical movement of aquatic plants caused by climatic factors such as wind helps keep the soil surface open for water percolation. The growth and decay of roots and rhizomes loosen the soil, creating tubular pores and channels that increase hydraulic conductivity and prevent clogging, especially in the VSSFCW [16, 127]. The roots and rhizomes also stabilize the sediment bed, facilitating the physical filtration process in the wetland system.

Nutrients are another vital component necessary for the growth and development of macrophytes [51]. Due to the high productivity of wetland plants, a significant amount of nutrients is bound within their biomass, affecting the rate of nutrient uptake. For example, Eichhornia crassipes (also known as Water Hyacinth), a highly productive aquatic plant, has a nutrient uptake rate of 230 g/Nm^2 for nitrogen (N) and 30 g/Nm^2 for phosphorus (P) [132]. The nutrient uptake rate by wetland plants is particularly significant in FSFCWs. The large surface area of leaves and stems of submerged vegetation provides colonization sites for photosynthetic algae, bacteria, and protozoa. These attached microbes oxygenate the water and take up nutrients from the water column, aiding in contaminant degradation and nutrient removal processes. However, extensive microbial growth on submerged plant tissues can also hinder plant growth [5, 16] that provides a brief description of the various crucial roles of aquatic plants (macrophytes) in wastewater treatment.

The exchange of gases within aquatic, flood-prone, and wetland environments is essential for plant survival [31]. Aquatic plant species utilize two pathways for oxygen (O_2) transfer: internal diffusion and convective flow [16]. Gas diffusion primarily occurs in the root system, while through-flow, a pressurized mechanism, takes place in the rhizomes and stems of wetland plants [12, 16]. O_2 moves from the atmosphere, starting from the upper parts of the plants (leaves), to the lower parts (roots). From the roots, O_2 is transferred to the rhizosphere, then to the surrounding sediment, facilitating microbial decomposition and pollutant uptake [132]. However, direct transfer of O_2 between the environment and submerged plant species is limited [31]. Diffusion of O_2 in wetland plants is influenced by environmental conditions such as temperature and humidity, as well as physiological (e.g., root and shoot biomass), anatomical (e.g., stems, leaves, and root system), and morphological (e.g., structure, shape, and size) characteristics [31]. In SSFCWs, the movement of O_2 to the rhizosphere enhances pollutant degradation through aerobic processes and promotes nitrification [12, 159]. This highlights the importance of wetland plant functions in different types of CWs. Besides their role in wastewater treatment, wetland plants also provide habitats for vertebrates and other wildlife [148]. Wetland plants are vital components of the food chain, particularly in FSFCWs and HSSFCWs. Moreover, when properly managed in CWs, wetland plants contribute to the aesthetic value of the environment [132].

4.5.2 Sedimentation and Filtration

Sedimentation is the process of settling of suspended solids and associated pollutants under gravity. Sedimentation can remove emerging pollutants that are either particulate-bound or have high density and low solubility. Sedimentation can be enhanced by increasing the hydraulic retention time, decreasing the flow velocity, and optimizing the inlet and outlet structures of CWs. Sedimentation can also be influenced by the characteristics of the wastewater, such as turbidity, organic matter content, pH, and salinity. Sedimentation can remove a significant fraction of emerging pollutants such as microplastics [82], nanomaterials [152], and some pharmaceuticals [35] from wastewater.

Filtration is the process of trapping of suspended solids and associated pollutants by the porous media (such as gravel, sand, or soil) in CWs. Filtration can remove emerging pollutants that are either particulate-bound or have low molecular size and high hydrophobicity. Filtration can be enhanced by increasing the surface area, decreasing the pore size, and optimizing the depth and configuration of the porous media in CWs. Filtration can also be influenced by the characteristics of the wastewater, such as turbidity, organic matter content, pH, and salinity. Filtration can remove a significant fraction of emerging pollutants such as microplastics [60], nanomaterials [118], hormones, and some pharmaceuticals [152] from wastewater. Due to its natural, environmentally friendly, and economical advantages, an increasing number of countries and regions are beginning to use wetlands as an effective means of wastewater treatment. In the future, wetland technology will play an even more important role in the field of pollution treatment [109].

Sedimentation and filtration are important physical processes that can reduce the concentration and load of emerging pollutants in wastewater. However, sedimentation and filtration alone may not be sufficient to achieve complete removal of emerging pollutants in CWs. Therefore, other mechanisms such as adsorption, biodegradation, photodegradation, plant uptake, and volatilization should also be considered and integrated in the design and operation of CWs.

4.6 Prospects

CWs are artificial wetland systems that mimic the natural processes of wetlands to treat wastewater, stormwater, agricultural runoff, and other sources of water pollution. They can provide many benefits for the environment, society, and economy, but they also have some challenges and limitations. However, it is hard to quantify the contributions of each mechanism process to specific pollutants removal under different operating conditions and determine the most important process. In addition, at present, the knowledge of degradation pathways of different types of pollutants in different substrates (plants, water and substrates), identification of transformation products, and specific functional microorganisms responsible for the pollutant's removal are very limited. Therefore, further research is needed to provide more convincing evidence for the potential removal mechanism in CWs. Therefore, there are many opportunities for future research and development in this field. Here are some possible research directions about CWs:

1. Improving the performance and reliability of CWs under extreme environmental conditions, especially in the area where have negative affect on the plant growth and microbial activities.
2. Developing multifunctional CWs that can have a wide range adaptive to remove multiple emerging pollutants in the future.
3. Exploring the potential of CWs for economic and ecological values, such as bioenergy production, fertilizer production, or reused for irrigation.
4. Investigating the ecological and social impacts of CWs on the surrounding environment and communities, such as the effects on biodiversity, ecosystem services, human health, and well-being.
5. Optimizing the design and operation of CWs using advanced modeling and simulation tools, such as kinetics models, hydrological models, mechanistic models, or artificial intelligence models.

References

1. Aiello R, Cirelli GL, Consoli S, Licciardello F, Toscano A (2013) Risk assessment of treated municipal wastewater reuse in Sicily [Article]. Water Sci Technol 67(1):89–98. https://doi.org/10.2166/wst.2012.535
2. Alimi OS, Farner Budarz J, Hernandez LM, Tufenkji N (2018) Microplastics and nanoplastics in aquatic environments: aggregation, deposition, and enhanced contaminant transport. Environ Sci Technol 52(4):1704–1724
3. Álvarez JA, Ruíz I, Soto M (2008) Anaerobic digesters as a pretreatment for constructed wetlands. Ecol Eng 33(1):54–67. https://doi.org/10.1016/j.ecoleng.2008.02.001
4. Andrew CS, Daniel S, William AJ, Khoiviet H, David EC (2003) Impact of the plant rhizosphere and augmentation on remediation of polychlorinated biphenyl contaminated soil. Environ Toxicol Chem 22

5. Ansari NA, Ram J, Nawab A (2015) Structure and composition of Butterfly (Lepidoptera: Rhopalocera) fauna in Surajpur wetland, National Capital Region, India. Asian J Conserv Biol 4(1):43–53
6. Azaizeh H, Linden KG, Barstow C, Kalbouneh S, Tellawi A, Albalawneh A, Gerchman Y (2013) Constructed wetlands combined with UV disinfection systems for removal of enteric pathogens and wastewater contaminants. Water Sci Technol 67(3):651–657. https://doi.org/10.2166/wst.2012.615
7. Babatunde AO, Zhao YQ (2009) Forms, patterns and extractability of phosphorus retained in alum sludge used as substrate in laboratory-scale constructed wetland systems. Chem Eng J 152(1):8–13. https://doi.org/10.1016/j.cej.2009.03.020
8. Bai X, Wang B, Yu M, Nie M (1999) Development of constructed wetland wastewater treatment technology and its application in China. 32(6):4. 1006-6780(1999)06-0088-05
9. Balabanič D, Hermosilla D, Merayo N, Klemenčič AK, Blanco Á (2012) Comparison of different wastewater treatments for removal of selected endocrine-disruptors from paper mill wastewaters. J Environ Sci Health Part A 47(10):1350–1363. https://doi.org/10.1080/10934529.2012.672301
10. Bao T, Chen T, Liu H, Chen D, Qing C, Frost RL (2014) Preparation of magnetic porous ceramsite and its application in biological aerated filters [Article]. J Water Process Eng 4:185–195. https://doi.org/10.1016/j.jwpe.2014.10.004
11. Bavor HJ, Roser DJ, Adcock PW (1995) Challenges for the development of advanced constructed wetlands technology [Article Proceedings Paper]. Water Sci Technol 32(3):13–20. https://doi.org/10.1016/0273-1223(95)00600-1
12. Białowiec A, Sobieraj K, Pilarski G, Manczarski P (2019) The oxygen transfer capacity of submerged plant Elodea densa in wastewater constructed wetlands. Water 11(3):575
13. Bier RL, Bernhardt ES, Boot CM, Graham EB, Hall EK, Lennon JT, Nemergut DR, Osborne BB, Ruiz-González C, Schimel JP, Waldrop MP, Wallenstein MD (2015) Linking microbial community structure and microbial processes: an empirical and conceptual overview. FEMS Microbiol Ecol 91(10):fiv113. https://doi.org/10.1093/femsec/fiv113
14. Boonyaroj V, Chiemchaisri C, Chiemchaisri W, Theepharaksapan S, Yamamoto K (2012) Toxic organic micro-pollutants removal mechanisms in long-term operated membrane bioreactor treating municipal solid waste leachate. Bioresour Technol 113:174–180. https://doi.org/10.1016/j.biortech.2011.12.127
15. Bowen L, Weiliang Z, Dezhi X, Shibin L, Jinyang H, Xiulei F (2022) Influence of different metals on production of sewage sludgeΓÇôBased biochar and its application for ammonium and phosphate adsorption removal from wastewater. J Environ Eng 148(9)
16. Brix H (1994) Functions of macrophytes in constructed wetlands. Water Sci Technol 29(4):71–78
17. Brix H (2003) Plants used in constructed wetlands and their functions. In: Dias V, Vymazal J (eds) 1st international seminar on the use of aquatic macrophytes for wastewater treatment in constructed wetlands. Lisboa, Portugal
18. Camacho-Muñoz D, Martín J, Santos JL, Alonso E, Aparicio I, De la Torre T, Rodriguez C, Malfeito JJ (2012) Effectiveness of three configurations of membrane bioreactors on the removal of priority and emergent organic compounds from wastewater: comparison with conventional wastewater treatments. J Environ Monitoring 14(5):1428–1436. https://doi.org/10.1039/C2EM00007E
19. Cavani F, Trifiro F, Vaccari A (1991) Hydrotalcite-type anionic clays: preparation, properties and applications. Catalysis Today 11(2):173–301. https://doi.org/10.1016/0920-5861(91)80068-K
20. Chen C, Liang H, Gao D (2019) Community diversity and distribution of ammonia-oxidizing archaea in marsh wetlands in the black soil zone in North-east China. Front Environ Sci Eng 13(4). https://doi.org/10.1007/s11783-019-1146-z
21. Chen J, Li X, Jia W, Shen S, Deng S, Ji B, Chang J (2021) Promotion of bioremediation performance in constructed wetland microcosms for acid mine drainage treatment by using organic substrates and supplementing domestic wastewater and plant litter broth. J Hazard Mater 404(Pt A):124125. https://doi.org/10.1016/j.jhazmat.2020.124125

22. Chen M, Zhu M, Zhu Y, Wang D, Li Z, Zeng G, Zhang C, Huang J, Xu P (2019) Collision of emerging and traditional methods for antibiotics removal: taking constructed wetlands and nanotechnology as an example. NanoImpact 15.https://doi.org/10.1016/j.impact.2019.100175
23. Chen XC, Kong HN, Wu DY, Wang XZ, Lin YY (2009) Phosphate removal and recovery through crystallization of hydroxyapatite using xonotlite as seed crystal. J Environ Sci 21(5):575–580. https://doi.org/10.1016/S1001-0742(08)62310-4
24. Chen YC, Banks MK, Schwab AP (2003) Pyrene degradation in the rhizosphere of tall fescue (*Festuca arundinacea*) and switchgrass (*Panicum virgatum* L.) [Article]. Environ Sci Technol 37(24):5778–5782. https://doi.org/10.1021/es030400x
25. Chen ZM, Chen B, Zhou JB, Li Z, Zhou Y, Xi XR, Lin C, Chen GQ (2008) A vertical subsurface-flow constructed wetland in Beijing. Commun Nonlinear Sci Numer Simul 13(9):1986–1997. https://doi.org/10.1016/j.cnsns.2007.02.009
26. Cheng J, Yuan J, Li S, Yang X, Lu Z, Xu J, He Y (2021) Promoted reductive removal of chlorinated organic pollutants co-occurring with facilitated methanogenesis in anaerobic environment: a systematic review and meta-analysis. Crit Rev Environ Sci Technol 52(14):2582–2609. https://doi.org/10.1080/10643389.2021.1886890
27. Cheng L, Yang W, Liang H, Nabi M, Li Y, Wang H, Hu J, Chen T, Gao D (2023) Nitrogen removal from mature landfill leachate through enhanced partial nitrification-anammox process in an innovative multi-stage fixed biofilm reactor. Sci Total Environ 877, Article 162959. https://doi.org/10.1016/j.scitotenv.2023.162959
28. Christenson L, Sims R (2011) Production and harvesting of microalgae for wastewater treatment, biofuels, and bioproducts. Biotechnol Adv 29(6):686–702. https://doi.org/10.1016/j.biotechadv.2011.05.015
29. Cirelli GL, Consoli S, Licciardello F, Aiello R, Giuffrida F, Leonardi C (2012) Treated municipal wastewater reuse in vegetable production. Agric Water Manag 104:163–170. https://doi.org/10.1016/j.agwat.2011.12.011
30. Colares GS, de Souza Celente G, da Silva FP, de Loreto AC, Lutterbeck CA, Kist LT, Machado ÊL (2019) Combined system for the treatment and reuse of urban wastewater: the efficiency of anaerobic reactors + hybrid constructed wetlands + ozonation. Water Sci Technol 80(2):254–264. https://doi.org/10.2166/wst.2019.270
31. Colmer T (2003) Long-distance transport of gases in plants: a perspective on internal aeration and radial oxygen loss from roots. Plant, Cell Environ 26(1):17–36
32. Compant S, Clément C, Sessitsch A (2010) Plant growth-promoting bacteria in the rhizo- and endosphere of plants: their role, colonization, mechanisms involved and prospects for utilization. Soil Biol Biochem 42(5):669–678. https://doi.org/10.1016/j.soilbio.2009.11.024
33. Conrad A, Kontro M, Keinänen MM, Cadoret A, Faure P, Mansuy-Huault L, Block J-C (2003) Fatty acids of lipid fractions in extracellular polymeric substances of activated sludge flocs. Lipids 38(10):1093–1105
34. Couillard D (1992) Appropriate waste-water management technologies using peat [Review]. J Environ Syst 21(1):1–20
35. David G, Rana MS, Saxena S, Sharma S, Pant D, Prajapati SK (2023) A review on design, operation, and maintenance of constructed wetlands for removal of nutrients and emerging contaminants. Int J Environ Sci Technol 20(8):9249–9270. https://doi.org/10.1007/s13762-022-04442-y
36. Deng S, Chen J, Chang J (2021) Application of biochar as an innovative substrate in constructed wetlands/biofilters for wastewater treatment: performance and ecological benefits. J Cleaner Prod 293.https://doi.org/10.1016/j.jclepro.2021.126156
37. Dissanayake CB, Weerasooriya SVR (1981) Peat as a metal trapping material in the purification of industrial effluents. Int J Environ Stud 17(3–4):233–238. https://eurekamag.com/research/006/076/006076492.php
38. Dong D, Aleta P, Zhao X, Choi OK, Kim S, Lee JW (2019) Effects of nanoscale zero valent iron (nZVI) concentration on the biochemical conversion of gaseous carbon dioxide (CO_2) into methane (CH_4). Biores Technol 275:314–320. https://doi.org/10.1016/j.biortech.2018.12.075

39. DuPoldt CA Jr (1996) Managing stormwater runoff in urban watersheds. Opflow 22(10):9–9
40. Gao D-W, Hu Q (2012) Bio-contact oxidation and greenhouse-structured wetland system for rural sewage recycling in cold regions: a full-scale study. Ecol Eng 49:249–253. https://doi.org/10.1016/j.ecoleng.2012.08.012
41. Gao H, Wang S, Song Y, Peng J (2013) Sewage purification effect of tidal flow subsurface flow hybrid constructed wetland. Chin J Environ Eng 7(10):3743–3748
42. Gong L, Chen G, Li J, Zhu G (2020) Utilization of rural domestic sewage tailwaters by Ipomoea aquaticain different hydroponic vegetable and constructed wetland systems. Water Sci Technol 82(2):386–400. https://doi.org/10.2166/wst.2020.373
43. Gunes K (2007) Restaurant wastewater treatment by constructed wetlands. Clean-Soil Air Water 35(6):571–575. https://doi.org/10.1002/clen.200700142
44. Guo M, Gong Z, Miao R, Jia C, Rookes J, Cahill D, Zhuang J (2018) Enhanced polycyclic aromatic hydrocarbons degradation in rhizosphere soil planted with tall fescue: bacterial community and functional gene expression mechanisms. Chemosphere 212:15–23. https://doi.org/10.1016/j.chemosphere.2018.08.057
45. Gupta S, Nayak A, Roy C, Yadav AK (2021) An algal assisted constructed wetland-microbial fuel cell integrated with sand filter for efficient wastewater treatment and electricity production. Chemosphere 263:128132. https://doi.org/10.1016/j.chemosphere.2020.128132
46. Gupta S, Srivastava P, Yadav AK (2020) 13–Integration of microbial fuel cell into constructed wetlands: effects, applications, and future outlook. In: Abbassi R, Yadav AK, Khan F, Garaniya V (eds) Integrated microbial fuel cells for wastewater treatment. Butterworth-Heinemann, pp 273–293. https://doi.org/10.1016/B978-0-12-817493-7.00013-8
47. Gupta S, Srivastava P, Patil SA, Yadav AK (2021) A comprehensive review on emerging constructed wetland coupled microbial fuel cell technology: potential applications and challenges. Bioresour Technol 320(Pt B):124376. https://doi.org/10.1016/j.biortech.2020.124376
48. Hartman K, Tringe SG (2019) Interactions between plants and soil shaping the root microbiome under abiotic stress. Biochem J 476(19):2705–2724. https://doi.org/10.1042/bcj20180615
49. He CF, Zhang FJ, Liang XQ, Wang ZY, Tian YL, Zhang HY (2018) Effect of acid-activated red mud on the simulated constructed wetland ecosystem. Environ Eng 36(07):51–55+61. https://doi.org/10.13205/j.hjgc.201807011
50. Headley JV, Gandrass J, Kuballa J, Peru KM, Gong Y (1998) Rates of sorption and partitioning of contaminants in river biofilm. Environ Sci Technol 32(24):3968–3973. https://doi.org/10.1021/es9804991
51. Hoagland CR, Gentry LE, David MB, Kovacic DA (2001) Plant nutrient uptake and biomass accumulation in a constructed wetland. J Freshw Ecol 16(4):527–540
52. Hu Y, Xiao R, Kuang B, Hu Y, Wang Y, Bai J, Wang C, Zhang L, Wei Z, Zhang K, Jorquera MA, Acuna JJ, Pan W (2022) Application of modified biochar in the treatment of pesticide wastewater by constructed wetland. Water 14(23), Article 3889. https://doi.org/10.3390/w14233889
53. Huang X-F, Chaparro JM, Reardon KF, Zhang R, Shen Q, Vivanco JM (2014) Rhizosphere interactions: root exudates, microbes, and microbial communities. Botany 92(4):267–275
54. Jacoby RP, Koprivova A, Kopriva S (2021) Pinpointing secondary metabolites that shape the composition and function of the plant microbiome. J Exp Bot 72(1):57–69. https://doi.org/10.1093/jxb/eraa424
55. Ji Z, Tang W, Pei Y (2022) Constructed wetland substrates: a review on development, function mechanisms, and application in contaminants removal. Chemosphere 286(Pt 1):131564. https://doi.org/10.1016/j.chemosphere.2021.131564
56. Jiang O, Li L, Duan G, Gustave W, Zhai W, Zou L, An X, Tang X, Xu J (2023) Root exudates increased arsenic mobility and altered microbial community in paddy soils. J Environ Sci (China) 127:410–420. https://doi.org/10.1016/j.jes.2022.05.036
57. Jin L, Shen K, Fang G, Liu S, Pan A, Deng Y, Gao G (2011) Different substrates in analog vertical flow constructed wetlands to dispose sbr effluent of papermaking mill. In: Proceeding of international mechanical pulping conference 2011

58. Jin L, Zhang G, Tian H (2014) Current state of sewage treatment in China. Water Res 66:85–98. https://doi.org/10.1016/j.watres.2014.08.014
59. Keiding K, Nielsen PH (1997) Desorption of organic macromolecules from activated sludge: effect of ionic composition. Water Res 31(7):1665–1672
60. Khan S, Nawab J, Waqas M (2020) Constructed wetlands: a clean-green technology for degradation and detoxification of industrial wastewaters. In: Bharagava RN, Saxena G (eds) Bioremediation of industrial waste for environmental safety: volume II: biological agents and methods for industrial waste management. Springer, Singapore, pp 127–163. https://doi.org/10.1007/978-981-13-3426-9_6
61. Kongthale G, Sotha S, Michu P, Madloh A, Wetchapan P, Chaijak P (2023) Electricity production and phenol removal of winery wastewater by constructed wetland—microbial fuel cell integrated with ethanol tolerant yeast. Biointerface Res Appl Chem 13(2), Article 157. https://doi.org/10.33263/briac132.157
62. Koprivova A, Kopriva S (2022) Plant secondary metabolites altering root microbiome composition and function. Curr Opin Plant Biol 67:102227. https://doi.org/10.1016/j.pbi.2022.102227
63. Kotoky R, Rajkumari J, Pandey P (2018) The rhizosphere microbiome: significance in rhizoremediation of polyaromatic hydrocarbon contaminated soil. J Environ Manage 217:858–870. https://doi.org/10.1016/j.jenvman.2018.04.022
64. Lancheros JC, Madera-Parra CA, Caselles-Osorio A, Torres-López WA, Vargas-Ramírez XM (2019) Ibuprofen and naproxen removal from domestic wastewater using a horizontal subsurface flow constructed wetland coupled to ozonation. Ecol Eng 135:89–97. https://doi.org/10.1016/j.ecoleng.2019.05.007
65. Langenbach K, Kuschk P, Horn H, Kästner M (2009) Slow sand filtration of secondary clarifier effluent for wastewater reuse. Environ Sci Technol 43(15):5896–5901. https://doi.org/10.1021/es900527j
66. Lawrence JR, Kopf G, Headley JV, Neu TR (2001) Sorption and metabolism of selected herbicides in river biofilm communities. Can J Microbiol 47(7):634–641. https://doi.org/10.1139/w01-061
67. Leverenz HL, Tchobanoglous G, Darby JL (2009) Clogging in intermittently dosed sand filters used for wastewater treatment. Water Res 43(3):695–705. https://doi.org/10.1016/j.watres.2008.10.054
68. Li C, Dong Y, Lei Y, Wu D, Xu P (2015) Removal of low concentration nutrients in hydroponic wetlands integrated with zeolite and calcium silicate hydrate functional substrates. Ecol Eng 82:442–450. https://doi.org/10.1016/j.ecoleng.2015.05.003
69. Li H, Xu H, Yang Y-L, Yang X-L, Wu Y, Zhang S, Song H-L (2019) Effects of graphite and Mn ore media on electro-active bacteria enrichment and fate of antibiotic and corresponding resistance gene in up flow microbial fuel cell constructed wetland. Water Res 165, Article 114988. https://doi.org/10.1016/j.watres.2019.114988
70. Li J, Han X, Brandt BW, Zhou Q, Ciric L, Campos LC (2019) Physico-chemical and biological aspects of a serially connected lab-scale constructed wetland-stabilization tank-GAC slow sand filtration system during removal of selected PPCPs. Chem Eng J 369:1109–1118. https://doi.org/10.1016/j.cej.2019.03.105
71. Liu D, Ge Y, Chang J, Peng C, Gu B, Chan GY, Wu X (2009) Constructed wetlands in China: recent developments and future challenges. Front Ecol Environ 7(5):261–268
72. Liu H, Fang HH (2002) Extraction of extracellular polymeric substances (EPS) of sludges. J Biotechnol 95(3):249–256
73. Liu H, Brettell LE, Qiu Z, Singh BK (2020) Microbiome-mediated stress resistance in plants. Trends Plant Sci 25(8):733–743
74. Liu H, Sun L, Chen X, Jing J, Shen X, Li J (2015) Treatment of dairy wastewater by oil separation and regulation/air flotation/UASB/CASS. China Water Wastewater 31(8):93–95, Article 1000-4602(2015)31:8<93:Gytjqf>2.0.Tx;2-v
75. Liu J, Li C, Jing J, Zhao P, Luo Z, Cao M, Ma Z, Jia T, Chai B (2018) Ecological patterns and adaptability of bacterial communities in alkaline copper mine drainage. Water Res 133:99–109. https://doi.org/10.1016/j.watres.2018.01.014

76. Liu W, Chen X, Li W, Yu Y, Yan K (2014) Environmental assessment, management and utilization of red mud in China. J Cleaner Prod 84
77. Liu XH, Wang YQ, Lu SY, Liu Y, Zhao B, Xi BD, Guo XC, Guo W, Zhang J (2020) Intensified sulfamethoxazole removal in an electrolysis-integrated tidal flow constructed wetland system. Chem Eng J 390, Article 124545. https://doi.org/10.1016/j.cej.2020.124545
78. Liu X, Liu Y, Guo X, Lu S, Wang Y, Zhang J, Guo W, Xi B (2020) High degree of contaminant removal and evolution of microbial community in different electrolysis-integrated constructed wetland systems. Chem Eng J 388.https://doi.org/10.1016/j.cej.2020.124391
79. Lu S, Wan Z, Li F, Zhang X (2016) Ammonia nitrogen adsorption and desorption characteristics of twenty-nine kinds of constructed wetland substrates. Res Environ Sci 29(8):1187–1194, Article 1001-6929(2016)29:8<1187:2zsdtl>2.0.Tx;2-d
80. Ma Y (2018) Black-odorous water governance by a combined process of fabricated PB, AF BAF and ICW. China Water Wastewater 34(20):76–81. https://doi.org/10.19853/j.zgjsps.1000-4602.2018.20.017
81. Maltais-Landry G, Maranger R, Brisson J, Chazarenc F (2009) Nitrogen transformations and retention in planted and artificially aerated constructed wetlands. Water Res 43(2):535–545
82. Matamoros V, Bayona JM (2008) Behavior of emerging pollutants in constructed wetlands. In: Barceló D, Petrovic M (eds) Emerging contaminants from industrial and municipal waste: removal technologies. Springer, Berlin, pp 199–217. https://doi.org/10.1007/698_5_096
83. Matamoros V, Salvadó V (2012) Evaluation of the seasonal performance of a water reclamation pond-constructed wetland system for removing emerging contaminants. Chemosphere 86(2):111–117. https://doi.org/10.1016/j.chemosphere.2011.09.020
84. Matos MP, Sperling MV, Matos AT, Miranda ST, Souza TD, Costa LM (2017) Key factors in the clogging process of horizontal subsurface flow constructed wetlands receiving anaerobically treated sewage. Ecol Eng 106:588–596. https://doi.org/10.1016/j.ecoleng.2017.06.013
85. McKay G, Poots VJP (1980) Kinetics and diffusion-processes in color removal from effluent using wood as an adsorbent. J Chem Technol Biotechnol 30(6):279–292
86. Meng P, Pei H, Hu W, Shao Y, Li Z (2014) How to increase microbial degradation in constructed wetlands: influencing factors and improvement measures. Bioresour Technol 157:316–326. https://doi.org/10.1016/j.biortech.2014.01.095
87. Min Z, Bin GUO, Qinglin FU, Jinfu J, Hua LI, Nengfei D, Chen LIU, Yicheng LIN (2011) Comparison of removal efficiency of phosphorus and ammonium in solutions by five substrates. J Soil Water Conserv 25(3):63–67, Article 1009-2242(2011)25:3<63:Btjzqc>2.0.Tx;2-3
88. Mittal Y, Dash S, Srivastava P, Mishra PM, Aminabhavi TM, Yadav AK (2022) Azo dye containing wastewater treatment in earthen membrane based unplanted two chambered constructed wetlands-microbial fuel cells: a new design for enhanced performance. Chem Eng J 427:131856. https://doi.org/10.1016/j.cej.2021.131856
89. Mizukawa K, Takada H, Ito M, Geok YB, Hosoda J, Yamashita R, Saha M, Suzuki S, Miguez C, Frias J, Antunes JC, Sobral P, Santos I, Micaelo C, Ferreira AM (2013) Monitoring of a wide range of organic micropollutants on the Portuguese coast using plastic resin pellets. Mar Pollut Bull 70(1):296–302. https://doi.org/10.1016/j.marpolbul.2013.02.008
90. Mlih R, Bydalek F, Klumpp E, Yaghi N, Bol R, Wenk J (2020) Light-expanded clay aggregate (LECA) as a substrate in constructed wetlands—a review. Ecol Eng 148.https://doi.org/10.1016/j.ecoleng.2020.105783
91. Mu C, Wang L, Wang L (2021) Removal of Cr(VI) and electricity production by constructed wetland combined with microbial fuel cell (CW-MFC): influence of filler media. J Cleaner Prod 320.https://doi.org/10.1016/j.jclepro.2021.128860
92. Muñoz MA, Rosales RM, Gabarrón M, Faz A, Acosta JA (2016) Effects of the hydraulic retention time on pig slurry purification by constructed wetlands and stabilization ponds. Water Air Soil Pollut 227(9):293. https://doi.org/10.1007/s11270-016-2993-x
93. Peterson SB, Teal JM (1996) The role of plants in ecologically engineered wastewater treatment systems. Ecol Eng 6(1–3):137–148

94. Qualls RG, Richardson CJ (1995) Forms of soil phosphorus along a nutrient enrichment gradient in the Northern Everglades. Soil Sci 160(3):183–198. https://doi.org/10.1097/00010694-199509000-00004
95. Rana S, Viraraghavan T (1987) Use of peat in septic tank effluent treatment—column studies. Water Qual Res J 22(3):491–504. https://doi.org/10.2166/wqrj.1987.039
96. Rane NR, Tapase S, Kanojia A, Watharkar A, Salama ES, Jang M, Kumar Yadav K, Amin MA, Cabral-Pinto MMS, Jadhav JP, Jeon BH (2022) Molecular insights into plant-microbe interactions for sustainable remediation of contaminated environment. Bioresour Technol 344(Pt B):126246. https://doi.org/10.1016/j.biortech.2021.126246
97. Ravanbakhsh M, Sasidharan R, Voesenek LA, Kowalchuk GA, Jousset A (2018) Microbial modulation of plant ethylene signaling: ecological and evolutionary consequences. Microbiome 6:1–10
98. Rochman CM, Lewison RL, Eriksen M, Allen H, Cook A-M, Teh SJ (2014). Polybrominated diphenyl ethers (PBDEs) in fish tissue may be an indicator of plastic contamination in marine habitats. Sci Total Environ 476–477, 622–633. https://doi.org/10.1016/j.scitotenv.2014.01.058
99. Ruiz AM, Maerz JC, Davis AK, Keel MK, Ferreira AR, Conroy MJ, Morris LA, Fisk AT (2010) Patterns of development and abnormalities among tadpoles in a constructed wetland receiving treated wastewater. Environ Sci Technol 44(13):4862–4868. https://doi.org/10.1021/es903785x
100. Saeed T, Sun G (2012) A review on nitrogen and organics removal mechanisms in subsurface flow constructed wetlands: dependency on environmental parameters, operating conditions and supporting media. J Environ Manage 112:429–448. https://doi.org/10.1016/j.jenvman.2012.08.011
101. Santos-Medellín C, Liechty Z, Edwards J, Nguyen B, Huang B, Weimer BC, Sundaresan V (2021) Prolonged drought imparts lasting compositional changes to the rice root microbiome. Nat Plants 7(8):1065–1077. https://doi.org/10.1038/s41477-021-00967-1
102. Sanusi A, Wortham H, Millet M, Mirabel P (1996) Chemical composition of rainwater in Eastern France. Atmos Environ 30(1):59–71. https://doi.org/10.1016/1352-2310(95)00237-S
103. Sardans J, Lambers H, Preece C, Alrefaei AF, Penuelas J (2023) Role of mycorrhizas and root exudates in plant uptake of soil nutrients (calcium, iron, magnesium, and potassium): has the puzzle been completely solved? Plant J. https://doi.org/10.1111/tpj.16184
104. Schorer M, Eisele M (1997) Accumulation of inorganic and organic pollutants by biofilms in the aquatic environment. Water Air Soil Pollut 99(1):651–659. https://doi.org/10.1007/BF02406904
105. Seidel K (1964) Abbau von bacterium coli durch höhere wasserpflanzen. Naturwissenschaften 51(16):395–395
106. Seidel K (1966) Reinigung von Gewässern durch höhere Pflanzen. Naturwissenschaften 53(12):289–297
107. Seidel K, Happel H, Graue G (1978) Contributions to revitalisation of waters. Stiftung Limnologische Arbeitsgruppe
108. Sharma PK, Minakshi D, Rani A, Malaviya P (2018) Treatment efficiency of vertical flow constructed wetland systems operated under different recirculation rates. Ecol Eng 120:474–480
109. Sharma R, Malaviya P (2022) Constructed wetlands for textile wastewater remediation: a review on concept, pollutant removal mechanisms, and integrated technologies for efficiency enhancement. Chemosphere 290:133358. https://doi.org/10.1016/j.chemosphere.2021.133358
110. Shelef O, Gross A, Rachmilevitch S (2013) Role of plants in a constructed wetland: current and new perspectives. Water 5(2):405–419
111. Shen Z, Zhou Y, Liu J, Xiao Y, Cao R, Wu F (2015) Enhanced removal of nitrate using starch/PCL blends as solid carbon source in a constructed wetland. Bioresour Technol 175:239–244. https://doi.org/10.1016/j.biortech.2014.10.006

112. Shi X, Fan J, Zhang J, Shen Y (2017) Enhanced phosphorus removal in intermittently aerated constructed wetlands filled with various construction wastes. Environ Sci Pollut Res 24(28):22524–22534. https://doi.org/10.1007/s11356-017-9870-z
113. Shirdashtzadeh M, Chua LHC, Brau L (2022) Microbial communities and nitrogen transformation in constructed wetlands treating stormwater runoff. Front Water 3.https://doi.org/10.3389/frwa.2021.751830
114. Song L, Yao Y, Lin L, Gao W, Cai T, Liang H, Gao D (2019) The potential source of nitrous oxide in the pristine riparian marsh during freeze-thaw cycles, case study in Northeast China. Ecol Eng 134:18–25. https://doi.org/10.1016/j.ecoleng.2019.05.004
115. Srivastava P, Abbassi R, Garaniya V, Lewis T, Yadav AK (2020) Performance of pilot-scale horizontal subsurface flow constructed wetland coupled with a microbial fuel cell for treating wastewater. J Water Process Eng 33:100994. https://doi.org/10.1016/j.jwpe.2019.100994
116. Srivastava P, Abbassi R, Yadav AK, Garaniya V, Lewis T, Zhao Y, Aminabhavi T (2021) Interrelation between sulphur and conductive materials and its impact on ammonium and organic pollutants removal in electroactive wetlands. J Hazard Mater 419:126417. https://doi.org/10.1016/j.jhazmat.2021.126417
117. Su X, Yuan J, Lu Z, Xu J, He Y (2022) An enlarging ecological risk: Review on co-occurrence and migration of microplastics and microplastic-carrying organic pollutants in natural and constructed wetlands. Sci Total Environ 837:155772. https://doi.org/10.1016/j.scitotenv.2022.155772
118. Sudarsan JS, Annadurai R, Mukhopadhyay M, Chakraborty P, Nithiyanantham S (2018) Domestic wastewater treatment using constructed wetland: an efficient and alternative way. Sustain Water Resour Manag 4(4):781–787. https://doi.org/10.1007/s40899-017-0164-x
119. Suzuki M (1997) Role of adsorption in water environment processes. Water Sci Technol 35(7):1–11. https://doi.org/10.1016/S0273-1223(97)00108-X
120. Tabrizian P, Ma W, Bakr A, Rahaman MS (2019) PH-sensitive and magnetically separable Fe/Cu bimetallic nanoparticles supported by graphene oxide (GO) for high-efficiency removal of tetracyclines. J Colloid Interface Sci 534:549–562. https://doi.org/10.1016/j.jcis.2018.09.034
121. Tao ZK, Jing ZQ, Tao MN, Chen RJ (2022) Recycled utilization of ryegrass litter in constructed wetland coupled microbial fuel cell for carbon-limited wastewater treatment. Chemosphere 302, Article 134882. https://doi.org/10.1016/j.chemosphere.2022.134882
122. Tong X-n, Wang X-z, He X-j, Wang Z, Li W-x (2020) Effects of antibiotics on microbial community structure and microbial functions in constructed wetlands treated with artificial root exudates. Environ Sci Process Impacts 22(1):217–226. https://doi.org/10.1039/C9EM00458K
123. Trivedi P, Leach JE, Tringe SG, Sa T, Singh BK (2020) Plant-microbiome interactions: from community assembly to plant health. Nat Rev Microbiol 18(11):607–621. https://doi.org/10.1038/s41579-020-0412-1
124. Tsalkatidou M, Gratziou M, Kotsovinos N (2009) Combined stabilization ponds–constructed wetland system. Desalination 248(1):988–997. https://doi.org/10.1016/j.desal.2008.10.015
125. Valentin FHH (1986) Peat beds for odour control recent developments and practical details. Filtr Sep 23:224–226
126. Verduzo Garibay M, Fernández del Castillo A, de Anda J, Senés-Guerrero C, Gradilla-Hernández MS (2022) Structure and activity of microbial communities in response to environmental, operational, and design factors in constructed wetlands. Int J Environ Sci Technol 19(11):11587–11612. https://doi.org/10.1007/s13762-021-03719-y
127. Verhofstad M, Poelen Mv, Van Kempen M, Bakker E, Smolders A (2017) Finding the harvesting frequency to maximize nutrient removal in a constructed wetland dominated by submerged aquatic plants. Ecol Eng 106:423–430
128. Vistanty H, Crisnaningtyas F (2021) Integration of upflow anaerobic sludge blanket and constructed wetlands for pharmaceutical wastewater treatment. IOP Conf Series Earth Environ Sci 623(1):012082. https://doi.org/10.1088/1755-1315/623/1/012082
129. Vymazal J (2007) Removal of nutrients in various types of constructed wetlands. Sci Total Environ 380(1–3):48–65

130. Vymazal J (2010) Constructed wetlands for wastewater treatment. Water 2(3):530–549
131. Vymazal J (2011) Plants used in constructed wetlands with horizontal subsurface flow: a review. Hydrobiologia 674(1):133–156
132. Vymazal J, Kröpfelová L (2008) Wastewater treatment in constructed wetlands with horizontal sub-surface flow, vol 14. Springer Science & Business Media
133. Wan Z, Zhang X, Lu S (2015) The adsorption and desorption of phosphorus by nineteen constructed wetland substrates. Technol Water Treat 41(4):35–39, 44
134. Wang J, Bai Z (2017) Fe-based catalysts for heterogeneous catalytic ozonation of emerging contaminants in water and wastewater. Chem Eng J 312:79–98. https://doi.org/10.1016/j.cej.2016.11.118
135. Wang J, Yuan H, Kang X, Lu W (2010) Critical success factors for on-site sorting of construction waste: a china study. Resour Conserv Recycl 54(11):931–936. https://doi.org/10.1016/j.resconrec.2010.01.012
136. Wang Q, Liu Q, Hu Y, Ding J, Ma Q, Zong K, Yang Z (2019) Effect of carbon source derived from macrophytes on microbial denitrification in constructed wetlands: role of plant species. Bioresour Technol Rep 7:100217. https://doi.org/10.1016/j.biteb.2019.100217
137. Wang R, Zhao X, Wang T, Guo Z, Hu Z, Zhang J, Wu S, Wu H (2022) Can we use mine waste as substrate in constructed wetlands to intensify nutrient removal? A critical assessment of key removal mechanisms and long-term environmental risks. Water Res 210:118009. https://doi.org/10.1016/j.watres.2021.118009
138. Wang T, Wang L, Chen Q, Kalogerakis N, Ji R, Ma Y (2020) Interactions between microplastics and organic pollutants: effects on toxicity, bioaccumulation, degradation, and transport. Sci Total Environ 748:142427
139. Wang W, Wang W, Zhang X, Wang D (2002) Adsorption of p-chlorophenol by biofilm components. Water Res 36(3):551–560. https://doi.org/10.1016/S0043-1354(01)00267-6
140. Wang Xl, Cao Wp, Wang Ym, Zhang Hh, Guo Yf (2013) Application and development of rhizosphere microorganisms in environmental pollution control. Water Saving Irrig(06):4–7
141. Wang X, Jiang C, Wang H, Xu S, Zhuang X (2023) Strategies for energy conversion from sludge to methane through pretreatment coupled anaerobic digestion: potential energy loss or gain. J Environ Manag 330.https://doi.org/10.1016/j.jenvman.2022.117033
142. Wang YA, Zhang XH, Lin H (2022) Removal of Cr(vi) and p-chlorophenol and generation of electricity using constructed wetland-microbial fuel cells based on Leersia hexandra Swartz: p-chlorophenol concentration and hydraulic retention time effects. RSC Adv 12(24):15123–15132. https://doi.org/10.1039/d2ra01828d
143. Wang YF, Li DP, Song XS, Cao X, Xu ZS, Huang W, Wang YH, Wang ZW, Sand W (2022) Intensifying anoxic ammonium removal by manganese ores and granular active carbon fillings in constructed wetland-microbial fuel cells: metagenomics reveals functional genes and microbial mechanisms. Bioresour Technol 352, Article 127114. https://doi.org/10.1016/j.biortech.2022.127114
144. Wang Y, Cai Z, Sheng S, Pan F, Chen F, Fu J (2020) Comprehensive evaluation of substrate materials for contaminants removal in constructed wetlands. Sci Total Environ 701:134736. https://doi.org/10.1016/j.scitotenv.2019.134736
145. Wang Z, Dong J, Liu L, Zhu G, Liu C (2013) Screening of phosphate-removing substrates for use in constructed wetlands treating swine wastewater. Ecol Eng 54:57–65. https://doi.org/10.1016/j.ecoleng.2013.01.017
146. Wen HY, Zhu H, Yan BX, Banuelos G, Shutes B, Wang XY, Cao SJ, Cheng R, Tian LP (2022) High removal efficiencies of antibiotics and low accumulation of antibiotic resistant genes obtained in microbial fuel cell-constructed wetlands intensified by sponge iron. Sci Total Environ 806, Article 150220. https://doi.org/10.1016/j.scitotenv.2021.150220
147. Wen ZD, Wu WM, Ren NQ, Gao DW (2016) Synergistic effect using vermiculite as media with a bacterial biofilm of Arthrobacter sp. for biodegradation of di-(2-ethylhexyl) phthalate. J Hazard Mater 304:118–125. https://doi.org/10.1016/j.jhazmat.2015.10.060
148. Wetzel R (2001) Fundamental processes within natural and constructed wetland ecosystems: short-term versus long-term objectives. Water Sci Technol 44(11–12):1–8

149. Wetzel RG (2001) Limnology: lake and river ecosystems. Gulf Professional Publishing
150. Wingender J, Strathmann M, Rode A, Leis A, Flemming H (2001) Isolation and biochemical characterization of extracellular polymeric substances from *Pseudomonas aeruginosa*. Microb Growth Biofilms, Pt A. Academic Press Inc., San Diego
151. Wolfe RL (1990) Ultraviolet disinfection of potable water. Environ Sci Technol 24(6):768–773. https://doi.org/10.1021/es00076a001
152. Wu H, Wang R, Yan P, Wu S, Chen Z, Zhao Y, Cheng C, Hu Z, Zhuang L, Guo Z, Xie H, Zhang J (2023) Constructed wetlands for pollution control. Nat Rev Earth Environ 4(4):218–234. https://doi.org/10.1038/s43017-023-00395-z
153. Xie JY, Li HX, Wang SY, Chen H, Jiang W, Zhang L, Wang LJ, Wu YF, Li LR, Lu XQ (2022) Removal and fate of carbamazepine in the microbial fuel cell coupled constructed wetland system. Environ Eng Res 27(3), Article 210119. https://doi.org/10.4491/eer.2021.119
154. Xing C, Xu X, Xu Z, Wang R, Xu L (2021) Study on the decontamination effect of biochar-constructed wetland under different hydraulic conditions. Water 13(7). https://doi.org/10.3390/w13070893
155. Xiong J, Du C, Zheng Y, Ge Y, Wang Y, Wang X (2015) Purification of highly polluted river water using surface horizontal subsurface flow hybrid constructed wetlands. Chin J Environ Eng 9(11):5167–5172
156. Xu L, Zhao Y, Fan C, Fan Z, Zhao F (2017) First study to explore the feasibility of applying microbial fuel cells into constructed wetlands for COD monitoring. Bioresour Technol 243:846–854. https://doi.org/10.1016/j.biortech.2017.06.179
157. Yadav AK, Dash P, Mohanty A, Abbassi R, Mishra BK (2012) Performance assessment of innovative constructed wetland-microbial fuel cell for electricity production and dye removal. Ecol Eng 47:126–131. https://doi.org/10.1016/j.ecoleng.2012.06.029
158. Yang R, Liu MH, Yang Q (2022) Microbial fuel cell affected the filler pollution accumulation of constructed wetland in the lab-scale and pilot-scale coupling reactors. Chem Eng J 429, Article 132208. https://doi.org/10.1016/j.cej.2021.132208
159. Yang X, He Q, Guo F, Sun X, Zhang J, Chen M, Vymazal J, Chen Y (2020) Nanoplastics disturb nitrogen removal in constructed wetlands: responses of microbes and macrophytes. Environ Sci Technol 54(21):14007–14016
160. Yang Y, Zhao Y, Liu R, Morgan D (2018) Global development of various emerged substrates utilized in constructed wetlands. Bioresour Technol 261:441–452. https://doi.org/10.1016/j.biortech.2018.03.085
161. Yang Z, Yang L, Wei C, Wu W, Zhao X, Lu T (2018) Enhanced nitrogen removal using solid carbon source in constructed wetland with limited aeration. Bioresour Technol 248(Pt B):98–103. https://doi.org/10.1016/j.biortech.2017.07.188
162. Ye F, Li Y (2009) Enhancement of nitrogen removal in towery hybrid constructed wetland to treat domestic wastewater for small rural communities. Ecol Eng 35(7):1043–1050. https://doi.org/10.1016/j.ecoleng.2009.03.009
163. Yin Q, Zhang B, Wang R, Zhao Z (2017) Biochar as an adsorbent for inorganic nitrogen and phosphorus removal from water: a review. Environ Sci Pollut Res 24(34):26297–26309. https://doi.org/10.1007/s11356-017-0338-y
164. Zhang B (2019) Research on phosphorus and nitrogen removal via prefabricated stepped wetland in river channels and experiment of its filler regeneration, Tianjin University
165. Zhang B, Ning D, Yang Y, Van Nostrand JD, Zhou J, Wen X (2020) Biodegradability of wastewater determines microbial assembly mechanisms in full-scale wastewater treatment plants. Water Res 169:115276. https://doi.org/10.1016/j.watres.2019.115276
166. Zhang H, Wang XC, Zheng Y, Dzakpasu M (2023) Removal of pharmaceutical active compounds in wastewater by constructed wetlands: performance and mechanisms. J Environ Manage 325(Pt A):116478. https://doi.org/10.1016/j.jenvman.2022.116478
167. Zhang K, Yang SQ, Wang W, Luo HB, Chen W, Zhang XX, Ma DD, An XC, Chen FH, Cheng L, Chen J (2022) Bioelectrochemical processes and cellulosic carbon source enhance the autotrophic and heterotrophic denitrification of low C/N ratio wastewater in tidal flow constructed wetland—microbial fuel cells. J Cleaner Prod 363, Article 132368. https://doi.org/10.1016/j.jclepro.2022.132368

168. Zhang T, Xu D, He F, Zhang Y, Wu Z (2012) Application of constructed wetland for water pollution control in China during 1990–2010. Ecol Eng 47:189–197. https://doi.org/10.1016/j.ecoleng.2012.06.022
169. Zhang X, Liu XT, Xu L, Luo Q, Chen J, Hu L, Jin J (2013) Purification effect of vertical flow constructed wetlands using modified substrates coated by MgFe-LDHs. China Environ Sci 33(08):1407–1412
170. Zhang X, Peng S, Zhu C, Li J, Wang P, Teng F (2015) Treatment of hydrogen sulphide gas by iron oxide porous ceramsite biotrickling filter. Chin J Environ Eng 9(2):841–846, Article 1673-9108(2015)9:2<841:Tyhwdk>2.0.Tx;2-k
171. Zhang Y, Liu F, Lin YD, Sun L, Guo XR, Yang S, He JL (2022) Enhanced swine wastewater treatment by constructed wetland-microbial fuel cell systems. Water 14(23), Article 3930. https://doi.org/10.3390/w14233930
172. Zhao LF, Xue N, Lu ZR, Xue LY (2023) Pyrrhotite-based constructed wetland-microbial fuel cell: reactive brilliant red X-3B removal performance and microbial communities. J Environ Eng 149(1), Article 04022086. https://doi.org/10.1061/joeedu.Eeeng-6958
173. Zhao L, Fu GP, Pang WC, Tang J, Guo ZP, Hu ZL (2022) Biochar immobilized bacteria enhances nitrogen removal capability of tidal flow constructed wetlands. Sci Total Environ 836, Article 155728. https://doi.org/10.1016/j.scitotenv.2022.155728
174. Zhao L, Zhu W, Tong W (2009) Clogging processes caused by biofilm growth and organic particle accumulation in lab-scale vertical flow constructed wetlands. J Environ Sci 21(6):750–757. https://doi.org/10.1016/S1001-0742(08)62336-0
175. Zhao XH, Zhao YQ, Kearney P (2011) Transformation of beneficially reused aluminium sludge to potential P and Al resource after employing as P-trapping material for wastewater treatment in constructed wetland. Chem Eng J 174(1):206–212. https://doi.org/10.1016/j.cej.2011.09.001
176. Zheng H, Liao Y, Chai H, Zhao L, Cao X, Feng L, Ji F (2023) Performance and mechanism of falling water enhanced tidal flow constructed wetlands (F-TFCW) for rural grey water treatment. J Cleaner Prod 136969
177. Zhou XY (2018) Study on the design of new type spiral composite constructed wetland and its purification effect. Vol. Master of Engineering, University of Jinan
178. Zhu X, Shen CF, Huang JX, Wang LM, Pang QQ, Peng FQ, Hou J, Ni LX, He F, Xu B (2022) The effect of sulfamethoxazole on nitrogen removal and electricity generation in a tidal flow constructed wetland coupled with a microbial fuel cell system: microbial response. Chem Eng J 431, Article 134070. https://doi.org/10.1016/j.cej.2021.134070

Chapter 5
Fate and Transport of Emerging Pollutants in Aquatic Environment

5.1 Introduction

With the rapid development of heavy chemical industries and the massive use of various chemicals, emerging pollutants have started to be frequently detected in the environment and are threatening human health and ecological environment quality, thus becoming a hot spot of concern. Emerging pollutants (EPs) are pollutants that have been newly discovered or are of concern, pose risks to the ecological environment and human health, and haven't been taken into account by controls yet or the management measures in place are insufficient to properly avoid and control their hazards [12]. Such substances are not necessarily new chemicals but are usually pollutants that have existed in the environment for a long time, but whose presence and potential hazards have only recently been discovered due to their low concentration in the environment. These are primarily unregulated man-made chemicals that are present in trace concentrations in air, soil, water, food, and human/animal tissues, persist in the environment, and are capable of disrupting the physiological functions of receptors.

One view is that EPs are substances, whether synthetic or naturally occurring, that are not usually monitored in the environment yet may enter there and cause harm to the ecosystem and human health. It is called emerging not because they are new, but because of the increasing level of concern. Given the diverse behavior and production sources of these contaminants, their detection and removal can be problematic and challenging [30].

At present, several EPs of international attention include environmental endocrine disruptors (EDCs), persistent organic pollutants (POPs), microplastics, antibiotics, personal care products, etc. Different kinds of EPs have one or more kinds of biological toxicity, such as organotoxicity, embryotoxicity, reproductive toxicity, neurotoxicity, endocrine disrupting effect, carcinogenicity, teratogenicity, mutagenicity, and some of the EPs have environmental persistence or bioaccumulation. ESCs refer to exogenous substances or mixtures that, after entering the body, can interfere with the

G. Dawen and M. Nabi, *Novel Approaches Towards Wastewater Treatment*,
Springer Water, https://doi.org/10.1007/978-3-031-55189-5_5

normal endocrine function in the body, thus destroying the stability and regulation of the body, and have adverse effects on the organism itself and its offspring or population. Endocrine disruptors enter the biological body through a variety of ways, such as food, water, etc., their chemical properties are stable, can be enriched through the food chain, long-term accumulation in the body, and difficult to discharge, therefore, even low concentration of intake may cause harm to the organism.

EPs that occur in the air may include volatile organic solvents and bioaerosols. Humans may be exposed to different EPs in the air through inhalation, exposure, or ingestion. The extent to which these EPs are harmful to humans depends on the concentration, frequency and duration of exposure. These pollutants can have a direct impact on people in a variety of ways, including DNA mutations, cancer, coronary heart disease, headaches, vomiting, skin irritation, and liver damage [31].

Emerging contaminants in water are primarily wastes from agricultural, industrial and domestic water use, which then enter groundwater, municipal wastewater and drinking water. These emerging contaminants can enter the human body through the food chain by enriching drinking contaminated water. The most common emerging pollutant in water is bioplastics, with large amounts of bioplastic waste being released into the ocean each year. Furthermore, these microplastics absorb a variety of contaminants in the water cycle, including antibiotics, polycyclic aromatic hydrocarbons, and organic chemicals. This makes microplastics carriers of various pollutants. These dangerous interactions accumulate in sediments and thus are absorbed by some aquatic organisms, and then through enrichment in the food chain, microplastics can bioaccumulate in the human body and lead to dangerous complications [108].

Numerous chemicals are utilized in pharmaceutical and personal care products (PPCPs) for both therapeutic and consumer uses. Pharmaceuticals can be prescription and over-the-counter drugs used for veterinary and human purposes to prevent or treat animal or human diseases, and personal care products are used primarily to improve the quality of daily life. The PPCP list covers a variety of commonly used chemicals, such as painkillers, antidepressants, antimicrobials, fever reducers, cosmetics, disinfectants, perfumes, steroids, and stimulants. It is common for pharmaceutical products to be used in human health and animal husbandry. These substances possess biological activity and are specifically formulated to generate precise effects. Human medicine encompasses a diverse array of substances. In the case of drugs utilized by humans or veterinary drugs administered to animals, patients serve as the primary source of contamination. Drugs and their metabolites enter the aquatic environment through discharge from sewage treatment plants, surface water through runoff, leaching from soil, or the application of manure onto land. Wastewater containing PPCPs compounds, discharged into streams without treatment. Generally, wastewater treatment facilities are designed to remove nitrogen, phosphorus and other pollutants from wastewater. Therefore, the presence of PPCPs in wastewater may affect the quality of wastewater after treatment. The EPs mainly come from industrial production, daily life and agricultural activities. As the production and utilization of chemical substances increase, the number and diversity of EPs also grow. Currently, there are over 20 categories of EPs identified globally, with each category consisting of dozens or even hundreds of compounds. Consequently, it is crucial to gather data for the purpose of assessing the risks associated with the presence and impacts of trace environmental pollutants.

A variety of sources contribute to PPCP contamination of water systems: drainage, illegal disposal of human excreta, industry, landfill leeches, and sewage. The unintended physiological effects of PPCPs in environmental waters on wildlife and humans are currently unknown. However, some studies suggest that continued exposure to sub-toxic concentrations of PPCPs may lead to unintended outcomes and unintended effects that may be detrimental to humans and ecosystems.

In recent years, residues of emerging contaminants have been removed or degraded using a wide range of physical, chemical, and biological techniques. Biological treatment technologies are currently the most widely used emerging contaminant removal technologies, including activated sludge, membrane bioreactor (MBR), and biosorption. Some organic micropollutants that are non-biodegradable cannot be sufficiently eliminated by biological treatment methods. In addition to using conventional oxidation techniques like ozonation and photolysis, chemical treatment technologies are also frequently used to degrade these micropollutants. Such as photo-Fenton, photocatalysis, solar-driven processes, ultrasonic processes, and electro-Fenton processes.

Nowadays, with more than 700 EPs and their metabolites and transformation products, EPs are described as artificial or naturally occurring substances with the potential to enter the environment and have known or suspected negative ecological and/or health consequences on humans. These chemicals are not often monitored in the environment. In some cases, the release of EPs into the environment may have occurred over a long period of time but may not have been detected until new detection methods have been developed.

An overview of emerging contaminants in various environmental matrices is provided in this chapter, the types and characteristics of some typical and widely noticed emerging contaminants, several biotechnological and abiotic techniques for the removal and degradation of residues of emerging contaminants, and the transport of emerging contaminants in wastewater treatment plants (WWTPs). Biological treatment of EPs includes aerobic sludge, anaerobic sludge, isolated strain, genetically engineered bacteria and biological activated carbon. Abiotic treatments of EPs include adsorption, coagulation, membrane separation and chemical oxidation. This will help to raise awareness, recognition and action in environmental monitoring and regulation.

5.2 Emerging Pollutants

Natural aquatic systems are the final settlement sites for EPs, such as rivers and oceans. In many developing countries, wastewater is used for agricultural purposes without pretreatment. The use of untreated or substandard treated agricultural wastewater can enrich many contaminants in the food chain, posing a potential risk to human health.

When considering the sources of pollutants, it is important to note that numerous chemicals are extensively used in products that are directly connected to people's daily lives. As a result, emerging pollutants have a broad range of release sources. Municipal WWTPs and landfills, as places for the treatment and disposal of municipal wastewater and waste, have also become important collection points for EPs in cities. There have been numerous studies showing that many EPs are not removed efficiently in WWTPs; it is worth pointing out that some EPs such as nonylphenols (NP) and perfluoroalkyl acids have even higher concentrations in the effluent than in the influent. This may be due to the incomplete degradation of precursors, making WWTPs an important point source of emerging pollutants in the urban water environment.

With the study of EPs, including drugs, understanding the environmental effects of hospital wastewater is an important topic. Numerous investigations have shown that both treated and untreated hospital wastewater contain numerous medications to understand the effects on WWTPs and the environment. In China, wastewater from large hospitals and cancer hospitals specializing in cancer treatment were the main sources of the Methotrexate, Azathioprine, Cyclophosphamide, Ifosfamide and Etoposide, suggesting that hospital wastewater is an important source of some cell suppressor drugs.

Air pollution sources are mainly atmospheric emissions from human activities related to urbanization and industrialization. A variety of natural sources can also contribute to the formation of dust, such as wind-blown dust, wildfires, and volcanoes. Regardless of the source, various air pollutants (such as particulate matter (PM2.5 and PM10), volatile organic compounds (VOCs), nitrogen dioxide (NO_2), etc.) are introduced into indoor and outdoor environments by atmospheric emissions. Some of the pollutants emitted are well known, while others (including butadiene, black carbon, formaldehyde, ultrafine particles, several transition metals, and others) are commonly regarded as "EPs" because they are known or believed to be detrimental to human health.

5.2.1 Phthalates (PAEs)

Phthalates (PAEs), also known as phthalates, are the collective name for esters formed from phthalic acid, which are generally viscous liquids with very low volatility. They have a peculiar odor, insoluble or insoluble in water, soluble in most organic solutions, low melting point and low toxicity. For different PAEs, as their side chains grow, their lipid solubility and boiling point are also higher. In addition, PAEs have the properties of general esters, such as they can be hydrolyzed in acidic or alkaline environments, can undergo ester exchange reactions, can be reduced by organic metals, etc. Because PAEs and the chemical makeup of their substrate material don't have a covalent bond. PAEs are easily discharged into the environment during manufacturing, use, and disposal, which leads to their ubiquity in a variety of environmental media. This includes household products (furniture, clothing, cosmetics, children's toys, nutritional supplements/food packaging, etc.), construction and transportation materials, industrial areas (paints and varnishes, adhesives, lubricants, waxes, cleaning materials, electronics, inks), agricultural activities (pesticides, insecticides, fertilizers, mulch plastics) or others (e.g. powders, medical devices, etc.) [111]. It is thus important to note that PAEs are found in a variety of environmental matrixes, including the atmosphere, lithosphere (soil, sediment) and hydrosphere (surface water, wastewater, etc.). Some PAEs such as Di (2-ethylhexyl) phthalate (DEHP), Dimethyl phthalate (DMP), and Dinoctyl phthalate (DnOP) have been reported as endocrine disrupting chemicals and listed as priority pollutants by the United States Environmental Protection Agency (USEPA) and the China Environmental Monitoring Center [36]. The toxicity and bioaccumulative potential of PAEs have raised concerns about their impact on the environment. The growth of China's economy and population have led to it becoming an important manufacturer and consumer of PAEs and poses related pollution threats to several environmental subdistricts.

DEHP is a type of PAEs, because DEHP can increase the elasticity and toughness of plastics, it is widely used in the plastics industry, and is the main constituent substance of plasticizers, and its output ranks 1st in plastic additives. Due to the expansion of agricultural production in recent years, agricultural mulch is widely used in China. DEHP, as the main constituent of mulch, is gradually released into the soil, affecting soil temperature, soil microorganisms and soil physicochemical properties. DEHP is a typical endocrine disruptor that has toxic effects on crops when it enters the crop body, mainly in terms of inhibition of growth, development and physiological properties, etc. DEHP has bleaching properties and reduces chlorophyll content in plant leaves. It was found that DEHP can damage the mitochondrial and chloroplast membranes of cucumber, and has an inhibitory effect on chlorophyll synthesis, which can cause damage to cucumber. After the soil where peppers were grown was contaminated with DEHP, the content of vitamin C and capsaicin in peppers was significantly reduced, which decreased the quality of peppers. When DEHP reaches a certain dose, it inhibits the germination rate of wheat seeds, affects the normal development of wheat leaves, and has an inhibitory effect on the growth of its root system (Fig. 5.1).

DMP is commonly used in the production of plastic films such as cling film because of its good characteristics such as film-forming, water repellency and adhesion, and is also the main raw material for food packaging materials and personal living products. It can be detected in perfume solvents, plastic oil packs for instant noodles, pickle wrappers, yogurt cups, aged vinegar, white wine and other care products and food packaging materials or food, which shows that DMP pollution has penetrated into human. China has included several phthalate esters such as DMP in the list of priority pollutants for control. Studies have reported that exposure to DMP during pregnancy can cause fetal death and malformation, while exposure of the body to DMP can cause liver toxicity and reproductive toxicity; in vitro tests have found that DMP can affect cellular immunity causing immunotoxicity and induce cancer cell proliferation (Fig. 5.2).

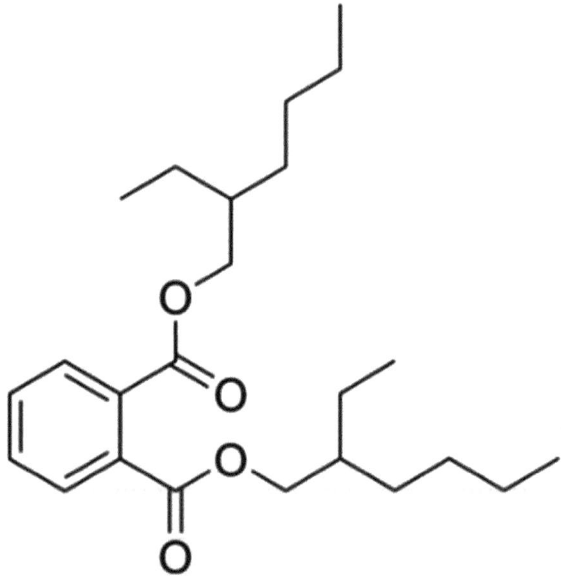

Fig. 5.1 Structural formula of DEHP

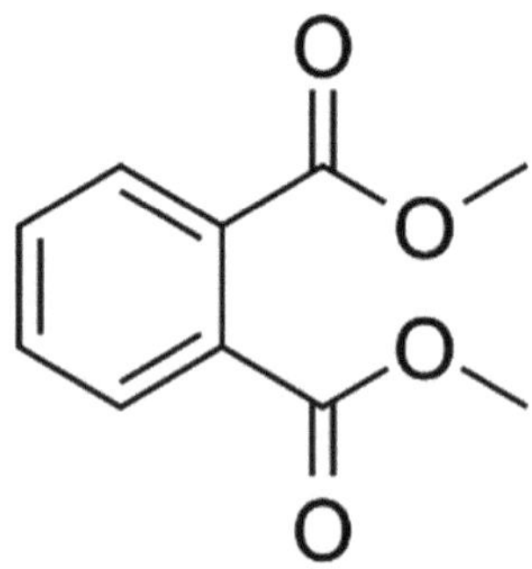

Fig. 5.2 Structural formula of DMP

DEP, a new plasticizer, has been widely used in various industries. Because it is produced and used industrially in large quantities, DEP is constantly detected in water, soil, and atmospheric environmental media. Because of the lack of covalent bonds with the main material and relatively high water solubility, DEP is more likely to enter the aquatic environment and affect the ecological environment and human body. Studies have reported that DEP exposure can reduce antioxidant properties in organisms, cause apoptosis, interfere with the endocrine system and immune system, and affect the intellectual development of children, with "triptogenic" toxicity (Fig. 5.3).

DBP is a common phthalate ester compound, mainly used as plasticizer, dye solvent, rubber additive and so on. In recent years, due to atmospheric deposition, sewage irrigation, sludge farming and the massive use of fertilizers, pesticides and agricultural films, DBP has become one of the more easily detected and high levels

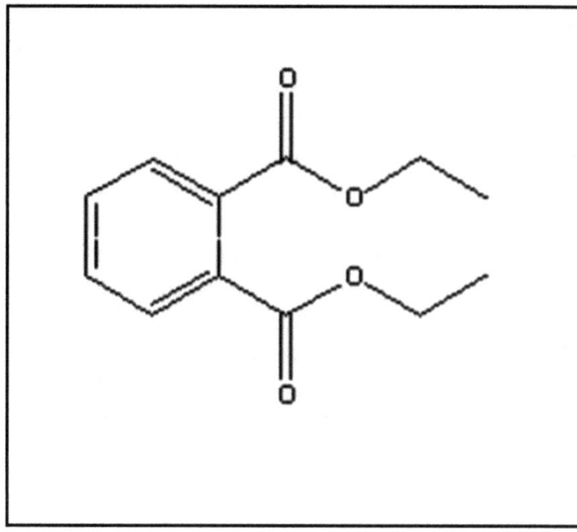

Fig. 5.3 Structural formula of DEP

of phthalate pollutants in agricultural soils. The high content of DBP in agricultural soil will not only affect the growth of crops and quality of agricultural products, but also endanger human health through the food chain, leading to endocrine disruption and reproductive malfunction. The U.S. National Environmental Protection Agency has listed DBP as a priority pollutant and endocrine disruptor, and China has also listed it as a priority pollutant for control (Fig. 5.4).

PAEs are widely used as plasticizers to e improve the flexibility and elasticity of polyvinyl chloride (PVC) [79]. In China, PAEs is widely used in daily necessities, food packaging, medical equipment, industrial pipes, interior decoration, agricultural plastic film and so on. According to estimates from 2009, 1.3 million tons of PAEs were used each year, mainly dibutyl phthalate (2-ethylhexyl) ester (DEHP) and dibutyl phthalate (DnBP) [40]. When phthalate-containing items are exposed to the aging and decomposition processes associated with plastics, PAEs readily migrate into the environment because they are not covalently bound to polymer substrates in soft PVC and exist in a leachable form. The misuse of these compounds has led to occasional residues in different environmental zones, indicating the potential

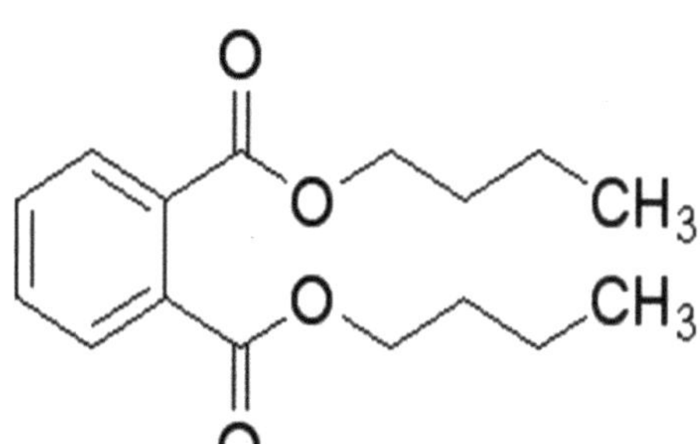

Fig. 5.4 Structural formula of DEP

for PAEs contamination. Due to the prevalence and high concentration of PAEs, researchers have expressed strong concern about PAEs in recent years. Due to its wide application and diversity of transport routes, large amounts of PAEs residues even reach remote environments through atmospheric transport, landfill leaching, agricultural runoff, and sewage discharge [46].

Moreover, there is an increasing body of evidence indicating that numerous PAEs have detrimental impacts on aquatic and terrestrial organisms, including humans, both in the short term and over prolonged periods [106]. The typical symptoms of PAEs toxicity in aquatic embryos are death, tail curvature, necrosis, cardiac edema, and no touch response. BBP, DnBP and DEHP have estrogen endocrine disrupting effects on intact organisms. Human reproductive, liver, and renal systems may be negatively impacted by exposure to PAEs [55]. There is growing concern about the uncertainty of PAEs exposure and the risks PAEs may pose to human health and the environment.

PAEs' effects on ecosystems and the environment It has been demonstrated that aquatic hazardous PAEs can cause genetic mutations and have an adverse effect on reproduction in annelids (both aquatic and terrestrial), mollusks, crustaceans, insects, fish, and amphibians. A large number of studies have focused on the endocrine disrupting effects of PAEs in different species of fish and mammals. A recent study reported that prolonged exposure to DEHP disrupts endocrine function in Marine fish, with males being more sensitive than females [132]. The aquatic toxicity of PAEs varies widely among different types of compounds, with only lower alkyl chain lengths (C_1 to C_4) consistently toxic. Literature data indicate that DMP, DEP, DnBP, and BBP have acute and chronic effects on algae, invertebrates, and fish in both fresh and salt water. PAEs (DEHP, DnOP) with alkyl chain length C_6 have no acute toxicity to aquatic organisms, and acute toxicity has been reported only at concentrations significantly higher than their water solubility.

Humans are inevitably exposed to environmental PAEs through contaminated surface water and creatures, which raises severe concerns about the possible health implications. Effects of PAEs on reproductive development have been observed in humans, including reduced anogenital distance in newborns, shortened gestation period in adults, reduced sex and thyroid hormone levels, and reduced sperm quality. As a result, due to the potential harm they could cause to the environment and public health, PAEs in ecosystems have garnered a lot of attention.

5.2.2 *Endocrine Disrupting Chemicals (EDCs)*

There are many different endocrine-disrupting chemicals (EDCs) in the environment. By changing the endocrine system's hormonal balance, these EDCs have an impact on hormone production or receptor binding. EDCs can lead to reproductive, developmental and sexual behavioral disfunction, resulting in deleterious outcomes in animals and humans. Most endogenous steroid hormones, such as estradiol or androgens, contain structures that are similar to those of natural or synthesized EDCs.

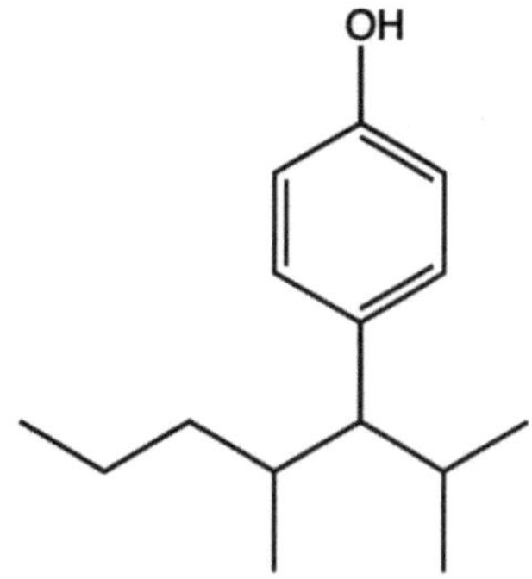

Fig. 5.5 Structural formula of alkylphenol

Therefore, they tend to interfere with the action of steroid hormones by binding to the corresponding hormone receptors [7].

Alkylphenols (AP) and alkylphenol ethoxylates (APEO) have good emulsification, wetting, penetration, foaming, washing, decontamination, etc. They are widely used in antioxidants, textile printing and dyeing auxiliaries, lubricant additives, pesticide emulsifiers, resins and rubber stabilizers, etc. AP and APEO have endocrine disrupting effects and belong to EDC, which can be enriched through the food chain and are easily harmful to the ecological environment. They can be enriched through the food chain, which is harmful to the ecological environment. Isothiazolinone fungicides, a new class of efficient broad-spectrum antiseptic fungicides, widely used in cosmetics, coatings, adhesives, pulp and paper industries. Isothiazolinones have allergic reactions to eyes and skin, and have neurotoxicity and cytotoxicity (Fig. 5.5).

Alkylphenols [63] are one of the major degradation products of the nonionic surfactants alkylphenol ethoxylates (APEOs). APS are used in a wide range of industrial applications, such as plastic plasticizers, industrial detergents, lubricant additives, oil-soluble phenolic resins, pesticide emulsifiers, textile finishing agents, etc., resulting in a wide range of degradation products and generally high contamination levels. The ecotoxic effects of the degradation products of APS are strongest for octylphenol (OP), followed by NP [135], while the other APs are weaker in order of strength according to the shortest to longest hydrophilic end chains. Therefore, OPs are included as persistent toxic chemical pollutants by UNEP.

Polybrominated diphenyl ethers (PBDEs) have the chemical formula $C_{12}H_{(0-9)}Br_{(1-10)}O$. There are 209 congeners of PBDEs depending on the number and position of bromine atoms on the benzene ring, and PBDEs are often used as an important class of flame retardants and insulation materials in industrial production because of their good flame retardancy and thermal stability and low cost. Since 1965, PBDEs have been used worldwide as flame retardants and are widely added to textiles, home furnishing materials, polyurethane foam, thermoplastics and electronics. Pentabromodiphenyl ether (penta-BDE or BDE-99), octabromodiphenyl ether (octa-BDE) and decabromodiphenyl ether (deca-BDE or BDE-209) are the most prominent PBDEs commercially available (Fig. 5.6).

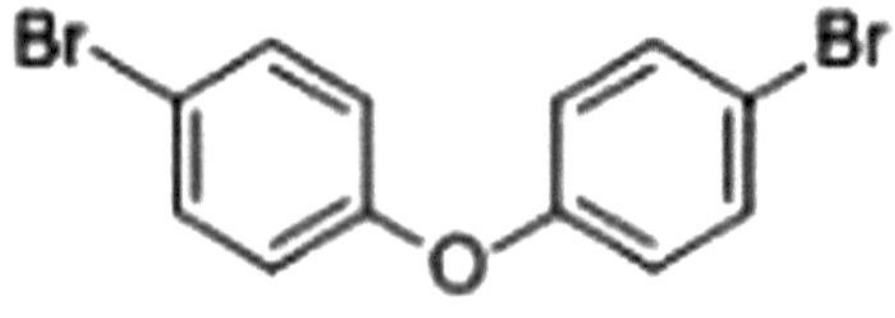

Fig. 5.6 Structural formula of poly brominated diphenyl ethers (PBDEs)

The production and demand of PBDEs have been increasing over the past decades. As the health hazards of PBDEs to human populations have become apparent with increased use, concerns about PBDEs have gradually shifted from product applications to environmental and human health impacts. Because PBDEs are not chemically bonded to various substances, a small percentage escapes into the environment during production, use, disposal and recycling. Over the past few decades, scientists have detected the presence of PBDEs in indoor and outdoor air, dust, sediment, sludge, birds, fish and mammalian tissues [67]. Exposure to PBDEs has been detected at different concentrations in our environment such as atmosphere, soil, water, organisms, and human body. PBDEs can be enriched in the food chain due to their special chemical structure with high bioaccumulation, low degradability, lipophilicity, hydrophobicity, and semi-volatility, which allows PBDEs in the environment to eventually enter the human body through enrichment and biomagnification in the food chain, thus posing a serious threat to human health. Due to the typical characteristics of PBDEs, PentaBDE and OctaBDE have been listed on the Stockholm POPs list in 2009, and BDE-209 has been banned in the EU and some countries. However, due to the low price, superior performance and low acute toxicity of BDE-209, it is still widely used in China, and China has some of the highest levels of both flame retardant manufacture and usage worldwide.

Bisphenol A (BPA), one of the EDCs, is one of the most widely produced chemicals in the world and is widely used in people's daily lives. BPA exposure affects follicular growth and development, reduces oocyte quality, interferes with granulosa cell steroid hormone secretion, and causes ovarian disfunction. BPA, with a similar chemical structure to estrogen and thyroid hormone, is the first EDCs that can bind to thyroid hormone receptors and affect thyroid hormone signaling and transduction in vitro. It is the first EDCs found to bind to thyroid hormone receptors and affect thyroid hormone signaling and transduction in vitro. Even at low concentrations, BPA can alter endogenous hormone regulation and cause many adverse effects on human health, such as abnormal genital development, decreased reproductive function, abnormal placental function, abnormal thyroid function and neurobehavior. Due to the special physical and chemical properties of BPA, it can produce raw materials such as polycarbonate and epoxy resin with excellent performance. These materials have the characteristics of high temperature resistance, impact resistance, transparency and corrosion resistance, and are therefore widely used in the manufacturing of electronic products, medical devices, food packaging, plastic bottles, can linings, children's toys, etc. With the change of people's lifestyle in recent years, the demand for express delivery and food takeaway has increased enormously, and BPA is more widely present in daily life and industrial supplies.

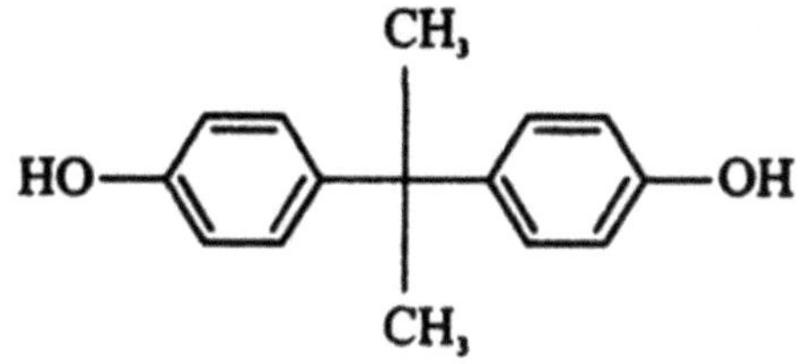

Fig. 5.7 Structural formula of BPA

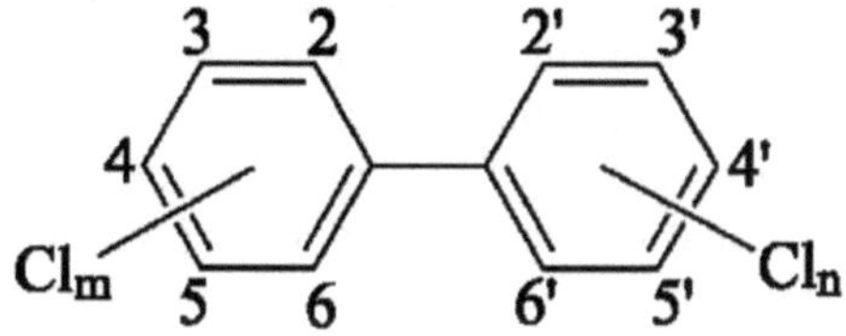

Fig. 5.8 Structural formula of PCBs

Polychlorinated biphenyls (PCBs) are a class of synthetic chlorinated aromatic hydrocarbons, also known as chlorinated biphenyls. PCBs have the molecular formula $C_{12}H_{10-m-n}Cl_{m+n}(1 \leq m + n \leq 10)$ and the structural formula shown in the Fig. 5.7 CAS No.:1336-36-3. In theory, the number of chlorine atoms substituted can be up to 10, but in practice, common PCBs contain only 3–6 chlorine atoms. Due to the different substitution positions and numbers of chlorine atoms on the benzene ring, PCBs can be combined into 209 congeners. Different molecular structures make these congeners differ greatly in terms of toxicity, physicochemical properties and environmental behavior (Fig. 5.8).

At room temperature, the pure PCBs compounds are crystalline, while the mixtures are oily liquids. Generally, commercially produced PCBs are mixtures, and low-chlorinated biphenyls (such as di-, tri- and tetrachlorinated biphenyls) are mostly liquid with good fluidity. PCBs have a high octanol–water partition coefficient [98] and a strong affinity for lipids, and are highly soluble in fats and organic solvents. PCBs have very low solubility in water, and solubility decreases with the increase in the number of chlorine atoms; conversely, the viscosity of PCBs increases with the increase in the number of chlorine atoms. As PCBs have excellent properties such as alkali resistance, acid resistance, heat resistance, corrosion resistance, oxidation resistance and electrical insulation, they have been widely used in electric power, chemical and mechanical industries.

PCBs are physicochemically stable, persistent, bioaccumulative, semi-volatile and highly toxic in the environment, and very hazardous to living organisms. Based on their toxicological characteristics, PCBs can be classified as either dioxin-like (dl-PCBs) or non-dioxin-like (ndl-PCBs). Dioxin-like PCBs include 12 monomers of co-planar structure, with typical "triptogenic" toxicity (mutagenic, carcinogenic and teratogenic), acute lethal toxicity, reproductive toxicity, dermal toxicity, immunotoxicity and neurological toxicity. The toxicity of non-dioxin PCBs is relatively weak, but can cause a series of complex adverse reactions, in the case of long-term accumulation and high concentration intake, its toxicity manifested as memory loss,

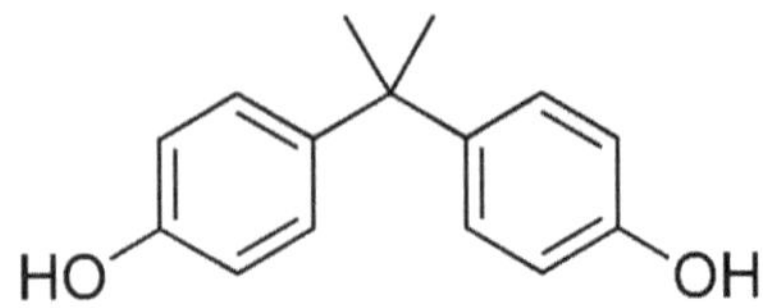

Fig. 5.9 Chemical structure of EDCs

cognitive disfunction, endocrine disorders, liver damage, immune impairment, etc. Due to the lack of awareness of the toxicity of PCBs in the early days, humans have suffered from them. In addition to the world-shocking rice bran oil incident, there are also studies that found that the concentration of PCBs in tissue sections of white women suffering from breast cancer exceeds that of normal people by more than 50%, and workers using PCBs to produce capacitors are more likely to suffer from lymphatic and digestive tract cancers. These cases show that PCBs have strong toxic effects on the human body. Therefore, the International Agency for Research on Cancer (IARC) classifies PCBs as human class I carcinogens.

In addition to interfering with endocrine function, EDCs affect the immune system as well, which is found in almost all animals (including humans, fish, amphibians, birds, snails, crustaceans, and other species). Any chemical that disrupts the endocrine system's operation is considered an EDC. The endocrine system controls behavior, growth, reproduction, and development through the hormones it produces. EDCs don't fall under any category of substances. Concerns about endocrine active chemicals in the environment have been raised since studies have revealed that certain EPs have effects on the endocrine system at a very low quantity (Fig. 5.9).

EDCs are thought to be the most significant water pollutants. Hormones are among the substances on the EDC list, along with pesticides, disinfection byproducts, bisphenols, and fluorinated compounds. The presence of EDCs in water has become a public health concern because these chemicals have been linked to harmful health and reproductive disorders in humans and animals [48]. Even in low doses, these substances have the potential to interfere with some species' endocrine systems. EDCs pose potential risks to humans and animals because they are thought to cause diseases such as nephrotoxicity, cancer, and metabolic disorders [70].

EDCs have a variety of adverse effects on the endocrine system in humans and animals that may be cumulative and may only occur in future generations, and then produce effects that may be irreversible and threaten the sustainability of human development. The majority of EDCs are synthetic organic substances that have released into the environment, such BPA. But they can also be naturally occurring estrogens such as estrone and 17b estradiol and are therefore prevalent in aquatic environments that receive wastewater discharges.

Manure produced by livestock is an important source of hormones in surface water and soil, yet information on global hormone consumption is limited. For decades, hormone levels found naturally in food have been deemed safe. However, hormone toxicity, including estrone, estradiol and estriol, is severe and poses serious risks to soil, plants, water resources and humans. The uncontrolled release of hormones into the environment is linked to this toxicity, which has negative and harmful impacts on human and other creatures' immune, neurological, and reproductive systems. Recent epidemiological and experimental research on humans and animals have connected human exposure to hormones and other endocrine-disrupting pollutants to health issues.

5.2.3 *Microplastics*

Microplastics generally refer to particles smaller than 5 mm in size, depending on their source, particles can be classified as primary or secondary particles. Most microplastics in the ocean are secondary products of degradation and fragmentation of medium plastics or larger debris; Primary microplastics introduced directly into the ocean by runoff are made into micron-scale particles, common applications include exfoliators in synthetic garments, textile applications, industrial abrasives.

Because of low cost, dependability, light weight, and good ductility, plastics are widely used in all facets of our daily life. In recent years, production of global plastic has dramatically increased, and the limited recycling of plastic waste and its difficulty in natural degradation is now considered a serious threat to the marine environment. Plastic pollution is ubiquitous, a growing number of marine species are known to be affected by plastic pollution, and it is now a top priority for research to do a more thorough evaluation of these materials' ecological risk. In addition, microplastics can also serve as carriers of various toxic pollutants, including heavy metals, polychlorinated biphenyls (PCBs), polycyclic aromatic hydrocarbons (PAHs), and other pollutants. Microplastics present major ecological dangers when they are present in large quantities. They can also be transmitted and enriched through the food chain, endangering human health.

There are numerous varieties of plastic polymers and additives that can be mixed to create products with certain characteristics. The most common polymers are polyethylene (PE), polypropylene [50], polystyrene [63], PVC, polyamide (PA), polyethylene terephthalate (PET), polyvinyl alcohol (PVA). Polymer density, which impacts buoyancy, placement in the water column, and the possibility of a subsequent interaction with biota, plays a significant role in determining an object's environmental fate once it is discharged into the ocean. Polymers that are denser than seawater (such as PVC) will sink, while polymers that are less dense (such as PE and PP) will tend to float in the water column. Processes such as biosiltation and biological settlement on plastic surfaces increase the weight of particles and thus accelerate their settlement on the bottom sediment [131]. The object's density and distribution throughout the water column are also changed by fragmentation, degradation, and leaching of additives (Fig. 5.10).

Microplastics are widely present in various environmental matrices, such as air, fresh water, wastewater, seawater, bottled water, tap water, and food. There are also multiple pathways for the release of environmental microplastics, mainly from human sources (e.g., synthetic textiles, personal care products), transportation sources (e.g.,

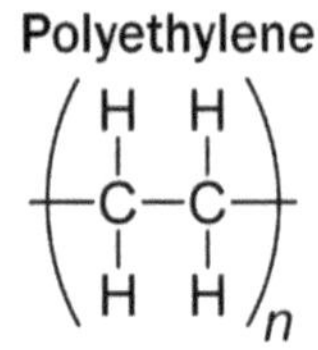

Fig. 5.10 Chemical structure of microplastics

synthetic rubber tires), and industrial sources (e.g., plastic pellets), most of which migrate through surface water or are discharged directly into the ocean. Following their release into the environment, these microplastics will go through a variety of exposure pathways, including ingestion, inhalation, and skin contact, which will have a negative impact on human health. They will also accumulate, degrade, and migrate under various environmental conditions before eventually entering the human body.

The threat of contamination of the aquatic environment by microplastics is often associated with organisms in the aquatic environment, and microplastics can serve as particulate matter for ingestion, solid supports for contaminant transport, and a significant opportunity for physical injury during exercise. Compared to aquatic systems, the environmental impact of microplastic pollution in agricultural soils may be underestimated. Thus, research are urgently needed to draw attention to parliamentarians in the agricultural system. However, only a few recent studies have examined pollution levels and possible sources of microplastics in agricultural environments and their harmful effects on soil biota. According to recent findings, microplastics are common and persistent in the soil environment and have negative effects on soil biota's growth, reproduction, nutrition, survival, and immune systems (animals, plants and microorganisms) [65].

5.2.4 *Antibiotic*

Antibiotics are secondary metabolites produced by microorganisms, higher plants and animals during the life process that have anti-pathogenic or other activity and can be used to treat a variety of bacterial infections in humans and animals. Antibiotics are frequently utilized in aquaculture and animal husbandry as well as in the prevention and treatment of human diseases. Most antibiotics enter the environment through excretion as a result of incomplete metabolism in living organisms.

Antibiotics are classified according to their mechanism of action and spectrum, route of administration and chemical structure. β-lactams (including carbapenems and monoamides), aminoglycosides, chloramphenicol, glycopeptides, lincomycin, macrolides, polyenes, polypeptides, quinolones and fluoroquinolones, rifamycin, sulfonamides, and tetracyclines are the different structural categories for these compounds [49].

Since the discovery of penicillin, β-lactam antibiotics have been widely used in medicine, biology, biochemistry, life science, agriculture and other fields. Due to the high consumption and low metabolism of antibiotics by humans or animals, a large amount of β-lactam antibiotics are directly released into the environment after passing through the human body. The antibiotics in the environment mainly come from point sources such as human and animal medication, industrial production, animal husbandry, aquaculture, hospitals, sewage treatment plants, etc., and then spread to the environment through agriculture, urban and other runoff. Currently, trace antibiotics are often detected in surface water, groundwater, seawater, and soil, and antibiotics have become one of the emerging pollutants in the environment, and their large release may lead to the production of antibiotic resistant bacteria (ARBs) and antibiotic resistant genes (ARGs), thus posing potential environmental risks. β-lactam antibiotics are the most widely produced and used class of antibiotics, mainly including penicillin and its derivatives, cephalosporins They include penicillin and its derivatives, cephalosporins, monoamide cyclines, carbapenems and penicillin enzyme inhibitors (Fig. 5.11).

Macrolide antibiotics (MA) are basic lipophilic antibiotics consisting of a large central ring of 12–16 carbon atoms and one or more glycosidic bonds together. MA can display strong lethality against Gram-positive bacteria and has been widely used in the antimicrobial treatment of respiratory and tissue infections. However, the uptake of macrolide antibiotics by organisms is limited, and a large proportion of MA is released into the environment before it is broken down in the organism, resulting in exposure to MA in the environment. The presence of MA has been detected in the waters surrounding several municipal WWTPs.

MA is usually a colorless, weakly basic compound with a large octanol–water partition coefficient and is readily soluble in organic solvents. MA is poorly soluble in water but can become soluble in water by forming salts with acids (gluconates, lactates, dodecyl sulfates). These substances can be persistent in the environment

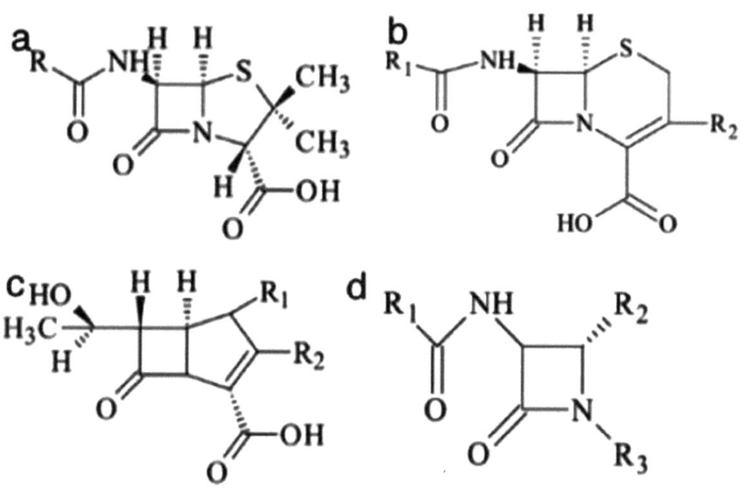

Fig. 5.11 Structural of **a** penicillin, **b** cephalosporins, **c** carbapenems and **d** monocyclic-β-lactams

and are somewhat bioconcentrated. Structurally, MA is a class of large 14- to 16-membered rings linked to monosaccharides or polysaccharides by glycosidic bonds, which are prone to glycosidic bond hydrolysis under acidic conditions, and their lactones are prone to ring exchange rupture under alkaline conditions.

Aminoglycoside antibiotics (AGs) are broad-spectrum antibiotics that are mainly used to inactivate Gram-positive and Gram-negative bacteria. Since its discovery, it has been widely used in livestock and poultry farming because of its antibacterial effect. Most of the aminoglycoside antibiotics use 2-deoxystreptamine (2-DOS) as the core structure, modified by sugar substituents at C4, C5 and C6 positions, and their structures are shown in Fig. 5.12. Since AGs themselves have relatively serious ototoxicity and nephrotoxicity, and are influenced by a variety of cephalosporin antibiotics, a series of semi-synthetic products such as amikacin, netilmicin, dibekacin, sisomicin, etc. have been produced, and so far, aminoglycoside antibiotics are still one of the most used antibiotics in clinical practice.

With the increasing production of aminoglycoside antibiotics, the amount of antibiotic bacterial residue produced is also increasing. Antibiotic residue is a solid waste produced during the fermentation of microorganisms to produce antibiotics, whose main components are mycelium of fermenting bacteria, metabolites from the fermentation process, degradation of the culture medium and a small amount of antibiotics. The residual antibiotics inside the antibiotic residue may migrate and transform into the ecological environment, and once induced to produce a large number of drug-resistant bacteria, it may induce soil microorganisms to develop resistance at the same time (Fig. 5.13).

Quinolones are an important class of synthetic antibiotics that can be used as antibacterial drugs. Target bacterial DNA and effectively inhibit bacterial DNA replication by hindering the activity of bacterial DNA gyrase. As a result, they are widely employed in the medical field for their quick and effective action against

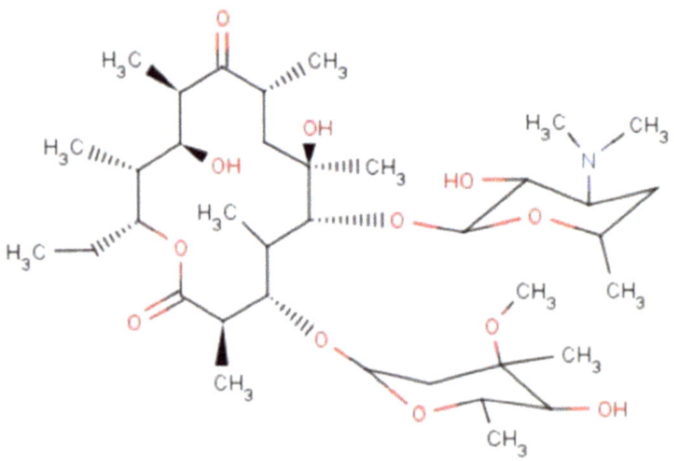

Fig. 5.12 Structural formula of macrolide antibiotics

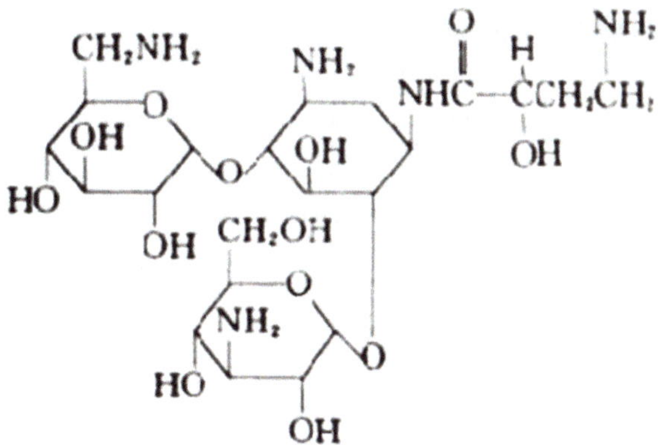

Fig. 5.13 Structural formula of aminoglycoside antibiotics

bacteria. Currently, the most used products are 3rd generation products, including ciprofloxacin, levofloxacin, lomefloxacin, enrofloxacin, sparfloxacin, norfloxacin and ofloxacin. However, their heavy use has led to a series of negative environmental impacts.

More than 70% of quinolone antibiotics are ingested by organisms and enter the environment through urine and excretion, to the extent that they are detected in the environment such as drinking water, surface water and soil. In addition, the production wastewater from pharmaceutical plants contains many residues. These residues can be toxic to humans and animals, and residues of quinolone antibiotics in different environmental matrices may affect environmental processes, damage ecosystems, and ultimately pose a threat to human health and ecological environment (Fig. 5.14).

Antibiotics are frequently detected as micropollutants in natural and man-made environments, including surface water, groundwater, wastewater, and treated wastewater at the lowest concentrations. Antibiotics are consequently a serious global threat as micropollutants. Antibiotics are successful drugs that protect human health and are widely used around the world. Since the discovery of penicillin in 1929, antibiotics have been used to prevent and treat human diseases and are widely used in areas

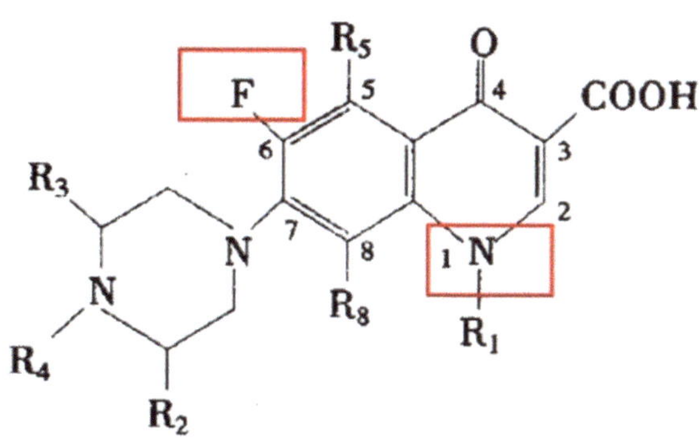

Fig. 5.14 Structural formula of quinolone antibiotics

Table 5.1 Types of emerging pollutants

Types	Definition	Source
PAEs	PAEs, also known as phthalates, are the collective name for esters formed from phthalic acid	Household products (furniture, clothing, cosmetics, children's toys, nutritional supplements/food packaging, etc.), construction and transportation materials, industrial areas (paints and varnishes, adhesives, lubricants, waxes, cleaning materials, electronics, inks), agricultural activities (pesticides, fertilizers, mulch plastics)
EDCs	Endocrine disruptors, endocrine-disrupting substances that mimic human hormones (hormones)	Manure produced by livestock is an important source of hormones in surface water and soil
Microplastics	Microplastics, defined as plastic particles less than 5 mm in diameter, are a major carrier of pollution	Mainly from human sources (e.g., synthetic textiles, personal care products), transportation sources (e.g., synthetic rubber tires), and industrial sources (e.g., plastic pellets)
Antibiotic	Antibiotics are secondary metabolites produced by microorganisms, higher plants and animals during the life process that have anti-pathogenic or other activity	Antibiotics as micropollutants are commonly found in natural and man-made environments such as surface water, groundwater, wastewater, and treated wastewater

such as animal husbandry and aquaculture. In 2013, about 162,000 tons of antibiotics were used in China; of which, 48% is consumed by humans and the rest by animals [134].

According to many studies, WWTPs are often relatively inefficient in removing antibiotic drugs from wastewater. WWTPs As a result, large amounts of antibiotic drugs and metabolites enter the surface water directly [74]. Due to the continuous transmission of antibiotics in the environment, concentrations of antibiotics in aquatic environments are very low, ranging from nanograms per liter to micrograms per liter. In addition, trace amounts of antibiotics remaining in the environment can induce antibiotic resistant bacteria (ARBs). These are produced by the emergence of antibiotic resistance genes (ARGs) [50]. Aquatic bioaccumulation may result in ecological concerns, antibiotic residues in aquatic habitats may maintain or promote the spread of ARGs. As a result, antibiotics have emerged as a significant contaminant in the aquatic environment. As a result, ecological hazards have gained attention, particularly the potential threat that antibiotic residues in freshwater systems provide to human health.

China has become a major producer and consumer of antibiotics. Antibiotics are commonly found as micropollutants in the natural and man-made environment, such as surface water, groundwater, wastewater and treated wastewater, in the lowest concentrations. Generally, antibiotics are routinely released into sediments and water bodies because WWTPs cannot remove them. Antibiotics are therefore a serious global threat as micropollutants (Table 5.1).

5.3 Occurrence of Emerging Pollutants in Environment

The distribution of EPs in the environment is a growing concern because most of the chemicals in emerging pollutants are unregulated and many of their potential impacts are ignored or overlooked by many monitoring and surveillance programs. Among these EPs, we are likely to find many different chemicals, including PAEs, EDCs, microplastics, antibiotic, PPCPs, consumer products (food additives, detergents, etc.), industrial compounds (perfluoroalkyl substances (PFAS), plasticizers, etc.), pesticides, and engineering (nano) materials. Although many of these pollutants can be found in WWTPs effluent, agricultural surface runoff, and industrial discharges, the effects of the former category (WWTP effluent) are particularly significant in urban areas [93]. Environmental monitoring in particularly sensitive areas such as estuaries has become a challenge due to the complexity of analysis endpoints and potential impacts [84].

With the rapid development of industry and the mass production and use of various chemical substances, the danger of some EPs to public health and the ecological environment is gradually emerging. Antibiotics, painkillers, active ingredients in sunscreen quinolones, ibuprofen, aspirin and other substances that should only exist in medicines and cosmetics are often detected in the natural environment such as rivers, soil and offshore in recent years. The Yangtze River Delta, the Pearl River Delta, and Beijing-Tianjin-Hebei have all experienced significant increases in EP pollution. From the national pattern, the concentration levels of EPs generally show the distribution characteristics of increasing from west to east, and there are more types of EPs in coastal areas than in central and western regions. Endocrine disruptors are mainly distributed in coastal and central regions, among which Bohai Bay, Yangtze River Estuary and Pearl River Estuary are more densely distributed.

The occurrence of environmental pollution can be caused by point (mainly urban and industrial) or diffuse (agricultural) pollution. The characteristics of EPs play a major role in the transmission of EPs from diffuse sources to aquatic bodies, such as volatility, polarity, sorption properties, persistence, and other properties. The environmental fate of EPs from municipal or industrial WWTPs discharged directly into rivers is of concern (degradation, sediment adsorption, aqueous phase transport). Significant biodegradation of EPs in surface water and groundwater after contamination with sewage. The ability of a community of organisms to degrade pollutants through metabolic pathways is necessary for the biodegradability of EPs, as is the bioavailability of pollutants, particularly in sediments and soils. Realistic biodegradation simulations are best performed using experimental systems that are as close as possible to real environmental conditions.

Until they reach water bodies, EPs in rural regions are extensively dispersed throughout the environment and are transferred by air, runoff, erosion, or leaching.

For example, properties such as adsorption behavior of drugs may vary considerably in different soil types because they are present in ionized and non-ionized forms, which can affect their interaction with different compounds in the soil [69]. The behavior of engineered nanomaterials (ENMs) in the environment is mainly unknown.

Exploring the transformation of ENMs before and after they enter the environment, such as their interaction with common cations or their dissolution under natural conditions, may be important for controlling transport and fate, and further studies are needed. Their behavior differs from that of non-particulate pollutants and therefore new paradigms are needed to study ENMs in water, soil, sediment and biota. Research in this area is needed with a view to providing information that can be used to develop, refine and calibrate models used to study EPs.

5.3.1 Emerging Pollutants in Surface Water

The main routes through which pharmaceutical products enter the environment are through industrial processes, human and animal waste, hospital wastewater, improper treatment of pharmaceutical products, agricultural wastewater, and irrigation with contaminated water. Surface and ground water both include a range of pharmaceutical compounds, and these drug wastes are transferred to the water cycle via several pathways. Among other things, WWTPs act as a conduit for these drug products to enter the water cycle, as many of these compounds are not completely removed during the treatment process. The presence of these compounds in the environment has not been a widespread concern because of their random distribution and their low concentrations in the environment.

Northeast China's Songhua River is used for irrigation as well as domestic water supplies. The river originates from the merger of the Second Songhua River and Nen River, before falling into Heilongjiang, covering a total catchment area of 557,200 km^2. The productivity, economics, and living conditions of Northeast China are greatly impacted by the 76.20 billion m^3 of yearly runoff from the Songhua River basin. The Second Songhua River originates in the Changbai Mountain Range, where it forms the southern source of the Songhua River that covers 958 km, with a catchment area of 73,400 km^2. The Second Songhua River acts as the source of the Songhua River, which is 1897 km long, with only the southern source of the Songhua River watershed being investigated in this study. The Songhua River watershed encompasses many lakes and numerous tributaries, including the Yitong River, Yinma River, Hulan River, Lalin River, Woken River, Mudan River, and Tangwang River.

A total of 28 surface water sampling locations were chosen throughout the Songhua River watershed based on the location and tributaries of the cities there, and samples were taken in July 2016 for examination of PAE concentration and CDOM light absorption. These sampling sites cover many national monitoring sections, provincial monitoring sections, and inflow points of a number of important tributaries. Among these sites, 12 samples were collected in the mainstream and 16 samples in tributaries, as detailed in Fig. 5.15. Major cities and WWTPs present within this watershed are shown in Fig. 5.15. The ambient temperature during the sampling period ranged from 25 to 30 °C, with water sample temperatures ranging from 19.1 to 23.8 °C, wind speed ranging from 1.6 to 4.5 m/s, with a water pH range of 6.69–7.81. Concentrations of dissolved organic carbon in these water samples ranged from 3.53 to 14.90 mg L^{-1}. Surface water samples (approximately 2 L) were collected in acetone-rinsed amber glass bottles with polytetrafluoro ethylene (PTFE)-lined caps at a depth of 0.5–1 m.

Most PAEs compounds exhibit stable chemical properties due to their low volatility, low water solubility, and high lipophilicity. The contribution of hydrolysis and photolysis to the overall environmental degradation of PAEs (including DnBP and BBP) in water is usually very low, and DEHP and DnOP do not hydrolyze in water. Therefore, abiotic hydrolysis is not expected to play an important role in determining the environmental fate of PAEs. The hydrolysis rate of PAEs depends on its structural characteristics and environment, usually ranging from about 4 months for DMP to about 100 years for DEHP [124]. The photooxidation process of PAEs is also relatively slow, with half-lives of 2.4–12 years for DEP and DnBP and 0.12–1.5 years for DEHP (Figs. 5.16 and 5.17).

Surface water is considered to be the component of the natural world most affected by human activities. The occurrence of PAEs in surface waters has been monitored around the world, including rivers, lakes, groundwater, and even tap water. In China, the occurrence of PAEs in surface water has been more widely recorded in southern cities than in northern cities [123]. PAEs pollution in the Songhua River, the largest tributary of Northeast China's Heilongjiang basin, is a concern. All six species of PAEs were detected in the water in the Harbin reach. The concentrations of DnBP and DEHP both exceeded 10 μg/L, with most of the highest concentrations occurring downstream from WWTPs [41]. This study suggests that WWTP emissions are a potential source of surface water contamination by PAEs.

PAEs have also been detected in drinking water sources, ranging from 0 to 6.6 μg/L, in Mopanshan Reservoir (a source of drinking water for Harbin) and the Yangtze River (a source of drinking water for Nanjing) [125]. DnBP and DEHP are the main pollutants in the Yangtze River, with concentrations as high as 35.65 μg/L and 54.73 μg/L respectively [119]. The general trend of PAEs in surface water in China is shown in Fig. 5.18, with average levels of DnBP and DEHP higher than other levels, which is consistent with the widespread use of plasticizers.

Surface water is considered to be the component of the natural world that is most affected by human activity. The occurrence of PAEs in surface water has been monitored around the world, including in rivers, lakes, groundwater, and even tap water (Table 5.2). In China, the occurrence of PAEs in surface water has been documented more widely in southern cities than northern cities, with the highest levels, 41 μg/L DnBP and 101 μg/L DEHP, being found in the Haihe river.

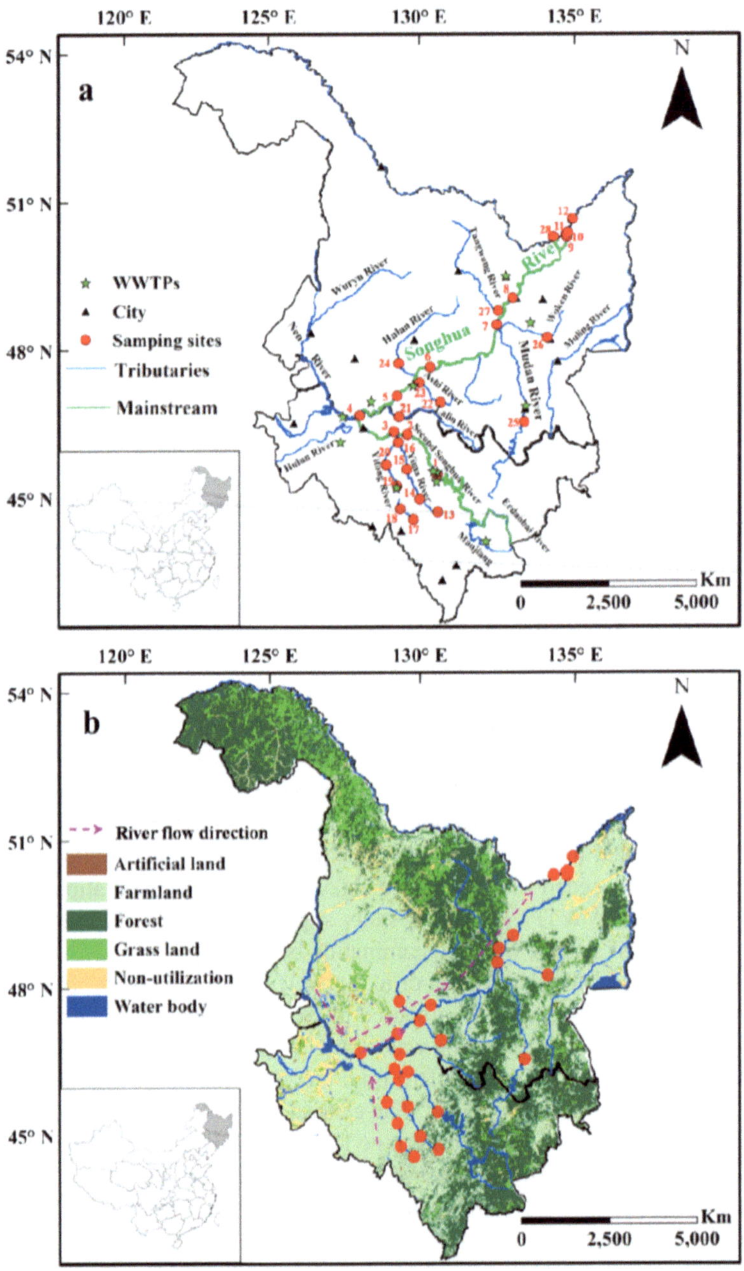

Fig. 5.15 Study area location and sampling station distribution **a** sampling station distribution and **b** land use types in the Songhua River watershed [123]

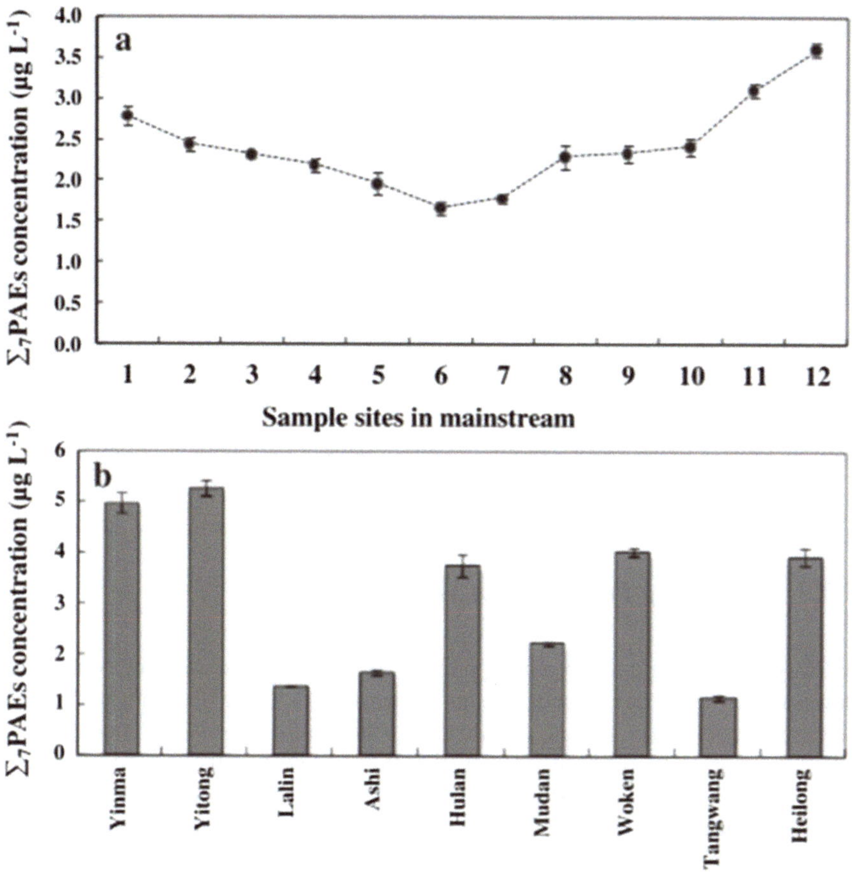

Fig. 5.16 Distribution of ∑PAEs in surface water of the Songhua River watershed, **a** distribution along the direction of the Songhua River mainstream flow, **b** distribution in all the surveyed tributaries [123]

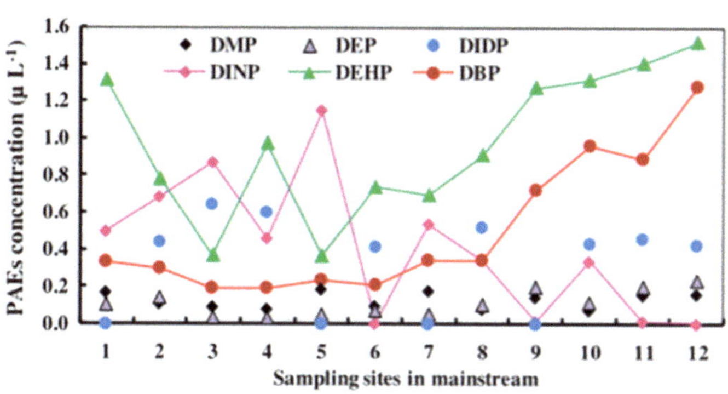

Fig. 5.17 PAE distribution in surface water of the Songhua River mainstream along the direction of the river flow [123]

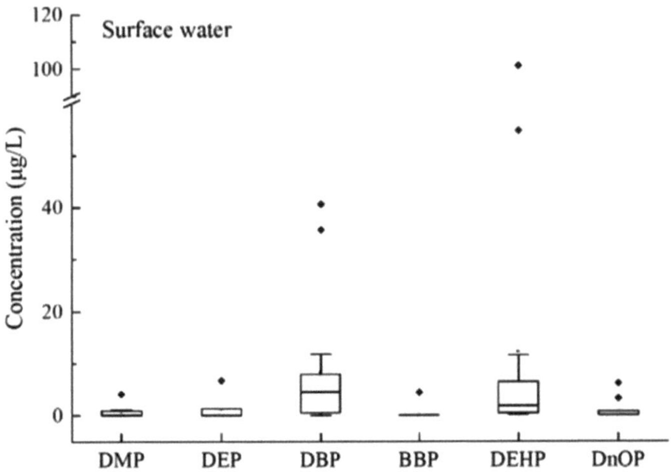

Fig. 5.18 PAEs in surface water

5.3.2 *Emerging Pollutants in Wastewater*

Industry, agriculture and the general public use water and release many compounds in wastewater every day. In actuality, human activity, industrial discharges, and agricultural practices all contribute significantly to the issue of wastewater contaminants. All of these practices produce a variety of pollutants and alter the water cycle, which is of global concern in relation to their ultimate impact on wildlife and human health.

EPs are new products or chemicals with no regulatory status, but its effects on the environment and human health are unknown. They come from industry (PAEs, polychlorinated biphenyls) or wastewater discharges (e.g., pharmaceuticals). Human medications such painkillers, antibiotics, beta-blockers, anticonvulsants, lipomodulators, contrast agents, anticancer agents, hormones, and disinfectants are among the drug molecules found in the environment. Wastewater treatment is necessary to eliminate potentially toxic compounds, but its efficiency is not known. The problem with EPs is a lack of understanding of their medium—and long-term effects on human health, the environment and the aquatic environment.

PAEs have been in use for many years and millions of tons are produced worldwide each year. They are widely utilized as plasticizers in polymers like PVC to give them flexibility and improve their impact resistance and cold resistance. They can be found in many consumer products. Cosmetics are another application area for PAEs. They are used as fixatives to increase the penetration of products into the skin or to prevent nails from cracking. The most commonly used is DEHP, especially for food containers, blood bags, catheters, or bibs [11].

Polychlorinated biphenyls (PCBs) are 209 chlorinated aromatic compounds. Depending on their chlorine content, they are more or less sticky or resinous, insoluble in water, colorless or yellowish, and have a strong odor. They are part of the

Table 5.2 Occurrence of PAEs in surface water (range/mean value, ng/mL)

Location	Types	DMP	DEP	DnBP	BBP	DEHP	DnOP	Reference
Harbin, China	Mopanshan Reservoir	ND–0.042	ND–0.055	0.053–4.498	n.d	0.129–6.571	ND–0.448	[76]
Tianjin, China	Haihe River	–	–	0.35–40.68	–	3.54–101.1	–	[21]
Nanjing, China	Yangtze River	0.003–0.499	0.0073–0.051	0.01–6.165	ND–0.044	1.197–6.205	–	[125]
Nanjing, China	Taihu lake	0.037–0.097	0.017–0.021	3.35–7.88	ND–0.025	0.446–0.752	–	[44]
Wuxi, China	Wujing River	ND	ND	0.528–5.067	ND–0.07	0.513–1.483	–	[136]
Hefei, China	Chaohu Lake	0.123–1.106	0.133–0.157	0.924–5.43	0.066–0.076	0.162–0.231	0.032–0.04	[56]
Wuhan, China	Yangtze River	ND–0.1	ND	ND–35.65	–	3.9–54.73	0.07–3.2	[121]
Other countries								
Sweden	Surface water	–	0.01–0.03	0.02–0.06	–	0.01–0.04	–	[13]
Germany	Surface water	–	–	0.12–8.8	–	0.33–97.8	–	[33]
Netherlands	Surface water	< 0.005–0.19	< 0.07–2.3	< 0.066–3.1	< 0.01–1.8	0.9–5	< 0.002–0.078	[115]
France	Marne River	0.030–0.050	0.025–0.092	0.140–0.220	0.023–0.035	0.307–0.708	ND	[26]
U.S.	Eleven Point River	–	–	0.16–1.36	0.07–0.14	–	–	[103]
Nigeria	Ogun river catchments	ND	1480–1755	2080–2705	–	255–480	–	[3]
South Africa	Port Elizabeth harbor	0.03–350.8	0.03–398.3	1–1028	–	2.1–2306	–	Fatoki and Noma (2002)

biocumulative pollutants found in certain fatty tissues in the body. They are produced worldwide, due to their low solubility in water, they are often discharged into the environment and stored in sediments [92].

The most common groups of PAEs, DEHP and DnBP, are prevalent in WWTP effluent. DEP and DMP also occur frequently, while DnOP and BBP are less frequently reported (Table 5.3). In Harbin, the largest provincial capital of northeast China, PAEs were detected in infected patients with detection frequency of 42% to 100%, and the concentrations of DEHP and DnBP reached 30.99 μg/L and 24.46 μg/L, respectively [41]. The three studied sewage treatment plants had limited efficiency in removing part of the PAE, implying a potential risk of PAE from sewage discharge into the Songhua River. The average concentrations of DnBP, DEP and DEHP in Beijing sewage were 27 μg/L, 21 μg/L and 15 μg/L, respectively (Fig. 5.19).

5.3.3 Emerging Pollutants in Sediments

PAEs pollution in soil is mainly caused by the application of plastic agricultural films, urban biosolids, agricultural chemicals and wastewater irrigation. Some PAE compounds transfers through soil, reach groundwater, or enter the food system through accumulation in crops or plants, according to studies on PAE uptake and metabolism in plants. PAEs have been detected in several soil types in China at levels ranging from one thousandth of a milligram to several mg/kg [58].

In total, Gao et al. [117] collected twenty-seven sediment samples from 9 different sites, including three wetland types: a Deyeuxia angustifolia wetland (DAW), a Carex lasiocarpa wetland (CLW) and a Phragmites australis wetland (PAW) (Fig. 5.20). Three 5 m × 5 m sampling plots were established in each wetland type. In this study, all PAW is perennial wetland, DAW is drained wetland with less surface water, and CLW is seasonally flooded. In each sample plot surface sediments (0–20 cm depth) and vertical (20–40 and 40–60 cm depth) sediments were collected with a pre-cleaned, stainless steel sediment sampler, and transferred into solvent-rinsed aluminum containers. Any pieces of vegetation or roots were removed. Five sub-samples were taken from each site for each sediment sample, which was then cooled and brought straight to the lab. After being sieved through a stainless steel 100-mesh sieve, samples were air dried, powdered, and homogenized before being kept at − 20 °C until extraction.

PAEs pollutants in soil mainly come from human agricultural activities. However, urbanization may also exacerbate PAE pollution. The presence of PAEs in natural wetland sediments with limited human disturbance indicates that PAEs are widespread in Qixinghe natural wetland and other ecosystems. The concentration of PAEs in the sediments of Qixinghe Wetland is one order of magnitude lower than that in the facilities, ranging from 1.37 to 4.90 mg kg^{-1} in northeast China [117]. In this study [117], the concentration of PAEs showed significant differences at different locations. The highest concentration was detected in WS1, which received direct

Table 5.3 Occurrence of PAEs in wastewater (range/mean value, ng/mL)

Location	Mdia	DMP	DEP	DnBP	BBP	DEHP	DnOP	Reference
Harbin, China	Influent (3 STPs) Effluent	4.05–6.449 ND–152	2.25–16.86 ND–1.37	8.73–24.46 3.47–4.13	ND–5.67 ND–17.03	2.42–30.99 1.7–25.4	4.63–12.84 1.11–14.15	[42]
Harbin, China	Influent Effluent	0.779 0.111	1.977 0.022	6.411 0.674	ND ND	ND ND	7.152 2.953	[68]
Tianjin, China	Influent Effluent			0.21–0.583 0.14–0.32		1.12–2.382 0.14–0.32		[133]
Other countries								
Finland	Influent (4 STPs) Effluent	ND–1 ND–< 1	< 1–74 ND–4	3.0–9.0 ND–6	< 1–5 ND–< 1	28–122 2–8	ND–2 ND–1	[83]
Norway	Influent (5 STPs) Effluent		1.3–4.9 < 0.5–9.9			< 0.5–21 < 0.5–34		Vogelsang et al. (2006)
Netherlands	Untreated municipal sewage Sewage Effluent (2 STPs)	0.39–6.2 < 0.002–1.3 < 0.003–0.32	0.0041–44 < 0.35–5.2 < 0.3–0.93	< 0.38–51 < 0.69–21 < 0.42–0.48	0.56–4.9 < 0.17–1.3 < 0.07–0.29	< 13–100 1–150 < 0.47–2.4	0.26–2.4 0.012–2.8 < 0.002–0.19	
Denmark	Influent Effluent	0.269–4.31 ND–0.237		15.34–24.67 1.83–2.73	9.41–80.74 1.99–4.33	52.23–84.1 2.08–9.93		Roslev et al. (2007)
Austria	Influent (2 STPs) Effluent	0.26–0.81 ND	1.2–2.7 ND–< 0.1	< 0.1–0.47 ND	ND–0.26 ND	4.1–13 < 0.2–1.3	ND–< 0.1 ND	Clara et al. (2010)
France	Influent Effluent	0.82 ± 1.13 ND	7.71 ± 5.2 0.78 ± 0.22	11.10 ± 0.37 0.15 ± 0.12	1.12 ± 0.54 0.30 ± 0.12	22.46 ± 13.22 5.02 ± 1.53	0.10 ± 0.1 ND	[26]
Spain	Influent Effluent	0.6 ± 0.5 0.13 ± 0.02	50.7 ± 19 49.8 ± 2.8	46.8 ± 15 ND	0.67 ± 0.8 0.01 ± 0.03	47.9 ± 25 1.28 ± 0.5		Sanchez-Avila et al. (2009)
Canada	Influent Effluent					70 ± 9.1 54 ± 7		Barnabe et al. (2008)

(continued)

Table 5.3 (continued)

Location	Mdia	DMP	DEP	DnBP	BBP	DEHP	DnOP	Reference
U.S.A	Influent Effluent		ND–10 < 1	ND–3.6 < 0.36	14 0.84	33 2.9	4.2 0.39	Jackson and Sutton (2008)
Australia	Influent Effluent		1.08 0.0049	0.201 0.0344	0.134 0.0757	0.716 0.589		Tan et al. (2007)

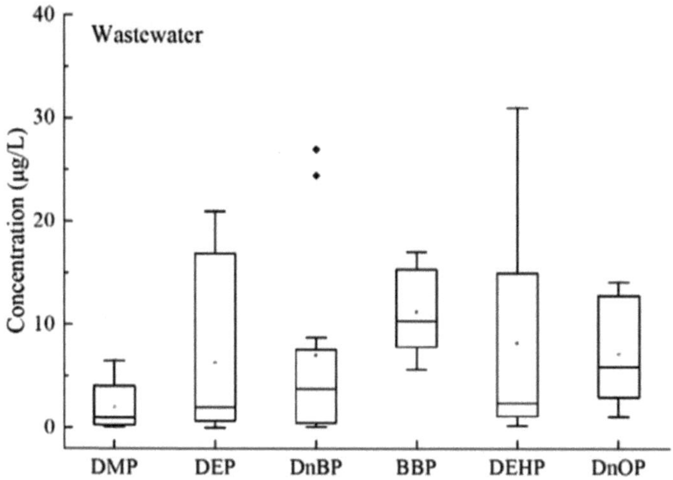

Fig. 5.19 PAEs in wastewater [40]

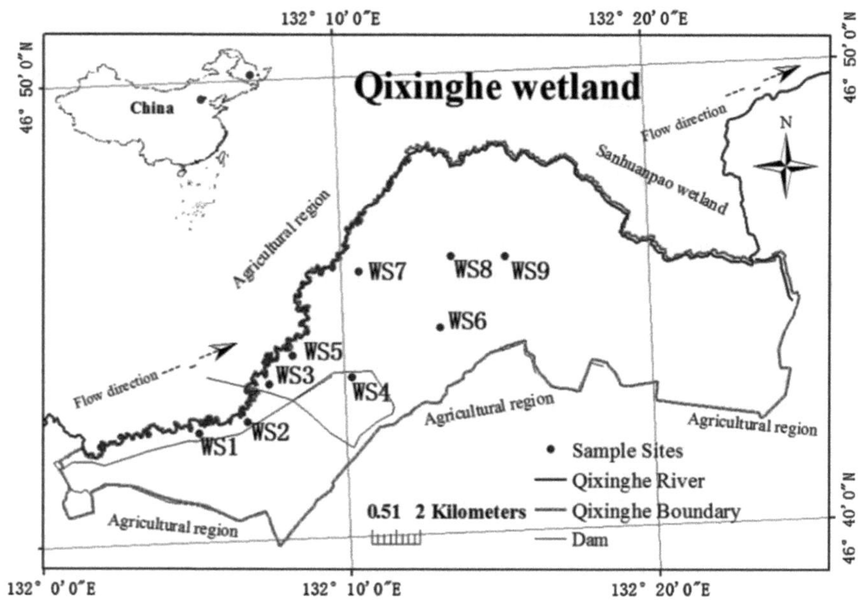

Fig. 5.20 Map of study area

agricultural runoff in the test area. Two other sites with relatively high concentrations were located near the ecotourism route at WS2 (376.93 lg kg^{-1}) and near the pumping station at WS4 (394.9 lg kg^{-1}). The core area with the least amount of non-point source pollution, WS9, has the lowest concentration. Additionally, all sediment

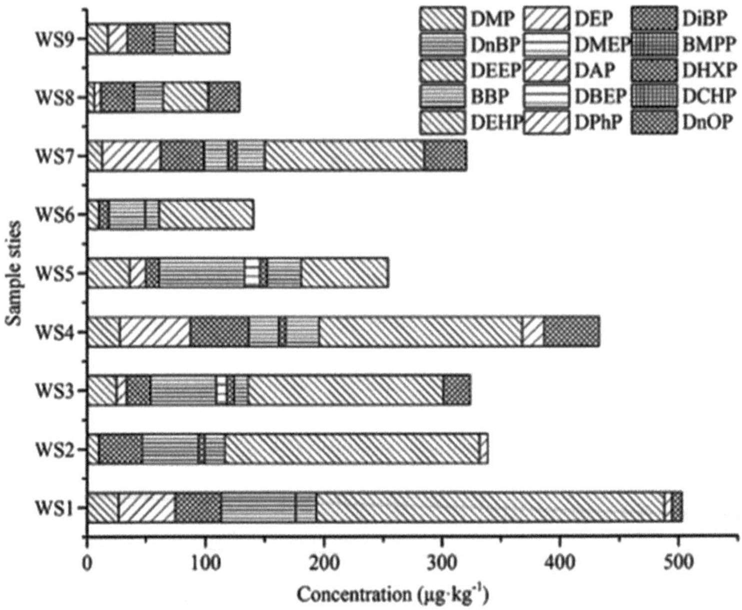

Fig. 5.21 PAEs concentration in three types of wetland sediments in Qixinghe wetland, Heilongjiang province, China [117]

samples contained DEHP, which was definitely the dominating PAEs. Farming activities in the vicinity of Qixinghe Wetland are very intensive, and more anthropogenic pollution sources are used as insecticides and herbicides through aerial spraying and over application of fertilizers and fertilizers. It is well known that the use of agricultural plastic films, fertilizers, and pesticides is attributable to the large amount of PAEs pollutants in cultivated land. During the growing season, excess water is drained from arable land to natural wetlands; PAE can be transferred to wetlands by irrigation water, resulting in heavy use of soil amendments (Fig. 5.21).

EPs are distributed between sediment and water phases during wastewater treatment, as well as in stream environments. The chemical, biological, and physical characteristics of the material are used to determine the allocation ranges, environmental factors (such as pH, temperature), and sediment composition. Adsorbed compounds can linger in the environment and are typically only discovered years later.

There are two ways that EPs might contaminate soil and sediment: (1) point source pollution, and (2) diffuse contamination. The first one is a single identifiable source, such as industrial wastewater or septic tanks. Diffuse contamination is difficult to identify as coming from a discrete site, for example, runoff. Some direct access routes to the soil include large-scale cattle operations, irrigation with treated or untreated wastewater, intentional disposal, landfill leachate, sewage spills, and phosphogypsum modification. Indirect paths involve atmospheric air masses moving long distances and depositing them on plants, animals, and soil (Fig. 5.22).

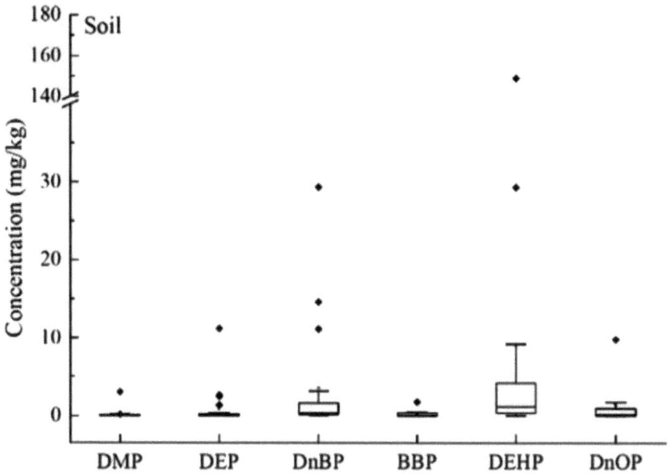

Fig. 5.22 PAEs in soil

DMP, DiBP, DnBP and DEHP were detected in all wetland sediments. The detection rates of DEP, DMEP, DHXP, BBP, DPHP and DnOP were 77.8%, 22.2%, 55.6%, 77.8%, 33.3% and 55.6%, respectively. BMPP, DEEP, DAP, DBEP, and DCHP were not detected. The distribution of PAEs in wetland sediments was uneven, and the concentration was in his order: DHXP < DPHP < DMEP < DMP < BBP < DnOP < DEP < DiBP < DnBP < DEHP.

This might be as a result of its substantial octanol–water partition coefficient, low vapor pressure, limited solubility in water, and comparatively high molecular weight, resulting in increased binding to organic matter in sediments. Among the PAEs, DEHP is the main homologue, with concentrations ranging from 37.62 to 294.90 lg kg^{-1} [117].

In the depth of sediment profile, the concentration of PAEs is much lower than that of surface sediments. The detection rate showed an uneven downward trend. No DPHP was detected in sediments from 20 to 40 cm and 40 to 60 cm. DHXP and DMEP were not detected in the 20–40 cm layer but were present in the 40–60 cm sediment. DiBP, DnBP and DEHP were 100% detected in deep sediments, and their concentrations showed a fluctuating trend. Anaerobic biodegradation conditions vary with the property of the vertical sediment. Therefore, this confirms that the vertical distribution of PAEs is not only affected by adsorption and elution processes, but also by the anaerobic degradation of PAEs and groundwater migration [43] (Tables 5.4 and 5.5).

Table 5.4 Detection rates and concentrations of individual PAEs (ND: not detected) [117]

	Depth	0–20 cm					20–40 cm					40–60 cm			
	Detection rate (%)	Mean	SD	Minimum (μg kg^{-1})	Maximum (μg kg^{-1})	Detection rate (%)	Mean	SD	Minimum (μg kg^{-1})	Maximum (μg kg^{-1})	Detection rate (%)	Mean	SD	Minimum (μg kg^{-1})	Maximum (μg kg^{-1})
DMP	100	19.13	10.12	ND	36.04	77.8	10.57	7.63	ND	21.49	55.6	8.87	6.81	ND	19.08
DEP	77.8	28.88	23.55	ND	60.13	77.8	4.02	3.71	ND	9.72	33.4	5.02	4.35	ND	11.32
DiBP	100	29.12	13.35	8.15	48.78	100	10.19	6.54	6.24	20.08	100	13.24	10.81	NS	27.6
DnBP	100	39.55	19.92	18.03	71.28	100	9.09	6.07	7.28	19.73	100	17.06	7.63	ND	24.83
DMEP	22.2	11.01	4.98	ND	13.24	0	ND	ND	ND	ND	33.4	5.73	3.61	ND	11.49
BMPP	0	ND	ND	ND	ND	0	ND	ND	ND	ND	0	ND	ND	ND	ND
DEEP	0	ND	ND	ND	ND	0	ND	ND	ND	ND	0	ND	ND	ND	ND
DAP	0	ND	ND	ND	ND	0	ND	ND	ND	ND	0	ND	ND	ND	ND
DHXP	55.6	6.08	3.19	ND	7.13	0	ND	ND	ND	ND	22.3	7.06	5.41	ND	13.04
BBP	77.8	19.73	10.6	ND	28.4	55.6	5.67	2.57	ND	10.85	33.4	10.03	5.37	ND	22.15
DBEP	0	ND	ND	ND	ND	0	ND	ND	ND	ND	0	ND	ND	ND	ND
DCHP	0	ND	ND	ND	ND	0	ND	ND	ND	ND	0	ND	ND	ND	ND
DEHP	100	135.28	85.74	37.62	294.9	100	60.05	54.02	35.97	125.07	100	71.22	41.57	28.02	113.74
DPhP	33.3	10.66	6.33	ND	18.55	0	ND	ND	ND	ND	0	ND	ND	ND	ND
DnOP	55.6	27.82	17.76	ND	46.39	55.6	5.81	3.06	ND	11.04	33.4	5.68	3.47	ND	11.3
$\sum_{PAEs}$ 15	100	284.61	120.35	128.41	502.79	100	98.25	40.61	52.06	137.03	100	118.58	50.91	80.27	192.56

Table 5.5 Occurrence of PAEs in sediments (range/mean value, mg/kg)

Location	Types	DMP	DEP	DnBP	BBP	DEHP	DnOP	Reference
Northeast China	Arable soils (3 cities)	ND	0.18–1.36	0.16–1.56	–	0.51–2.15	–	Hu et al. (2003)
Northwest China	(2 cities)	ND	0.18–0.25	0.38–0.39	–	1.62–2.17	–	
Northern China	(4 cities)	ND–0.2	0.15–2.61	0.14–0.98	–	0.51–2.18	–	
Eastern China	(7 cities)	ND	ND–1.29	0.21–1.38	–	0.2–5.98	–	
Southern China	(4 cities)	ND	ND–0.17	ND–0.26	–	0.54–3.42	–	
Southwest China	(3 cities)	ND	ND–0.37	0.51–00.64	–	1.02–2.08	–	
Xinjiang, China	Cotton fields	ND–3.01	ND–2.42	11.15–57.7		103.5–149		Guo and Wu (2011)
Sanjiang Plain, China	Topsoil	0.0266	0.0349	0.0285	–	0.0279	–	Wang et al. (2013)
Harbin, China	Black soils	–	–	2.75–14.62	–	0.49–4.20	–	Xu et al. (2008)
Handan, China	Fluvo-aquic soils	–	–	3.18–29.37	–	1.15–7.99	–	
Beijing, China	Greenhouse soils	< 0.01–0.02	< 0.01–0.05	0.34–1.66	–	0.22–0.74	< 0.09	Ma et al. (2003)
Tianjin, China	Farmland	0.003–0.088	0.003–0.081	0.007–0.189	0–1.79	0.039–2.37	ND–0.647	Kong et al. (2012)
Yellow River Delta, China	Urban soil	0.002–0.060	ND–0.004	0.245–2.058	ND	1.465–6.320	ND–0.044	Yang et al. (2013)
Yangtze River Delta, China	Agriculture soil	0.0002–0.071	0.0005–0.0906	ND–1.5	ND–0.0122	ND–9.19	ND–0.273	Sun et al. (2016)
Nanjing, China	Vegetable soils	ND–0.012	ND–0.007	ND–0.046	ND–0.002	0.204–0.704	0.002–0.019	Wang et al. (2013)
Guangzhou, China	Agricukture soil	0.001–0.157	0.001–0.178	0.009–2.74	ND–0.18	0.107–29.37	ND–0.084	Zeng et al. (2008)

(continued)

Table 5.5 (continued)

Location	Types	DMP	DEP	DnBP	BBP	DEHP	DnOP	Reference
Other countries								
Denmark	Agriculure soil	ND	0.009	0.061	ND	0.182	ND	Laturnus and Gron (2007)
U.K.	Brickearth	0.0001	0.0002	0.008		0.022	0.012	Gibson et al. (2005)
Netherlands				0.006				Peijnenburg and Struijs (2006)

5.4 Biotic Treatments of Emerging Pollutants

Despite their effectiveness, physical and chemical approaches used to remove EPs have certain drawbacks including high costs, low efficiency, and the potential for generating secondary pollutants [87]. On the other hand, biological treatment stands as an environmentally friendly technique for EPs removal, as it eliminates the need for chemical additives and has minimal production of harmful by-products [4]. The primary objective of biological treatment is to establish an efficient system to break down and decompose pollutants effectively [52]. Biological treatment employs mechanisms like biodegradation, bioaccumulation, and biotransformation, with the involvement of organisms such as nematodes, bacteria, or other microorganisms in organic waste decomposition. Furthermore, the growth of microorganisms is hindered by toxic pollutants, but co-metabolic mechanisms often augment their proliferation and facilitate the degradation of EPs [52].

Besides the utilization of microorganisms, plants can serve as a highly efficient method of bioremediation, referred to as phytoremediation [6]. Several plant species have demonstrated the ability to degrade EPs. Biological treatment methods have the potential not only to eliminate pollutants but also to enhance the condition of polluted environments. Ornamental plants, in particular, can enhance soil quality, stimulate plant growth, and contribute to environmental sustainability. Utilizing phytoremediation can aid in preventing soil contamination, while fungal remediation in agricultural land can offer alternatives to pesticides [57].

Extensive research has been conducted on the efficacy of aerobic and anaerobic sludge in the treatment of wastewater. Both aerobic and anaerobic sludge have been employed to eliminate EPs from wastewater. Numerous studies have demonstrated that various microbial communities possess the capability to degrade distinct types of pollutants. Biosorption and biodegradation are the prevalent biological techniques utilized for the removal of EPs [4].

Yang et al. [129] studied the isolated strains to improve the degradation efficiency of antibiotics in wastewater sludge. Batch experiments showed that the combination of two strains enhanced the biodegradation of antibiotics in sludge under aerobic and anaerobic conditions better than the use of a single strain. By introducing bacteria capable of degrading antibiotics, not only were the antibiotics themselves directly degraded, but there was also an alteration in the microbial community structure within the sludge. Consequently, there was a modification in the pathway responsible for the degradation of xenobiotics and an enhancement in the capacity to degrade antibiotics [129].

5.4.1 *Aerobic Sludge*

In the removal of EPs from soil and water, the utilization of a mixed flora composite aerobic sludge has proven to be more efficient [4]. The effective removal of EPs can be attributed to the diversity of microbial species and their respective abundance in the sludge. Aerobic granular sludge offers numerous potential benefits, including a high capacity to retain biomass, excellent settling ability, resilience against contamination, and a composition comprising numerous strains of microorganisms [4]. Using enriched activated sludge for atrazine removal, it was hypothesized that the main players in atrazine degradation might be *Thiobacillus* and *Ferrobacter* based on the changes in microbial community composition and relative abundance of different species in the activated sludge [137]. Ni et al. [88] investigated the removal of PBDEs by aerobic granular sludge. The results showed that the removal of decabromodiphenyl ether (BDE-209) from wastewater was mainly by biosorption of aerobic granular sludge. The adsorption effect is directly influenced by temperature, contact time, and the dosage of sludge, while the absorption capacity is inversely related to the pH of the solution. Microbial activity is notably impacted by temperature, which subsequently affects the removal of EPs in wastewater treatment facilities [4]. By conducting microbial community analysis, we can investigate the impact of PPCPs on the environment. This analysis reveals that certain genera exhibit no response to PPCPs, while others display sensitivity towards them [4].

Shukla and Ahammad [101] showed that the aerobic zone in the new improved trickling filter was better than the anoxic zone in removing PPCP, the new improved trickling filter was more effective in eliminating macrolide resistance genes, while the conventional activated sludge process was more effective in eliminating TC resistance genes. Microplastics polylactic acid (DMP-PLA), an EP, can also enhance the particle stability of aerobic granular sludge, thus enhancing the pollutant removal capacity [104]. Karim and Shriwastav [64] studied the removal efficiency of Sequencing Batch Reactor Activated Sludge Process (SBR) on CIP, TTC and AMX. The results showed that the removal efficiency of CPX was below 65%, TTC was better than 60–90%, and AMX was between 40 and 90%.

5.4.2 Anaerobic Sludge

Similar to aerobic granular sludge, anaerobic granular sludge harbors a diverse range of microbial strains and possesses the capability to degrade various pollutants. While certain EPs may resist degradation by aerobic sludge, they can be efficiently broken down by anaerobic sludge [4]. The removal of different EPs relies on both the sludge type and the concentration of co-substrates. Anaerobic ammonia oxidation plays a significant role in nitrogen removal and the enrichment process of this anaerobic reaction can effectively biodegrade azole compounds. However, further investigations are necessary to identify the responsible bacteria for degradation. The degradation of persistent compounds is contingent upon the presence of co-substrates, and additional research is required to optimize the co-substrate concentration in the degradation environment. A crucial aspect is to classify pollutants based on their aerobic or anaerobic degradability, as this knowledge facilitates the design of strategies for EP removal without compromising microbial systems [4].

Previous studies have considered that the removal of Fluoxetine (FLX) is mainly adsorption, not biodegradation, so its biological removal efficiency is low. Palma and Costa [89] found that the biodegradability of sulfate-reducing bacteria enriched from anaerobic sludge was studied for the first time, and FLX was used as the sole carbon source, and good removal effect was detected. Anaerobic biodegradation seems to play an important role in the removal of FLX. Mares-Carbajal et al. [82] investigated the biotransformation of the SARS-CoV-2 antiviral drugs ribavirin and remdesivir in a methane-producing bioreactor. They also evaluated the impact of the recovered iron-rich minerals from metallurgical wastewater on the biotransformation process. Compared with the control bioreactor lacking minerals, the removal efficiency of two antiviral drugs in the reactor enriched with recycled minerals was improved. 22 kinds of EPs in 43 reclaimed water treatment plants in Beijing were investigated [85]. Including 11 kinds of medicines and PPCPs, 10 kinds of organophosphorus flame retardants and 3 kinds of ARGs. The anaerobic/anoxic/aerobic system combined with the membrane bioreactor process achieved a relatively high water removal rate of PPCP (87%).

5.4.3 Isolated Strain

Because the biological treatment process can effectively remove some EPs, researchers have also conducted research on separating and identifying microorganisms responsible for degrading these compounds [47]. Madukasi et al. [80] isolated *Rhodopseudomonas globosa z08*, a wild strain of purple non-sulfur bacteria, can degrade micro-pollutants in pharmaceutical factory wastewater. *Bacillus cereus*, a bacterium isolated from petroleum sludge, has been shown to effectively degrade promethazine, vincristine, and sulfamethoxazole (SMX) [15].

The removal rate of diclofenac (DCF) by *Bacillus subtilis* and *Bacillus lateralis* strains reached 20% after 17 h in the culture medium at 100°C [51]. Chopra and Kumar [22] isolated *Enterococcus lactis strain DSD-1*(MN744331), *Bacillus paraanthrax strain DYD-1*(MN744332), *Citrobacter strain PPD-2*(MN744333) and *Enterococcus faecium strain DKD-1*(MN744334) from wastewater samples. These four strains can degrade 20% of DCF in 90 days. Ren and Ni [95] selected different MPs in 11 as carbon sources to domesticate the microbial community in the sediments of mangrove areas in Shenzhen. It was found that *Pseudomonas*, *Pandora* and *Dysenteria* could grow well in the mixed strains. Thirty new acesulfame-degrading strains related to shinella were isolated [60]. Examination and sequencing showed that the isolated strain could use acesulfame (ACE) as the only carbon source under aerobic and anoxic conditions, which indicated that ACE could be degraded under anoxic conditions. Bacterial strains capable of degrading paracetamol were isolated from sewage sludge produced by WWTP [113]. Two strains, *Pseudomonas stutzer CSW02* and *Pseudomonas extremophila CSW01*, which can degrade paracetamol with very high concentration in solution as the only carbon and energy source, were isolated. None of them has been reported as acetaminophen degrading agent before, and it is much faster than any other paracetamol degrading agent strain reported in previous literature. A new type of bacteria, *Bacillus megaterium strain ISO-2*, was isolated from wastewater collected from polycarbonate industry, which can degrade BPA and can tolerate high total dissolved solids and alkaline environment [107]. The *aminoglycoside strain nyz550* was isolated as a metformin decomposing strain [72]. *Bacillus strain DSI-1* can effectively remove IBP from wastewater [23]. Paracetamol-degrading bacteria were isolated from sewage samples by enrichment culture, belonging to *Bacillus licheniformis strain PPy-2* [24]. *Bacillus cereus* and *Pseudomonas guarico* were isolated from petroleum sludge [9]. Among them, *Bacillus cereus* not only effectively degraded various aromatic dyes, but also found that *Bacillus cereus* could degrade various EPs. Sangeetha Devi et al. [99] found that *Rhodococcus rhodochrous* alone could not degrade phthalate plasticizers, but the addition of cetane stimulated *R. rhodochrous* to degrade phthalate plasticizers. However, there were differences in the degradation ability when different types of phthalate plasticizers were used.

While there are certain pollutants that resist degradation by aerobic and anaerobic sludge, they can be broken down by specific isolated strains. However, it is important to note that isolated strains alone cannot eliminate all pollutants. Therefore, a combination of isolated strains and aerobic or anaerobic sludge is necessary to achieve complete pollutant removal, with the isolated strains playing a vital stimulatory role [4].

5.4.4 Genetically Engineered Bacteria

Some natural strains do not have the ability to degrade highly toxic pollutants, which leads to the accumulation of pollutants. However, strains with degradable ability may not be suitable for living in the specified environment, so EPs can be degraded by genetically engineered bacteria with biological enhancement. Genetic engineering is a method to change the genetic content of microbial communities, which involves the change of gene sequence of target microbial cells [100]. With the growing attention towards environmental conservation, the utilization of genetically engineered bacteria emerges as a highly efficient and eco-friendly approach. Genetic engineering holds immense research potential in the field of pollutant removal. Researchers have successfully isolated certain strains from the natural environment that possess the capability to degrade diverse compounds. However, the degradation rate exhibited by these strains is lower than the expectations. To address this issue, target genes responsible for pollutant degradation are carefully selected and inserted into rapidly reproducing bacteria, resulting in the effective breakdown of POPs. Comparative studies conducted between natural strains and genetically engineered strains validate the superior degradation ability of the latter [4].

The primary aim of utilizing genetically engineered bacteria is to enable them to adapt to the environment and efficiently degrade pollutants. One feasible approach to enhance the genetic potential of genetically engineered bacteria in breaking down persistent pollutants by creating degradation pathways. Introducing genetically engineered bacteria into natural communities can also enhance the degradation capabilities of these communities. For instance, when bio-enhanced activated sludge was introduced, atrazine degradation started within 11 days of operation, whereas in the absence of bio-enhanced activated sludge, atrazine degradation began after approximately one month [4]. The survival of transgenic organisms is very important for the success of bioremediation process. The survival of genetically modified organisms is affected by various factors, such as environmental conditions, growth and survival rate, and competition between genetically modified organisms and existing protozoa. Studies have shown that genetically modified organisms have survived in wild microbial communities for about 6 years [100].

The long-term existence of genetically engineered organisms will change or harm the existing natural environment after the pollutants are removed. Horizontal gene transfer and survival are the main potential problems faced by genetically engineered organisms after they are released into the environment. Thus the concern authorities should put forward appropriate laws and regulations to contain, evaluate risks and problems and continuously monitor genetically modified organisms after they are put into the environment [100]. Although there have been no reports on the widespread implementation of genetically engineered bacteria, the escalating levels of chemical and biological pollution might compel scientists to consider the utilization of engineered bacteria in environmental remediation [4].

5.4.5 *Biological Activated Carbon*

Under appropriate temperature and nutrient conditions, accumulation or artificial immobilization of microorganisms on the surface of activated carbon (AC) produces bioactive carbon [5]. AC (mostly in granular form) can act as a carrier for both adsorption and biodegradation [63]. The mechanism is the interaction of granular activated carbon (GAC) particles, microorganisms, pollutants and dissolved oxygen in solution. In most cases, ozone oxidation is followed by a bioactivated carbon process to remove contaminants. Bioactivated carbon is effective in removing nitrogen and organic carbon and can be used as part of a tertiary treatment process for reclamation [5]. Before ozone oxidation, the bioactivated carbon process was more effective in the removal of EPs, especially for pesticides (e.g., atrazine and triclosan), β-blockers (e.g., atidolol), and pharmaceuticals (painkillers, antibiotics, lipid regulators, and antidepressants). The bioactive carbon process showed low removal of EDCs such as E3, BPA, and octylphenol, but 99% removal of E1 [5]. Therefore, the combination of bioactivated carbon process and ozone oxidation process can significantly improve the removal efficiency of EPs. The order of removal of EPs by the bioactive carbon process is: pesticides > β-blockers > drugs > EDCs > PPCPs. Since the bioactive carbon process is less effective in removing EDCs and PPCPs (including some drugs), the process should be mainly applied to mixed systems [5].

5.5 Abiotic Treatments of Emerging Pollutants

5.5.1 Adsorption

Adsorption, in a broad sense, refers to the phenomenon of adsorption of a gas or liquid by a solid surface or a liquid surface, and the adsorbed substance is called adsorbent. The principles of pollutant adsorption and desorption are illustrated in the Fig. 5.23. According to the nature of the intermolecular bond between the adsorbent and the adsorbent surface, it can be divided into physical adsorption and chemical adsorption. The adsorbent has a high specific surface area, but the adsorbent mass needs to be removed from the adsorbent before it can be used. AC is often used for adsorbent activation [5]. The removal capacity of sorbents depends on density (material, particle and volume), porosity (particle and volume), external surface area, internal surface area, pore size distribution (between macro-, meso- and micropores), surface chemistry and operating parameters (temperature, pH, contact time) [5]. The adsorption mechanisms of the different materials include hydrogen bonding, van der Waals forces, electrostatic interactions and ion exchange. And the adsorption of different EPs was influenced by factors such as pH, temperature, and ionic strength. Adsorption is effective for chemicals that show resistance to chemical, physical or biological degradation [105].

The mechanism of adsorption dictates that it can be combined with other wastewater treatment technologies to better remove EPs [5]. For different compound classes, adsorption can remove up to 21%, but biodegradation can increase the removal of some compounds up to 98% [96]. Microplastics and EPs extracted from different systems are widely used as sorbents to improve the environmental sustainability of

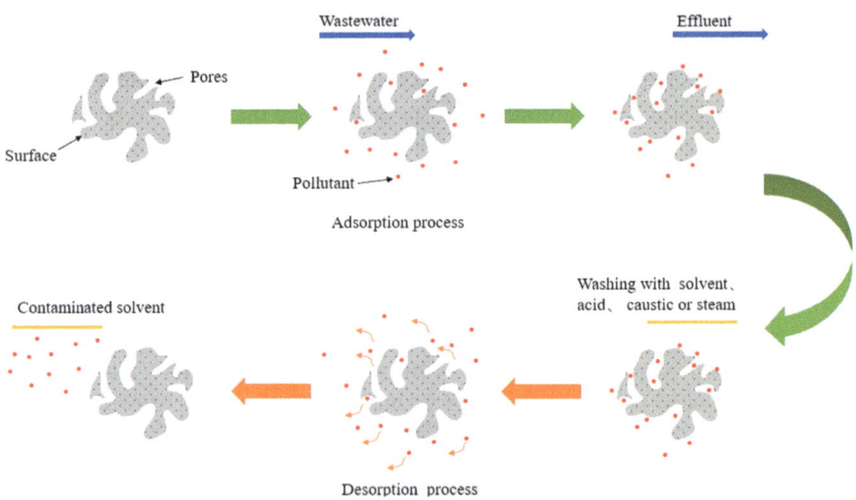

Fig. 5.23 Wastewater treatment techniques of adsorption and desorption

the sorption process. The sorption capacity of other materials for EPs is also worth exploring [6].

AC has high porosity and specific surface area, so AC has strong adsorption properties [96]. AC has been widely used in wastewater treatment, and its role in the adsorption of EPs is being investigated [75]. AC is divided into powdered activated carbon (PAC) and GAC by particle size, and macroporous activated carbon ($\geq$ 50 nm), mesoporous activated carbon (2–50 nm) and microporous activated carbon (2 $\geq$ 0.8 nm) by pore size. Mesoporous AC was found to be the most suitable for the removal of EPs because it reduces the interference of organic matter on the adsorption active sites [27].

The nature of EPs, material conditions, environmental conditions and type of AC, and source of raw material are all factors that affect the adsorption effectiveness. The removal rate of diclofenac using GAC was greater than 90%, while the removal effect using PAC was not efficient due to the carbon structure of the raw material. The nature of the wastewater and the operating conditions of the WWTP also affect the removal of EPs. The presence of more organic matter in the wastewater interferes with the adsorption active site of AC, which affects the removal efficiency of EPs. The advantage of using AC-mediated adsorption of EPs over ozonation is that no by-products are formed and the energy consumption of the WWTP is reduced.

AC adsorption usually takes the form of a packed bed filter with PAC or GAC. PAC is mainly applied directly to the aeration tank or the new flocculation tank after the secondary sedimentation tank, which requires a subsequent separation step, but is less expensive and has a shorter contact time than GAC. GAC filters are generally placed in the secondary sedimentation tank in conventional WWTP processes. After the settling tank, both adsorption and filtration are possible, and it can be operated continuously without a subsequent separation step. Figure 5.24 shows the three most common PAC dosing methods, all three of which can achieve stable removal of EPs. The process in Fig. 5.24a is the simplest and has the lowest investment cost, and the subsequent filtration tank can be set up to ensure the effluent quality and reduce PAC loss, but this process has the highest PAC dosing; Fig. 5.24b, c processes can effectively save 30–50% of PAC dosing, but the process is relatively complex and the overall cost is higher.

Adsorption can be used in combination with other wastewater treatment processes to achieve better treatment results. Acero et al. [2] proposes to combine three different treatment methods, AC and coagulation, for the removal of EPs, which can result in better removal rates.

5.5.2 *Coagulation*

Chemical coagulation is a commonly used conventional treatment process for drinking water treatment and to improve phosphorus removal from wastewater [18]. Park et al. [90] used bench-scale activated sludge systems to investigate the effect of flocculated aluminum (Al) on the effect of Al flocculation on activated sludge

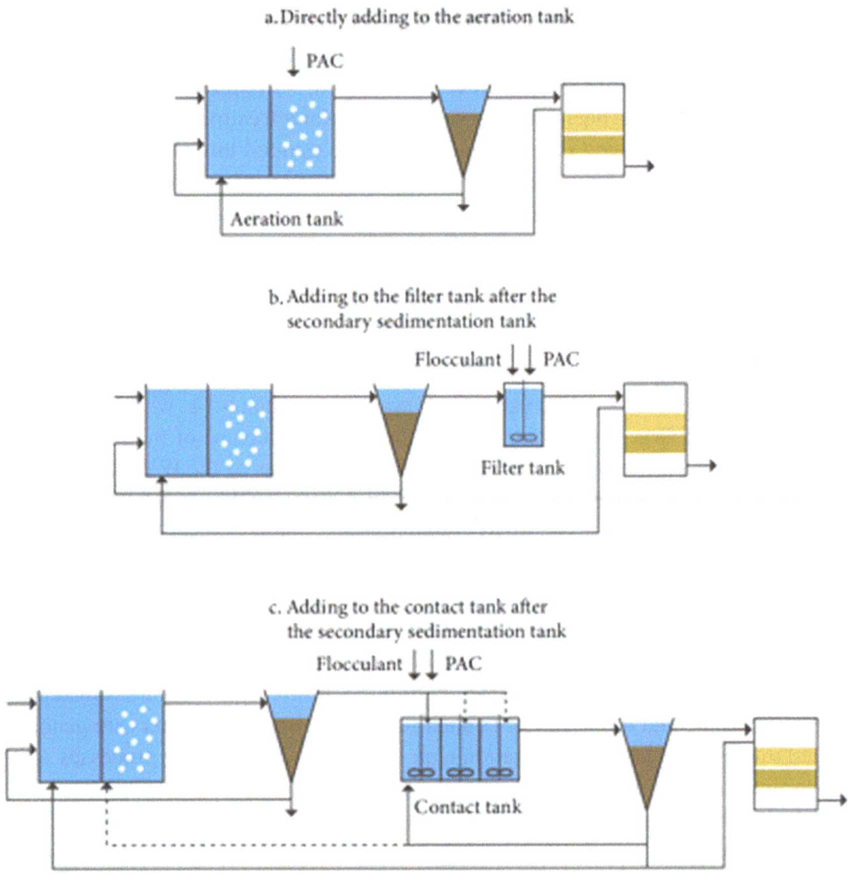

Fig. 5.24 Three ways of PAC addition

performance and removal of 17-α-ethinylestradiol was investigated using a bench-scale activated sludge system. The results showed that higher Al dosing resulted in better settling, dewatering and wastewater quality, and better 17-α-ethinylestradiol removal.

Electroflocculation is a method to produce metal hydroxide flocs by applying electric current to metal electrodes to produce metal ions [91]. Al-Qodah et al. [8] studied the combined use of electrocoagulation and anaerobic fixed membrane fixed bed reactor (AFFBR) reactor for treatment of high strength pharmaceutical wastewater and investigated the effect of various operating parameters such as pH, current density and electrolysis time on removal of pollutants. The results showed that increasing the current density could shorten the electrolysis time while improving the degradation efficiency. Also, the results showed that the combined treatment method can effectively treat high strength wastewater. Yang et al. [130] found that microwaves can produce more Fe(c) than the thermostatic method under the same temperature conditions, and Fe(c) can promote the net rolling removal of organic molecules in the coagulation process and improve the removal efficiency of the whole reaction system.

Zhang and Wang [133] studies have shown that adding iron or aluminum coagulants to DOM separated from leachate can reduce DOC values by 20–25%. Among the coagulants (ferric chloride, alum and PAC), PAC was the most effective in the removal of DOM and PAEs from leachate.

5.5.3 *Membrane Separation*

Membrane technology is a physical treatment mechanism that filters out pollutants through membranes of different pore sizes depending on the size and nature of the pollutant, as shown in Fig. 5.25 [6]. Membrane processes are based on using hydrostatic pressure to remove suspended solids and high molecular weight solutes and allowing water and low molecular weight solutes to pass through [96].

As shown in Fig. 5.25, depending on the size of the pores on the membrane, membrane technology or filtration can be classified into various types, including: ultrafiltration (UF), nanofiltration (NF), microfiltration (MF), forward osmosis (FO), and reverse osmosis (RO). The pore size of MF is between 0.001 and 0.1 μm; UF pore size is less than 0.001 μm; NF pore size is 1–10 nm, etc. [6].

A disadvantage of the membrane process compared to other processes is that the contaminants are not removed directly, but are enriched in the filtrate. The filtrate requires further treatment. However, the membrane process also has many advantages, such as modularity, low energy consumption, and no need for many chemicals [109]. The membrane process also has a serious disadvantage, which is the problem

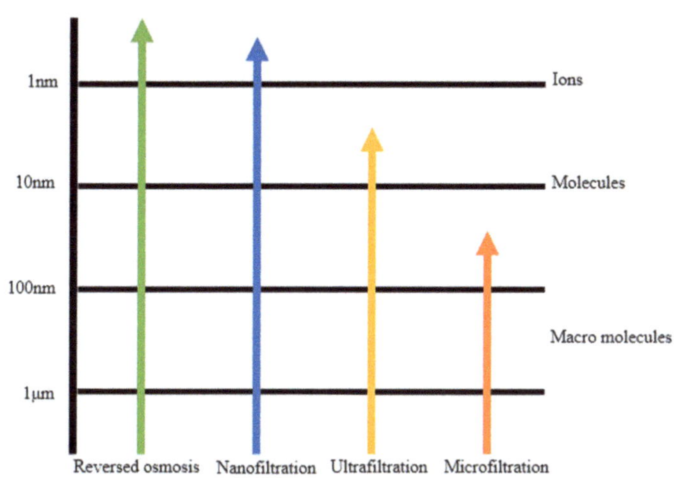

Fig. 5.25 Wastewater treatment using membrane technology

of membrane fouling. This problem can be solved by adding a pretreatment unit before the membrane process. Common pretreatment processes include coagulation, adsorption, and advanced oxidation [135].

Malaeb and Ayoub [81] concluded that the retention rate of EPs by membranes depends on the physicochemical properties (molecular size and weight, charge and hydrophobicity), membrane properties (membrane material, pore size, etc.) and operating conditions of EPs. and operating conditions. The ideal membrane material should be resistant to chemical corrosion, stable mechanical structure and long life. Different membrane materials have different retention effects on the same EPs, and the same membrane has different retention effects on different EPs. Further study on membrane separation mechanism and its relationship with the chemical structure of organic compounds is essential for the design and control of membrane processes.

In EPs removal, membrane separation technology has received more attention, among which NF and RO have the best effect on EPs removal. NF and RO technologies both use high-pressure membranes, which can retain smaller molecular weight substances due to their extremely small pore size and are generally applied to the retention of inorganic ions and organic pollutants. In comparison, small molecules are relatively less effectively retained by NF membranes due to their slightly larger pore size (ranging from 5 to 10 nm) compared to RO membranes. Low-pressure membranes such as MF and UF membranes are not sufficient to retain most EPs due to their larger pore size, but because the secondary effluent of WWTPs still contains relatively high concentrations of organic matter, which can adversely affect the retention effect of NF/RO, MF/UF is often used as a pretreatment process for NF/RO. Organic matter is the primary factor responsible for membrane fouling. Extended operation of NF/RO membranes can lead to reduced membrane flux and decreased retention efficiency. Therefore, incorporating MF/UF as a pretreatment process in the treatment of secondary effluent is essential to ensure long-term stable operation of NF/RO systems. This pretreatment helps maintain optimal membrane performance by effectively removing contaminants and minimizing the impact of organic matter on the NF/RO membranes. As the pore size decreases, the removal efficiency of EPs increases significantly. FO and RO have been reported to have high removal rates in treating water contaminated with various EPs [96]. NF has been shown to have higher efficiency than UF in the removal of certain EPs. For example, UF has a removal rate of 221% for caffeine, while NF has a removal rate of 46–84%, with the same trend for ketone tributylamine [96]. Bolong et al. [14] investigated the removal of BPA by laboratory fabricated hollow fiber polyethersulfone (PES) NF membranes for the removal of endocrine-disrupting compounds (i.e., BPA, BPA). It was shown that the modified PES hollow fibers of cSMM were twice as effective in removing BPA compared to unmodified PES. The ability of RO treatment to remove 10 drugs and their metabolites from seasonally generated secondary separately treated wastewater from Spanish WWTPs has been studied. All three low-energy polyamide membranes removed compounds ranging from 30% (nicotine in March) to 100% (caffeine in August) [91]. Acero et al. [1] tested UF and NF membranes for secondary municipal wastewater filtration. The exclusion coefficients of 28 pharmaceutical compounds were determined and one of the NF membranes showed a removal rate higher than

75%, indicating that NF membranes may be an excellent choice for the removal of pharmaceuticals from municipal wastewater. The main retention mechanism of UF membranes is adsorption, while size exclusion and electrostatic rejection of NF membranes at high pH are considered to be responsible for the retention of microcontaminants. Heo et al. [58] tested the removal of BPA from water using UF membranes made of polysulfone and polyvinylidene chloride. The former was able to achieve 75% removal, while the latter was able to eliminate up to 98% of the load in the experimental feed water. Electrostatic spinning technology is a simple and suitable method for preparing fibers, which are effective in the removal of EPs because of their large specific surface area, high porosity, high surface roughness, low weight, and the ability to couple with other functional materials [17].

High removal rates for most compounds can be achieved when membranes or combinations of membranes are used to remove emerging contaminants [75]. Gur-Reznik et al. [53] conducted a study to evaluate the rejection of carbamazepine (CBZ) and complex amphoteric acid by three RO materials and two NF materials, it was found that all three ROs were very efficient in removing these two contaminants, but the two NFs were not as efficient as the ROs, and one of the NFs with larger pore size was seasonal in removing CBZ. Valladares Linares et al. [112] investigated the removal of 13 selected micropollutants from wastewater in the clean and fouled state of FO membranes. Results showed that, in addition to hydrophilic neutral compounds, the removal of hydrophobic and ionic compounds was increased due to the presence of the fouling layer. And, when combined with RO, the removal rates in both states increased to more than 96%. The coupled forward osmosis-low pressure reverse osmosis (FO-LPRO) system is an effective double barrier to selected microcontaminants.

5.5.4 Chemical Oxidation

Chemical oxidation, as a modern technique in the realm of chemical treatment technologies, involves the use of chemical oxidants to transform pollutants into controlled and harmless forms. A significant parameter employed in the chemical oxidation treatment process is chemical oxygen demand (COD), particularly useful for treating wastewater with relatively low COD levels. Within the chemical oxidation process, pollutants are eliminated through the utilization of oxidants, ultimately reducing them into benign by-products such as water and carbon dioxide. Two primary categories exist within chemical oxidation treatment processes: conventional oxidation methods and advanced oxidation methods [6].

5.5.4.1 Advanced Oxidation Processes

Advanced oxidation processes (AOPs) are processes that oxidize pollutants using highly reactive oxidants or oxidant-induced reactive radicals [109]. Advanced oxidation methods are characterized by the utilization of different sources of reactive radicals. Simple or conventional oxidation processes do not always suffice to decompose organic compounds present in wastewater. Hence, advanced oxidation methods have emerged, wherein the oxidation process involves the simultaneous utilization of highly reactive hydroxyl molecules as free radicals. Several advanced oxidation techniques employed for wastewater treatment include photo Fenton, photocatalytic, and solar-driven processes [6].

The implementation of advanced oxidation techniques necessitates the presence of both oxidants and catalysts. Typical oxidants used include ozone (O_3) and hydrogen peroxide (H_2O_2), while common catalysts consist of substances such as Fe^{2+} and TiO_2, among others. Catalyst modification can be employed to address challenges encountered during the oxidation process. Additionally, external energy sources like UV radiation can be utilized to degrade certain EPs. The susceptibility of these pollutants to photocatalytic degradation is determined by their molecular structures [15]. Bracamontes-Ruelas et al. [16] investigated the ability of laboratory level Fenton to oxidize mosquito repellent, caffeine, CBZ, potassium acetyl sulfonate, IBP, and triclosan. It was shown that the removal rates of several emerging contaminants were close to 100% except for mosquito repellent (85.21%), and the higher removal rates in this study than in other studies may be due to the different hydraulic retention times and the dosing of oxidants and catalysts. It indicates that Fenton oxidation can effectively remove EPs from wastewater.

The generation of hydroxyl radicals (HO·) is a fundamental characteristic of AOPs, making the type and experimental conditions of hydroxyl radical generation crucial for pollutant removal. For instance, difluoxacillin has been found to be unresponsive to UV radiation and only slightly degraded through photocatalysis. However, it exhibits high sensitivity to acoustochemical conditions. This highlights the importance of utilizing longer wavelength UV radiation, as shorter wavelengths may not induce any changes in the chemical structure of environmentally persistent compounds. Introduction of semiconductor photocatalysts, like titanium oxide, can enhance hydroxyl radical production and optimize the degradation process. Nonetheless, inadequate or excessive photocatalyst loading can result in slow degradation rates. Numerous studies have elucidated that the concentration of oxidant, pH of the reaction mixture, chemical structure and initial concentration of the targeted pollutant, as well as the wavelength and intensity of the radiation source (e.g., sun), significantly influence the overall outcomes [96]. Vela et al. [114] used persulfate in sunlight conditions to generate sulfate radicals. These sulfate radicals can oxidize and mineralize pesticides to remove them from agricultural wastewater. Falås et al. [32] used MnO_x-coated coconut shell fibers to remove biologically active compounds from hospital wastewater, and of the 52 organic micropollutants detected 29 were removed by more than 90% after 24 h reaction period.

The photo Fenton process necessitates the utilization of hydrogen peroxide (H_2O_2), iron salts, and UV–visible light. The presence of light expedites the reaction, requiring a wavelength of $\lambda < 580$ nm for effective implementation of the photo Fenton AOP. In order to keep iron in solution, an acidic environment is crucial during the photo Fenton process. Various studies have indicated that the optimal pH range for photo Fenton operation is between 3 and 3.5, mirroring conventional Fenton AOP conditions. However, the Ph may vary depending on the type of pollutant and the composition of the wastewater being treated. Furthermore, research has explored the use of chelating agents to expand the Ph range and analyze their feasibility in the photo Fenton process [15].

Peracetic acid (PAA) has a high oxidation potential and has the potential to remove EPs. However, because of its low homolytic cleavage, catalysts are required to activate the generation of its strong oxidation [25]. UV, metal catalysts, non-metal catalysts, synergistic Fe^{2+}/UV, phosphate buffer, and thermal catalysts all contribute to the catalysis of PAA. Notably, the UV/PAA process, metal catalyst/PAA process, and synergistic Fe^{2+}/UV/PAA process have garnered significant attention. Extensive research conducted on these processes has demonstrated their efficacy in eliminating EPs. However, further exploration is required to fully comprehend the catalytic methods, oxidation conditions, mechanisms, and by-products associated with PAA.

The application of AOP treatment in water and wastewater systems has shown potential in enhancing the detoxification of treated water. To evaluate the overall efficacy of the treatment process, bioassays involving invertebrates, plants, microbes, and fish have been employed in various studies. These bioassays have provided valuable insights into assessing the effectiveness of AOPs. Additionally, alternative techniques like excitation emission matrix (EEM) fluorescence spectroscopy have been proposed as a means to quantify the impact of AOPs on the degradation of pharmaceuticals and PPCPs [75].

It should be noted that the use of AOP methods requires the use of chemicals and energy, resulting in increased operating costs. In addition, the main difficulties in using the AOP method for decentralized systems located in rural areas may be the handling and transportation of chemicals and the complex operational methods of AOP. More importantly, AOP may generate toxic or antioxidant secondary contaminants during oxidation of EPs that require further treatment in other steps [61].

5.5.4.2 Electrochemical Oxidation

Electro-oxidation is another oxidation process that has been evaluated for the treatment of emerging contaminants [91]. Electrochemical oxidation (EO) includes anodic oxidation, electro-Fenton oxidation and photo-electric Fenton oxidation. The advantages of EO are high efficiency, good controllability and mild operating conditions, and disadvantages are the need to add electrolytes to increase electrical conductivity, the formation of degradation intermediates, and easy electrode deactivation. EO electrodes include inert electrodes (such as Pt, IrO_2 and RuO_2, etc.) and active electrodes (such as PbO_2, SnO_2 and boron-containing diamond), etc. Anodic oxidation mainly includes 2 processes: (1) direct electron transfer; (2) release of reactive oxygen groups (e.g. –OH) from the electrode, chemisorption of reactive oxygen, etc.

EO occurs in a reactor called an electrolytic cell, which consists of a tank, an electrolyte, a pair of electrodes, an anode (where the expected pollutants are oxidized), a cathode (where heavy metals, and usually organic matter, are reduced), and a power source [35]. The main goal of EO is to remediate wastewaters contaminated with EPs through in situ production of reactive oxygen species such as $HO^{\bullet}$. One of the extensively researched electrochemical AOPs is anodic oxidation. In this process, a high current density (J) is applied to the anode (M) of a cell, resulting in the generation of physically adsorbed $HO^{\bullet}$ on the surface of the anode M. However, the oxidation capacity of an electrochemical AOP will depend on the electrochemical system used, the type of cell, the electrodes and the other equipment used [15].

In electrochemical AOP, the hydrodynamic parameters and the energy used for pumping must be considered, as it may lead to a large part of the energy consumption in these processes. If a lower current intensity is used in electrochemical AOP to increase the production of $HO^{\bullet}$ since the treatment duration, but the pumping time will be prolonged, this will significantly increase the cost [15].

On the one hand, EO is extremely costly due to the long reaction time required and the expensive electrodes. On the other hand, the generation of toxic disinfection and degradation by-products complicates the reuse and/or discharge of EO-treated effluents. Therefore, various studies have recommended the use of EO in combination with biological processes. EO is commonly used as a pretreatment to improve the biodegradability of industrial wastewater. However, sufficient organic matter needs to be left for the subsequent biological treatment process and the toxic by-products generated by EO can also affect the subsequent biological treatment process. Therefore, the most promising approach for the combination of EO and biological treatment processes seems to be EO as a post-treatment, since the organics and nutrients are eliminated in the biological stage, thus concentrating the OH-radicals generated by EO and mineralizing the more persistent pollutants. By using EO as a post-treatment, the production of disinfection by-products (e.g. THMs) can also be reduced due to the reduction of their precursors (e.g. organics or nutrients).Another benefit of EO as a post-treatment is the reduction of the formation of polymeric layers on the electrode surface, thus reducing their fouling [35].

A non-radical oxidation process with natural air diffusion cathode (ADC) and RuO_2-Ti as the anode was constructed [71]. A self-rebreathing high-efficiency hydrogen peroxide cathode with good catalytic performance without aeration was prepared using BP2000. The hydrogen peroxide was then activated by ClO^- from the anode to generate $^1O_2$, which efficiently degraded the pollutants by the non-free radical oxidation process without adding any catalyst. It was found that the system has good degradation effect on many EPs such as AN, BPA, TC and SMX.

de Amorim et al. [10] used a boron-containing diamond electrode to oxidatively degrade SMX (1 m M), and the results showed that the TOC removal rate was 78% when the solution was 0.1 M $NaClO_4$ and 69% when the solution was 0.1 M NaCl. Perez et al. studied the removal of 10 PPCPs from RO brine and WWTP effluent and found that the current density in the range of 20–100 A/m^2 had no effect on the removal, probably due to the mass transfer resistance in the process when the oxidized solute was present at such a low concentration. Domínguez et al. [28] used boron doped diamond and platinum electrodes to study the ketoprofen. Mineralization of ketoprofen via anodic oxidation was studied using boron-doped diamond and platinum electrodes. The redox behavior of ketoprofen molecules, contamination of the electrode, oxygen and active chloride production were studied by cyclic voltammetry. The effects of electrolyte, pH of aqueous medium and applied current density on the mineralization behavior of ketoprofen were investigated. It was observed that the physisorption of $HO^{\bullet}$ onto the boron-doped diamond electrode initiated the mineralization process of ketoprofen. However, the mineralization efficiency was found to be low when using sodium chloride as the supporting electrolyte. This was attributed to the formation of chlorinated organic compounds that are difficult to degrade. In contrast, when sodium sulfate was used as the supporting electrolyte, complete mineralization of ketoprofen was achieved for both boron-doped diamond and platinum anodes. Radjenovic et al. [94] studied RO concentrates used for electrochemical oxidation of 28 drugs and pesticides and concluded that electrodes can oxidize most drugs and pesticides, but may form by-products that may require additional treatment. In another study, Cheng et al. [20] studied microbially mediated iodination in an electrolytic cell contrast agent dehalogenation. Biogenic palladium nanoparticles in the cathode promoted the direct electrochemical reduction of the enantiomerate, and complete deiodination was achieved after 24 h of cycling. The higher cell voltage increased the rate of dehalogenation, suggesting the potential of this combined process for drug dehalogenation in hospital wastewater. The novel combination of anode and cathode is being tested for its ability to provide tertiary treatment of wastewater to reduce the concentration of target organic compounds, total estrogens and total organic matter. For the treatment of secondary effluent from a municipal sewage treatment plant containing 29 pharmaceuticals and pesticides, Garcia-Segura et al. [45] utilized an anodic oxidation technique employing a boron-doped diamond (BDD) anode. The compound was completely degraded within 7 min at 2.1 mA/cm^2 and removed within seconds by the addition of 0.1 mol/L NaCl.

5.5.4.3 Ozone

Ozone can react directly with EPs or indirectly through secondary oxidants. Ozone reacts with specific kinds of wastewater organics (e.g. phenols or amines) in a side reaction to form $HO^{\bullet}$. The combination of O_3, H_2O_2 and UV improves the $HO^{\bullet}$ produced by ozone decomposition. Ozone is potentially selective and therefore usually attacks electron-rich EPs and EPs with deprotonated amine groups at low pH, whereas the action of $HO^{\bullet}$ is non-selective, so the method can remove a wide range of EPs. Many studies have reported that the use of ozone oxidation for the removal of PPCPs and pharmaceuticals is very effective, as most products can be eliminated [6].

It is worth mentioning that in order for ozone to produce $HO^{\bullet}$, a significant amount of hydroxide ions must first be present, thus facilitating the production of these radicals. The pH for the production of $HO^{\bullet}$ is a key factor and the pH of the wastewater must be maintained at approximately above 8, but this may vary depending on the nature of the contaminants to be treated and the substrate in which they are found. In UV ozone AOP, a wavelength of $\lambda < 300$ nm is to be used to cause the dissolved ozone to cleave and atomic oxygen (O) to occur with water (H_2O), thermally excited to produce hydrogen peroxide (H_2O_2) [15].

An important issue in ozonation is the formation of oxidation by-products. The mechanism of ozonation depends on pH, temperature and ozone dose, which can interfere with the degradation of EPs. If insufficient ozone dose is applied, transformation products or oxidation byproducts will be formed instead of complete decomposition. Sometimes, the temporary increase in toxicity after ozone oxidation is attributed to the formation of toxic oxidation byproducts. After ozonation, toxicity can be further reduced by a subsequent biological treatment step, typically a sand filter or biofilter [5]. Luo et al. [78] reported that the problem of ozone oxidation is the high energy consumption that formation of some oxidation by-products and interference of free radical scavengers. Therefore, these are the areas that need to be considered for future ozonation research. Mišík et al. [86] also showed that ozone oxidation of the effluent from tertiary wastewater reduced its genotoxic activity. The effect of ozonation on secondary wastewater treatment has also been shown to have a positive impact on reducing ecotoxicity, as evidenced by a series of studies evaluating Gammarus fossarum (leaf-cutting invertebrates); the researchers concluded that this effect was the result of a reduction in micropollutant loading, although no specific compounds were measured [75].

5.6 Fate of Emerging Pollutants in Wastewater Treatment Plants

5.6.1 Phthalate Esters

The processes affecting the fate of organic contaminants in sewage treatment plants include volatilization, solids settling, hydrolysis, biodegradation, and adsorption [97]. Biodegradation and adsorption are considered to be the main pathways for the removal of PAEs. Both mechanisms may be influenced by several factors, including the nature of the treatment process, temperature, and the initial concentration of the target compound [40].

Wang et al. [118] studied the removal efficiency of PAEs in five WWTPs and found that the removal of PAEs mainly occurred in primary and secondary treatment, and the influences of influent PAE concentration and treatment process were key factors affecting the degradation efficiency of PAEs in primary and secondary treatment, respectively. In primary treatment, the removal of long-chain PAEs was found to be more efficient irrespective of the influent PAE concentration. On the other hand, during secondary treatment, short-chain PAEs were effectively removed regardless of the specific treatment process employed. Anaerobic/Anoxic/Oxic (A/A/O) advantage in the removal of long-chain PAEs may be related to better control of dissolved oxygen for biological growth and favorable particle adsorption. At the laboratory scale, Huang et al. [59] investigated the fate of DBP in an A/A/O treatment process and found that only 2.4% was discharged with the effluent. However, such good results were not found in the full-size A/A/O treatment process. Yan et al. [128] provides a performance comparison of the recirculating activated sludge technology [32] process, Anaerobic/Oxic (A/O) and A/A/O processes. A/O and A/A/O treatment have limited effectiveness in removing PAEs, while CAST treatment is more effective. This may be due to differences in ambient temperature and recirculation methods. In the CAST process, biodegradation was the main pathway for PAEs removal, whereas in the A/A/O process, more PAEs were removed by sludge adsorption. Yan et al. [128] also found that short alkyl chain groups (DMP, DEP, and DBP) tended to be removed by biotransformation rather than adsorbed to excess sludge, while long alkyl chain groups (BBP, DEHP, and DOP) on the contrary. Marttinen et al. [83] also found that at 20 °C, aqueous-phase DEHP removal was up to 95% and average biotransformation removal was up to 42% in an SBR with an optimal aerobic/anoxic cycle. In contrast, only 14% of the DEHP load in the influent was removed by biotransformation during the full biological treatment at 18.5 °C, with an overall removal efficiency of 97 and 68% by adsorption to the primary sludge and residual sludge.

In summary, biodegradation and adsorption are the main removal mechanisms of PAEs from WWTPs. the proportion of PAEs adsorbed on sludge increases with their alkyl chain length and the lipophilicity of the substance. These studies suggest that PAEs can be removed by adsorption without being completely metabolized or degraded, which is a crucial finding considering the potential environmental risk posed by pollutants discharged from WWTPs [40].

5.6.1.1 Cyclic Activated Sludge Technology

Gao et al. [37, 42] conducted a qualitative assessment of the fate of PAEs in two different types of wastewater treatment plants, CAST and A/A/O. The results indicated that the removal of PAEs is a result of both biotransformation and adsorption, with the former being particularly important in the CAST process, and the latter being more significant in the A/A/O process.

WWTP#1 operates an indoor Cyclic Activated Sludge Technology [32] process, where the influent sewage is treated by primary sedimentation before being passed through a sequence of biological selectors and finally entering the CAST bioreactors. The CAST tanks employ intermittent aeration with a cycle of 6 h (1.5 h feeding, 3 h aeration, and 1.5 h settlement). The mean daily inflow during the sampling collection period was 11 × 104 $m^3\ d^{-1}$ for WWTP#1. Figure 5.26 shows the process flow and sampling point distribution of WWTP#1.

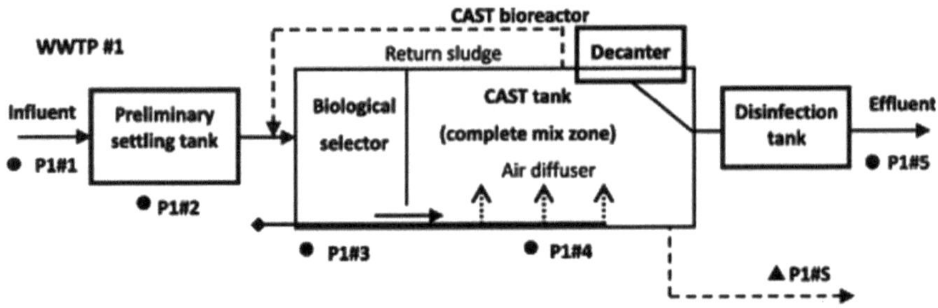

Fig. 5.26 Schematic diagrams of WWTP#1. Samling points (wastewater): P1#1: influent to WWTP#1, P1#2: effluent from the preliminary settling tank, P1#3: effluent from the biological selector, P1#4: effluent from the CAST tank, P1#5: effluent from WWTP#1, P1#S: excess sludge from WWTP#1

The PAE removal effect of WWTP#1 with CAST as the biological treatment unit was investigated, Fig. 5.26 shows the PAE concentrations at different sampling points in WWTP#1 [37, 42]. The aqueous concentrations of low molecular weight (LMW) and intermediate molecular weight (IMW) PAEs reach their peak after leaving the preliminary settling tank (P1#2). In particular, the amount of DMP and DBP discharged from the preliminary settling tank is much higher. The levels of high molecular weight (HMW) PAEs were moderately lower at this stage, and the DEHP concentration was reduced by about 30%. Most PAEs exhibited a faster decrease in concentration in the CAST bioreactor than in the other processes, especially for less refractory compounds such as LMW and IMW PAEs. Concentrations of PAEs appeared to decrease slightly in the bioselector (P1#3) and subsequently in the aerated CAST tank (P1#4), indicating a further decrease in PAEs due to biodegradation. Data from the effluent samples (P1#5) indicated that LMW-PAEs had been effectively removed, while DEHP levels were barely affected, an observation that underscores the effective removal of LMW-PAEs compounds under nitrification conditions. Aerobic tanks are essential for the biodegradation of some PPCPs.

The CAST process performed better than either the A/O or A/A/O processes, the mean removal efficiency of ΣPAEs being 72%, 30%, and 32%, respectively (Fig. 5.27d). However, the CAST process produced poor elimination efficiencies for DEHP (< 40%). BBP was not detected in the CAST process. Similarly, relatively high removal efficiencies for di-n-octyl phthalate (DOP) (> 70%) were achieved by the CAST process (Table 5.6).

Because the $\log K_{ow}$ of BBP, DEHP and DOP were 4.8, 7.5 and 8.1, respectively, indicating the poor degradation of these compounds, which reflects their alkyl chain length and their tendency to adsorb into the sludge. Vieno et al. [116] considered biodegradation and adsorption as the main elimination of PAEs removal mechanisms, and since both mechanisms are temperature dependent, the better performance of the CAST process may be due to the indoor operation of the WWTP#1 treatment process. In this study, the ambient temperature dropped below − 10 °C during sampling and the wastewater temperature in the treatment facility was approximately 4–6 °C.

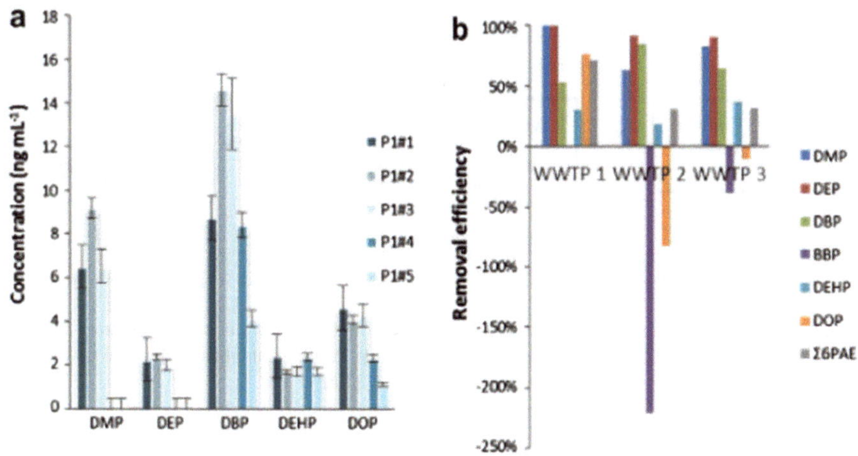

Fig. 5.27 Occurrence and removal of PAE during different WWTP processes. **a** WWTP#1-CAST process; and **b** removal efficiency of three different WWTPs (WWTP 1-CAST process (WWTP#1), WWTP 2-A/O process (WWTP#2), WWTP 3-A/AO process (WWTP#3))

Table 5.6 Concentrations (ng/mL) of PAEs in WWTP#1

	PAEs concentrations (ng/mL)
DMP	6.49 ± 0.96
DEP	< 5
DBP	8.74 ± 1.05
BBP	0
DEHP	< 5
DOP	< 5

Data were taken from Gao et al. [37, 42]

Under these conditions, which are not conducive to microbial activity or growth, the low biodegradation rate resulted in poor elimination of PAE from the wastewater at higher concentrations.

5.6.1.2 Anaerobic/Oxic Process

The PAE removal effect of WWTP#2 with A/O as the biological treatment unit was investigated [37, 42]. Figure 5.28 shows the process flow and sampling point distribution of WWTP#2. The basic design parameters of WWTP#2 are shown in Table 5.7.

Figure 5.29 shows the PAE concentrations at different sampling points in WWTP#2 and its removal efficiency. The A/O process (WWTP#2) mainly focused on nitrogen removal. In the A/O process, the concentrations of LMW-PAEs (DMP,

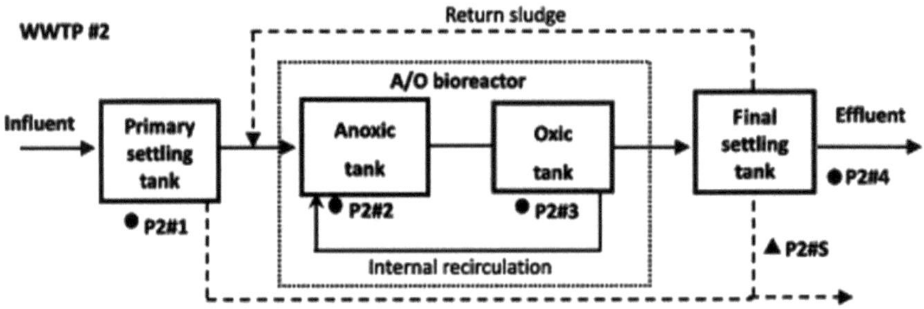

Fig. 5.28 Schematic diagrams of WWTP#2. Samling points (wastewater): P2#1: effluent from the primary settling tank, P2#2: effluent from the anoxic tank, P2#3: effluent from the oxic tank; P2#4: effluent from WWTP#2, P2#S: excess sludge from WWTP#2

Table 5.7 Operational and design parameters of investigated WWTP#2

Parameter	Details
WWTP#2	
Technology	A/O
Location	Harbin
Average flow rate (m^3/d)	325,000
HRT (hours)	8
SRT (days)	19
Treatment processes	Pst → A/O → Sst

Data were taken from Gao et al. [37, 42]

Pst Primary settling tank, *A/O* Anoxic/Oxic, *Sst* Second settling tank

DEP and DEHP) were significantly reduced under denitrification conditions in the anoxic cell (P2#2). The aerobic tank (P2#3) and the final settling tank (P2#4) played a minor role in the removal of LMW PAEs. The reason may be mainly that LMW-PAEs are relatively soluble and easily decomposed by microorganisms, and dilution of the primary settling effluent (P2#1) by internal recirculation and return sludge may also be a possible explanation. DBP accumulates in the anoxic tank, while the removal effect was highest in the aerobic tank. Regarding HMW-PAEs, BBP and DOP concentrations remained relatively constant throughout the process but increased significantly in the final effluent, while DEHP was effectively removed in the anoxic tank but also increased in the final effluent. DEP, DMP and DBP were effectively removed from the wastewater by the A/O process. However, the elimination efficiency of the A/O process for DEHP was low (< 40%). Concentration of BBP increased during the course of the treatment. Similarly, relatively high DOP removal efficiency (> 70%) was observed.

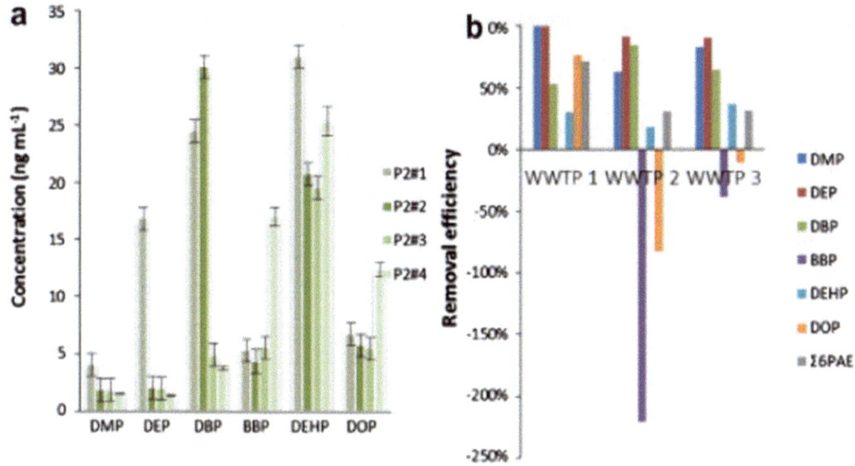

Fig. 5.29 Occurrence and removal of PAE during different WWTP processes. **a** WWTP#2-A/O process; and **b** removal efficiency of three different WWTPs (WWTP 1-CAST process (WWTP#1), WWTP 2-A/O process (WWTP#2), WWTP 3-A/AO process (WWTP#3))

5.6.1.3 Anaerobic/Anoxic/Oxic Process

The removal efficiency of PAE in WWTP#3 with A/A/O treatment as the biological treatment unit was investigated [37, 42]. Figure 5.30 shows the process flow and sampling point distribution of WWTP#3. The basic design parameters of WWTP#3 are shown in Table 5.8. Figure 5.31 shows the PAE concentrations and removal efficiency at each sampling point in WWTP#3.

The A/A/O process of WWTP#3 was optimized to remove both nitrogen and phosphorus. Comparison of different sampling points P2#3 and P2#4 and P3#3 and P3#4 showed that in the A/A/O process, the concentration of HMW-PAE increases after the wastewater passes through the final settling tank. The low solubility and high logK_{ow} values suggest that HMW-PAEs rapidly adsorb onto suspended organic matter in the

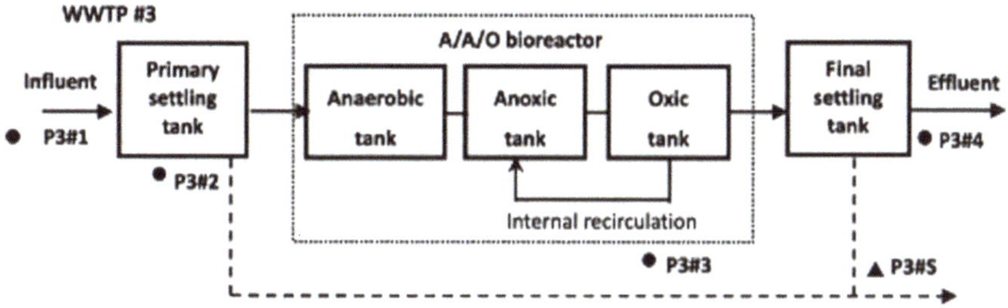

Fig. 5.30 Schematic diagrams of WWTP#3. Samling points (wastewater): P3#1: influent to WWTP#3, P3#2: effluent from the primary settling tank, P3#3: effluent from the A/A/O tank, P3#4: effluent from WWTP#3, P3#S: excess sludge from WWTP#3

Table 5.8 Operational and design parameters of investigated WWTP#3, WWTP#4 and WWTP#5

Parameter	Details
WWTP#3	
Technology	A/A/O
Average flow rate (m^3/d)	6400
HRT (h)	9.5
SRT (days)	17
Location	Harbin
Treatment processes	Pst → A/A/O → Sst
WWTP#4	
Technology	A/A/O
Average flow rate (m^3/d)	53,000
Wastewater source	Domestic + Industrial
Location	Qingdao
Service area	All over the Chengyang district administration between the east of the Moshui River and south of the Baisha River
WWTP#5	
Technology	Multiple-influent inverted A/A/O
Average flow rate (m^3/d)	198,200
Wastewater source	Industrial
Location	Qingdao
Service area	124 km^3, covers the Licun River and Zhangcun River basins

Data were taken from Gao et al. [37, 42] and Wu et al. [126]
Pst Primary settling tank, *A/A/O* Anaerobic/Anoxic/Oxic, *Sst* Second settling tank

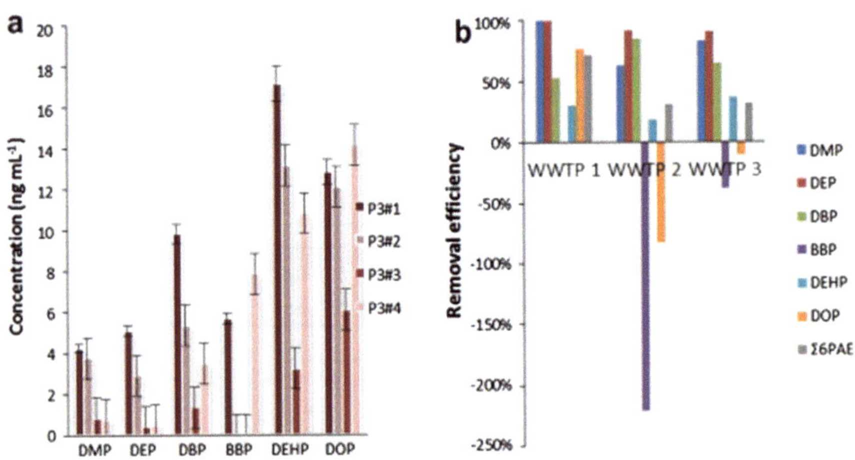

Fig. 5.31 Occurrence and removal of PAE during different WWTP processes. **a** WWTP#3-A/A/O process; and **b** removal efficiency of three different WWTPs (WWTP 1-CAST process (WWTP#1), WWTP 2-A/O process (WWTP#2), WWTP 3-A/AO process (WWTP#3))

effluent, which may carries the compounds for further degradation. However, during biological processes, suspended organic matter may undergo structural changes such as degradation and hydrolysis, so that the eventually adsorbed PAEs may be released in the wastewater. Previous studies have shown that the biodegradability of PAEs seems to be related to the length of alkyl side chains and alkyl branched chains [37, 42]. In this study, the degradation efficiency of short-chain PAEs such as DMP, DEP, and DBP was higher than that of long-chain PAEs such as DOP and DEHP. In addition, alkyl-branched PAEs, such as BBP, showed relatively lower efficiency than straight-chain PAEs such as DBP and DOP, respectively. DEP, DMP and DBP were effectively removed from the wastewater by the A/A/O process. However, the elimination efficiency of DEHP by the A/A/O process was low (< 40%). BBP was found to accumulate in the wastewater treated by the A/O and A/A/O processes.

The removal efficiency of PAE in WWTP#4 and WWTP#5 with A/A/O treatment as the biological treatment unit was investigated [126]. The basic design parameters of WWTP#4 and WWTP#5 are shown in Table 5.8. Table 5.9 shows the influent and effluent concentrations and removal efficiencies for WWTP#4 and WWTP#5. The average removal rate of $\sum$PAE in WWTP#4 with A/A/O wastewater treatment process reached 70.31%. The removal rates studied were as follows: DMP: 86.47%, DEP: 81.69%, DBP: 58.58%, BBP: 73.91%, DCHP: 69.38%, DEHP: 73.12%, and DOP: 100%. The removal rates of dimethylglycol phthalate (DMGP), DEEP, and dipentyl phthalate (DPP) were less than 1.76%. No DOP was detected in the effluent compared to PAE in the influent. The average removal efficiency of $\sum$PAE by the multiple influent inversion A/A/O wastewater treatment process of WWTP#5 was similar to WWTP#4. DMP (95.59%), DEP (85.31%), DBP (57.49%), BBP (74.86%), DCHP (68.60%), DEHP (90.08%) and DOP (73.87%) were effectively removed. The results showed that DEHP removal was relatively good and the concentration of DEHP in the influent was relatively low. DMP was the most easily removed pollutant, while the poorly removed pollutants were DMGP and DEEP. Similar results from biodegradation studies showed that the removal of DMP and DEP was 75–99%.

5.6.1.4 Sequencing Batch Reactor

The removal efficiency of PAE in WWTP#6 with SBR treatment as the biological treatment unit was investigated [34]. The basic design parameters of WWTP#6 are shown in Table 5.10. The study data are shown in Table 5.11. It was found that the removal of four phthalates by SBR was 80% (31.18 ng/mL) with 6% (2.33 ng/mL) by adsorption and 74% (28.85 ng/mL) by biotransformation. The removal of DEP, DBP, BBP and DEHP by SBR biotransformation was 84, 70, 88 and 70%. Two reasons were considered for the high removal of phthalates by SBR treatment. One is the effect of SBR treatment configuration on the biotransformation of phthalates. SBR has a pretreatment chamber before the aeration tank, which is often called a biological selector. The pretreatment chamber is in anoxic condition due to recirculation of nitrification solution. Two different redox conditions (anoxic and aerobic) in the reactor provided two metabolic degradation pathways, which enhanced the

Table 5.9 Concentrations (ng/mL) of PAEs in WWTP#4 and WWTP#5

PAEs concentrations (ng/mL)	WWTP#4						WWTP#5					
	Influent				Effluent (ng/mL)	Removal efficiency (%)	Influent				Effluent (ng/mL)	Removal efficiency (%)
	Min (ng/mL)	Max (ng/mL)	Mean (ng/mL)	Percentage (%)			Min (ng/mL)	Max (ng/mL)	Mean (ng/mL)	Percentage (%)		
DMP	34.73	43.67	40.59	30.20	5.49	86.47	38.55	46.50	43.80	23.50	1.93	95.59
DEP	11.43	20.63	15.62	11.60	2.86	81.69	16.68	24.86	19.47	10.50	2.86	85.31
DBP	16.79	24.09	21.10	15.70	8.74	58.58	49.05	51.15	46.55	25.00	19.79	57.49
DMGP	8.13	8.21	8.17	6.08	8.16	0.12	7.98	8.66	8.37	4.50	8.28	1.08
DEHP	3.84	4.10	3.83	2.85	1.03	73.12	2.04	5.60	3.83	2.06	0.38	90.08
DEEP	6.44	6.49	6.46	4.81	6.42	0.62	6.59	7.16	6.80	3.66	6.69	1.62
DPP	2.24	2.29	2.27	1.69	2.23	1.76	nd	nd	nd	nd	nd	–
BBP	3.98	4.76	4.37	3.25	1.14	73.91	8.96	9.48	9.19	4.94	2.31	74.86
DCHP	11.46	14.47	12.54	9.33	3.84	69.38	9.96	14.46	12.04	6.47	3.78	68.60
DOP	16.68	24.86	19.47	14.5	nd	100	30.55	40.77	35.97	19.30	9.40	73.87
DHXP	nd	nd	nd	nd	nd	–	nd	nd	nd	nd	nd	–
Total PAEs	115.72	153.57	134.42		39.91	70.31	170.36	208.64	186.02		55.42	70.21

Data were taken from Wu et al. [126]
nd not detected, meaning the concentration is below limit of detection (LOD)

degradation of phthalate esters. The second influencing factor is the larger SRT run time (12–27 days) of the SBR. It has been shown that large SRTs are expected to increase biodegradation and adsorption by increasing sludge biodiversity and suspension in the mix, respectively. The seasonal variation of phthalate removal in SBR was observed by studying the seasonal variation of wastewater samples analyzed with a collection period of one year. The removal rate in winter was lower than the removal rate in summer, with less than 67% removal in winter (November–February) when the average temperature was 17 °C, and higher than 75% removal in summer (June and August) when the average temperature was 30 °C. The minimum removal of low molecular weight phthalates DEP and DBP was observed in November at 30% and 19%, respectively. No significant changes in BBP removal were observed, except in March (51% removal). The removal rate of DEHP in SBR was better in summer, with the highest removal rate of 100% in April.

Table 5.10 Operational and design parameters of investigated WWTP#6 and WWTP#7

Parameter	Details
WWTP#6	
Technology	SBR
Design capacity (m^3/d)	27,000
Operating capacity (m^3/d)	27,000
HRT (h)	11
SRT (days)	12–27
Mixed liquor suspended solid (mg/L)	5200 ± 1800
Mixed liquor volatile suspended solid (mg/L)	1640 ± 500
Cycle time (hours)	3
Filling and aeration time	1.5 h per cycle
Settling time	0.75 h per cycle
Decanting time	0.75 h per cycle
Treatment processes	Scr → Gc → Pst → SBR
Location	India
Effluent disposal	River Ganga
WWTP#7	
Technology	Modified SBR
Average flow rate (m^3/d)	147,473
Wastewater source	Domestic + Industrial
Location	Qingdao
Service area	The south of the Sifang district and the north of the Changle River

Data were taken from Gani and Kazmi [34] and Wu et al. [126]
Scr Screening, *Gc* Grit chamber, *Pst* Primary settling tank, *SBR* Sequencing Batch Reactor

Table 5.11 Concentrations (ng/mL) of PAEs in WWTP#6 and WWTP#7

PAEs concentrations	WWTP#6			WWTP#7					
	Influent (ng/mL)	Effluent (ng/mL)	Removal efficiency (%)	Influent				Effluent (ng/mL)	Removal efficiency (%)
				Min (ng/mL)	Max (ng/mL)	Mean (ng/mL)	Percentage (%)		
DMP				41.25	49.46	44.87	33.10	5.07	88.70
DEP		0.93		31.68	39.86	34.47	25.40	8.35	75.78
DBP		3.30		nd	nd	nd	nd	4.32	–
DMGP				8.24	8.31	8.27	6.10	8.27	0.00
DEHP		3.30		3.56	4.14	3.84	2.83	1.32	65.63
DEEP				9.48	9.68	9.60	7.07	9.80	–
BBP		0.41		nd	nd	nd	nd	nd	–
DCHP				7.98	8.66	8.34	6.17	0.17	97.97
DOP				18.84	24.65	22.49	16.60	3.28	85.42
DHXP				3.84	4.10	3.83	2.82	3.79	1.04
Total PAEs	38.85	7.85	80	124.87	148.86	135.74		44.37	67.31

Data were taken from Gani and Kazmi [34] and Wu et al. [126]
nd not detected, meaning the concentration is below limit of detection (LOD)

The removal efficiency of PAE in WWTP#7 with MSBR treatment as the biological treatment unit was investigated [126]. The basic design parameters of WWTP#7 are shown in Table 5.11. The MSBR process is a new wastewater treatment technology that can remove biological nitrogen and phosphorus. Moreover, the MSBR process can be considered as the combination of the A/A/O and sequencing batch reactor processes. The inlet and outlet PAE concentrations as well as the removal efficiency of WWTP#7 are shown in Table 5.11. The average PAE removal rate of MSBR process in WWTP#7 was 67.31%. The effective removal rates of DMP, DEP, DEHP, DCHP and DOP were 88.70%, 75.78%, 65.63%, 97.97% and 85.42%, respectively. The removal efficiency of DMP, DEP, DCHP and DOP was relatively high, while that of DHXP was relatively high. While the removal efficiency of DHXP was relatively low. In addition, the MSBR process had poor removal efficiency for DMGP and DEEP. No DBP was detected in the influent samples, but DBP was detected in the treated effluent.

5.6.1.5 Upflow Anaerobic Sludge Bed

The removal efficiency of PAE in WWTP#8 via Upflow Anaerobic Sludge Bed (UASB) treatment as the biological treatment unit was investigated [34]. The basic design parameters of WWTP#8 are shown in Table 5.12.

The study data of WWTP#8 are shown in Table 5.13. The total removal of phthalates in UASB was 53% (17.47 ng/mL), with 37% (12.05 ng/mL) in the biodegradation fraction and 16% (5.39 ng/mL) in the adsorption fraction. The lowest removal rate was achieved by biotransformation in UASB. The biotransformation removal rates were 44% (DEP), 61% (DBP), 27% (BBP) and 30% (DEHP), respectively. DEHP is the most common phthalate in wastewater compared to other low molecular weight phthalates such as DEP. The high concentrations observed in the final

Table 5.12 Operational and design parameters of investigated WWTP#8

Parameter	Details
WWTP#8	
Technology	UASB + polishing pond
Design capacity (m^3/d)	38,000
Operating capacity (m^3/d)	38,000
HRT in UASB (h)	10.2
HRT in polishing pond (h)	24
Treatment processes	Scr → Gc → UASB → Pp
Location	India
Effluent disposal	River Dhamola

Data were taken from Gani and Kazmi [34]
Scr Screening, *Gc* Grit chamber, *UASB* Upflow Anaerobic Sludge Bed, *Pp* Polishing pond

Table 5.13 Concentrations (ng/mL) of PAEs in WWTP#8

PAEs concentrations (ng/mL)	WWTP#8			
	Influent (ng/mL)	Effluent (ng/mL)		Removal efficiency (%)
		UASB	Polishing pond	
DEP			0.79	
DBP			1.03	
DEHP			4.45	
BBP			0.42	
Total PAEs	33.03	15.55	6.68	53

Data were taken from Gani and Kazmi [34]

effluent of the combined UASB and pond may be a result of the lower removal of DEHP from wastewater in the anaerobic process UASB. A study reported from laboratory-scale anaerobic degradation experiments that water solubility is the main factor limiting the anaerobic degradation of phthalates [34]. The water solubility of DEHP (3 ng/mL) is lower than that of DEP, DBP and BBP (11.2–50 ng/mL), with lower bioavailability and weaker degradation capacity. The UASB effluent was further treated in the polishing pond. The removal of the four phthalates in the polishing tank was 63% (9.82 ng/mL), and the total removal with UASB treatment was 83% (27.29 ng/mL). Seasonal variations in the removal of phthalates in the UASB and polishing ponds were observed by studying the seasonal variations in the analysis of effluent samples collected with a one-year period. The removal rate in winter was lower than the removal rate in summer, with a removal rate below 67% in winter (November–February) when the average temperature was 17 °C, and above 75% in summer (June and August) when the average temperature was 30 °C. A minimum removal of 22% was observed in November for the low molecular weight phthalate DEP and 17% and 13% for the low molecular weight phthalate DBP at UASB and pond, respectively. BBP removal similarly dropped below 35% in the winter months (November–March), while in the other months of the year, BBP removal was higher than 70%.

5.6.2 Nonylphenol Ethoxylates (NPnEOs)

The concentrations of short-chain nonylphenol ethoxylates (NPnEO, n = 1, 2) and nonylphenol (NP) were found to be relatively high in wastewater. Extensive research has shown that short-chain NPnEO exhibits estrogenic properties and is more toxic to aquatic organisms compared to long-chain NPnEO [102]. Due to their hydrophobic nature, these compounds are prone to adsorption onto solid particles. As a result, higher concentrations can be detected in wastewater sludge and sediments. Some research findings indicate that removal rates of NPnEO in conventional activated sludge reactors can range from 74 to 97%, and membrane-assisted treatments can be even more effective and reliable than traditional biological processes [110].

Under dry weather conditions, the total concentration of nonylphenol compounds (NP-c) was correlated with the domestic and the concentrations of NP-c in the influent water of the WWTPs receiving both domestic and industrial wastewater were lower than those of the WWTPs receiving only domestic wastewater [73]. The content of long-chain NPnEO in the influent water of four WWTPs in Beijing, China, was only one-third of the total NP-c, clearly different from the data from other countries but consistent with the data from China [73]. This may be due to the long route of wastewater discharge to WWTPs in China, during which the long-chain NPnEO has been converted to other metabolites. The study concluded that the existing wastewater treatment processes can meet the need for removal of nonylphenols, and the NP-c [73]. The main removal processes were biodegradation and sludge adsorption. NP, nonylphenol monoethoxylate (NP1EO), nonylphenol diethoxylate (NP2EO) and carboxylated derivatives including carboxylated NPnEOs (NPECs) are not only biodegraded but also produced during the wastewater treatment process, while NPnEOs ($n \geq 3$) are mainly biodegraded, mineralized or converted into their metabolites, so the removal rate of NPnEOs ($n \geq 3$) can reflect the degree of biodegradation. The affinity of NPnEOs for solids resulted in a fraction of NPnEOs being adsorbed by sludge flocs. Compared with other metabolites, NP and NP1EO and NP2EO tend to aggregate more on the sludge flocs due to their higher hydrophobicity. Lian et al. [73] found that the concentration of nonylphenols was slightly higher in summer than in winter, probably due to more washing events in summer and Surfactant dosage increased and the higher ambient temperature contributed to the biodegradation of long-chain NPnEO by biochemical treatment.

5.6.2.1 Anaerobic/Oxic Process

Lian et al. [73] investigate behaviours of nonylphenol polyethoxylates and their metabolites in different wastewater treatment processes, the basic design parameters of WWTP#9 are shown in Table 5.14. The study data of WWTP#9 are shown in Table 5.15. The results pointed out that the ratio of biodegradation to sludge adsorption for WWTP#9 was 1.3. This may be due to the fact that the P2 influent was first settled in the primary settling tank and a large amount of NP-c were removed by adsorption on the primary sludge before entering the biological treatment unit. Therefore, the primary settling tank plays an important role in the adsorption output pathway of nonylphenols in the wastewater treatment process.

Gao et al. [37, 42] investigated the ability of WWTP#10 to treat NPs using the A/O process as a biological treatment unit, the basic design parameters of WWTP#10 are shown in Table 5.15. The wastewater treatment process and the distribution of sampling points for WWTP#10 are shown in Fig. 5.32. The concentrations of NP, NP1EO and NP2EO detected in wastewater samples collected from different locations of the A/O process are shown in Fig. 5.33. The operation of the A/O process achieved effective removal of NP and NPnEO from the effluent. Influent

Table 5.14 Operational and design parameters of investigated WWTP#9 and WWTP#10

Parameter	Details
WWTP#9	
Technology	A/O
Average flow rate (m^3/d)	750
HRT (h)	6–9
SRT (days)	20
Treatment process	Scr → Gc → Pst → A/O → Sst
Wastewater source	Domestic + Industrial
Location	Beijing
WWTP#10	
Technology	A/O
Capacity of A/O (m^3)	160,000
Treatment process	Scr → Agc → Pst → A/O → Sst
Wastewater source	Domestic + Industrial
Location	Harbin

Data were taken from Lian et al. [73] and Gao et al. [37, 42]
Scr Screening, *Gc* Grit chamber, *Agc* Aerated grit chamber, *Pst* Primary settling tank, *A/O* Anoxic/Oxic, *Sst* Second settling tank

concentrations of NP, NP1EO, and NP2EO were similar, ranging from 1.8 to 2.0 ng/mL, but effluent concentrations of NP, NP1EO, and NP2EO measured from the sedimentation tanks increased from approximately 1.9 ng/mL to 2.5, 4.4, and 2.1 ng/mL, respectively. One possible reason for the increase in concentration was the return flow of sludge from the primary sedimentation tank to the sand settling tank. After the A/O treatment, the concentrations of NP, NP1EO, and NP2EO decreased below the concentrations in the influent water, and after the secondary settling unit, the concentrations of NP, NP1EO, and NP2EO decreased to 0.44, 0.28, and 0.2 ng/mL, respectively. In the A/O process, the removal rates of NP, NP1EO, and NP2EO in the wastewater were 78%, 84%, and 89% respectively. The comparison of the removal rates of different treatment units is shown in Fig. 5.34. The results show that the removal rates of NP, NP1EO and NP2EO in the anaerobic cell were 39%, 46% and 53%, respectively, while the removal rates of NP, NP1EO and NP2EO in the aerobic cell were 37%, 26% and 20%, respectively. The aerobic unit of the A/O process had highly active microorganisms due to the presence of aeration in the unit, which increased the degradation of pollutants. However, the degradation of NP1EO and NP2EO occurred only in some aeration units of both processes. Therefore, the removal efficiency of target compounds in individual units is not comparable to other units. NP removal by the A/O process is good due to the large microbial biomass and long HRT time in the A/O process, which gives the microorganisms enough time to adapt to the environmental conditions. A large amount of NP, NP1EO and NP2EO was biodegraded when the wastewater passed through the secondary sedimentation tank, and the relative removal rates were measured to be about 60% for NP and

Table 5.15 Concentrations (ng/mL) of NPs in WWTP#9

Sampling time	T (°C)	Samples	Concentration (μmol/L)					
			NP	NPnEOs (n = 1.2)	NPnEOs (n ≥ 3)	NPECs	Sum	Removal percentage (%)
Sep, 27th	20.7	Inf.	0.037	0.064	0.016	0.032	0.149	89.9
		Eff.	0.001	0.001	0.001	0.012	0.015	
Dec, 17th	13.8	Inf.	0.029	0.071	0.013	0.023	0.136	89.7
		Eff.	0.001	0.002	0.002	0.009	0.014	
Average ± S.D		Inf.	0.033 ± 0.005	0.067 ± 0.006	0.014 ± 0.002	0.028 ± 0.007	0.142 ± 0.008	89.8 ± 1.4
		Eff.	0.001 ± 0.000	0.001 ± 0.001	0.002 ± 0.001	0.010 ± 0.002	0.014 ± 0.001	

Data were taken from Lian et al. [73]

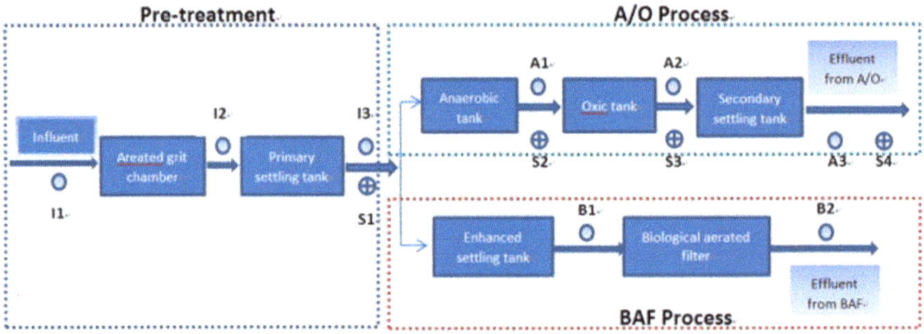

Fig. 5.32 Sampling points along the process flow of WWTP#10 and WWTP#14. Sampling points: wastewater: I1: influent to the WWTP, I2: effluent from the aerated grit chamber, I3: effluent from the primary settling tank, A1: effluent from the anaerobic tank (A/O), A2: effluent from the oxic tank (A/O), A3: effluent from the secondary settling tank (A/O), B1: effluent from the enhanced settling tank (BAF), B2: effluent from the biological aerated filter (BAF). Sludge: S1: sludge from the primary settling tank, S2: sludge from the anaerobic tank (A/O), S3: sludge from the oxic tank (A/O), S4: sludge from the secondary settling tank (A/O)

more than 70% for both NP1EO and NP2EO. High concentrations of NP, NP1EO and NP2EO were observed in the sludge collected from the A/O process, suggesting that the removal of NP, NP1EO and NP2EO in the A/O process may be related to the adsorption of sludge. For compounds with minimal adsorption potential and $\log K_{ow}$ less than 2.5, such as long-chain NPnEO, the main removal mechanism is degradation. For compounds with $\log K_{ow}$ values greater than 4, such as NP and short-chain NPnEO, were not easily biodegradable and removal by binding to solids may be the main removal mechanism due to their high adsorption potential.

The seasonal variation of nonylphenol ethoxylate removal by A/O process WWTP#11 was investigated [38]. The WWTP#11 effluent treatment process and sampling points are shown in Fig. 5.35, and the basic design parameters and seasonal variation of NPnEOs concentration for WWTP#11 are shown in Table 5.16. NP and NP1EO and NP2EO concentrations differed significantly between treatment units and seasons (Fig. 5.36). In the effluent samples extracted from the main sedimentation tank, NP, NP1EO, and NP2EO concentrations were highest in winter compared to other seasons (Fig. 5.36b). NP, NP1EO, and NP2EO concentrations were relatively similar in summer and autumn. The spring NP, NP1EO, and NP2EO concentrations were maintained at lower levels, all less than 2000 ng/L. Among the three short-chain NPEO compounds studied, NP2EO exhibited more significant month-to-month variation. The results indicated that seasonal differences in NP, NP1EO and NP2EO concentrations in the main sedimentation tank effluent were related to the influent concentration. Anaerobic and aerobic tanks were rich in microorganisms, and certain microorganisms used long-chain NPnEO as a carbon source, resulting in short-chain NPnEO compounds. Therefore, changes in environmental conditions could theoretically affect the concentration and fate of NPnEO compounds in the tanks by altering the structure and physiological characteristics of the microbial community. After passing through the anaerobic and aerobic tanks, NPnEO compounds were removed

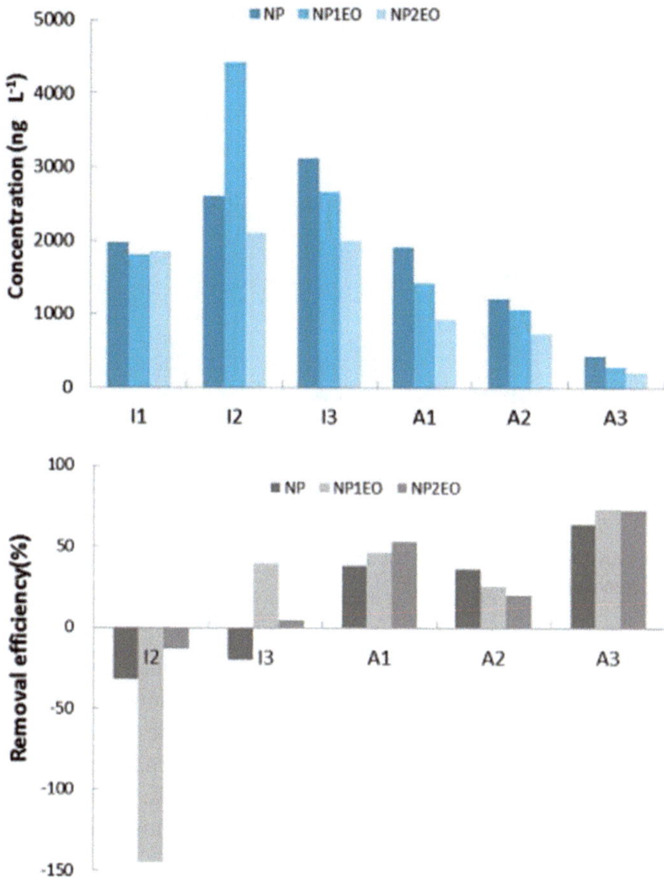

Fig. 5.33 Concentration and removal efficiency of NP, NP1EO and NP2EO for the A/O process (WWTP#10)

in large quantities and the concentrations of NP, NP1EO and NP2EO in the effluent of the two units were relatively close (Fig. 5.36c, d). The concentration values of the three short-chain NPnEO compounds were lower in spring and summer than in autumn and winter, and all the concentration values were relatively similar. The average concentrations of the three NPnEO in the effluent of the anaerobic and aerobic cells are shown in Table 5.16. One possible reason for this phenomenon is that the low temperatures in spring and winter had a negative impact on the microbial metabolic activity, resulting in lower concentrations of short-chain NPnEO compounds after the degradation of long-chain NPnEO compounds. Comparing the average NP, NP1EO, and NP2EO concentrations in the effluent of the two different units, only the NP concentration showed a similar trend. Actual measurements of NP, NP1EO, and NP2EO concentrations in the effluent of the secondary sedimentation tank indicated that seasonal variations in NP1EO and NP2EO levels were more pronounced than for

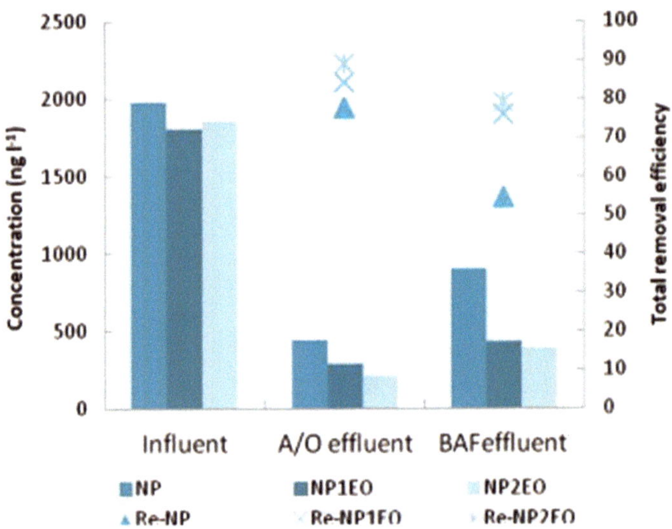

Fig. 5.34 Comparison of A/O(WWTP#10) and BAF(WWTP#14) processes for the total removal efficiency. Re: removal efficiency

NP. NP1EO and NP2EO concentrations were highest in winter, with mean values of 0.870 ng/mL and 0.920 ng/mL, respectively, followed by autumn, with NP1EO and NP2EO concentrations of 0.550 ng/mL. On the contrary, the concentrations of these two compounds measured in spring and summer were much lower than those in the cold season, 0.160–0.280 ng/mL and 0.390–0. 200 ng/mL, respectively. Seasonal variations of NP in the effluent of the secondary sedimentation tank were relatively insignificant.

The total removal efficiency of short-chain NPnEO is shown in Fig. 5.37 [38]. It can be seen that the removal of NPnEO varied slightly from season to season. The removal efficiencies of NP and NP2EO were relatively similar, and generally the removal efficiencies observed in spring were much lower than those in other seasons. However, the removal efficiency of NP1EO varied smoothly across seasons, with the best performance observed in spring. The average removal of NP across seasons ranged from 90 to 60%, with the highest in winter, followed by summer and autumn

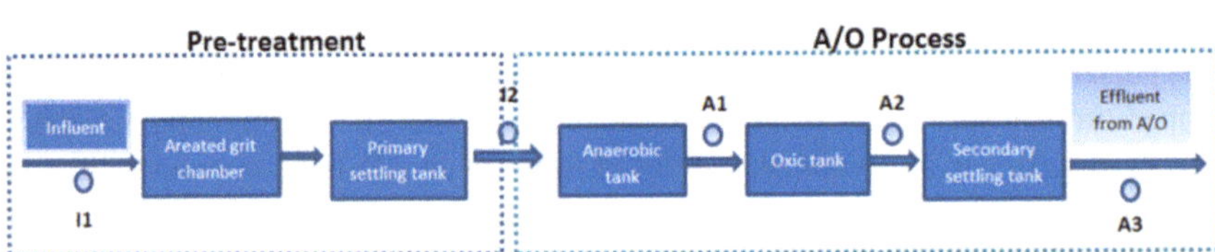

Fig. 5.35 Sampling points along process of WWTP#11. Sampling points: I1: influent to WWTP#11; I2: effluent from the primary settling tank; A1: effluent from the anaerobic tank (A/O); A2: effluent from the oxic tank (A/O); A3: effluent from the secondary settling tank

Table 5.16 Average concentrations of the three NPnEO in the effluent of the anaerobic and aerobic tanks

Parameter	Details			
WWTP#11				
Technology	A/O			
The corridors of A/O process	6			
The corridors of anoxic condition	2			
The corridors of oxic condition	4			
HRT (h)	10			
SRT (days)	19			
Wastewater source	Domestic			
Location	Harbin			
Temperature	Summer		Winter	
	20.2–28.8 °C		− 13.2 to − 24.8 °C	
Concentration of ΣNPnEO (ng/mL)	Spring	Summer	Autumn	Winter
	0.25–1.50	0.50–1.60	0.50–3.50	1.00–2.15

Data were taken from Gao et al. [38]

(both over 86%), and the lowest in spring. the average removal of NP2EO ranged from 90 to 70% across seasons in the order of: summer > autumn > winter > spring. The removal rate of NP1EO was slightly greater than that of the other two compounds, and the average removal rate of NP1EO ranged between 80 and 90%, decreasing smoothly in the order of spring, summer, autumn and winter. These results suggest that the differences in removal efficiency of NP, NP1EO and NP2EO may be related to their respective biodegradation rates. The concentrations of short-chain NPnEO were highest in winter. The concentrations decreased sequentially from winter to autumn and then to summer, with the lowest concentrations in spring. Although the removal efficiency of NP, NP1EO and NP2EO showed different trends in different seasons, all showed better removal performance in summer. Higher temperature accelerates microbial degradation.

The removal efficiency of NPnEO by different wastewater treatment units in WWTP#11 is shown in Fig. 5.38 [38]. In the A/O process, anaerobic treatment plays a major role in the removal of short-chain NPnEO, indicating that the main pathway for the removal of NP, NP1EO and NP2EO in the water column is anaerobic digestion. Alternatively, NP and NP1EO accumulated in the main sedimentation tank after treatment. These results help to elucidate the relationship between seasonal variations and the effectiveness of NP, NP1EO and NP2EO elimination during A/O treatment.

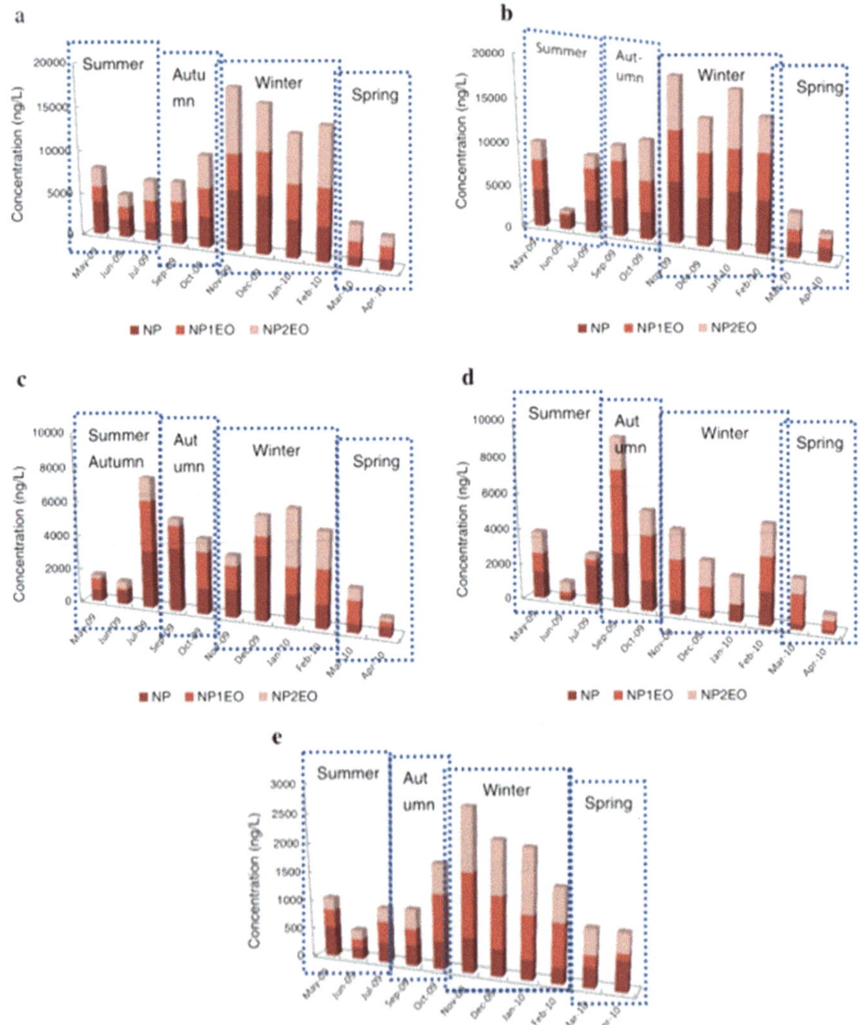

Fig. 5.36 NP, NP1EO and NP2EO concentrations in different sampling points during the whole year (WWTP#11). **a** I1; **b** I2; **c** Al; **d** A2; **e** A3

5.6.2.2 Anaerobic/Anoxic/Oxic Process

Lian et al. [73] investigate behaviours of nonylphenol polyethoxylates and their metabolites in different wastewater treatment processes. The NPnEO concentrations of the studied WWTP#12 and WWTP#13 are shown in Table 5.17 [73]. WWTP#12 was investigated for the removal of nonylphenols in the sedimentation tank, and the sand adsorbed NP at a concentration of 0.029 mmol/kg. Considering the sand discharge, it can be calculated that the NP released through the sand discharge is

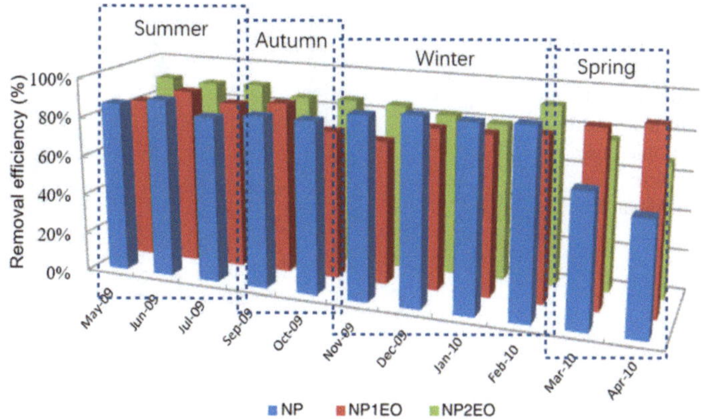

Fig. 5.37 Total removal efficiencies of NP, NP1EO and NP2EO during different seasons (WWTP#11)

0.3 mmol/d, which is 0.004% of the NP input to the WWTP. This percentage is too small to be considered and other NPnEO metabolites eliminated by sand discharge are also neglected because they are less likely to be adsorbed to sand particles than NP. Therefore, the sand chamber was considered to have no effect on the removal of NPs. the ratios of biodegradation to sludge adsorption in WWTP#12 and WWTP#13 were 6.9 and 3.0, respectively, and biodegradation played a major role in the removal of NPs. The reason for the ratio of WWTP#12 being greater than that of WWTP#13 may be due to the fact that the WWTP#13 influent was first settled in the primary settling tank and a large amount of nonylphenols were removed by adsorption on the primary sludge before entering the biological treatment unit.

The response of WWTPs under different weather conditions was investigated [73]. The results showed that since this WWTP is open, rainfall diluted the concentration of nonylphenols in the influent water by about 60% of that in dry weather, but had no significant effect on the removal rate of nonylphenols. This may be the result of the combined effect of rainfall dilution, microorganisms and long hydraulic retention time (HRT) of the wastewater treatment system.

5.6.2.3 Biological Aerated Filter

Gao et al. [37, 42] investigated the comparison of NPs removal efficiency between Biological Aerated Filter (BAF) process and A/O process, WWTP#14 of the wastewater treatment process and sampling points are shown in Fig. 5.32. Table 5.18 shows the basic operating and design parameters of WWTP#14. The concentrations of NP, NP1EO and NP2EO detected in wastewater samples collected from different locations of the BAF process are shown in Fig. 5.39. The pretreatment measures of the BAF process were the same as those of the A/O process, so the concentrations of NP, NP1EO and NP2EO in the effluent from the aeration and abrasion chamber

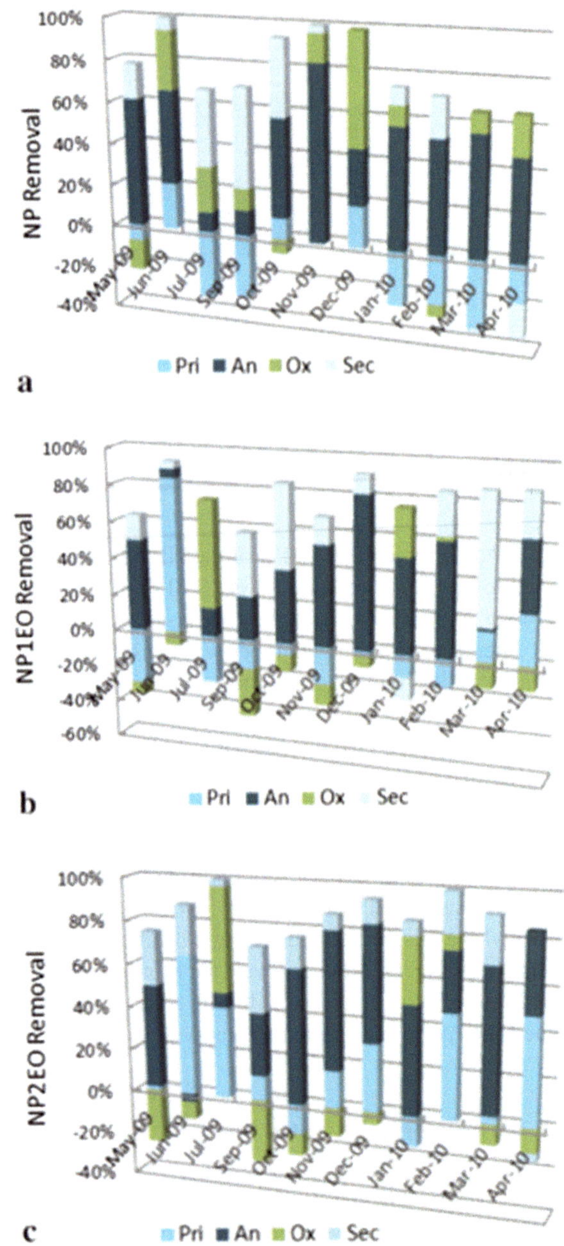

Fig. 5.38 Contribution ratio of the individual treated units of the A/O process (WWTP#11) during different seasons. Pri: primary settling tank; An: anaerobic tank; Ox: oxic tank; Sec: secondary settling tank

Table 5.17 Concentrations (ng/mL) of NPs in WWTP#12 and WWTP#13

Sampling time	T (°C)	Samples	Concentration					
			NP	NPnEOs (n = 1.2)	NPnEOs (n ≥ 3)	NPECs	Sum	Removal percentage (%)
WWTP#12								
June, 14th	25.7	Inf.	0.029	0.067	0.009	0.021	0.126	88.0
		Eff.	nd	0.004	0.001	0.010	0.015	
July, 8th	23.8	Inf.	0.035	0.044	0.005	0.031	0.115	75.7
		Eff.	0.001	0.008	0.001	0.018	0.028	
July, 24th	25.3	Inf.	0.032	0.053	0.007	0.038	0.130	80.8
		Eff.	0.002	0.002	0.001	0.020	0.025	
Sep, 11th	22.9	Inf.	0.042	0.052	0.008	0.043	0.145	81.4
		Eff.	0.002	0.001	0.001	0.023	0.027	
July, 11th*	20.5	Inf.	0.019	0.032	0.004	0.021	0.076	82.9
		Eff.	0.001	0.001	0.001	0.010	0.013	
July, 14th*	24.1	Inf.	0.021	0.036	0.005	0.014	0.076	85.5
		Eff.	0.002	0.001	0.001	0.007	0.011	
July, 18th*	25.3	Inf.	0.040	0.055	0.008	0.018	0.121	86.0
		Eff.	0.002	0.004	0.001	0.010	0.017	
Average ± S.D. (in dry weather)		Inf.	0.036 ± 0.005	0.054 ± 0.008	0.007 ± 0.001	0.030 ± 0.011	0.127 ± 0.011	82.4 ± 4.8
		Eff.	0.001 ± 0.001	0.004 ± 0.002	0.001 ± 0.000	0.016 ± 0.005	0.022 ± 0.005	

(continued)

Table 5.17 (continued)

Sampling time	T (°C)	Samples	Concentration					
			NP	NPnEOs (n = 1.2)	NPnEOs (n ≥ 3)	NPECs	Sum	Removal percentage (%)
WWTP#13								
Oct, 26th	20	Inf.	0.038	0.067	0.110	0.132	0.347	90.8
		Eff.	nd	nd	0.001	0.031	0.032	
Dec, 17th	14.8	Inf.	0.085	0.106	0.016	0.060	0.267	90.3
		Eff.	nd	nd	0.002	0.024	0.026	
Average ± S.D		Inf.	0.062 ± 0.033	0.086 ± 0.027	0.063 ± 0067	0.096 ± 0.051	0.308 ± 0.057	90.6 ± 3.5
		Eff.	nd	nd	0.002 ± 0.001	0.028 ± 0.005	0.030 ± 0.004	

Data were taken from Lian et al. [73]
nd not detected, meaning the concentration is below limit of detection (LOD)
*The samplings times were in 2008, while others in 2007

Table 5.18 Operational and design parameters of investigated WWTP#14

Parameter	Details
WWTP#14	
Technology	BAF
Average flow rate (m^3/d)	40,000
Treatment process	Scr → Agc → Pst → Est → BAF
Wastewater source	Domestic + Industrial
Location	Harbin

Data were taken from Gao et al. [37, 42]
Scr Screening, *Agc* Aerated grit chamber, *Pst* Primary settling tank, *Est* Enhanced settling tank, *BAF* Biological Aerated Filter

and primary settling tank were the same. After BAF treatment, NP, NP1EO and NP2EO were eliminated, with the concentrations of NP being about 0.9 ng/mL and NP1EO and NP2EO being about 0.4 ng/mL. The BAF process showed relatively low removal rates of 55%, 76% and 79% for NP, NP1EO and NP2EO, respectively. This may be related to the shorter HRT in the pool, which reduces the time required for microorganisms to synthesize the relevant enzymes for biodegradation of long-chain NPnEO. In addition, the high intensity of aeration coupled with the fast upflow rate limited the adsorption of filters and biofilms. Due to the presence of aeration in the unit, the biofilter of the BAF process had highly active microorganisms, which increased the degradation of contaminants. However, the degradation of NP1EO and NP2EO occurred only in some aeration cells. Therefore, the removal efficiency of target compounds in individual units is not comparable to other units.

Gao et al. [39] studied the seasonal changes of NPnEOs in Sewage Treatment Plant with BAF Process. The WWTP#15 wastewater treatment process and sampling points are shown in Fig. 5.40. According to the climatic characteristics of Harbin city and the results of previous studies, the samples collected in November, December, January and February are called "winter" samples, the samples collected in March and April are called "spring" samples, the samples collected in May, June and July are called "summer" samples, and the samples collected in September are called "summer" samples. No samples were collected in August due to WWTP maintenance. The NP0 ~ 2EO concentrations used to represent each season were determined by averaging all samples collected in the relevant months of each season.

Figure 5.41 shows the concentrations of NP, NP1EO and NP2EO at each sampling site in WWTP#15 [39]. Figure 5.41a shows the total influent concentrations of NP, NP1EO and NP2EO. From Fig. 5.41a, it can be seen that there are obvious changes in the concentrations of NP, NP1EO and NP2EO in different seasons. The total concentrations of NP, NP1EO and NP2EO observed in winter were 16 ng/mL, which were much higher than those in other seasons. There was less variation in autumn, summer and spring, with concentrations decreasing to 8.9, 6.7 and 4.1 ng/mL, respectively. Seasonal differences may be due to the different water volumes associated with winter washing events, which resulted in increased concentrations of NPnEO detergent in the wastewater. In spring, the raw WWTP#15 effluent was diluted by melted water,

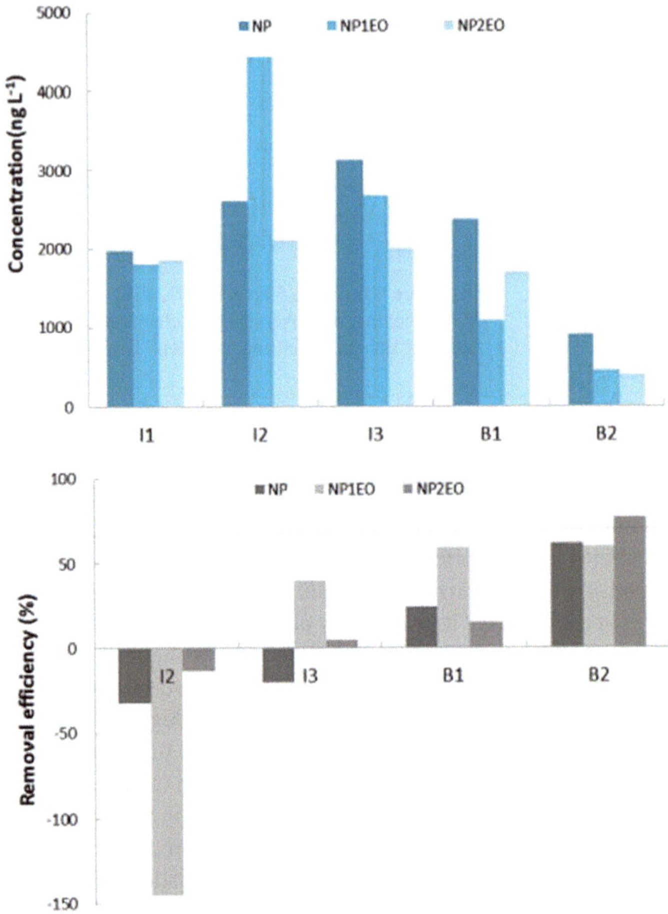

Fig. 5.39 Concentration and removal efficiency of NP, NP1EO and NP2EO for the BAF process (WWTP#14)

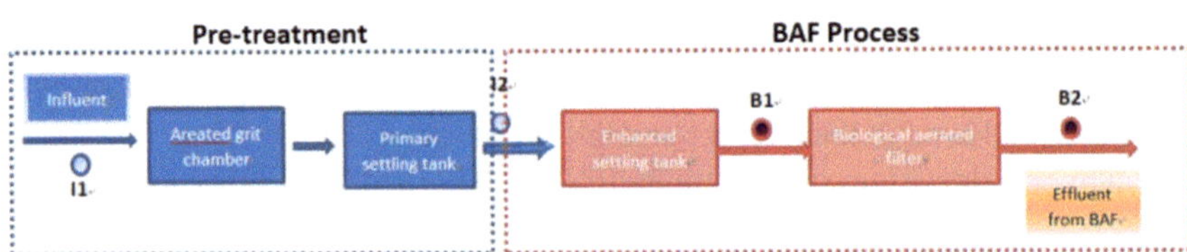

Fig. 5.40 Sampling points along the process flow of WWTP#15. Sampling points: Il: influent to WWTP#15, I2: effluent from the primary settling tank, B1: effluent from the enhanced settling tank, B2: effluent from the biological aerated filter (BAF)

so that the winter influent NP, NP1EO and NP2EO concentrations were more than three times higher than in spring. The increased use of water and NPnEO detergents in summer and autumn may be a factor in the accumulation of NP, NP1EO, and NP2EO in the influent. However, a clear relationship between seasonal characteristics (e.g., temperature) and concentrations of NP, NP1EO, and NP2EO compounds in raw wastewater needs to be further investigated. There were significant differences in the concentrations of NPnEO compounds between treatment units and seasons. As shown in Fig. 5.41b, the distribution trends of NP, NP1EO and NP2EO concentrations in the effluent of the main sedimentation tank (I2) in different seasons were similar to those of the total influent. The total concentrations of NP, NP1EO and NP2EO were higher than the total concentrations in summer, autumn and winter. NP compounds were found to accumulate after passing through the primary sedimentation tanks, with the largest increase observed in spring, reaching approximately 50% compared to the influent. A significant increase of 23% and 21% in NP concentrations was also measured in the autumn and summer, respectively. NP1EO concentrations increased by 40%, 25% and 19% in the autumn, summer and winter, respectively. NP2EO concentrations decreased by 19%, 35% and 20% in the spring, autumn and winter, respectively. Among the three short-chain NPnEOs compounds, NP and NP1EO concentrations increased more during the transitional periods of spring, summer, autumn and winter. In summary, the seasonal differences in NP, NP1EO and NP2EO concentrations in the primary sedimentation tank effluent were relative to the total influent concentration. This phenomenon may be the result of physical sedimentation associated with the primary sedimentation tank, which contains few microorganisms and therefore little biodegradation occurs in this unit. The effluent was pretreated in an intensified sedimentation tank before entering the BAF reactor. The concentrations of NP, NP1EO and NP2EO in the effluent of the intensive sedimentation tank are shown in Fig. 5.41c. Compared with the primary settling tank, the NP, NP1EO and NP2EO concentrations decreased significantly in winter and summer, by about 47% and 40%, respectively. In addition, there was an increase of 33% and 17% in autumn and spring, respectively. the distribution of NP, NP1EO and NP2EO in different seasons was generally consistent with that of total NP, NP1EO and NP2EO, and the concentrations in the colder season (autumn and winter) were higher than those in the warmer season (spring and summer). The results indicate that the distribution of NP, NP1EO and NP2EO compounds correlated with seasonal variations; whereas the removal behavior of NP, NP1EO and NP2EO in the enhanced sedimentation tank did not follow this seasonal dependence. A preliminary interpretation is that the presence of flocculants and alkaline conditions have an effect on the conversion of NP, NP1EO, and NP2EO. After the BAF process treatment (Fig. 5.41d), NP, NP1EO and NP2EO compounds were effectively removed, especially in autumn. The total concentrations of NP, NP1EO and NP2EO measured in different seasons were lower than 4 ng/mL. The differences in NP, NP1EO and NP2EO concentrations were also reduced in the cold and warm seasons compared to the intensive settling tank, indicating that the ambient temperature had little effect on the distribution and removal of NP, NP1EO and NP2EO in the BAF unit. It can be inferred that the removal of

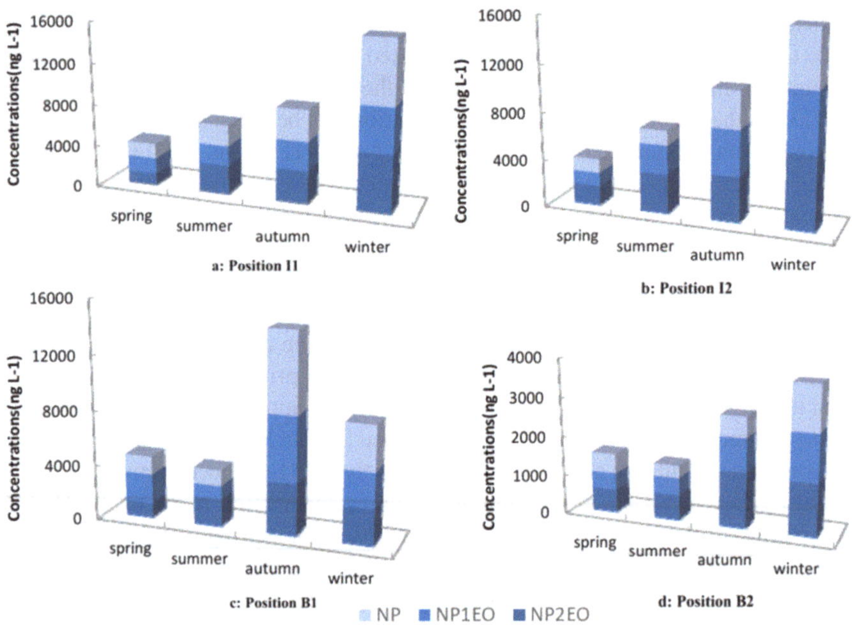

Fig. 5.41 Seasonal variations of NP, NP1EO, and NP2EO concentrations in water (WWTP#15)

NP, NP1EO and NP2EO in the BAF reactor is related to the uptake by the biofilter/biofilm rather than to the degradation by microorganisms.

The total removal efficiencies of the three short-chain NPnEOs compounds in different seasons are shown in Fig. 5.42 [39]. The BAF process was effective in removing NP, NP1EO and NP2EO, with an average removal rate of 78% in summer and 76% in winter. In contrast, the average removal rates of NP, NP1EO and NP2EO were 69% and 60% in autumn and spring, respectively. The highest removal rate for NP2EO was about 80% in summer, autumn and winter. The seasonal variation in the reduction of NP1EO concentration was not significant from the results obtained. The NP1EO removal rate observed in summer was about 76%, while the NP1EO removal rates observed in autumn, spring and winter were all between 69 and 72%. NP showed relatively poor removal performance. The removal rates of NP were above 75% in both summer and winter, while the removal rates in spring and autumn were about 20%.

The contribution of individual treatment units to NP removal is shown in Fig. 5.43a [39]. The enhanced sedimentation tank plays an important role in the removal of NP from WWTP#15; however, its contribution to NP removal is minimal in the fall. In the autumn, the BAF process was effective in removing NP. In the summer, the contribution of the BAF process was almost equal to that of the enhanced settling tank. The contribution of the primary settling tank was negative compared to the enhanced settling tank and BAF, and its performance was in this order: spring > autumn > summer > winter. As shown in Fig. 5.43b, the contribution of each treatment unit

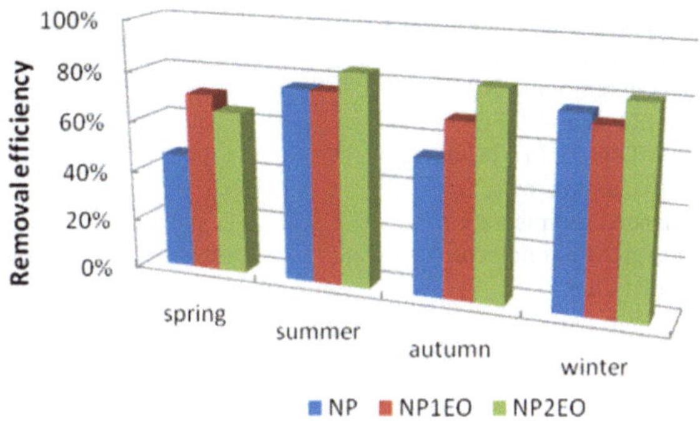

Fig. 5.42 Seasonal variations in the total removal efficiency of NP, NP1EO, and NP2EO compounds (WWTP#15)

to NP1EO removal varied by season. The contribution of enhanced sedimentation tanks to NP1EO removal was more significant in summer and winter. However, accumulation of NP1EO in the enhanced sedimentation tank occurred in spring and autumn when the BAF unit played a major role in NP1EO removal. As can be seen in Fig. 5.43c, the primary sedimentation tanks showed significant effects in the removal of NP2EO, except in autumn. The contribution of the enhanced sedimentation tanks to NP2EO removal had a similar seasonal distribution to that of NP1EO. In summary, there were seasonal differences in the removal of NP, NP1EO and NP2EO compounds by the treatment units, with the BAF units typically performing best in the fall and relatively poorly in the winter.

During the BAF of WWTP#15, the concentrations and distributions of NP, NP1EO and NP2EO metabolites had significant seasonal variations. The concentrations of short-chain NPnEO were significantly higher in winter compared to any other season in the influent, and exhibited a decreasing trend from autumn to summer to spring. The concentrations of NP, NP1EO and NP2EO compounds in the primary settling tank effluent had a certain seasonal distribution pattern with the total influent concentration. The seasonal distribution of NP, NP1EO and NP2EO changed after passing through the enhanced settling tank. Higher concentrations were measured in autumn and the lowest concentrations in summer. The biological treatment process of NP, NP1EO, and NP2EO in the BAF reactor was minimally influenced by ambient temperature. This suggests that the removal of NP, NP1EO, and NP2EO in the BAF reactor primarily occurred through uptake by the biofilter/biofilm, rather than microbial degradation. Although the removal rates of NP, NP1EO and NP2EO showed different trends in different seasons, the removal rates were better in summer and winter. The highest removal rate of NP2EO and the relatively poor removal rate of NP indicated that the ethoxylation unit might have contributed to the instability of NPnEO. In the BAF process, the BAF unit played a major role in the elimination of short-chain NPnEO compounds; however, the removal efficiency did not increase with increasing temperature. Intensive sedimentation tanks also played an important role in the removal of NP and NP1EO, while the treatment of the main sedimentation tank resulted in the accumulation rather than the removal of NP and NP1EO. These results suggest that the presence of flocculants and alkaline conditions may have

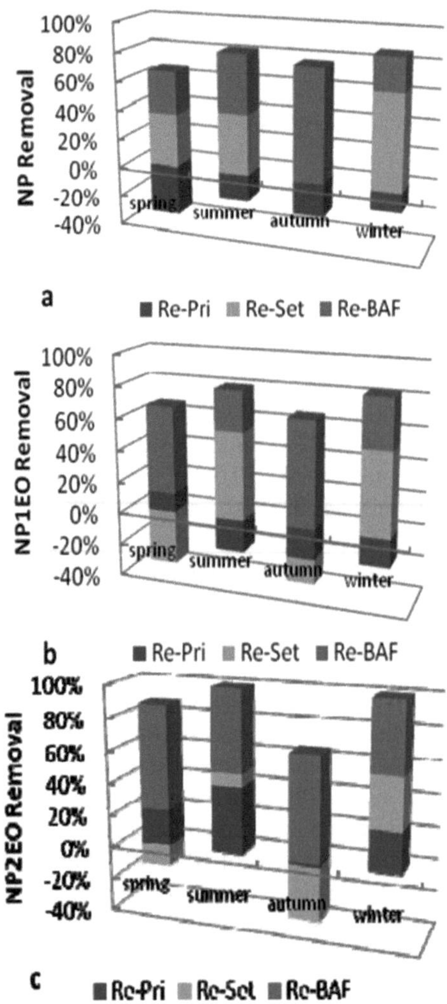

Fig. 5.43 Contribution rates of the individual treatment units in removing NP, NP1EO, and NP2EO (WWTP#15). Pri: primary sedimentation tank; Set: enhanced sedimentation tanks; BAF: biological aerated filter

some effect on the conversion of NP and NP1EO. These results help to elucidate the relationship between seasonal changes and the removal of NP, NP1EO and NP2EO compounds during BAF treatment.

5.6.3 *Antibiotics*

The removal of 12 typical antibiotics (three CEs, three MCs, three FQs, and three HAs), including cefazolin (CFA), cefmetazole (CFM), cefotaxime (CTX), azithromycin [6], clarithromycin (CLA), roxithromycin (ROX), OFL, NOR, flumequine (FLU), sulfadiazine (SD), sulfapyridine (SMPD), and SMX in numerous treatment plants treated with CASS, A/A/O, A/O and CAST as bioreactors was investigated [120]. In general, the removal of antibiotics from wastewater treatment systems occurred mainly in the biological treatment stage. The removal rates of typical antibiotics varied widely, with an average of 30–80%. CEs and HAs antibiotics were mainly removed by biological treatment, while MCs and FQs antibiotics were removed by sludge adsorption. The removal of antibiotics was positively correlated with organic substrate concentration, and together with high SRT, low substrate loading may lead to increased biodiversity of active biomass, resulting in a wider range of degradation pathways [54]. Four different treatment processes, Cyclic Activated Sludge System (CASS), A/A/O, A/O, and CAST, removed antibiotics without significant differences, indicating that the conventional wastewater treatment processes were not sufficient to remove antibiotics. In the desilting stage (sedimentation tank) of the four treatment processes, CASS, A/A/O, A/O, and CAST, only a small amount of the target antibiotics was removed from the influent water, probably because the desilting stage is based on gravity separation, which only precipitates and removes a relatively high percentage of inorganic particles from the wastewater. This stage has no retention time and therefore has less effect on the target antibiotics [120]. Yan et al. [128] studied the removal effect of antibiotics in four WWTPs, and the results showed that negative removal of antibiotics occurred in these WWTPs. There may be several reasons for this phenomenon. There are several issues related to sampling in the context of antibiotic concentration analysis. Firstly, diurnal variation can pose a challenge as antibiotic concentrations may fluctuate throughout the day. Secondly, some antibiotics could potentially be reconverted during the wastewater treatment process, leading to inaccuracies in measured concentrations. Lastly, changes in the adsorption of antibiotics on particles could also contribute to variations in observed concentrations.

The three cephalosporin antibiotics in the study, CFA, CFM, and CTX, had only 0.2–5.4% removal in the sedimentation and primary sedimentation stages in the four WWTPs [120]. And the highest removal rates of the three antibiotics were found in the biological treatment stage, ranging from 54.4 to 90%. The results of this study showed that the biological treatment stage was the most effective stage for the removal of cephalosporin antibiotics. However, it has also been suggested that the hydrolysis of cephalosporin antibiotics in water is mainly due to abiotic factors [62]. Therefore, the removal mechanism of cephalosporin antibiotics needs to be further investigated.

Furthermore, SD, SMPD, and SMX had the highest removal efficiencies of 18.4–74.1% in the biological treatment stages of the four WWTPs studied [120]. There was no absolute difference in the removal of sulfonamide antibiotics among the four different treatment processes of CASS, A/A/O, A/O, and CAST in the study. In general, the removal rate of AO was higher than the other three treatment processes, which may be due to the relatively long hydraulic retention time of the AO process. The residual concentration of sulfonamide-type antibiotics in the sludge was extremely low, indicating that the removal of HAs from the WWTP relied on biodegradation rather than sludge adsorption. Several studies confirmed that HAs are highly water soluble and have negligible adsorption to sludge biomass [77]. Additionally, Göbel et al. [54] proposed that the higher concentrations of SD and SMX in the effluent, compared to the influent, could indicate the conversion of the metabolites of SD and SMX back into their parent compounds.

For the three fluoroquinolone antibiotics of OFL, NOR, and FLU, each treatment unit had some removal effect. Although the biological treatment stage was dominant, the removal rates were significantly higher in the primary and secondary sedimentation stages, ranging from 3.6% to 11.8% and 10% to 42.9%, respectively. Due to the relatively high solid-water partition coefficients of NOR and OFL, they were more easily adsorbed on the surface of solid particles, resulting in lower concentrations in the deposition stage. On the other hand, the high fluoroquinolone antibiotic residues in the dewatered sludge indicated that activated sludge adsorption was the main removal mechanism of fluoroquinolone antibiotics. The adsorption process is reversible, and the adsorption performance is optimal when the concentration of fluoroquinolone antibiotics is 100–300 ng/mL. The adsorption capacity depends mainly on the amphiphiles of fluoroquinolone antibiotics. There was no significant difference in the removal of these antibiotics between the four treatment processes CASS, A/A/O, A/O and CAST, and the level of antibiotics in the sludge was generally high. This indicates that these antibiotics were mainly removed by sludge adsorption, which was related to the operating temperature. Yan et al. [128] showed that FQs accounted for most of the antibiotics removed by adsorption.

The removal rate of macrolide antibiotics was higher in the primary and secondary sedimentation stages, while the removal rate was lower in the biological treatment stage, and may even be negative [120]. The overall removal rate was lower. The removal rate of A/O treatment process was slightly higher, which may be related to the difference of biological treatment units. The longer hydraulic retention time of A/O treatment process resulted in higher removal efficiency of FQs and macrolide antibiotics, which were removed by adsorption. The large variation in removal efficiency of different antibiotics in different WWTPs may be due to imperfect analytical methods for antibiotics in wastewater and sludge. The quality of wastewater also varies from region to region. Even in the same treatment process, the removal of target antibiotics varies from one treatment unit to another due to differences in WWTP management. The study showed that macrolide antibiotics accounted for most of the antibiotics removed by adsorption [128]. MCs tend to be adsorbed by sludge, however, conjugated metabolites and adsorption behavior can be easily altered by changing physicochemical parameters during the treatment process, leading to secondary release of MCs from sludge into water or uncoupling [77].

5.6.3.1 Cyclic Activated Sludge Technology (CAST)

The removal of 12 typical antibiotics in a treatment plant with CAST as the biological tank reaction treatment was investigated [120], the basic design parameters of

WWTP#16 are shown in Table 5.19. The wastewater treatment process and the distribution of sampling points for WWTP#16 are shown in Fig. 5.44.

Figure 5.45 shows the concentration of antibiotics in the effluent of each wastewater treatment unit in WWTP#16. It can be seen in the Fig. 5.45. That the ROX in the influent water of WWTP#16 is extremely high, over 1.5 ng/mL, and the average antibiotic concentration in the influent water of the WWTP is 1.6 ng/mL. This was due

Table 5.19 Operational and design parameters of investigated WWTP#16 and WWTP#17

Parameter	Details
WWTP#16	
Technology	CAST
HRT (h)	6
SRT (days)	15
Treatment process	Scr → Ggc → CAST → ClD
Wastewater source	Domestic + Industrial
Location	Harbin
WWTP#17	
Technology	CAST
Average flow rate (m^3/d)	40,000
HRT (h)	3
SRT (days)	15
Treatment process	Scr → Ggc → CAST → Sst → ClD
Service area (km^2)	36
Population served	110,000
Location	Chongqing

Data were taken from Wang et al. [120] and Yan et al. [128]
Scr Screening, *Ggc* Grease and grit chamber, *CAST* Cyclic Activated Sludge Technology, *Sst* Second settling tank, *ClD* ClO_2 disinfection

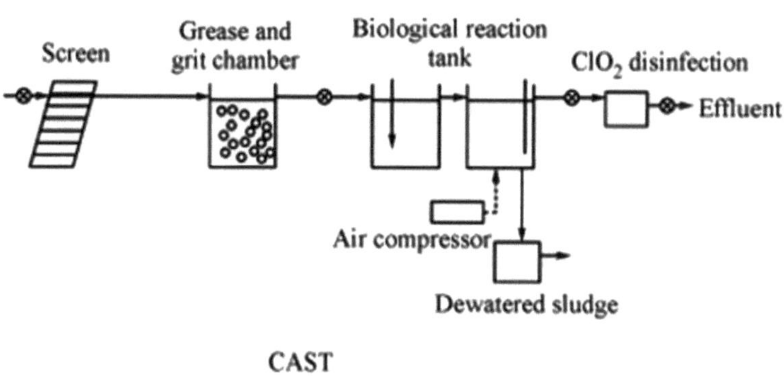

Fig. 5.44 Schematic diagrams of WWTP#16

to the fact that this area of WWTP is a populated with livestock, and the main pollutants were MCs, FQs, and sulfonamide antibiotics. Only a small amount of antibiotics was removed before biological treatment, and most of the antibiotic removal occurred in the biological treatment stage. Some ROX was also removed after the biological treatment stage, but the concentration of ROX in the final effluent was still very high (> 0.700 ng/mL). It indicates that the basic CAST process cannot meet the requirement of reducing antibiotic contamination in WWTPs.

The removal effect of 10 antibiotics in WWTP#17 was investigated [128]. The basic design parameters of WWTP#17 are shown in Table 5.19. Table 5.20 shows the antibiotic concentrations in the influent and effluent water in WWTP#17. The results showed that the effluent concentrations of SD and SMX in WWTP#17 were higher than the influent concentrations, and the remaining eight antibiotics were better removal rate, especially NOR. Concentration of SD and erythromycin [79] increased in the disinfection link, but an overall decreasing trend in WWTP#17.

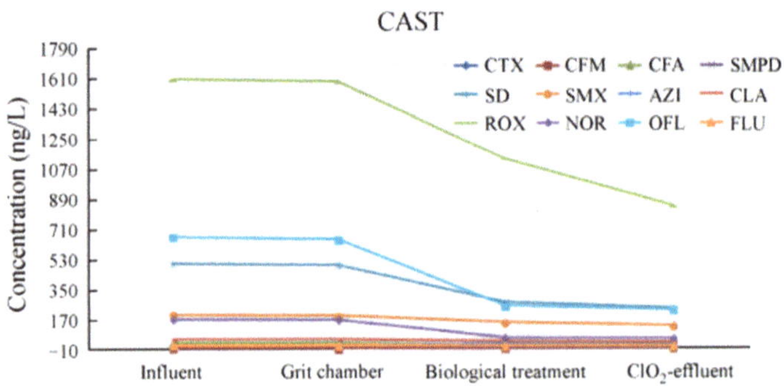

Fig. 5.45 Concentrations of antibiotics in effluents from different treatment units in the WWTP#16

Table 5.20 Concentrations (ng/mL) of antibiotics in WWTP#17

		Influent after screen (ng/mL)	Secondary effluent (ng/mL)	Disinfection effluent (ng/mL)
Sulfonamide antibiotics	SD	0.0229 ± 0.0047	0.0215 ± 0.0085	0.0223 ± 0.0054
	SMXZ	0.2038 ± 0.0219	0.2141 ± 0.0397	0.2166 ± 0.0287
	TMP	0.0449 ± 0.0096	0.0408 ± 0.0382	0.0332 ± 0.0209
	OFL	0.2430 ± 0.0199	0.2702 ± 0.0417	0.1599 ± 0.0471
Quinolone antibiotics	NOR	0.1916 ± 0.0401	0.1109 ± 0.0225	0.0575 ± 0.0136
	MOX	< LOQ	< LOQ	< LOQ
	ERY	0.2084 ± 0.0427	0.1218 ± 0.0024	0.1243 ± 0.0040
Macrolide antibiotics	ROX	0.3011 ± 0.0565	0.2396 ± 0.0305	0.2253 ± 0.0355
	AZIM	0.2974 ± 0.0301	0.2686 ± 0.0339	0.2432 ± 0.0326

Data were taken from Yan et al. [128]

5.6.3.2 Cyclic Activated Sludge System (CASS)

The removal of 12 typical antibiotics in a WWTP#18 with CASS as the biological tank reaction treatment was studied [120]. Table 5.21 shows the basic operating and design parameters of WWTP#18. The wastewater treatment process and the distribution of sampling points for WWTP#18 are shown in Fig. 5.46.

Figure 5.47 shows the concentration of antibiotics in the effluent of each wastewater treatment unit in WWTP#18. It can be seen in the Fig. that the ROX in the influent water of WWTP#18 was extremely high, over 1.5 ng/mL, and the average antibiotic concentration in the influent water of the WWTP is 1.6 ng/mL. This is due to the fact that this WWTP is a populated area with livestock, and the main pollutants are MCs, FQs, and sulfonamide antibiotics. Only a small amount of antibiotics was removed before biological treatment. Most of the antibiotic removal occurred in the biological treatment stage. Some ROX was also removed after the biological treatment stage, but the concentration of ROX in the final effluent was still very high (> 0.700 ng/

Table 5.21 Operational and design parameters of investigated WWTP#18

Parameter	Details
WWTP#18	
Technology	CASS
HRT (h)	6
SRT (days)	18
Treatment process	Scr → Ggc → Pst → CASS → Sst → UVD
Wastewater source	Domestic + Industrial
Location	Harbin

Data were taken from Wang et al. [120]
Scr Screening, *Ggc* Grease and grit chamber, *Pst* Primary settling tank, *CASS* Cyclic Activated Sludge System, *Sst* Second settling tank, *UVD* UV disinfection

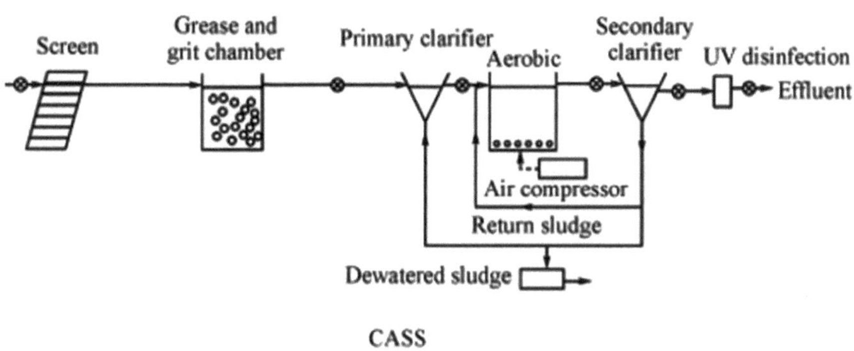

Fig. 5.46 Schematic diagrams of WWTP#18

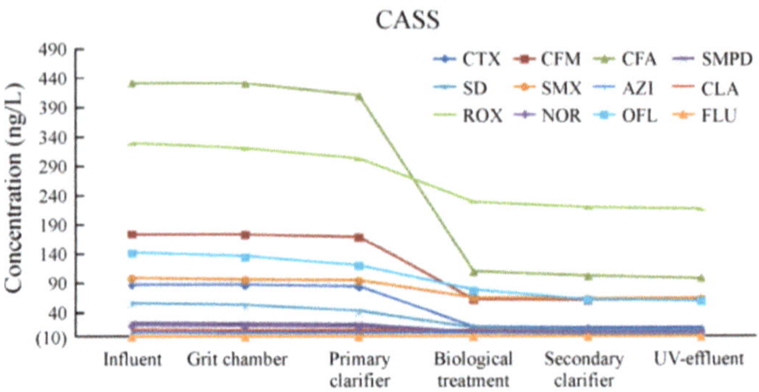

Fig. 5.47 Concentrations of antibiotics in effluents from different treatment units in the WWTP#18

mL). This indicates that the basic CASS process cannot meet the requirements of the WWTP to reduce antibiotic contamination.

5.6.3.3 Anaerobic/Oxic Process

Wang et al. [120] studied the removal of 12 typical antibiotics in numerous treatment plants treated with AO. Table 5.22 shows the basic operating and design parameters of WWTP#19. The wastewater treatment process flow and distribution of sampling points for WWTP#19 are shown in Fig. 5.48.

The effluent concentrations in each wastewater treatment unit in WWTP#19 are shown in Fig. 5.49. It can be seen that the concentration of antibiotics in WWTP#19 influent water is not very high, and the highest concentration of ROX < 0.120 ng/mL. This WWTP is located in a more densely populated area of Harbin city, so it can be seen that the population is not the main influencing factor of antibiotic content in the wastewater. The top three concentrations of antibiotics in the influent water were ROX, CFA, and OFL from high to low. The removal of antibiotics before biological treatment was not effective, and most of the antibiotic removal occurred in the biological treatment stage. In the biological treatment stage, the removal of ROX and CFA higher followed by OFL. The concentrations of ROX, CFA and OFL in the effluent were lower than 0.040 ng/mL, 0.020 ng/mL and 0.030 ng/mL, respectively.

Du et al. [29] studied the removal of TMP and SMX by WWTP#20 with > 50% and 29%, respectively, and the concentration of antibiotics in WWTP#20 is shown in Table 5.23.

5.6.3.4 Anaerobic/Anoxic/Oxic Process

The removal of 12 typical antibiotics in a treatment plant treated with A/A/O was studied [120]. The wastewater treatment process flow and the distribution of sampling

Table 5.22 Operational and design parameters of investigated WWTP#19 and WWTP#20

Parameter	Details
WWTP#19	
Technology	A/O
HRT (h)	16
SRT (days)	20
Treatment process	Scr → Ggc → Pst → A/O → Sst → UVD
Wastewater source	Domestic + Industrial
Location	Harbin
WWTP#20	
Technology	A/O
Average flow rate (m^3/d)	30,000
HRT (h)	17
Wastewater source	Domestic
Temperature (°C)	13–25
Location	Xuzhou

Data were taken from Wang et al. [120] and Du et al. [29]
Scr Screening, *Ggc* Grease and grit chamber, *Pst* Primary settling tank, *A/O* Anoxic/Oxic, *Sst* Second settling tank, *UVD* UV disinfection

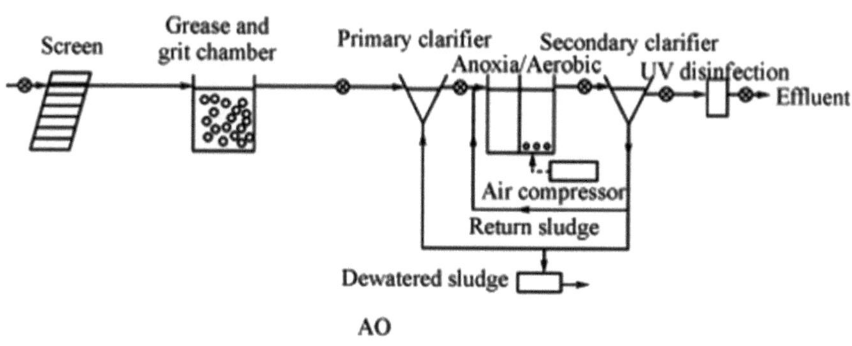

Fig. 5.48 Schematic diagrams of WWTP#19

points for WWTP#21 are shown in Fig. 5.50. Table 5.24 shows the basic operating and design parameters of WWTP#21.

The effluent concentration in each wastewater treatment unit in WWTP#21 is shown in Fig. 5.51. It can be seen in the Fig., in this WWTP influent that, except for FLU, 11 other antibiotics (CFA, CFM, CTX, AZI, CLA, ROX, OFL, NOR, SD, SMPD and SMX) were all present, and the total antibiotic concentration in the influent was 1.796 ng/mL. The highest concentration in the influent was SMX. The removal of antibiotics before biological treatment was not effective, and most of the antibiotic removal occurred in the biological treatment stage. The highest removal

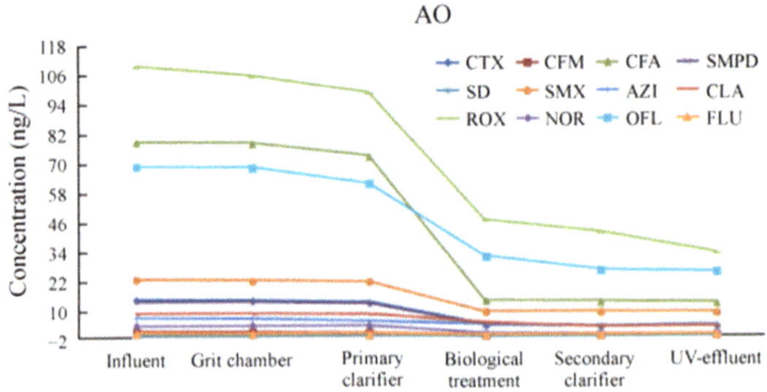

Fig. 5.49 Concentrations of antibiotics in effluents from different treatment units in the WWTP#19

Table 5.23 Concentrations (ng/mL) of antibiotics in WWTP#20

	Influent (ng/mL)*	Secondary effluent (ng/mL)*
NOR	0.271 ± 0.021	0.084 ± 0.013
TMP	0.368 ± 0.027	0.183 ± 0.012
ROX	0.093 ± 0.020	0.047 ± 0.008
SMX	0.612 ± 0.056	0.435 ± 0.039

Data were taken from Du et al. [29]
*Values are means ± standard deviation

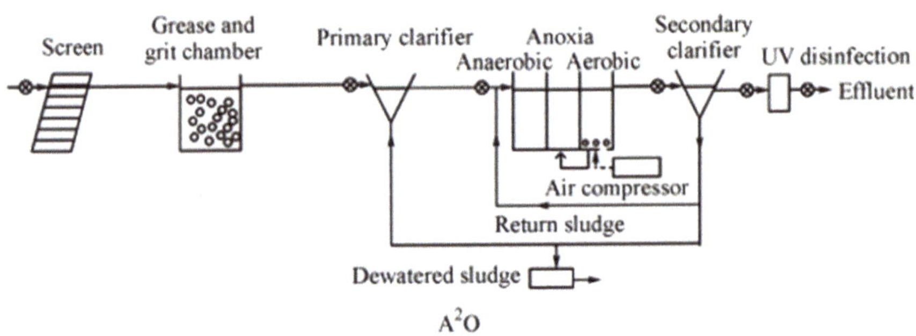

Fig. 5.50 Schematic diagrams of WWTP#21

efficiency in the biological treatment stage was CFA, but the concentration of ROX had increased in the biological treatment stage. The concentration of SMX in the effluent was still higher than 0.500 ng/mL, which still had a negative impact on the environment.

Table 5.24 shows the basic operating and design parameters of WWTP#22 and WWTP#23. The results of studying the effectiveness of two WWTPs for the removal of antibiotics with influent and effluent water concentrations are shown in Table 5.25 [128]. Which shows that the highest removal rates of SAs and QAs were observed in

Table 5.24 Operational and design parameters of investigated WWTP#21, WWTP#22 and WWTP#23

Parameter	Details
WWTP#21	
Technology	A/A/O
HRT (h)	6
SRT (days)	15–30
Treatment process	Scr → Ggc → Pst → A/A/O → Sst → UVD
Wastewater source	Domestic + Industrial
Location	Harbin
WWTP#22	
Technology	A/A/O
Average flow rate (m^3/d)	300,000
HRT (h)	8.9
SRT (days)	15
Treatment process	Scr → Ggc → Pst → A/A/O → Sst → ClD
Wastewater source	Domestic
Service area (km^2)	61.45
Population served	805,000
Location	Chongqing
WWTP#23	
Technology	A/A/O
Average flow rate (m^3/d)	600,000
HRT (h)	17
SRT (days)	11
Treatment process	Scr → Ggc → Pst → A/A/O → Sst → ClD
Wastewater source	Domestic
Service area (km^2)	125
Population served	1,540,000
Location	Chongqing
WWTP#24	
Technology	A/A/O
Average flow rate (m^3/d)	100,000
HRT (h)	12–15
Treatment process	Pst → A/A/O → Sst → SF
Wastewater source	Domestic + Industrial (8:2)

(continued)

Table 5.24 (continued)

Parameter	Details
Temperature (°C)	16–26
Location	Xuzhou

Data were taken from Wang et al. [120], Yan et al. [128] and Du et al. [29]
Scr Screening, *Ggc* Grease and grit chamber, *Pst* Primary settling tank, *A/A/O* Anaerobic/Anoxic/Oxic, *Sst* Second settling tank, *UVD* UV disinfection, *ClD* ClO_2 disinfection, *SF* Sand Filtration

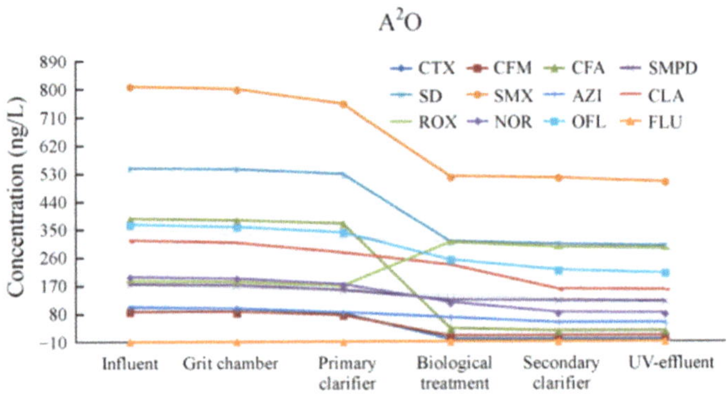

Fig. 5.51 Concentrations of antibiotics in effluents from different treatment units in the WWTP#21

WWTP#23. Also, the negative removal of CBZ, MTP and CA with higher effluent concentrations than influent was observed in WWTP#22 and WWTP#23 in this study.

The study analyzed the fate of antibiotics in A/A/O [29]. Table 5.24 shows the basic operating and design parameters of WWTP#24. The antibiotic concentrations in WWTP#24 are shown in Table 5.26. In the primary treatment, the removal of antibiotics was low, and only NOR had a high removal rate of up to 72%. This indicates that the removal of NOR is mainly through adsorption process. While in the primary treatment, the level of SMX was slightly increased. In the secondary biological treatment, the removal of different compounds ranged from − 21 to 77%. In comparison, the removal rates of NOR and ROX were lower, both at 38%. The removal rate of methicillin was relatively high at 47%. In contrast, the concentration of SMX was elevated. The removal rates of methicillin and SMX were reduced by 38% and 23%, respectively, in the tertiary treatment sand filter. For the other antibiotics NOR and ROX, their concentrations in the wastewater were not detected after tertiary treatment, so there is some uncertainty about their removal efficiency in tertiary treatment. Since the amount of suspended solids in the sand filtration tank was very small, the removal during sand filtration was mainly considered as biodegradation brought by biofilm on the surface of the sand particles.

Table 5.25 Concentrations (ng/mL) of antibiotics in WWTP#22 and WWTP#23

		WWTP#22				WWTP#23			
		Influent after screen (ng/mL)	Primary effluent (ng/mL)	Secondary effluent (ng/mL)	Disinfection effluent (ng/mL)	Influent after screen (ng/mL)	Primary effluent (ng/mL)	Secondary effluent (ng/mL)	Disinfection effluent (ng/mL)
Sulfonamide antibiotics	SD	0.1584 ± 0.0229	0.2161 ± 0.0648	0.1497 ± 0.0296	0.1445 ± 0.0289	0.2299 ± 0.0225	0.2097 ± 0.0149	0.1641 ± 0.0255	0.1550 ± 0.0256
	SM1	0.0167 ± 0.0048	0.0168 ± 0.0025	0.0088 ± 0.0075	0.0106 ± 0.0044	0.1502 ± 0.0201	0.1272 ± 0.0250	0.0398 ± 0.0076	0.0399 ± 0.0065
	SMX	0.8063 ± 0.1240	0.7209 ± 0.0999	0.5046 ± 0.0473	0.3440 ± 0.0178	2.9354 ± 0.3276	1.8063 ± 0.1652	1.1736 ± 0.0900	1.1479 ± 0.0651
	TMP	0.2270 ± 0.0388	0.2132 ± 0.0441	0.1532 ± 0.0276	0.1524 ± 0.0258	0.0774 ± 0.0227	0.0815 ± 0.0028	0.0473 ± 0.0117	0.0526 ± 0.0172
Quinolone antibiotics	OFL	0.5469 ± 0.0506	0.4193 ± 0.1113	0.2551 ± 0.0395	0.2207 ± 0.0356	0.3459 ± 0.0594	0.2726 ± 0.0361	0.1017 ± 0.0052	0.0579 ± 0.0184
	NOR	0.2484 ± 0.0307	0.2425 ± 0.0354	0.0692 ± 0.0073	0.0416 ± 0.0078	0.2030 ± 0.0594	0.1518 ± 0.0181	0.0296 ± 0.0056	0.0304 ± 0.0038
	MOX	0.0189 ± 0.0023	0.0215 ± 0.0062	0.0059 ± 0.0008	0.0060 ± 0.0005	0.0199 ± 0.0074	0.0187 ± 0.0010	0.0069 ± 0.0010	0.0066 ± 0.0009
Macrolide antibiotics	ERY	0.2762 ± 0.0634	0.2760 ± 0.0336	0.2030 ± 0.0452	0.2006 ± 0.0274	0.2542 ± 0.0154	0.2579 ± 0.0240	0.1454 ± 0.0148	0.1530 ± 0.0165
	ROX	0.5674 ± 0.0805	0.4796 ± 0.0435	0.4275 ± 0.0669	0.4391 ± 0.0453	0.4040 ± 0.0342	0.3857 ± 0.0448	0.3070 ± 0.0361	0.3475 ± 0.0354
	AZI	0.6090 ± 0.0466	0.6290 ± 0.0719	0.3942 ± 0.0506	0.3402 ± 0.0179	0.3625 ± 0.0217	0.2628 ± 0.0148	0.0886 ± 0.0253	0.0815 ± 0.0238

Data were taken from Yan et al. [128]
SM1 sulfamethazine

Table 5.26 Concentrations (ng/mL) of antibiotics in WWTP#24

	Influent* (ng/mL)	Primary effluent* (ng/mL)	Secondary effluent* (ng/mL)	Final effluent* (ng/mL)
NOR	0.023 ± 0.009	0.068 ± 0.012	0.042 ± 0.007	< 0.010
TRI	0.815 ± 0.058	0.640 ± 0.042	0.338 ± 0.027	0.210 ± 0.015
ROX	0.127 ± 0.021	0.105 ± 0.018	0.065 ± 0.013	< 0.020
SMX	0.370 ± 0.056	0.446 ± 0.035	0.510 ± 0.047	0.395 ± 0.026

Data were taken from Du et al. [29]
*Values are means ± standard deviation

5.6.3.5 Membrane Bioreactor

The removal effect of antibiotics in the MBR process was investigated. Table 5.27 shows the basic operating and design parameters of WWTP#25.

The antibiotic removal efficiency in WWTP#25 was shown in Table 5.28 [66]. The results showed that the removal of TC, 4-Epitetracycline, Doxycycline (DOX), NOR, CIP, and OFL were removed mainly by adsorption on the sludge. Trimethoprim (TRI), and CLA even showed negative removal. The previous study achieved a high removal efficiency for these two antibiotics, however, this may be attributed to a shorter solid retention time (SRT) of only 6–8 days in comparison to the previous study's SRT of 16–70 days. Therefore, the operating conditions have a great influence on the removal rate of antibiotics in MBR.

Table 5.27 Operational and design parameters of investigated WWTP#25

Parameter	Details
WWTP#25	
Technology	MBR
Average flow rate (m^3/d)	8800
HRT (h)	11
SRT (days)	6–8
Treatment process	Scr → Pst → MBR
Temperature (°C)	21
Population served	24,000
Location	Canada

Data were taken from Kim et al. [66]
Scr Screening, *Pst* Primary settling tank, *MBR* Membrane Bio-Reactor

Table 5.28 Removal efficiency of antibiotics in WWTP#25

Compounds	Total removal (%)	Fractions		
		Degradation/ transformation (%)	Sorption onto sludge (%)	Final effluent (%)
TC	97	0	98	2
4-ETC	95	0	98	2
DOX	90	0	97	3
NOR	90	0	98	2
CIP	89	0	98	2
AZI	78	49	20	31
SMX	66	65	1	34
OFL	56	0	86	14
ERY-H_2O	12	15	6	79
TRI	– 2	0	7	93
CLA	– 34	0	11	89

Data were taken from Kim et al. [66]

5.6.3.6 Upflow Anaerobic Sludge Bed

The removal effect of UASB as a pretreatment stage for 6-APA and amoxicillin (AMX) was investigated [19, 128]. 6-APA is recognized as one of the primary core structures or parent nuclei found in traditional antibiotics. It is commonly detected in wastewater originating from pharmaceutical production processes. Influent concentrations of 6-APA and AMX fluctuated between 0.143–0.315 ng/mL and 0.061–0.171 ng/mL. The removal rate of 6-APA ranged from 19 to 40% with an average removal rate of 26.3%, while the removal rate of AMX ranged from 13 to 47% with an average removal rate of 21.6%. Although the concentrations of 6-APA and AMX kept changing, the effluent concentrations of 6-APA and AMX remained stable at 0.160 ± 0.030 ng/mL and 0.075 ± 0.029 ng/mL, respectively. During the first stage of the study (day 91–day 150), the concentration of 6-APA in the effluent varied between 0.105 and 0.176 ng/mL, with an average value of 0.1434 ng/mL. The removal rates of 6-APA ranged from 19 to 37% during this period. Correspondingly, the concentrations of AMX in the effluent ranged from 0.059 to 0.093 ng/mL, with an average value of 0.0784 ng/mL. AMX The AMX removal rates ranged from 13 to 36%. In stage 2 (day 151–day 164), although the concentrations of 6-APA and AMX in the effluent and influent were higher, reaching 0.1785 ng/mL and 0.0856 ng/mL, respectively, the percentage reduction of 6-APA and AMX by UASB was higher than the other two stages, reaching 33% and 32%, respectively. It is worth mentioning that in the third stage (day 155–day 205), the concentrations of effluent 6-APA and AMX gradually decreased to 0.1303 ng/mL and 0.0647 ng/mL, respectively, indicating that UASB has potential resistance to shock loading of 6-APA and AMX.

5.7 Prospect

This section provides a comprehensive review of the classification, origin, environmental occurrence, migration routes to different environmental compartments, and various treatment technologies for EPs in WWTPs. However, it is evident from the literature that achieving significant removal of EPs using a single treatment technique is challenging. Integrated systems, on the other hand, show promise as an effective approach for mitigating this limitation. It should be noted that most of the proposed alternative treatment technologies are still in the laboratory-scale phase and lack comprehensive information on their feasibility for implementation in wastewater treatment plants. While conventional treatment processes significantly reduce the presence of commonly investigated environmental pollutants in effluents, some still persist. Moreover, certain EPs exhibit higher toxicity than their parent compounds, while a few even demonstrate negative removal efficiencies. To address these challenges, the generation of toxic oxidation byproducts during ozonation and their occurrence in both dissolved and particle phases necessitate biological reprocessing for effective remediation.

References

1. Acero JL, Benitez FJ, Leal AI, Real FJ, Teva F (2010) Membrane filtration technologies applied to municipal secondary effluents for potential reuse. J Hazard Mater 177(1):390–398. https://doi.org/10.1016/j.jhazmat.2009.12.045
2. Acero JL, Javier Benitez F, Real FJ, Teva F (2012) Coupling of adsorption, coagulation, and ultrafiltration processes for the removal of emerging contaminants in a secondary effluent. Chem Eng J 210:1–8. https://doi.org/10.1016/j.cej.2012.08.043
3. Adeniyi AA, Okedeyi OO, Yusuf KA (2011) Flame ionization gas chromatographic determination of phthalate esters in water surface sediments and fish species in the Ogun river catchments Ketu Lagos Nigeria. Environ Monit Assess 172(1–4):561–569. https://doi.org/10.1007/s10661-010-1354-2
4. Ahmad HA, Ahmad S, Cui Q, Wang Z, Wei H, Chen X et al (2022) The environmental distribution and removal of emerging pollutants, highlighting the importance of using microbes as a potential degrader: a review. Sci Total Environ 809:151926
5. Ahmed MB, Zhou JL, Ngo HH, Guo W, Thomaidis NS, Xu J (2017) Progress in the biological and chemical treatment technologies for emerging contaminant removal from wastewater: a critical review. J Hazard Mater 323:274–298. https://doi.org/10.1016/j.jhazmat.2016.04.045
6. Ahmed S, Mofijur M, Nuzhat S, Chowdhury AT, Rafa N, Uddin MA et al (2021) Recent developments in physical, biological, chemical, and hybrid treatment techniques for removing emerging contaminants from wastewater. J Hazard Mater 416:125912
7. Ahn C, Jeung EB (2023) Endocrine-disrupting chemicals and disease endpoints. Int J Mol Sci 24(6). https://doi.org/10.3390/ijms24065342
8. Al-Qodah Z, Al-Shannag M, Bani-Melhem K, Assirey E, Yahya MA, Al-Shawabkeh A (2018) Free radical-assisted electrocoagulation processes for wastewater treatment. Environ Chem Lett 16(3):695–714. https://doi.org/10.1007/s10311-018-0711-1
9. Alhefeiti MA, Athamneh K, Vijayan R, Ashraf SS (2021). Bioremediation of various aromatic and emerging pollutants by *Bacillus cereus* sp. isolated from petroleum sludge. Water Sci Technol 83(7):1535–1547. https://doi.org/10.2166/wst.2021.065

10. de Amorim KP, Romualdo LL, Andrade LS (2013) Electrochemical degradation of sulfamethoxazole and trimethoprim at boron-doped diamond electrode: performance, kinetics and reaction pathway. Sep Purif Technol 120:319–327. https://doi.org/10.1016/j.seppur.2013.10.010
11. Barnabé S, Beauchesne I, Cooper DG, Nicell JA (2007) Plasticizers and their degradation products in the process streams of a large urban physicochemical sewage treatment plant. Water Res. https://doi.org/10.1016/j.watres.2007.07.043
12. Bell KY, Wells MM, Traexler KA, Pellegrin ML, Morse A, Bandy J (2011) Emerging pollutants. Water Environ Res 83(10):1906–1984. https://doi.org/10.2175/106143011x13075599870298
13. Bendz D, Paxéus NA, Ginn TR, Loge FJ (2005) Occurrence and fate of pharmaceutically active compounds in the environment a case study: Höje River in Sweden. J Hazard Mater 122(3):195–204. https://doi.org/10.1016/j.jhazmat.2005.03.012
14. Bolong N, Ismail AF, Salim MR, Rana D, Matsuura T, Tabe-Mohammadi A (2010) Negatively charged polyethersulfone hollow fiber nanofiltration membrane for the removal of bisphenol A from wastewater. Sep Purif Technol 73(2):92–99. https://doi.org/10.1016/j.seppur.2010.01.001
15. Bracamontes Ruelas A, Ordaz-Diaz L, Bailon-Salas A, Rios-Saucedo J, Reyes-Vidal Y, Reynoso-Cuevas L (2022) Emerging pollutants in wastewater, advanced oxidation processes as an alternative treatment and perspectives. Processes 10:1–23. https://doi.org/10.3390/pr10051041
16. Bracamontes-Ruelas AR, Reyes-Vidal Y, Irigoyen-Campuzano JR, Reynoso-Cuevas L (2023) Simultaneous oxidation of emerging pollutants in real wastewater by the advanced fenton oxidation process. Catalysts 13(4):748. Retrieved from https://www.mdpi.com/2073-4344/13/4/748
17. Cao X, Luo J, Woodley JM, Wan Y (2016) Bioinspired multifunctional membrane for aquatic micropollutants removal. ACS Appl Mater Interfaces 8(44):30511–30522. https://doi.org/10.1021/acsami.6b10823
18. Celić M, Farré M, Lopez de Alda M, Perez S, Barceló D, Petrović M (2017) Chapter 15—environmental analysis: emerging pollutants. In: Fanali S, Haddad PR, Poole CF, Riekkola M-L (eds) Liquid chromatography, 2nd edn. Elsevier, pp 451–477
19. Chen Z, Wang H, Chen Z, Ren N, Wang A, Shi Y, Li X (2011) Performance and model of a full-scale up-flow anaerobic sludge blanket (UASB) to treat the pharmaceutical wastewater containing 6-APA and amoxicillin. J Hazard Mater 185(2):905–913. https://doi.org/10.1016/j.jhazmat.2010.09.106
20. Cheng H-Y, Hou Y-N, Zhang X, Yang Z-N, Xu T, Wang A-J (2017) Activating electrochemical catalytic activity of bio-palladium by hybridizing with carbon nanotube as "e−Bridge." Sci Rep 7(1):16588. https://doi.org/10.1038/s41598-017-16880-7
21. Chi J (2009) Phthalate acid esters in Potamogeton crispus L. from Haihe River China. Chemosphere 77(1):48–52. https://doi.org/10.1016/j.chemosphere.2009.05.043
22. Chopra S, Kumar D (2020) Characterization, optimization and kinetic study of diclofenac degradation by novel bacterial strains and their synthetic consortia. Bioremediat J 24(2–3):150–170. https://doi.org/10.1080/10889868.2020.1793723
23. Chopra S, Kumar D (2022) Characterization and biodegradation of ibuprofen by Bacillus siamensis strain DSI-1 isolated from wastewater. Rend Fis Acc Lincei 33(3):643–652. https://doi.org/10.1007/s12210-022-01085-6
24. Chopra S, Kumar D (2023) Characterization and biodegradation of paracetamol by biomass of Bacillus licheniformis strain PPY-2 isolated from wastewater. Rend Fis Acc Lincei 34(2):491–501. https://doi.org/10.1007/s12210-023-01140-w
25. Correa-Sanchez S, Peñuela GA (2022) Peracetic acid-based advanced oxidation processes for the degradation of emerging pollutants: a critical review. J Water Process Eng 49:102986. https://doi.org/10.1016/j.jwpe.2022.102986
26. Dargnat C, Teil M-J, Chevreuil M, Blanchard M (2009) Phthalate removal throughout wastewater treatment plant. Sci Total Environ 407(4):1235–1244. https://doi.org/10.1016/j.scitotenv.2008.10.027

27. Deblonde T, Cossu-Leguille C, Hartemann P (2011) Emerging pollutants in wastewater: a review of the literature. Int J Hyg Environ Health 214(6):442–448. https://doi.org/10.1016/j.ijheh.2011.08.002
28. Domínguez JR, González T, Palo P, Sánchez-Martín J (2010) Anodic oxidation of ketoprofen on boron-doped diamond (BDD) electrodes. Role of operative parameters. Chem Eng J 162(3):1012–1018. https://doi.org/10.1016/j.cej.2010.07.010
29. Du J, Fan Y, Qian X (2015) Occurrence and behavior of pharmaceuticals in sewage treatment plants in eastern China. Front Environ Sci Eng 9(4):725–730. https://doi.org/10.1007/s11783-014-0661-1
30. Egbuna C, Amadi CN, Patrick-Iwuanyanwu KC, Ezzat SM, Awuchi CG, Ugonwa PO, Orisakwe OE (2021) Emerging pollutants in Nigeria: a systematic review. Environ Toxicol Pharmacol. https://doi.org/10.1016/j.etap.2021.103638
31. Enyoh CE, Verla AW, Qingyue W, Ohiagu FO, Chowdhury AH, Enyoh EC et al (2020) An overview of emerging pollutants in air: method of analysis and potential public health concern from human environmental exposure. Trends Environ Analy Chem. https://doi.org/10.1016/j.teac.2020.e00107
32. Falås P, Wick A, Castronovo S, Habermacher J, Ternes TA, Joss A (2016) Tracing the limits of organic micropollutant removal in biological wastewater treatment. Water Res 95:240–249. https://doi.org/10.1016/j.watres.2016.03.009
33. Fromme H, Küchler T, Otto T, Pilz K, Müller J, Wenzel A (2002) Occurrence of phthalates and bisphenol A and F in the environment. Water Res 36(6):1429–1438. https://doi.org/10.1016/S0043-1354(01)00367-0
34. Gani KM, Kazmi AA (2016) Comparative assessment of phthalate removal and risk in biological wastewater treatment systems of developing countries and small communities. Sci Total Environ 569–570:661–671. https://doi.org/10.1016/j.scitotenv.2016.06.182
35. Ganzenko O, Huguenot D, van Hullebusch ED, Esposito G, Oturan MA (2014) Electrochemical advanced oxidation and biological processes for wastewater treatment: a review of the combined approaches. Environ Sci Pollut Res 21(14):8493–8524. https://doi.org/10.1007/s11356-014-2770-6
36. Gao D, Huang X, Tao Y (2015) A critical review of NanoSIMS in analysis of microbial metabolic activities at single-cell level. Crit Rev Biotechnol. https://doi.org/10.3109/07388551.2015.1057550
37. Gao D, Li Z, Guan J, Li Y, Ren N (2014) Removal of surfactants nonylphenol ethoxylates from municipal sewage-comparison of an A/O process and biological aerated filters. Chemosphere 97:130–134. https://doi.org/10.1016/j.chemosphere.2013.10.083
38. Gao D, Li Z, Guan J, Liang H (2017) Seasonal variations in the concentration and removal of nonylphenol ethoxylates from the wastewater of a sewage treatment plant. J Environ Sci (China) 54:217–223. https://doi.org/10.1016/j.jes.2016.02.005
39. Gao DW, Li Z, Guan JX, Liang H (2017) Seasonal changes and spatial distributions of nonylphenol ethoxylates in sewage treatment plant with BAF process. Environ Technol 38(4):406–412. https://doi.org/10.1080/09593330.2016.1196737
40. Gao D, Li Z, Wang H, Liang H (2018) An overview of phthalate acid ester pollution in China over the last decade: environmental occurrence and human exposure. Sci Total Environ 645:1400–1409. https://doi.org/10.1016/j.scitotenv.2018.07.093
41. Gao D, Li Z, Wen Z, Ren N (2013) Occurrence and fate of phthalate esters in full-scale domestic wastewater treatment plants and their impact on receiving waters along the Songhua River in China. Chemosphere. https://doi.org/10.1016/j.chemosphere.2013.08.009
42. Gao D, Li Z, Wen Z, Ren N (2014) Occurrence and fate of phthalate esters in full-scale domestic wastewater treatment plants and their impact on receiving waters along the Songhua River in China. Chemosphere 95:24–32. https://doi.org/10.1016/j.chemosphere.2013.08.009
43. Gao D-W, Wen Z-D (2015) Phthalate esters in the environment: a critical review of their occurrence, biodegradation, and removal during wastewater treatment processes. Sci Total Environ. https://doi.org/10.1016/j.scitotenv.2015.09.148

44. Gao X, Shi X, Cui Y, Li M, Zhang R, Qian X, Jiang Y (2011) Organic pollutants and ambient severity for the drinking water source of western Taihu Lake. Ecotoxicol 20(5):959–967. https://doi.org/10.1007/s10646-011-0681-6
45. Garcia-Segura S, Keller J, Brillas E, Radjenovic J (2015) Removal of organic contaminants from secondary effluent by anodic oxidation with a boron-doped diamond anode as tertiary treatment. J Hazard Mater 283:551–557. https://doi.org/10.1016/j.jhazmat.2014.10.003
46. Gasperi J, Garnaud S, Rocher V, Moilleron R (2009) Priority pollutants in surface waters and settleable particles within a densely urbanized area: case study of Paris (France). Sci Total Environ. https://doi.org/10.1016/j.scitotenv.2009.01.024
47. Geissen V, Mol H, Klumpp E, Umlauf G, Nadal M, van der Ploeg M et al (2015) Emerging pollutants in the environment: a challenge for water resource management. Int Soil Water Conserv Res 3(1):57–65.https://doi.org/10.1016/j.iswcr.2015.03.002
48. Gonsioroski A, Mourikes VE, Flaws JA (2020) Endocrine disruptors in water and their effects on the reproductive system. Int J Mol Sci. https://doi.org/10.3390/ijms21061929
49. Gothwal R, Shashidhar T (2015) Antibiotic pollution in the environment: a review. CLEAN Soil Air Water 43(4):479–489. https://doi.org/10.1002/clen.201300989
50. Graham DW, Olivares-Rieumont S, Knapp CW, Lima L, Werner D, Bowen E (2011) Antibiotic resistance gene abundances associated with waste discharges to the Almendares River near Havana, Cuba. Environ Sci Technol. https://doi.org/10.1021/es102473z
51. Grandclément C, Piram A, Petit M-E, Seyssiecq I, Laffont-Schwob I, Vanot G et al (2020) Biological removal and fate assessment of diclofenac using *Bacillus subtilis* and *Brevibacillus laterosporus* strains and ecotoxicological effects of diclofenac and 4′-hydroxy-diclofenac. J Chem 2020:9789420. https://doi.org/10.1155/2020/9789420
52. Grandclément C, Seyssiecq I, Piram A, Wong-Wah-Chung P, Vanot G, Tiliacos N et al (2017) From the conventional biological wastewater treatment to hybrid processes, the evaluation of organic micropollutant removal: a review. Water Res 111:297–317
53. Gur-Reznik S, Koren-Menashe I, Heller-Grossman L, Rufel O, Dosoretz CG (2011) Influence of seasonal and operating conditions on the rejection of pharmaceutical active compounds by RO and NF membranes. Desalination 277(1):250–256. https://doi.org/10.1016/j.desal.2011.04.029
54. Göbel A, McArdell CS, Joss A, Siegrist H, Giger W (2007) Fate of sulfonamides, macrolides, and trimethoprim in different wastewater treatment technologies. Sci Total Environ 372(2):361–371. https://doi.org/10.1016/j.scitotenv.2006.07.039
55. Hauser R, Calafat AM (2005) Phthalates and human health. Occup Environ Med 62(11):806. https://doi.org/10.1136/oem.2004.017590
56. He et al (2013) Spatio-temporal distributions and the ecological and health risks of phthalate esters (PAEs) in the surface water of a large shallow. Chinese lake Science of The Total Environment 461–462; 672–680. https://doi.org/10.1016/j.scitotenv.2013.05.049
57. Hejna M, Moscatelli A, Stroppa N, Onelli E, Pilu S, Baldi A, Rossi L (2020) Bioaccumulation of heavy metals from wastewater through a *Typha latifolia* and *Thelypteris palustris* phytoremediation system. Chemosphere 241:125018
58. Heo J, Flora JRV, Her N, Park Y-G, Cho J, Son A, Yoon Y (2012) Removal of bisphenol A and 17β-estradiol in single walled carbon nanotubes–ultrafiltration (SWNTs–UF) membrane systems. Sep Purif Technol 90:39–52. https://doi.org/10.1016/j.seppur.2012.02.007
59. Huang M-Z, Ma Y-W, Wang Y, Wan J-Q, Zhang H-P (2010) The fate of di-n-butyl phthalate in a laboratory-scale anaerobic/anoxic/oxic wastewater treatment process. Biores Technol 101(20):7767–7772. https://doi.org/10.1016/j.biortech.2010.05.028
60. Huang Y, Yu Z, Liu L, Che Y, Zhang T (2022) Acesulfame anoxic biodegradation coupled to nitrate reduction by enriched consortia and isolated *Shinella spp.* Environ Sci Technol 56(18):13096–13106. https://doi.org/10.1021/acs.est.2c03656
61. Hube S, Wu B (2021) Mitigation of emerging pollutants and pathogens in decentralized wastewater treatment processes: a review. Sci Total Environ 779:146545. https://doi.org/10.1016/j.scitotenv.2021.146545

62. Jiang M, Wang L, Ji R (2010) Biotic and abiotic degradation of four cephalosporin antibiotics in a lake surface water and sediment. Chemosphere 80(11):1399–1405. https://doi.org/10.1016/j.chemosphere.2010.05.048
63. Kalkan Ç, Yapsakli K, Mertoglu B, Tufan D, Saatci A (2011) Evaluation of biological activated carbon (BAC) process in wastewater treatment secondary effluent for reclamation purposes. Desalination 265(1):266–273. https://doi.org/10.1016/j.desal.2010.07.060
64. Karim AV, Shriwastav A (2023) Integrated sonophotocatalytic oxidation and SBR processes for the effective treatment of antibiotics with an emphasis on process optimization and microbial diversity. J Environ Chem Eng 11(2):109632. https://doi.org/10.1016/j.jece.2023.109632
65. Kim SW, Liang Y, Lozano YM, Rillig MC (2021) Microplastics reduce the negative effects of litter-derived plant secondary metabolites on nematodes in soil. Front Environ Sci. https://doi.org/10.3389/fenvs.2021.790560
66. Kim M, Guerra P, Shah A, Parsa M, Alaee M, Smyth SA (2014) Removal of pharmaceuticals and personal care products in a membrane bioreactor wastewater treatment plant. Water Sci Technol 69(11):2221–2229. https://doi.org/10.2166/wst.2014.145
67. Klinčić D, Dvoršćak M, Jagić K, Mendaš G, Herceg Romanić S (2020) Levels and distribution of polybrominated diphenyl ethers in humans and environmental compartments: a comprehensive review of the last five years of research. Environ Sci Pollut Res. https://doi.org/10.1007/s11356-020-07598-7
68. Kong X-J, Li D, Cao L-Q, Zhang X-M, Zhao Y, Lv Y, Zhang J (2008) Evaluation of municipal sewage treatment systems for pollutant removal efficiency by measuring levels of micropollutants. Chemosphere 72(1):59–66. https://doi.org/10.1016/j.chemosphere.2008.02.005
69. Ter Laak TL, Agbo SO, Barendregt A, Hermens JLM (2006) Freely dissolved concentrations of PAHs in soil pore water: measurements via solid-phase extraction and consequences for soil tests. Environ Sci Technol. https://doi.org/10.1021/es0514803
70. Lecomte S, Habauzit D, Charlier TD, Pakdel F (2017) Emerging estrogenic pollutants in the aquatic environment and breast cancer. Genes. https://doi.org/10.3390/genes8090229
71. Li Y, Xie S, Yao J (2023) Singlet oxygen generation for selective oxidation of emerging pollutants in a flow-by electrochemical system based on natural air diffusion cathode. Environ Sci Pollut Res 30(7):17854–17864. https://doi.org/10.1007/s11356-022-23364-3
72. Li T, Xu Z-J, Zhou N-Y (2023) Aerobic degradation of the antidiabetic drug metformin by *Aminobacter* sp. strain NyZ550. Environ Sci Technol 57(3):1510–1519. https://doi.org/10.1021/acs.est.2c07669
73. Lian J, Liu JX, Wei YS (2009) Fate of nonylphenol polyethoxylates and their metabolites in four Beijing wastewater treatment plants. Sci Total Environ 407(14):4261–4268. https://doi.org/10.1016/j.scitotenv.2009.03.022
74. Lindberg RH, Wennberg P, Johansson MI, Tysklind M, Andersson BAV (2005) Screening of human antibiotic substances and determination of weekly mass flows in five sewage treatment plants in Sweden. Environ Sci Technol. https://doi.org/10.1021/es048143z
75. Liu B, Zhang SG, Chang CC (2019) Emerging pollutants-part II: treatment. Water Environ Res 91(10):1390–1401. https://doi.org/10.1002/wer.1233
76. Liu et al (2013) Occurrence and Removal Characteristics of Phthalate Esters from Typical Water Sources in Northeast China. J Anal Methods Chem 20131–8. https://doi.org/10.1155/2013/419349
77. Liu X, Zhang G, Liu Y, Lu S, Qin P, Guo X et al (2019) Occurrence and fate of antibiotics and antibiotic resistance genes in typical urban water of Beijing, China. Environ Pollut 246:163–173. https://doi.org/10.1016/j.envpol.2018.12.005
78. Luo Y, Guo W, Ngo HH, Nghiem LD, Hai FI, Zhang J et al (2014) A review on the occurrence of micropollutants in the aquatic environment and their fate and removal during wastewater treatment. Sci Total Environ 473–474:619–641. https://doi.org/10.1016/j.scitotenv.2013.12.065

79. Mackintosh CE, Maldonado J, Hongwu J, Hoover N, Chong A, Ikonomou MG, Gobas FAPC (2004) Distribution of phthalate esters in a marine aquatic food web: comparison to polychlorinated biphenyls. Environ Sci Technol. https://doi.org/10.1021/es034745r
80. Madukasi EI, Dai X, He C, Zhou J (2010) Potentials of phototrophic bacteria in treating pharmaceutical wastewater. Int J Environ Sci Technol 7(1):165–174. https://doi.org/10.1007/BF03326128
81. Malaeb L, Ayoub GM (2011) Reverse osmosis technology for water treatment: state of the art review. Desalination 267(1):1–8. https://doi.org/10.1016/j.desal.2010.09.001
82. Mares-Carbajal FJ, Espinosa-Arzate MC, Ramírez-Montoya LA, Pat-Espadas AM, Ramírez JE, Rangel-Mendez JR et al (2022) Biocatalyst developed with recovered iron-rich minerals enhances the biotransformation of SARS-CoV-2 antiviral drugs in anaerobic bioreactors. J Water Process Eng 50:103337. https://doi.org/10.1016/j.jwpe.2022.103337
83. Marttinen SK, Ruissalo M, Rintala JA (2004) Removal of bis (2-ethylhexyl) phthalate from reject water in a nitrogen-removing sequencing batch reactor. J Environ Manage 73(2):103–109. https://doi.org/10.1016/j.jenvman.2004.05.011
84. Maruya KA, Dodder NG, Sengupta A, Smith DJ, Lyons JM, Heil AT, Drewes JE (2016) Multimedia screening of contaminants of emerging concern (CECS) in coastal urban watersheds in southern California (USA). Environ Toxicol Chem. https://doi.org/10.1002/etc.3348
85. Meng Y, Liu W, Fiedler H, Zhang J, Wei X, Liu X et al (2021) Fate and risk assessment of emerging contaminants in reclaimed water production processes. Front Environ Sci Eng 15(5):104. https://doi.org/10.1007/s11783-021-1392-8
86. Mišík M, Knasmueller S, Ferk F, Cichna-Markl M, Grummt T, Schaar H, Kreuzinger N (2011) Impact of ozonation on the genotoxic activity of tertiary treated municipal wastewater. Water Res 45(12):3681–3691. https://doi.org/10.1016/j.watres.2011.04.015
87. Morsi R, Bilal M, Iqbal HM, Ashraf SS (2020) Laccases and peroxidases: the smart, greener and futuristic biocatalytic tools to mitigate recalcitrant emerging pollutants. Sci Total Environ 714:136572
88. Ni S-Q, Cui Q, Zheng Z (2014) Interaction of polybrominated diphenyl ethers and aerobic granular sludge: biosorption and microbial degradation. Biomed Res Int 2014:274620. https://doi.org/10.1155/2014/274620
89. Palma TL, Costa MC (2021) Anaerobic biodegradation of fluoxetine using a high-performance bacterial community. ANAEROBE 68.https://doi.org/10.1016/j.anaerobe.2021.102356
90. Park C, Fang Y, Murthy SN, Novak JT (2010) Effects of floc aluminum on activated sludge characteristics and removal of 17-α-ethinylestradiol in wastewater systems. Water Res 44(5):1335–1340. https://doi.org/10.1016/j.watres.2009.11.002
91. Petrovic M, Farré M, Eljarrat E, Diaz S, Barceló D (2013) Chapter 14—environmental analysis: emerging pollutants. In: Fanali S, Haddad PR, Poole CF, Schoenmakers P, Lloyd D (eds) Liquid chromatography. Elsevier, Amsterdam, pp 389–410
92. Pham T-T, Proulx S (1997) PCBs and PAHs in the montreal urban community (Quebec, Canada) wastewater treatment plant and in the effluent plume in the St Lawrence River. Water Res 31(8):1887–1896. https://doi.org/10.1016/S0043-1354(97)00025-0
93. Prasse C, Stalter D, Schulte-Oehlmann U, Oehlmann J, Ternes TA (2015) Spoilt for choice: a critical review on the chemical and biological assessment of current wastewater treatment technologies. Water Res. https://doi.org/10.1016/j.watres.2015.09.023
94. Radjenovic J, Bagastyo A, Rozendal RA, Mu Y, Keller J, Rabaey K (2011) Electrochemical oxidation of trace organic contaminants in reverse osmosis concentrate using RuO2/IrO2-coated titanium anodes. Water Res 45(4):1579–1586. https://doi.org/10.1016/j.watres.2010.11.035
95. Ren S-Y, Ni H-G (2023) Biodeterioration of microplastics by bacteria isolated from mangrove sediment. Toxics 11(5). https://doi.org/10.3390/toxics11050432
96. Rodriguez-Narvaez OM, Peralta-Hernandez JM, Goonetilleke A, Bandala ER (2017) Treatment technologies for emerging contaminants in water: a review. Chem Eng J 323:361–380. https://doi.org/10.1016/j.cej.2017.04.106

97. Rogers HR (1996) Sources, behaviour and fate of organic contaminants during sewage treatment and in sewage sludges. Sci Total Environ 185(1):3–26. https://doi.org/10.1016/0048-9697(96)05039-5
98. Rutkowska M, Płotka-Wasylka J, Sajid M, Andruch V (2019) Liquid–phase microextraction: a review of reviews. Microchem J 149:103989
99. Sangeetha Devi R, Ramya R, Kannan K, Robert Antony A, Rajesh Kannan V (2019) Investigation of biodegradation potentials of high density polyethylene degrading marine bacteria isolated from the coastal regions of Tamil Nadu, India. Mar Pollut Bull 138:549–560. https://doi.org/10.1016/j.marpolbul.2018.12.001
100. Saravanan A, Kumar PS, Duc PA, Rangasamy G (2023) Strategies for microbial bioremediation of environmental pollutants from industrial wastewater: a sustainable approach. Chemosphere 313:137323. https://doi.org/10.1016/j.chemosphere.2022.137323
101. Shukla R, Ahammad SZ (2023) Performance assessment of a modified trickling filter and conventional activated sludge process along with tertiary treatment in removing emerging pollutants from urban sewage. Sci Total Environ 858:159833. https://doi.org/10.1016/j.scitotenv.2022.159833
102. Soares A, Guieysse B, Jefferson B, Cartmell E, Lester JN (2008) Nonylphenol in the environment: a critical review on occurrence, fate, toxicity and treatment in wastewaters. Environ Int 34(7):1033–1049. https://doi.org/10.1016/j.envint.2008.01.004
103. Solís ME, Liu CC, Nam P, Niyogi DK, Bandeff JM, Huang Y-W (2007) Occurrence of Organic Chemicals in Two Rivers Inhabited by Ozark Hellbenders (Cryptobranchus alleganiensis bishopi). Arch Environ Contam Toxicol 53(3):426–434. https://doi.org/10.1007/s00244-006-0208-y
104. Song X, Sun S, Gao Y, Zhang W, Zhou L et al (2022) Laboratory-scale study of a biodegradable microplastic polylactic acid stabilizing aerobic granular sludge system. Environ Pollut 306:119329. https://doi.org/10.1016/j.envpol.2022.119329
105. Sophia AC, Lima EC (2018) Removal of emerging contaminants from the environment by adsorption. Ecotoxicol Environ Saf 150:1–17.https://doi.org/10.1016/j.ecoenv.2017.12.026
106. Staples CA, Peterson DR, Parkerton TF, Adams WJ (1997) The environmental fate of phthalate esters: a literature review. Chemosphere 35(4):667–749. https://doi.org/10.1016/S0045-6535(97)00195-1
107. Suyamud B, Inthorn D, Panyapinyopol B, Thiravetyan P (2018) Biodegradation of Bisphenol A by a newly isolated Bacillus megaterium strain ISO-2 from a polycarbonate industrial wastewater. Water Air Soil Pollut 229(11):348. https://doi.org/10.1007/s11270-018-3983-y
108. Tang Y, Liu Y, Chen Y, Zhang W, Zhao J, He S et al (2020) A review: research progress on microplastic pollutants in aquatic environments. Sci Total Environ. https://doi.org/10.1016/j.scitotenv.2020.142572
109. Teodosiu C, Gilca A-F, Barjoveanu G, Fiore S (2018) Emerging pollutants removal through advanced drinking water treatment: a review on processes and environmental performances assessment. J Clean Prod 197:1210–1221. https://doi.org/10.1016/j.jclepro.2018.06.247
110. Terzic S, Matosic M, Ahel M, Mijatovic I (2005) Elimination of aromatic surfactants from municipal wastewaters: comparison of conventional activated sludge treatment and membrane biological reactor. Water Sci Technol 51(6–7):447–453. https://doi.org/10.2166/wst.2005.0667
111. Tuan Tran H, Lin C, Bui X-T, Ky Nguyen M, Dan Thanh Cao N, Mukhtar H et al (2021) Phthalates in the environment: characteristics, fate and transport, and advanced wastewater treatment technologies. Bioresour Technol. https://doi.org/10.1016/j.biortech.2021.126249
112. Valladares Linares R, Yangali-Quintanilla V, Li Z, Amy G (2011) Rejection of micropollutants by clean and fouled forward osmosis membrane. Water Res 45(20):6737–6744. https://doi.org/10.1016/j.watres.2011.10.037
113. Vargas-Ordóñez A, Aguilar-Romero I, Villaverde J, Madrid F, Morillo E (2023) Isolation of novel bacterial strains pseudomonas extremaustralis CSW01 and stutzerimonas stutzeri CSW02 from sewage sludge for paracetamol biodegradation. Microorganisms 11(1). https://doi.org/10.3390/microorganisms11010196

114. Vela N, Fenoll J, Garrido I, Pérez-Lucas G, Flores P, Hellín P, Navarro S (2019) Reclamation of agro-wastewater polluted with pesticide residues using sunlight activated persulfate for agricultural reuse. Sci Total Environ 660:923–930. https://doi.org/10.1016/j.scitotenv.2019.01.060
115. Vethaak AD, Lahr J, S. Marca, Schrap Angélique C., Belfroid Gerard BJ., Rijs Anton, Gerritsen Jacob, de Boer Astrid S., Bulder Guy CM., Grinwis Raoul V., Kuiper Juliette, Legler Tinka AJ., Murk Willie, Peijnenburg Henk JM., Verhaar Pim, de Voogt (2005) An integrated assessment of estrogenic contamination and biological effects in the aquatic environment of The Netherlands. Chemosphere 59(4):511–524. https://doi.org/10.1016/j.chemosphere.2004.12.053
116. Vieno NM, Tuhkanen T, Kronberg L (2005) Seasonal variation in the occurrence of pharmaceuticals in effluents from a sewage treatment plant and in the recipient water. Environ Sci Technol 39(21):8220–8226. https://doi.org/10.1021/es051124k
117. Wang H, Liang H, Gao D (2017) Occurrence and distribution of phthalate esters (PAEs) in wetland sediments. J For Res. https://doi.org/10.1007/s11676-017-0371-1
118. Wang C, Wang J, Gao W, Ning X, Xu S, Wang X et al (2023) The fate of phthalate acid esters in wastewater treatment plants and their impact on receiving waters. Sci Total Environ 873:162201. https://doi.org/10.1016/j.scitotenv.2023.162201
119. Wang F, Xia X, Sha Y (2007) Distribution of phthalic acid esters in Wuhan section of the Yangtze River, China. J Hazard Mater. https://doi.org/10.1016/j.jhazmat.2007.10.028
120. Wang WH, Zhang WF, Liang H, Gao DW (2019) Occurrence and fate of typical antibiotics in wastewater treatment plants in Harbin, North-east China. Front Environ Sci Eng 13(3). https://doi.org/10.1007/s11783-019-1118-3
121. Wang F, Xia X, Sha Y (2008) Distribution of Phthalic Acid Esters in Wuhan section of the Yangtze River China. J Hazard Mater 154(1–3): 317-324. https://doi.org/10.1016/j.jhazmat.2007.10.028
122. Water Air and Soil Pollution 140(1/4) 85–98. https://doi.org/10.1023/A:1020134707450
123. Wen Z, Huang X, Gao D, Liu G, Fang C, Shang Y et al (2017) Phthalate esters in surface water of Songhua River watershed associated with land use types, Northeast China. Environ Sci Pollut Res. https://doi.org/10.1007/s11356-017-1119-3
124. Wolfe NL, Steen WC, Burns LA (1980) Phthalate ester hydrolysis: linear free energy relationships. Chemosphere 9(7):403–408. https://doi.org/10.1016/0045-6535(80)90023-5
125. Wu B, Zhang X, Zhang X, Yasun A, Zhang Y, Zhao D et al (2009) Semi-volatile organic compounds and trace elements in the Yangtze River source of drinking water. Ecotoxicology. https://doi.org/10.1007/s10646-009-0331-4
126. Wu J, Ma TJ, Zhou ZF, Yu N, He ZQ, Li BS et al (2019) Occurrence and fate of phthalate esters in wastewater treatment plants in Qingdao, China. Hum Ecol Risk Assess 25(6):1547–1563. https://doi.org/10.1080/10807039.2018.1471341
127. Wu B, Zhang X, Zhang X, Yasun A, Zhang Y, Zhao D, Ford T, Cheng S (2009) Semi-volatile organic compounds and trace elements in the Yangtze River source of drinking water. Ecotoxicol 18(6):707–714. https://doi.org/10.1007/s10646-009-0331-4
128. Yan Q, Gao X, Chen Y-P, Peng X-Y, Zhang Y-X, Gan X-M et al (2014) Occurrence, fate and ecotoxicological assessment of pharmaceutically active compounds in wastewater and sludge from wastewater treatment plants in Chongqing, the Three Gorges Reservoir Area. Sci Total Environ 470–471:618–630. https://doi.org/10.1016/j.scitotenv.2013.09.032
129. Yang C-W, Liu C, Chang B-V (2020) Biodegradation of amoxicillin, tetracyclines and sulfonamides in wastewater sludge. Water 12(8):2147
130. Yang Y, Wang P, Liu Y (2010) Species distribution of ferric hydrolysates in microwave enhanced Fenton-like process and possible mechanism. J Hazard Mater 178(1):293–297. https://doi.org/10.1016/j.jhazmat.2010.01.076
131. Ye S, Andrady AL (1991) Fouling of floating plastic debris under Biscayne Bay exposure conditions. Mar Pollut Bull 22(12):608–613. https://doi.org/10.1016/0025-326X(91)90249-R

132. Yin X, Zeb R, Wei H, Cai L (2020) Acute exposure of di(2-ethylhexyl) phthalate (DEHP) induces immune signal regulation and ferroptosis in oryzias melastigma. Chemosphere. https://doi.org/10.1016/j.chemosphere.2020.129053
133. Zhang C, Wang Y (2009) Removal of dissolved organic matter and phthalic acid esters from landfill leachate through a complexation–flocculation process. Waste Manage 29(1):110–116. https://doi.org/10.1016/j.wasman.2008.02.023
134. Zhang Q-Q, Ying G-G, Pan C-G, Liu Y-S, Zhao J-L (2015) Comprehensive evaluation of antibiotics emission and fate in the river basins of China: source analysis, multimedia modeling, and linkage to bacterial resistance. Environ Sci Technol. https://doi.org/10.1021/acs.est.5b00729
135. Zhang J, Li G, Yuan X, Li P, Yu Y, Yang W, Zhao S (2023) Reduction of ultrafiltration membrane fouling by the pretreatment removal of emerging pollutants: a review. Membranes 13(1):77. Retrieved from https://www.mdpi.com/2077-0375/13/1/77
136. Zhang R, Wang W, Shi X, Yu X, Li M, Xiao L, Cui Y (2011) Health risk of semi-volatile organic pollutants in Wujin river inflow into Taihu Lake. Ecotoxicol 20(5):1083–1089. https://doi.org/10.1007/s10646-011-0682-5
137. Zhou Q, Chen L, Wang Z, Wang J, Ni S, Qiu J et al (2017) Fast atrazine degradation by the mixed cultures enriched from activated sludge and analysis of their microbial community succession. Environ Sci Pollut Res 24(28):22152–22157. https://doi.org/10.1007/s11356-017-9052-z

Chapter 6
Innovations in Pretreatment of Sewage Sludge

6.1 Introduction

The term "sewage sludge" refers to the heterogeneous compound with solid or semi-solid forms that left over from wastewater treatment plants (WWTPs), which contains considerable amounts of persistent organic pollutants, nutrients, and contaminants [151]. It is highly producing a foul smell and pathogenic organisms, and it contaminates water bodies once the tremendous amount of sewage sludge cannot be treated with suitable methods just in time. In addition to these contaminants, 70–80% of heavy metals would be accumulated during the sewage sludge treatments via physical sedimentation or physisorption and hard to separated. Thus, the cumulative effect of heavy metals is significantly, and it is difficult to find various environmentally friendly technologies to deal with [160].

The methods including sanitary landfill, incineration, pyrolysis, and anaerobic digestion have been applied to wastewater treatments. All these methods have potential risks endangered to human and environmental health while the vast amount of sewage sludge is generated throughout the world [78]. Sanitary landfill is application of treated sewage sludge to agriculture as soil fertilizers or soil conditioners because sewage sludge can provide N and P for plant growth. The leachate of sanitary landfill is an alarming threat causing the deterioration of groundwater quality and the vicinity of soil pollution. Therefore, the remains of potentially contaminants of landfill have become one of the most challenging issues of the world. For thermochemical conversion processes, it usually requires higher temperature to oxidize organic materials, which could produce many by-products, like the air pollutants NO_x and SO_x [69]. Moreover, the relatively low calorific value of incineration creates obstacles for its development.

G. Dawen and M. Nabi, *Novel Approaches Towards Wastewater Treatment*,
Springer Water, https://doi.org/10.1007/978-3-031-55189-5_6

Out of the various treatment methods available, anaerobic digestion of sewage sludge stands out as a highly efficient and environmentally-friendly technology. It is a chemo-biological process that could convert organic compounds to biogas under anaerobic condition [66]. The collaborative activities of diverse microbial communities play an important role in the conversion of organic matter into methane (CH_4) and carbon dioxide (CO_2) [121]. Therefore, anaerobic digestion of sewage sludge has a number of merits, especially from environmental sustainability perspective, because the chemicals and biological contaminants are eliminated.

However, the efficient AD has specific practical constraints related to complex organic matter composition. The presence of extracellular polymeric substances (EPS), recalcitrant cell walls, and other high molecular weight organic matter in sewage sludge obstacles hydrolysis of AD [78]. To further enhance the efficient of AD, different kinds of pretreatment have been provided to decompose and release inner constituents of the cells [110]. Sewage sludge pretreatment prior to AD includes physical and chemical pretreatment, ultrasonication, high-pressure homogenization, microwave irradiation, electro-kinetic disintegration, ball mill and so on. Regarding the chemical pretreatment (the unconventional method for complex organic waste destruction), the acids, alkalis, and oxidants have been used to hydrolyze the sludge, which could enhance anaerobic sludge digestion [78]. The promising techniques of combining physical and chemical pretreatment methods, such as free nitrous acid (FNA) and mechanical pretreatment, have shown significant effects on the disintegration of sewage sludge. For instance, the synergistic effect of the combination of high-pressure homogenization (HPH) and FNA pretreatment has demonstrated great potential in the application of anaerobic digestion [112].

In this chapter, we delve into the exploration and elucidation of innovative strategies for sewage sludge pretreatment and anaerobic digestion treatment. The focus lies on identifying and implementing suitable pretreatment methods, analyzing the energy consumption associated with pretreatment, examining the energy production potential derived from the anaerobic digestion of pretreated sludge, and envisioning the future prospects in this field. By thoroughly examining these aspects, we aim to enhance our understanding of effective pretreatment techniques, optimize energy efficiency during the process, harness renewable energy from anaerobic digestion, and pave the way for advancements and opportunities in the treatment of sewage sludge. This chapter serves as a comprehensive guide, providing valuable insights and paving the path for sustainable and efficient sludge management practices.

6.2 Sewage Sludge Production and Challenges in Its Anaerobic Treatment

6.2.1 Global Sludge Production

The increasing quantities of sewage sludge produced each year, particularly by the most developed countries, have led to much debate about the challenges of disposal and management [107]. On one hand, the expansion of the water treatment industry through the construction of new facilities and on the other hand, the necessity to meet more rigorous quality standards imposed by recent regulations, are factors that contribute to this phenomenon.

The heightened focus on treatment of wastewater has led to a significant environmental problem globally—the excessive production of sludge in WWTPs. To address this, researchers and plant operators are exploring technological solutions, such as upgrading existing technologies or using alternative methods, to reduce sludge production. Various strategies have been identified, including biological and physical–chemical methods, as well as emerging technologies. However, many of these methods have not been thoroughly implemented at full-scale. Therefore, it is crucial to understand the reduction mechanisms, advantages, disadvantages, and the economic and environmental impacts of each technology to select the most suitable system for reducing sludge [106].

From 1992 to 2010, the annual production of sewage sludge (on a dry-weight basis) continuously increased in most countries. For instance, the European Union produced 5.5 million tons annually in 1992, which increased to more than 10 million tons per year in 2010. The USA produced nearly 8 million tons per year, while China produced 9 million tons per year of sewage sludge in 2010. Apart from EU countries, Germany generated the largest amount of sewage sludge with 1.85 million tons per year on a dry-weight basis. The UK was the second-largest producer of sewage sludge with an annual production of 1.14 million tons, followed by Spain which produced 1.03 million tons annually [127]. Despite the decreasing trend among Asian countries, Japan still produces more sewage sludge compared to EU countries. However, it is important to note that China has generated a vast amount of sewage sludge in recent years, according to statistics from the Ministry of Housing and Urban–Rural Development, PRC. In 2019 alone, China generated approximately 60 million tons of sewage sludge, with an annual increase of 10%. Therefore, it is crucial to focus on addressing this issue in China [58].

Anthropogenic activities often correlate with sewage production and treatment rates. In China, the eastern and southern regions typically have higher population density, more developed industrialization, and better urbanization than their western and northern counterparts, as evidenced by available data. Consequently, the higher population density in the eastern and southern regions leads to increased sewage production, which is produced by a higher sewage treatment rate. The wastewater treatment processes used across different regions are generally similar, and may include physical, chemical, and biological treatments, among others [168].

6.2.2 Challenges in Sludge Anaerobic Digestion

The AD process involves the biological stabilization of a mixture of primary and secondary sludge by a microbial consortium, without the use of molecular oxygen, through four sequential stages: hydrolysis, acidogenesis, acetogenesis, and methanogenesis. This method is both environmentally and economically sustainable, as it yields methane-enriched biogas that can be used to produce electricity, or converted into renewable gas and transportation fuels (Fig. 6.1). By reducing dependence on fossil fuel energy and greenhouse gas emissions, this process is an eco-friendly option. Nevertheless, WAS is a less desirable substrate for the hydrolysis step, which is often the rate-limiting step in conventional digestion. This is due to its lower organic degradation efficiency, typically around 30–50%, which results in a longer sludge retention time (SRT) of 20–30 days, requiring a larger digester volume and leading to higher capital costs [112]. The intricate floc structure of WAS poses challenges for the breakdown of organic matter within microbial cells by hydrolytic enzymes, leading to a slower and limited anaerobic degradation rate. Therefore, there is a growing attention in using pretreatment approaches to enhance the conventional AD process. These technologies can help overcome the physical and chemical barriers present in the WAS floc structure and accelerate the degradation of organic matter.

Sewage sludge floc is a complex mixture composed mostly of microbial aggregates, inorganic and organic particles, EPS, filamentous bacterial strains, and a significant amount of water. Sewage sludge composition can vary depending on the wastewater origin and type. EPS, which come from cell lysis and microbial secretion (microorganisms' activity) or from the adsorption of organic matter in wastewater,

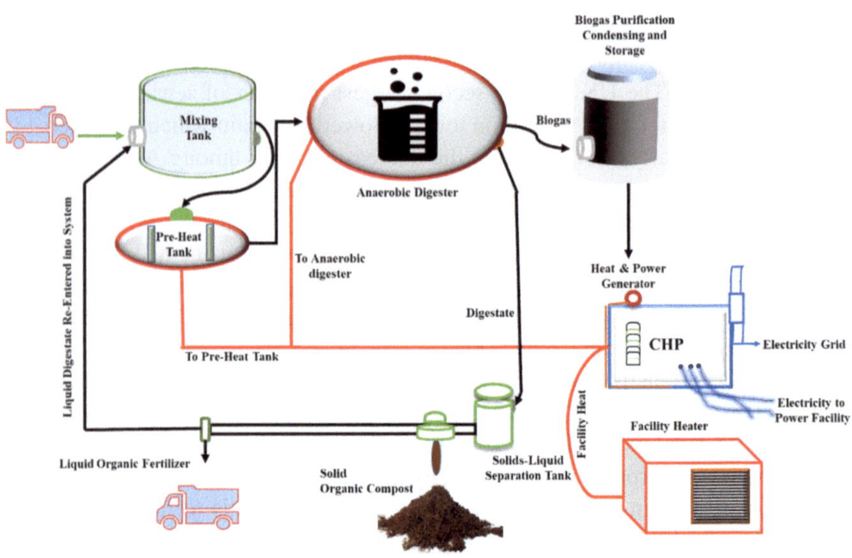

Fig. 6.1 Sludge anaerobic digestion process

make up the primary fraction sludge organic. EPS are composed of nucleic acids, proteins, polysaccharides, humic substances, and lipids, and are typically classified into three categories based on their distribution within the sludge floc: loosely bound EPS (LB-EPS), soluble EPS (S-EPS), and tightly bound EPS (TB-EPS). S-EPS are evenly dispersed in the aqueous phase, LB-EPS extend from TB-EPS and have a highly porous and dispersible structure, while TB-EPS adhere to the surface of bacterial cells within the sludge flocs. These three-dimensional, gel-like, and negatively charged biopolymers are responsible for the surface physicochemical properties of sludge matrixes, and provide protective shielding that prevents cell rupture and lysis, thereby affecting the functional integrity, strength, dewaterability, flocculation, and biodegradability of sludge. In addition to EPS, microbial cells themselves possess a hard cell envelope composed of glycan strands crosslinked by peptides, which presents physical and chemical barriers to direct anaerobic digestion. Therefore, sludge with high cell content and EPS have a rigid structure and is more difficult to digest and hydrolyze, resulting in low VS degradation (30–50%) in conventional mesophilic anaerobic system, even at lengthy retention time of 20–30 days. To overcome this challenge, pretreatment technologies can be employed to improve sludge digestion efficiency [182].

Despite the numerous advantages of the AD, instability is a likely issue that can cause the reactor to fail completely due to the accumulation of VFAs. These VFAs can surpass the digester's buffer capacity, causing acidification and ultimately leading to system failure. Alkalinity is consumed during the fermentation and hydrolysis stages of AD, while it is produced and acidification is compensated for during the methanogenic stage. The pH, which is one of the most important parameters affecting AD, is determined by VFAs. Fermentative bacteria require pH range of 4.0–8.5 for their growth, while the activities of methanogenic communities that thrive in the limiting range of 6.8–7.2 are affected by VFA concentrations and pH. Acetic acid is the primary VFA (volatile fatty acid) produced during the processes of acidogenesis and hydrolysis. Studies conducted earlier indicate that the undissociated VFAs significantly hindered the activity of methanogens in the single-stage anaerobic system [104].

The majority of organic compounds containing nitrogen are transformed into ammonia nitrogen (NH_4 and NH_3) during the process of anaerobic digestion. As the anaerobic digestion system operates, the concentration of ammonia nitrogen gradually rises. The total ammoniacal nitrogen (TAN) present in the anaerobic reactors serves a significant and multifaceted role. On one hand, ammonia nitrogen acts as a nutrient for anaerobic microorganisms, contributing to the alkalinity of the anaerobic digestion system. On the other hand, if the concentration of ammonia nitrogen exceeds a certain threshold, it can severely inhibit the anaerobic digestion system and even cause instability in the thermophilic anaerobic digestion system. Additionally, ammonia is the final product that results from the anaerobic digestion of urea, proteins, and nucleic acids in hydrolysis stage [14]. The high concentration of organic nitrogen in the raw material used for anaerobic digestion causes a reaction that leads to the production of ammonia, thereby increasing the concentration of ammonia

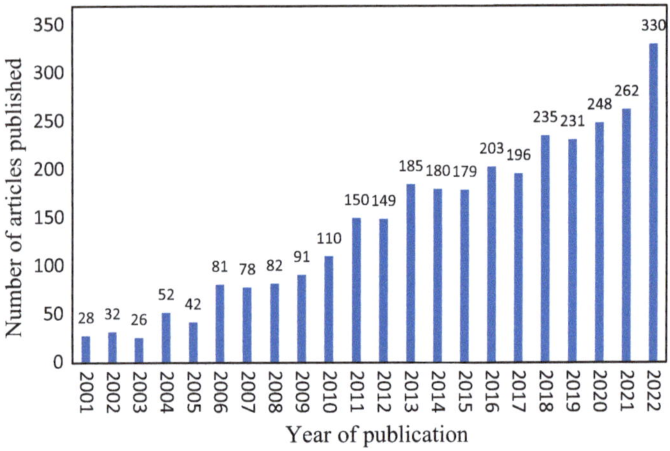

Fig. 6.2 The number of research articles published each year (2001–2022) pertaining to sludge pretreatment. *Source* https://www.engineeringvillage.com

nitrogen. However, an optimal level of ammonia concentration is necessary to maintain a sufficient buffer capacity of the methanogenic medium in the AD process. This helps to improve the stability of the anaerobic digestion process without inhibiting it [171].

The management of waste biomass presents a significant challenge for waste management authorities due to its vast production. Ensuring its safe disposal and effective utilization has been emphasized in studies by Rai and Rao [128] and Wei et al. [161]. In recent years, the pretreatment of waste sewage sludge has gained significant importance to improve anaerobic digestion. This is evidenced by the increase in published articles on sludge pretreatment, with only 28 articles in 2001 compared to 330 articles in 2022 as shown in Fig. 6.2 [109].

The pretreatment of sewage sludge facilitates the acceleration of the hydrolysis step in anaerobic digestion, leading to a decrease in the hydraulic retention time (HRT) required for the process. Moreover, this pretreatment significantly improves both methane production and the removal of organic compounds, as evidenced in Fig. 6.3 [111].

6.3 Pretreatment Technologies of Sewage Sludge to Improve Anaerobic Digestion

Anaerobic digestion, one of the promising treatments for sewage sludge disposal, can reduce the amount of sludge with lower capital costs of a WWTP. However, AD performance is limited by the complex floc structure (extracellular polymeric substances) and hard cell wall, because of the inefficient usage of organics. Thus,

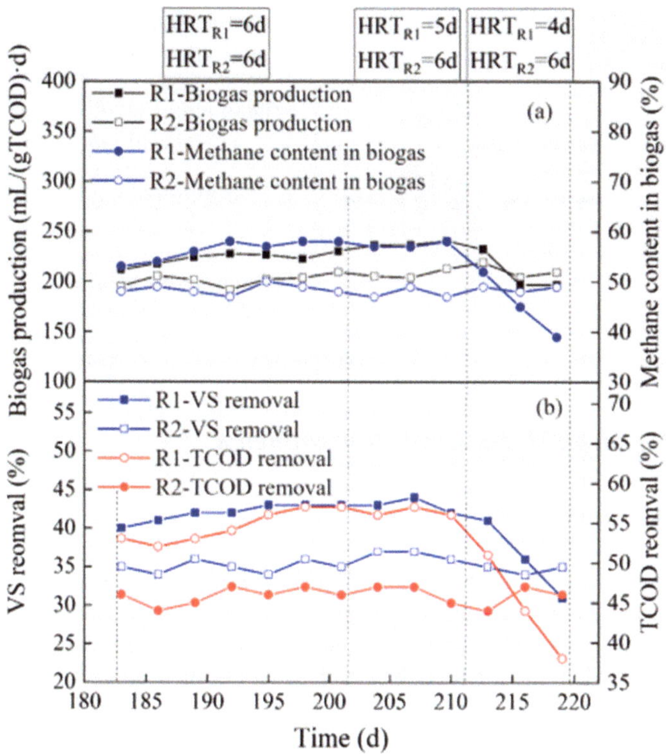

Fig. 6.3 Effect of sludge HPH pretreatment on EGSB operation performance: evaluating biogas production and methane content in biogas (**a**), and TCOD and VS (**b**). Removal (R1 (reactor 1) was fed with the HPH-pretreated sludge and R2 (reactor 2) was fed with the raw sludge as control)

various methods of pretreatments prior to anaerobic digestion have developed to accelerate the hydrolysis step and methane production, such as biological, thermal, chemical, mechanical, and emerging hybrid pretreatments [182]. Many experiments have provided that pretreatments can induce disintegration abilities by break of the cell wall and to make the available organics more accessible to microorganisms (Fig. 6.4). It also has benefits on pathogen elimination, methane production, solid organic disintegration, dewaterability, and bioavailability [156]. Meanwhile, the favorable characteristics of pretreatments with avoidable costs and mass reduction have been documented by numerous researches, but some of them found that the problems of input and greenhouse gas emissions and extra energy obstruct the development of pretreatment.

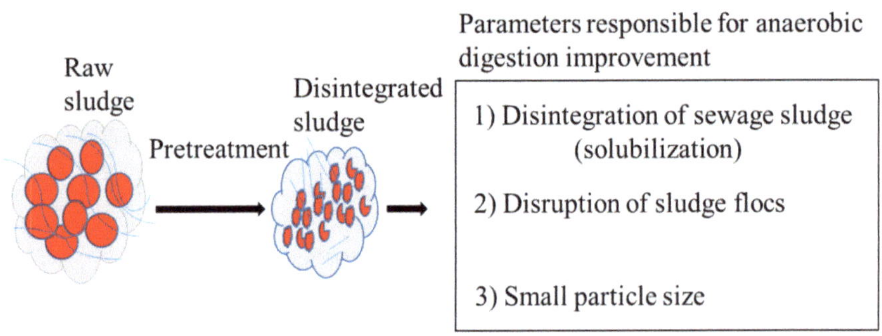

Fig. 6.4 Mechanisms involved in anaerobic digestion improvement via pretreatment

6.3.1 Physical/Mechanical Pretreatments

6.3.1.1 Ultrasonication

Ultrasonication use ultrasound waves at the frequency above 20 kHz to disrupt the floc matrix and sludge mechanically. It has been applied as mechanical technology for sludge decomposition. Microbubbles could be formed with high energy density ultrasound waves. When the microbubbles collapsing in milliseconds, periodic rarefactions and compressions situation will be created in aqueous medium, resulting in vigorous shear forces. The pressures could reach to 180 MPa and temperatures could go up to 5000 K on the interface between aqueous phases and gas [97]. This extreme condition could disrupt adjacent cells to release intercellular substances, like organic matters and enzymes, which would further increase biodegradation of sludge. In specific, reactive oxygen species (hydrogen, hydroperoxyl, hydroxyl, and hydrogen peroxide) could be produced under high temperature to breakdown recalcitrant compounds in oxidative way [86]. Ultrasonication improved AD efficiency by complete disaggregation of sludge flocs, which increase methane yield of sewage sludge.

Although there are many merits, high energy consumption will be the main obscure for widespread of ultrasonication. Celebi et al. [24] evaluated the energy generation and consumption in each step during pretreated wasted activated sludge. In this experiment, 24 kHz ultrasound at a rate of 0.73 W/mL for 10 min has been set up. And they found that soluble chemical oxygen demand (SCOD), methane production and biogas increased 32% and 15%, respectively. But the total energy consumption exceeded energy production. The reason could be that thermal drying, anaerobic digestion heating, and operating ultrasonic generator cause large amount of energy consumption. To make ultrasonication more energy-saving, lower frequency within tens of seconds and shorten operation time could be applied, but the degradation rate goes down. Chu et al. [32] pretreated sludge using 0.33 W/ml and 20 kHz for 20 min, which increased methane yield (143–292 gCH_4/kg dry solids). The possible reason is the insufficient ultrasound generate more hydrolyzed enzymes to degrade organic matters. Other research found that thickened sludge could increase methane production by 70%, but SCOD only increases by 12.8%, which indicates that biodegradable organic compounds may be adsorbed on solid phase [31]. Combined different frequency, treatment time, and energy density of ultrasounds would have different effects on AD. Ultrasonication with low-frequency waves, higher power, and short treatment time could be effectively for most pretreatments of AD.

In summary, ultrasonication can change the floc structure and reduce the particle size of sludge. Also, it has a high degree of decomposition by crushing the bacteria cell in the sludge, which cause the cell sap to seep out and easier for anaerobic bacteria to use. It is benefit for promoting anaerobic digestion, increasing methane production. The greater energy intensity could deconstruct sludge more easily to create smaller sludge particle size. Ultrasonication is good for sludge reduction with treatment time between 30 to 40 s and a frequency of 20–100 kHz. To make ultrasonication lower energy consumption and higher efficiency, the higher initial concentration of sludge would be used to achieve better sludge reduction effect.

6.3.1.2 Microwave Irradiation

Microwave irradiation is the electromagnetic wave in the range of 0.3–300 GHz with the wavelengths of 1 mm–1 m, which can penetrate tens of centimeters in the medium and heat inner substance [35]. Microwave irradiation is applicable used to destroy the cell structure in sludge treatment. The ultrafast friction between molecules cause thermal effect to destruct the cell wall. Tang et al. [145] claimed that microwave could have thermal effect on the vibration of molecular bonds to destruct cellular wall. Toreci et al. [148] conclude that the cell wall destruction is induced by increasing possibility of breakage of hydrogen bonds, the changing dipole orientation of polar molecules, and unfolding of complex biological molecules. Microwave pre-treatment has benefit both in solubilizing the organic matter and increasing biogas production. Appels et al. [9] did experiments of microwave pre-treatment on pilot scale semi-continuous AD and sludge solubilization. They discovered that the solubilization of organic matter in the sludge exhibited enhanced effectiveness, resulting in increase of biogas production (50%). Ahn et al. [1] used microwave irradiation (2450 MHz, 700 W for 15 min) to pretreat the sludge. They stated that the acidogenic potential improved up to 4.77 g COD/l and solubilization degree increased from 2 to 22%. In addition to more methane generated from the microwave pre-treatment, it can also deal with the pathogen present in sludge effectively. According to Hong et al. [68], a nearly 2.66 log reduction in fecal coliforms was observed when microwaves at approximately 2450 MHz were used for irradiation.

Along the time control evolved to temperature control in the operation, the microwave pretreatment has become a more effective method [119]. Alqaralleh et al. [2] set up the experiment with the microwave and the temperature reached 95, 135 and 175 °C. The result showed that he higher temperature could increase methane yield and the microwave would have better disintegrating effect on the thickened sludge. But there are some conflicting opinions on the microwave irradiation. Mehdizadeh et al. [105] compared the conditional heating technology and the microwave irradiation in the effect on dewatered sludge anaerobic digestion and solubilization. But there is no significant different in the biosolids solubilization and methane production. In other words, microwave can only enhance performance relay on the thermal effect. Therefore, it needs more effort to determine other factors that can influence sludge deconstruction beside the thermal effect.

In total, microwave irradiation has many potential benefits to improve the efficiency of AD, and it can generate more methane in the process because of the higher temperature. Cellular structure could be broken with the ultrafast friction between molecules. Microwave pretreatment gained lots of attention due to its short pretreating time and easy operation, but high cost to maintain the working temperature could obscure its industrial application. Therefore, reducing the energy consumption would improve the development of microwave irradiation.

6.3.1.3 Electrokinetic Disintegration

Electrokinetic disintegration is the high-voltage electric field methods, which could disrupt the cellular membranes and rigid sludge flocs. In disintegration process, the nutrients and intercellular materials would be easily released out [30]. Compared with traditional treatment methods, electrokinetic disintegration technology has the advantages with low cost, wide application range, strong controllability, rapid and thorough treatment, and no damage to the natural ecological environment. Lee and Rittmann [84] set up an electrokinetic with a specific energy input of 34 kWh/m^3, provoked soluble compounds up to110–460%, and increase CH4 production and TCOD removal by 33 and 18%. The net positive economic benefit could be generated by electrokinetic treatment, as the energy offsets from the increase in the production of biogas and reduction in biosolids disposal. Rittmann et al. [134] investigated the implementation of electrokinetic for sludge disintegration in a full-scale WWTP prior to anaerobic digestion. The researchers discovered that this electrokinetic treatment method was capable of handling 63% of the incoming waste sludge, resulting in a significant 40% increase in biogas production. Additionally, the application of electrokinetic treatment led to a notable 30% reduction in biosolids disposal.

The electrokinetic disintegration with positive and negative charges could destruct the cellular membrane of the substrates, and break ionic bond to enhances the AD process efficiency, which can increase biogas yield. The long-term observations showed that the electrokinetic disintegration was able to drastically reduce the amount of biological sludge produced by the plant, without affecting its treatment efficiency. The highest reduction was achieved when 100% return sludge flow rate was subjected to the disintegration process. The reduced sludge production gave rise to a considerable net cost saving for the company which manages the plant.

Totally, the electrokinetic pre-treatment technique, which treats biowastes through exposure to a high-voltage electric field, has been reported to increase biogas yield and soluble chemical oxygen demand in waste-activated sludge [182].

6.3.1.4 High-Pressure Homogenizer

HPH operates on the fundamental principle of utilizing intense turbulence, sudden pressure gradient, cavitation, and powerful shearing forces, all of which are generated through the rapid depressurization of extremely compressed sewage sludge organics (reaching pressures of up to 900 bar). In the course of this procedure, the cohesion of sludge flocs are disrupted, and the cell membranes undergo rupture, resulting in the release of intracellular substances. As a result, HPH has the potential to enhance both sludge biodegradation performance and disintegration. Figure 6.5 illustrates the HPH Disruption valve and its corresponding diagram. Currently, numerous studies are being conducted at both full-scale and laboratory scale demonstrations to examine the impact of HPH pretreatment on biogas production and sludge solubilization. Zhang et al. [179] conducted research to investigate how the solubilization of sludge was affected by homogenization cycle (ranging from 1 to 4) and pressure (ranging from 20 to 80 MPa). The findings indicated that increasing the number of homogenization cycles from 1 to 4 as well as pressure from 20 to 80 MPa, had a positive effect on sludge solubilization. The most energy-efficient approach for the sludge sample, which had a TS content of 2.48%, involved subjecting it to HPH pretreatment at 30 MPa with a single homogenization cycle. Rai and Rao [128] noted that the sludge's disintegrated and DD_{COD} exhibited increases of 4.5%, 10.7%, and 12.5% when subjected to treatments at 200, 300, and 400 bar, respectively. But, no additional enhancement was observed with higher pressures. Wahidunnabi and Eskicioglu [154] carried out a laboratory-scale semicontinuous experiment to compare the effectiveness of HPH pretreatment and two-phased anaerobic digestion (referred to as 2PAD + HPH) with conventional anaerobic digestion systems, including single-stage (control) and 2PAD approaches, for treating municipal sludge. The homogenization pressure was identified as the most influential factor in the solubilization of biopolymers and particulate COD in sludge. The HPH + 2PAD system, operated at 12,000 psi and using 0.009 g NaOH/g total solids (TS), demonstrated the highest levels of methane generation (0.61–1.32 l CH_4/l digester-d) and VS removal (43–64%). Moreover, this system exhibited positive energy balance and effectively removed pathogens. The HPH approach has gained significant popularity in recent years for its advantages, including low investment costs, ease of operation, high energy efficiency, and widespread adoption in large-scale implementations. Onyeche [119] conducted a demonstration project wherein a pressure of 150 bar and a flow rate of 2.7 m^3/h were utilized for pretreating concentrated sludge. The findings demonstrated a sludge reduction of approximately 23% and a remarkable increase of over 30% in biogas production. As a result, substantial savings were achieved in sludge disposal costs. In a separate demonstration project, Rabinowitz and Stephenson [126] implemented a patented MicroSludge™ unit at the Los Angeles WWTP. This process involved chemical pre-treatment to weaken cell membranes and high-pressure homogenization (HPH) to rupture the cells [182].

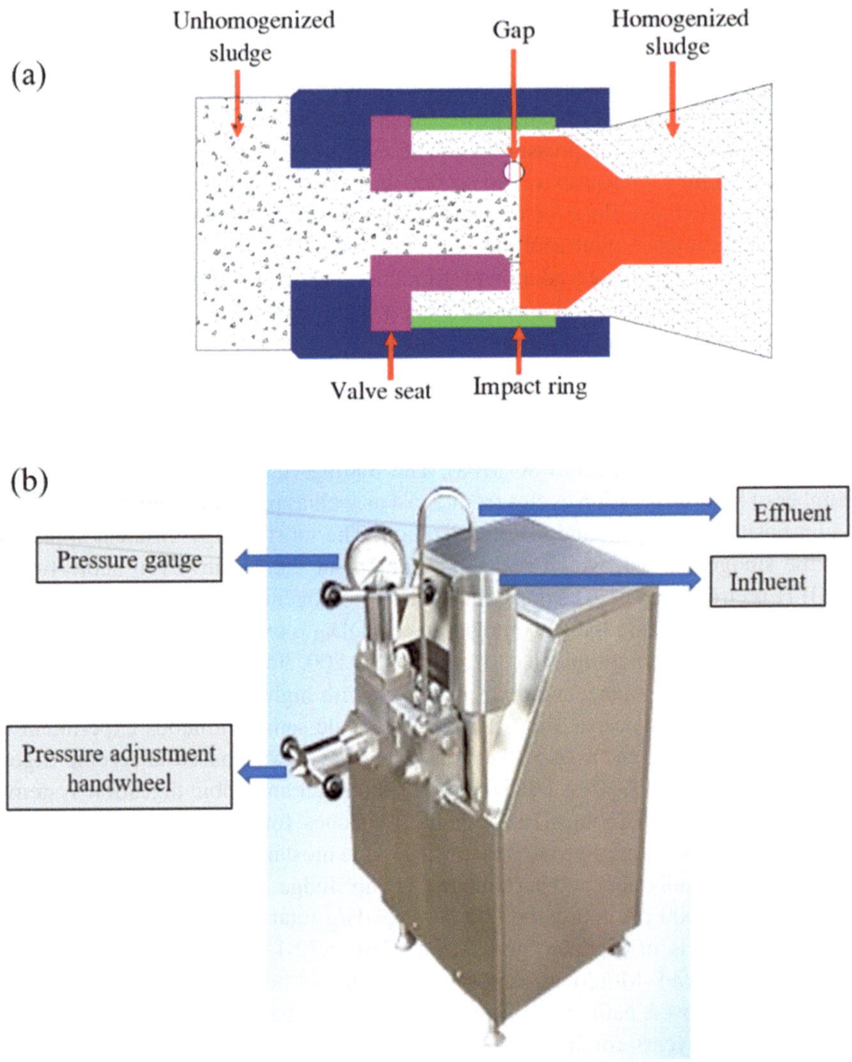

Fig. 6.5 Disruption valve (**a**) and diagram (**b**) of high pressure homogenizer

6.3.2 Chemical Pretreatments

Chemical pretreatment could sterilize and promote the hydrolysis of sludge particles by adding kind of reagents to precipitate and agglutinate sludge. Generally, chemical reagents include surfactants, metal salts, polymer polyelectrolytes, acids and so on, which could promote hydrolysis and sterilization by destructing EPSs and cells [114].

The pH condition is a crucial factor in chemical pretreatments, as it significantly influences their effectiveness. However, achieving the desired pH range can be challenging. Various pretreatment methods, such as alkaline/acid hydrolysis [163], free nitrous acid (FNA) [178], and free ammonia (FA) [155], rely on specific pH conditions. On the other hand, pH-independent pretreatment methods, like granular activated carbon [167], photocatalysis [7], ozonation [36], and nanobubble [169], can function effectively across a wide pH range. This classification allows for the optimal selection of a chemical pretreatment method based on the specific pH requirements of the wastewater treatment process [159].

6.3.2.1 Alkaline and Acidic Pretreatment

Alkaline and acidic pretreatment methods have great potential in sludge solubilization due to their numerous advantages, such as low cost, high efficiency in methane generation, and ease of operation and equipment [184]. Acidic hydrolysis involves the use of acids like H_3PO_4, HCl, HNO_3, and H_2SO_4, while alkali pretreatment utilizes alkaline solutions such as $Ca(OH)_2$, KOH, $Mg(OH)_2$, NaOH, and ammonia. Using acids or alkalis in sludge treatment, reduces the reaction requirements, enabling operation at ambient or moderate temperatures [8].

The efficiency of alkaline or acidic pretreatment may differ depending on their affinity to different organic components. Some studies have found that acidic pretreatment is particularly effective for lignocellulosic substrates by hydrolyzing hemicellulose, which releases monomeric sugars and soluble oligomers from the cell wall matrix, thus improving enzymatic digestibility [62]. However, acidic pretreatment has limitations in lignin hydrolysis, as lignin tends to precipitate and condense. Additionally, it can lead to the generation of toxic by-products, e.g., furfural and hydroxymethyl, which act as strong inhibitors in microbial fermentation system [62]. Acidic pretreatment also poses challenges due to its high toxicity and corrosiveness, requiring special materials for digester construction. To address these issues, the literature has presented the integration of acidic-thermal pretreatment. For example, Nielsen et al. [117] applied this method to waste activated sludge, Neyens et al. [116] utilized it for thickened sewage sludge, and Takashima and Tanaka [144] explored its application with anaerobically digested sludge [182]. This integrated approach aims to overcome the limitations of acidic pretreatment while improving its effectiveness in biomass solubilization.

6.3.2.2 Ozonation

Ozonation is widely recognized as a prominent chemical pretreatment method that effectively disintegrates cell structures and enhances sludge solubilization. It has found successful implementation in the recirculation of valuable products to improve the removal of biological nutrients in the activated sludge process, particularly under alternating anoxic/aerobic conditions. The efficiency of sludge solubilization exhibits a linear correlation and is dependent on the dose of ozone applied within a moderate range. However, this efficiency is closely associated with both the ozone reaction kinetics and mass transfer. Despite an increase in ozone dose, the kinetic reactions between the sludge mixed liquor and dissolved ozone display a lower average rate, which prevents continuous COD solubilization. Ozonation, the most commonly employed chemical method, partially solubilizes sludge, with the solubilization yield increasing with the dose of ozone applied. However, excessively high ozone doses can reduce solubilization due to oxidation of the solubilized components. Furthermore, ozone's oxidative nature may lead to increased destruction at the expense of methane yield. Several studies have identified an optimal ozone dose for enhancing anaerobic biodegradability. Initially, sludge ozonation was combined with the activated sludge process for wastewater treatment, demonstrating the feasibility of operating the process without physically wasting excess sludge. In one approach, thickened sludge was dosed with ozone (0.02 g O_3 g^{-1} TS) and reintroduced into the aerated tank, although the effluent displayed slight increases in nitrogen and suspended solids concentrations, which remained within authorized limits [136]. A recent review by Chu et al. [32] discussed various studies on the combination of ozonation with the activated sludge process, with around 30 installations being implemented commercially by the Japanese Kurita company. Deleris [39] proposed another industrial process known as Biolysis® O. It should be noted that excessive ozone levels can result in partial or complete mineralization of cellular materials, which significantly dropped methane production. The optimal sludge solubilization occurs within the ozone dose range of 0.05–0.5 g O_3/g TS, depending on the pretreatment conditions and sludge characteristics. However, sludge ozonation requires significant energy for ozone production (12.5 kWh/kg O_3) and liquid oxygen (0.5 kWh/Nm^3 O_2) as well as for ozone transfer to the sludge (2.5 kWh/kg O_3). To address the cost of sludge ozonation while maintaining high performance in pretreatment, Chu et al. [33] settled a system of microbubble ozonation. This system improved the efficiency of sludge solubilization from 15–30% to 25–40% at ozone dosage of 0.06–0.16 gO_3/g TSS and increased ozone application from 72 to 00%. The application of microbubble ozonation expedites the generation of hydroxyl radicals, accelerating sludge solubilization and mitigating the impact of high energy requirements [182].

6.3.2.3 Fe(II) Activated Persulfate Oxidation

Iron (II)-activated persulfate (Fe(II)-$S_2O_8{}^{2-}$) oxidation is a newly sludge pretreatment method first discovered by Zhen et al. [185]. Persulfate ($S_2O_8{}^{2-}$) is more easily activated by ultraviolet rays and heat, which may transform metal (Men +) into sulfate radicals ($SO_4{}^{-}$) the extremely strong oxidant with redox potential of 2.60 V. Under the optimum conditions of 1.2 mmol $S_2O_8{}^{2-}$/g VSS and 1.5 mmol Fe (II)/g VSS, the capillary aspiration time (CST) was eliminated by 88.8% in only 1 min at pH 3.0–8.5, which provided a strong oxidation effect for SO^4 to bacterial cells attack and EPS. SO^4 can destroy specific functional groups of fluorescent substances in EPS, such as tryptophan protein, humic acid, fulvic acid-like substances and aromatic protein. And it would cause bond breakage in polymer skeleton, while destroying bacterial cells, releasing water bound to EPS, hydrated water in cells and subsequent improvement of dehydration ability [185]. Zhen and his colleagues conducted additional research on the combined impact of Fe(II)-$S_2O_8^{2-}$ oxidation with heat treatment (mild temperature) [185] and even electrolysis [183]. In particular, they found that sulfate radicals possess a greater oxidation potential across a broader pH range of 3.0–8.5 when compared to hydroxyl radicals. Hydroxyl radicals, on the other hand, exhibit higher selectivity for oxidation under acidic and aqueous conditions. Therefore, it can be an alternative and more cost-effective method to improve dewatering capacity and to reduce sludge weight in the process. Fe (II)-$S_2O_8^{2-}$ oxidation has received unprecedented attention in recent years because of its superior properties [139]. Some studies have found that ZVI can be treated as a good catalyst in oxidation process [187], but compared with the performance of Fe(II), Fe0 needs to be dissolved before reacting with Peroxodisulfate ($S_2O_8^{2-}$). Different from dewatering experiments, the application of Fe(II)-$S_2O_8^{2-}$ oxidation in anaerobic digestion of sludge is still limited [182].

6.3.2.4 Fenton Oxidation

Fenton oxidation is a widely recognized technique for oxidation that involves the reaction between catalyst iron ions (Fe^{2+}) and hydrogen peroxide (H_2O_2), resulting in the generation of highly reactive hydroxyl radicals. Hydroxyl radicals are well-known for their superior oxidation potential i.e., + 2.80 V (under acidic conditions), surpassing that of ozone (+ 2.07 V) and hydrogen peroxide (+ 1.36 V) [116]. These hydroxyl radicals are particularly effective in breaking down sludge EPS and lysing microbial cells, leading to the release of bound water and intracellular materials. Many studies have applied Fenton oxidation to enhance sludge dewatering, reducing the volume of sludge and solids while increasing CH_4 production [115, 147, 184]. The effectiveness of this process depends on various factors, including the concentrations of reagents (H_2O_2 and Fe^{2+}), the Fe^{2+}/H_2O_2 ratio, initial temperature and pH, as well as the treatment time conditions [45]. Dewil et al. [40] showed the effectiveness of Fenton oxidation by using 0.07 g Fe^{2+}/g H_2O_2 and 50 g H_2O_2/kg DS at pH 3.

The pretreatment resulted in a significant improvement of 75% in biogas production and 5.0 times higher SCOD compared to the control group. In a recent study by Pilli et al. [122], they treated sludge through 60 g of H_2O_2 and 0.07 g of Fe^{2+} per gram of H_2O_2 per kilogram of total solids (TS) at pH 3. They observed a 15% higher methane production (increased from 430 to 496 m^3 CH4/Mg VS degraded), reduced greenhouse gas (GHG) emissions (0.128 Mg CO_2/Mg), and a 3.1-fold rise in net energy generation. Additionally, Şahinkaya et al. [135] explored and compared the enhancement of sludge disintegration and anaerobic biodegradability efficiency using conventional Fenton (Fe^{2+}/g H_2O_2, CFP) and Fenton-type (Fe°/g H_2O_2, FTP) processes. They found that CFP, utilizing Fe^{2+} in its dissolved form as the catalyst, was a more effective pretreatment method for sludge. Under optimal conditions of 40 g of H_2O_2 per kilogram of TS, 4 g of Fe^{2+}/° per kilogram of TS at pH 3, and 1 h of oxidation time, FTP and CFP improved methane production by 38.0% and 26.9%, respectively, compared to the control. Another experiment conducted by Zhou et al. [186] demonstrated a 10% improvement in methane production using 7 mg of Fe per gram of TS, 50 mg of H_2O_2 per gram of TS, at pH 2.0 for 30 min, where indigenous iron in the sludge served as the catalyst. A major drawback of Fenton reagent is its requirement of low pH values to avoid hydrolysis and precipitation of Fe^{3+} [12]. Moreover, pretreated sludge needs neutralization prior to digestion, and an excessive dosage of H_2O_2 or Fe^{2+} can scavenge hydroxyl radicals, thus reducing their concentration.

6.3.2.5 Thermal Hydrolysis

Thermal hydrolysis is a well-established and extensively adopted pre-treatment technique employed to improve the dewaterability of sludge [64, 116]. As a traditional technology, it has undergone extensive research efforts to improve sewage sludge digestion, with some notable studies conducted by Hii et al. [65] and Pilli et al. [122]. The effectiveness of the thermal hydrolysis process relies heavily on the time and temperature parameters employed during treatment. In the early 1990s, Li and Noike [85] conducted continuous and batch experiments and found that the optimum treatment time ranged from 15 to 60 min, with temperatures between 62 and 175 °C for the biological treatment of sludge. The results indicated that sludge solubilization increased with higher temperatures, and the thermal hydrolysis effect had a greater impact on proteins and carbohydrates compared to lipids. When incorporating thermal hydrolysis into an anaerobic sludge treatment system with a 5-day solids retention time (SRT), the optimal treatment time was found to be 30–60 min at a temperature of 170 °C, resulting in increased biogas production without any observed inhibitory effects at the studied temperature [186]. Consequently, the sludge thermal hydrolysis pretreatment has gained increasing attention from researchers due to its effectiveness. Carrère et al. [22] applied the thermal hydrolysis process at temperatures ranging from 60 to 210 °C, for sludge pretreatment. They found that sludge disintegration increased with the applied temperature. Within the investigated temperature range, both methane production and biodegradation were enhanced as a function of sludge COD solubilization through thermal hydrolysis. However, further increasing the thermal hydrolysis temperature (e.g., to 210 °C) had a reverse effect on biodegradability due to the formation of recalcitrant compounds through Maillard reactions, such as melanoidins, which are challenging to degrade and may inhibit the biodegradation of other organics to some extent [34]. Additionally, they concluded that sludge solubilization did not appear to differ significantly, even with longer treatment durations (e.g., 4 h compared to 30 min), indicating that the impact of thermal treatment time on sludge solubilization was less significant when compared to temperature. Conversely, if temperatures are not sufficiently high, longer heating periods may be required for the process [18]. In a study by Kinnunen et al. [80] examined how different thermal pretreatments impact sludge solubilization and methane production derived from paper wastewater. The researchers discovered that thermal pretreatments ranging from 105 to 134 °C resulted in a notable increase of 39–88% in sludge disintegration and biomethane potential. However, they observed that a pretreatment condition of 80 °C had no effect on biomethane potential. In a continuous mesophilic anaerobic process, thermal pretreatment at 121 °C reduced hydraulic retention time (HRT) to 10 days (organic loading rate (OLR) of 2.2 kgVS/m^3 day), resulting in a 77% increase in methane yield (138 NL CH_4/kgVS), while digestion of the untreated biosludge required longer HRT due to slower hydrolysis. Shorter HRT requirements lead to lower energy consumption and smaller reactor sizes. Climent et al. [34] found that thermal hydrolysis time.

6.3.3 Biological Pretreatments

Biological treatment encompasses a wide range of anaerobic and aerobic processes that can effectively facilitate sludge degradation, particularly in the context of anaerobic digestion. Biological pretreatment methods include various approaches such as pre-digestion, enzymatic hydrolysis, and the utilization of fungi or biosurfactants. One effective strategy is the implementation of temperature phased anaerobic digestion (TPAD), which involves subjecting sludge to a thermophilic pretreatment part before undergoing mesophilic anaerobic digestion. This approach enhances the hydrolysis process during the initial stages of digestion. TPAD represents a major form of biological pretreatment and has been extensively studied. Another notable biological treatment method is the microbial electrolysis cell, which is derived from the microbial fuel cell technology. It is an advanced technique for producing methane through a process known as electro methanogenesis. This section discusses the processes involved in these two biological pretreatment methods.

In TPAD, the integration of both thermophilic and mesophilic temperature regimes serves to enhance the hydrolysis of the feedstock and promote acid production during the thermophilic phase. This, in turn, facilitates increased syntrophic acetyl and methane production during the subsequent mesophilic stages [100]. Therefore, thermophilic hydrolytic activity including mesophilic digestion, single stage, dual digesters attributed to increase the efficiency of hydrolytic activity. Compared to the processes mentioned above. TPAD has several advantages, including better solid decomposition, use of low heat energy, higher methane production, low energy input [60] and pathogen sterilization [132]. Up to now, most studies have evaluated TPAD system. Bolzonella et al. [17] reported that the maximum specific methane yield of 370 mL/g VS added to TPAD (HRT 2–3 d, 70 and 37 °C) was 30–50% higher than that of moderate temperature and thermophilic single stage test. Ge et al. [51] conducted optimization studies on the thermophilic pretreatment conditions, including pH, temperature, and residence time, to enhance the overall performance of TPAD. They achieved methane production of up to 300 ml/g VS added (with a HRT of 14 days at 35 °C) after subjecting the sludge to thermophilic pretreatment at pH 6–7, for a duration of 1–2 days, at 65 °C. Skiadas et al. [140] observed that the volatile solids (VS) destruction of primary and secondary sludge in a TPAD system (with hydraulic retention times of 2 days at 70 °C and 55 °C) reached 55% and 43%, respectively. These values represented an increase of 43% and 6% compared to single-stage thermophilic anaerobic digestion at 55 °C. Furthermore, TPAD (at 55 °C with a hydraulic retention time of 3 days, and at 35 °C with a hydraulic retention time of 15 days) not only increased methane yield (0.62 L/g VS), but also resulted in the production of Class A biosolids with fecal coliform levels of 103 MPN/g TS and Salmonella levels of 1 MPN/4 g TS [133].

Furthermore, TPAD has been combined with various other pretreatment methods, such as ultrasonic treatment [131], microwave treatment [35], or ozonation applied between the thermophilic and mesophilic stages [177]. A recent study conducted by Coelho et al. [35] utilized ten semi-continuous reactors to examine the effects of digestion temperature, phase separation, and SRT on process efficiency. The integration of microwave treatment with staged digestion demonstrated several benefits, including increased biogas production, improved solid destruction, complete pathogen elimination, reduced hydraulic retention time (HRT), and enhanced dewatering capacity of the digested material. Despite the promising prospects, the TPAD process is still in the developmental stage. Further research is required to conduct energy balance analysis, particularly for system upgrades prior to large-scale implementation [100]. Additionally, it is crucial to investigate the composition of bacterial communities and the functional diversity involved in the TPAD process. This research focus aims to provide biodynamic parameters for optimizing the digester performance [177].

Microbial electrolysis cell process mainly includes bacteria (exoelectrogenic) in anode that required organic matters to release protons and electrons to anode and solution in the same time. In the cathode, microorganisms can accept electron by electrochemical activities with a small voltage input, which could drive methane formation. In addition, it has been proved that integrating MEC with anaerobic digester can significantly improve methane production and system stability [120, 137], and even repair AD system with process failure [37]. Recently, there have been various technical advancements aiming to integrate microbial electrolysis cells (MECs) with anaerobic digestion systems to enhance methane production. Encouraging results have been obtained in these endeavors. Guo et al. [57] conducted a study on the bioelectrochemical enhancement of methane and hydrogen production using a single-chamber MEC filled with sludge. They observed that at voltages of 1.4 and 1.8 V, methane and hydrogen production increased by 11.4–13.6 times and 1.7–5.2 times, respectively. In a separate study, Sasaki et al. [138] developed a bioelectrochemical reactor incorporating carbon fiber fabric for methane production of concentrated sewage sludge. They achieved a biogas production of 3.57 L/L d with a HRT of 4 days, which was 3.3 times higher than the control group yielding 0.83 L/L d. Similarly, Liu et al. [92] reported an increase in methane productivity to 91.8 g CH_4/m^3-reactor when combining microbial electrolysis with an anaerobic reactor, which was three times higher than the control group producing 30.6 g CH_4/m^3.reactor.d in anaerobic digestion system. Apart from methane production, MECs can also generate acetates, formates, bio-alcohols [81], and other valuable products, and can contribute to the bio-electrocatalytic reduction of CO_2. Furthermore, MECs can utilize external power sources such as solar energy and wind energy, enabling the simultaneous storage of solar energy/wind energy and bioenergy within this unique anaerobic system.

6.3.4 Hybrid Pretreatments

6.3.4.1 Thermal-Chemical Combination

The majority of pretreatment techniques either have high operation, installation, and O&M expenses or require significant amounts of energy. However, combining various chemical, physical, and mechanical pretreatments can offer a viable alternative for improving sludge liquification, reducing volatile solids and increasing methane production. This combined approach also has the potential to reduce energy consumption, minimize the need for chemical products, and save time in comparison to using single treatment methods [149, 153].

Compared to single chemical and thermal treatments of sludge, thermochemical pretreatment yielded much higher solubilization of volatile solids. In recent laboratory studies, a range of new combined pretreatments have been suggested and examined. Liu et al. [93] experimented with combining free ammonia (FA) and heat pretreatment to enhance the biochemical conversion of sludge to methane. The results showed a 25% increase in methane production and a 52% increase in hydrolysis rate after the pretreatment of 135 mgFA/l at 70 °C. Authors confirmed in another recent study that treating sludge with a nitrite pretreatment at − 5 °C freezing (600 mg NO_2^--N/L) for 4 h enhanced methane generation from 178 mlCH_4/gVS (untreated sludge) to 233 mlCH_4/gVS [96].

However, thermochemical treatments are mostly investigated merely on a laboratory scale when it comes to their applications in pretreatment methods. The Krepro process is a commercial facility that employs thermochemical method to process sludge. It was operated at a plant (demonstration) in Sweden for three years. The process involves adding H_2SO_4 to thickened sludge to lower its pH to between 1 and 2. The acidified sludge is then subjected to thermal hydrolysis in a pressure vessel at 3.5 bar and 140 °C for 30 to 40 min. This resulted in approximately 40% of particulate organics being solubilized [149].

The combination of alkalinity and pyrohydrolysis can effectively disintegrate sludge flocs and break down complex organic matter. For instance, Liu et al. [95] used hot alkali to pretreat WAS and observed improved degradation of PAM, as well as the prevention of acrylamide accumulation. However, during thermal hydrolysis pretreatment, the formation of maillard reaction products can resist degradation and even have adverse biological effects. Geng and Zhou [52] developed a method to decrease refractory maillard reaction products without affecting solubility by adding 10 mg/L $FeCl_3$ to the TH pretreatment, which increased the solubility and decreased refractory residues. This method also enhanced the biodegradability and methane production of WAS. The addition of Fe^{3+} triggers a Fenton-like reaction, leading to an increase in biodegradability and the enrichment of microbial communities related to protein degradation and methane production.

In addition to high temperature, recent studies have shown that freezing can also improve sludge flocs biodegradation performance. Liu et al. [93] found that nitrite-based freezing pretreatment can significantly promote the degradation of humic acid-like substrates. This is due to the destruction of EPS and cells through the formation of ice crystals during the nitrite-based freezing process. The freeze concentration effect also causes a dramatic decrease in pH and increase in nitrite concentration, which produces reactive derivatives of nitrite that act as oxidizing species to facilitate the degradation of refractory compounds such as humic substrates, pharmaceuticals, and PPCPs. Consequently, the bio-degradability of solubilized substrates is improved.

6.3.4.2 Mechanical-Chemical Combination Pretreatments

(a) **Ultrasound-chemical combination pretreatments**

Ultrasonic pretreatment proves to be highly effective in rapidly releasing intracellular substances by breaking down cell walls. However, it has a negative effect on breaking down larger molecules. This issue can be resolved by adding chemicals to the process. Combining ultrasound and alkaline processes to pretreat sludge has been shown to increase the soluble chemical oxygen demand (SCOD) through several times compared to using only ultrasound. Implementing a stepwise procedure of ultrasound-alkaline pretreatment enabled a significant 40% boost in methane production compared to a continuous procedure of equal duration. Additionally, the stepwise procedure required 20% less chemical input [31, 146].

The combination of ultrasound and alkali had a synergistic effect that involved two pathways. The initial pathway was attributed to the vigorous oscillation induced by ultrasound, enhancing the chances of alkaline reagents reacting with cellular walls and EPS. The second pathway relied on the alkaline environment, which facilitated the generation of radicals induced by ultrasound. According to Li et al. [87], the improvement in sludge biodegradability was higher with ultrasound combined with potassium ferrate compared to ultrasound with alkali.

However, the application of ultrasonic waves had an adverse effect on sludge dewaterability by creating eluted soluble extracellular polymeric substances (sEPS) and colloidal solids that hindered sludge filterability. Conversely, the presence of Fe (III) from potassium ferrate facilitated the reassembly of fragmented sludge particles, thereby improving dewaterability. Furthermore, the combined use of ultrasound and ozone pretreatment demonstrated the ability to eliminate antibiotic-resistant bacteria and degrade trace levels of levofloxacin, while concurrently reducing the presence of quinoline resistance genes-qnrA and qnrS in sludge [181]. Ushani et al. [152] discovered that the combination of surfactant and ultrasonic pretreatment yielded superior sludge disintegration and a higher hydrolysis constant compared to using ultrasound alone, leading to a remarkable six-fold reduction in energy demand.

(b) **Combination of HPH with other pretreatments**

Using a single pretreatment method, whether it's physical, chemical or biological, has certain drawbacks such as high energy input, poor sanitization, and high capital and operational costs. To mitigate these disadvantages, it is recommended to use hybrid pretreatment techniques. Many researchers have studied the use of hybrid pretreatment methods to improve bioenergy production [20, 173]. To increase sludge disintegration efficiency and decrease operational costs, HPH can be combined with other pretreatment methods like chemical pretreatment. However, there have been only a few studies conducted so far on combining HPH with other pretreatment methods, which requires further exploration.

The disintegration degree (DD_{COD}) of sewage sludge can be increased through a combination of alkaline and HPH pretreatment, according to Zhang et al. [180], Alkaline pretreatment weakens the cell wall, which makes HPH treatment more effective. In comparison to the individual alkaline and HPH treatments under the same operational conditions (pressure = 80 MPa, cycle number = 3 and NaOH dosage = 0.05 mol/L), the DD_{COD} of sludge after alkaline + HPH pretreatment was 59.25%, while the total DD_{COD} of the individual treatments was 40.07%. This suggests that other pretreatments may also have a synergistic effect when combined with HPH pretreatment. Fang et al. [50] discovered that when combined with an NaOH dosage of 0.04 mol/L, the DD_{COD} and SCOD increased by 26% and 93%, respectively, compared to HPH pretreatment alone at a pressure of 60 MPa, and cumulative biogas production increased by 60% after 14-day anaerobic digestion. The combination of NaOH and HPH pretreatment was also shown to improve energy efficiency when compared to single HPH pretreatment [180]. According to Fang et al. [49] through pretreatment at a pressure of 60 MPa, employing one homogenization cycle, and using an alkaline dosage of 0.04 mol/l, significant improvements were observed in a mesophilic anaerobic digestion process. Specifically, there was an 18.95% increase in sludge VS removal, a 24.68% increase in total TCOD, and a remarkable 95.81% increase in cumulative biogas production compared to single alkaline pretreatment. Enzymatic hydrolysis was significantly higher in lignocellulosic biomass pretreated with NaOH + HPH than with single NaOH pretreatment, according to Jiang et al. [72]. However, microbial activity may be inhibited by high alkaline dosages, necessitating close supervision during the practical application of such hybrid pretreatment methods [123].

In a different research, a combination of ethylenediaminetetraacetic acid (EDTA) and HPH was employed to disrupt bacterial cells and release intercellular proteins. The highest amount of proteins was observed at 13.8 MPa when using this approach, whereas using HPH alone required 34.5 MPa to achieve the same protein release. This low-pressure method resulted in a 60% reduction in energy consumption for cell disintegration. Combining EDTA pretreatment with HPH is an effective method for improving intercellular protein release at lower pressures while minimizing protein denaturation caused by chemical exposure [5].

The examples provided above demonstrate that combining HPH pretreatment with other methods positively impacts bioenergy production. When applying HPH pretreatment to substrates such as sewage sludge or lignocellulosic biomass, the resistance to disintegration can be addressed by combining it with other pretreatment methods. To further improve bioenergy production and reduce operating costs, there is a need for further research to identify the most suitable physical, biological, or chemical pretreatment methods to combine with HPH. The efficacy of FNA in disintegrating sewage sludge has been studied, with FNA being produced as a by-product through the nitration of anaerobic digestion liquor in WWTPs [158]. Given these advantages, it has been proposed to use a combined pretreatment of HPH and FNA for sewage sludge.

Recently, researchers have been utilizing free nitrous acid (FNA) pretreatment to treat sewage sludge. FNA is a by-product that can be produced in WWTP through the nitration of anaerobic digestion liquor [89]. In order to remove ammonium ($\sim$ 1 g N/L) from the liquor of anaerobic digestion, nitrite is an essential intermediate that needs to be removed. At pH $\sim$ 6, the nitrite concentration of this intermediate product is approximately 1 g N/L, which is equivalent to the ppm level concentration of FNA, thereby making FNA an effective option for treating sewage sludge. FNA present in the pretreated sewage sludge can be diluted and eliminated through denitrification without any adverse impact on the anaerobic digestion system.

Nabi et al. [110] employed a hybrid pretreatment method involving HPH and FNA to decrease digestion time and enhance methane production from sewage sludge. Their study aimed to examine the impact of combined HPH-FNA pretreatment on biogas generation during anaerobic digestion of sewage sludge. The researchers subjected the sludge to pretreatment using combined HPH-FNA (at 40 MPa and 2.49 mg N-HNO_2/L), FNA alone (at 2.49 mg N-HNO_2/L), and HPH alone (at 40 MPa). They evaluated and compared the sludge disintegration and subsequent enhancement in anaerobic digestion.

Both HPH and FNA pretreatments were found to impact biogas production during anaerobic digestion. From the 3rd to the 8th day, biogas production increased significantly. HPH-FNA pretreatment resulted in the highest biogas production rate of 80 ml/d, whereas raw sludge produced only 30 ml/d (Fig. 6.6a), indicating successful acceleration of the sludge's anaerobic digestion with pretreatment. Cumulative biogas production was 53.12, 28.81, and 182.81% higher in the reactor with HPH-FNA pretreated substrate than in those with FNA, HPH, and raw sludge, respectively (Fig. 6.6b). The increased biogas production was attributed to the disruption of flocs, release of extracellular substances, and destruction of cells, leading to the release of intercellular material into the aqueous phase. Zahedi et al. [176] reported an increase in biogas production with FNA pretreatment, while in this study, the synergistic effect of HPH and FNA pretreatments significantly enhanced biogas production.

The efficiency of anaerobic digestion can be evaluated by assessing the removal of VS and TCOD, which are highly correlated with biogas production. During anaerobic digestion, the microflora in the system converts organic matter through a complex process, ultimately producing biogas. As shown in Fig. 6.7, pretreatment had a significant impact on organic removal compared to raw sludge. The pretreatments of

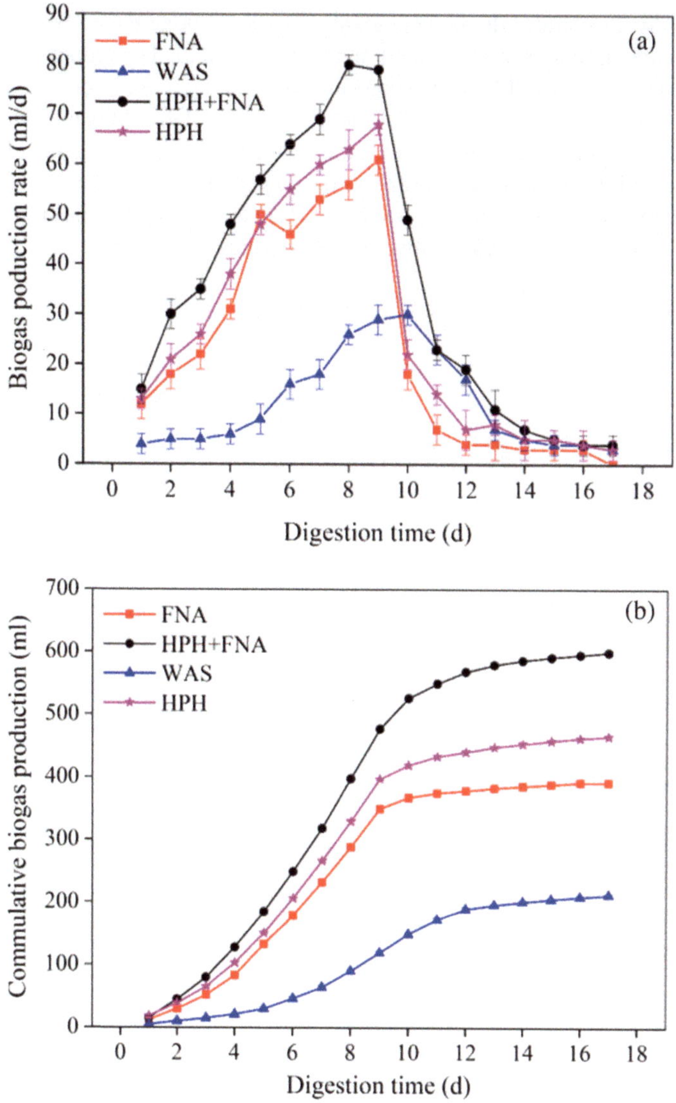

Fig. 6.6 Effect of pretreatment methods on **a** rate of biogas production and **b** cumulative biogas production

FNA, HPH, and HPH-FNA resulted in 40%, 47%, and 53% greater VS removal than raw sludge, respectively. Similarly, TCOD reduction was highest (58%) for the HPH-FNA pretreatment, which was 59% higher than raw sludge. The increase in organic removal can be attributed to the fragmentation and solubilization of sewage sludge, which increases vulnerability to enzymatic attacks [98]. This solubilization of organics supports microorganisms at the initial stage of anaerobic digestion and subsequently accelerates the biodegradation of solid sludge particles, as evidenced by the agreement between VS reduction and biogas yield. The reduction in particle size due to pretreatment likely also contributed to the increase in organic removal.

Figure 6.8 indicates a direct correlation between cumulative biogas production and DD_{COD}, suggesting that the pretreatment facilitated the release of organic matter from the sludge and its subsequent conversion to biogas through biochemical processes. This relationship is described by a linear Eq. (6.1) that highlights the significance of DDCOD as a predictor of cumulative biogas production.

$$y = 0.875x + 37.5 \tag{6.1}$$

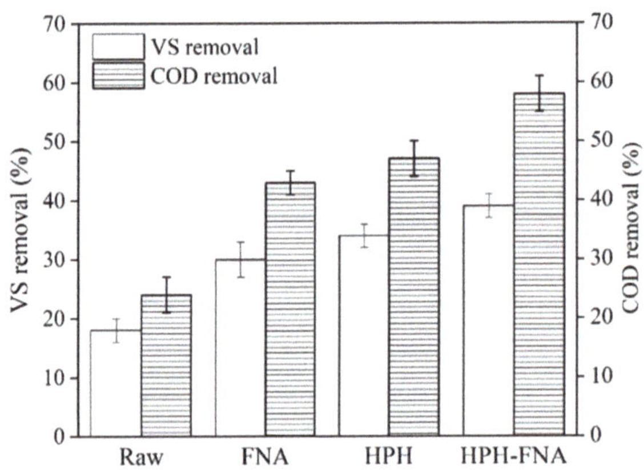

Fig. 6.7 Influence of pretreatment methods on the removal of TCOD and VS

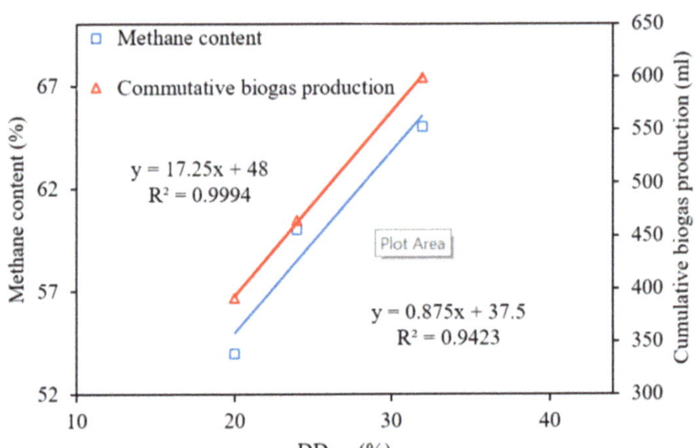

Fig. 6.8 Correlation between cumulative biogas production and methane content in biogas with DD_{COD} increase

where y and x represent cumulative biogas production (ml) and DDCOD (%), respectively.

With HPH-FNA pretreatment, the methane content in biogas increased by 8% and 20% compared to single HPH and FNA pretreatments, respectively. The release of organic matter to the liquid phase facilitated the conversion of methane by methanogens. A linear relationship between methane content and DD_{COD} is shown in Eq. 6.2, where x and y represent DDCOD (%) and methane content (%) in biogas. Figure 4.8 illustrates a strong correlation between the increase in DD_{COD} and the increase in methane content, as indicated by an R2 value of 0.9994.

$$y = 17.25x + 48 \quad (6.2)$$

(c) **Microwave-chemical combination pretreatments**

Numerous researches have shown that the VSS solubilization was notably greater when sludge underwent combined thermochemical pretreatment, as compared to when it underwent thermal or chemical treatment alone. Due to their time-consuming nature, traditional thermal or thermochemical treatments may not be the most practical options. As a result, the combination of chemical and microwave pretreatment presents an alternative approach for sludge treatment.

According to reports, the use of microwaves to release high molecular organic polymers such as proteins, polysaccharides and nucleic acids can lead to reduced dewaterability and high sludge viscosity. However, this problem can be solved through the use of alkali coupled microwave pretreatment, which effectively decomposes the polymers in a strong base. According to Kennedy et al. [76], the solubilization of COD in sludge increased significantly from 2% (control) to 21.7% after treatment with 2 g/L alkali (NaOH) at 85 °C, which was much higher than the increase seen with individual microwave treatment (7%) or alkaline treatment (16.3%). However, the combined alkali- microwave pretreatment did not have a significant effect on either the anaerobic degradability of pretreated sludge or biogas production. Eskicioglu et al. [46, 47] observed a 54% ultimate solubilization of sludge after microwave-alkali treatment of WAS at 175 °C with a 2-mol NaOH/L dose and a two-week reaction time. Qiao et al. [125] found that microwave heating alone reduced VSS by 40% at 170 °C, but adding 0.05 g NaOH per g dry solids increased the VSS dissolution ratio to 50% and achieved a 35% reduction in SS.

In a study conducted by Doğan and Sanin [44], a combination of alkali and microwave treatments at various pH levels was employed to enhance the SCOD/TCOD ratio, resulting in the highest methane yield observed at pH 12.0 and a notable 22% increase in dewaterability. The effectiveness of microwave-alkaline pretreatment relies on factors such as the dosage of NaOH, duration of heating, alkaline condition duration, and microwave power. However, it is important to note that excessive microwave intensity can lead to carbonization and gelatinization, subsequently reducing the dewaterability of the sludge. By utilizing a unique combination of alkaline-rhamnolipid with microwave technology, a significant net yield of 0.39 USD/ton was achieved at a large scale.

The Advanced Oxidation Process (AOP) is a type of eco-friendly technology for sludge treatment that can be considered as one of the "green" options available. AOP can be combined with microwave pretreatment for enhanced efficiency. This method is based on the generation of reactive free radicals, including the very potent hydroxyl radicals, with an oxidation potential of 2.33 V, resulting in a rapid reaction rate. Although AOP has many advantages, its operational cost is relatively high, as mentioned in source. According to the findings of Liao et al. [90], a 5-min microwave pretreatment (at a rate of 20 °C per minute) with a 7-ml/L hydrogen peroxide (30% by weight) dose resulted in over 96% of TCOD being dissolved into the solution. Similarly, Chan et al. [26] discovered that by subjecting sewage sludge to a 80 °C temperature along with a mixture of 34-mL/L hydrogen peroxide and 17-mL/L sulfuric acid, all of the COD was obtained in a soluble form, indicating complete solubilization of organic matter. Adding H_2O_2 into microwave-pretreated sludge can further increase the SCOD ratio, but excessive oxidation of partial organic substances at temperatures higher than 120 °C can lead to low methane production. According to a study by Ambrose et al. [3], enhancing the pH to 11.0 and employing stepwise heating can significantly enhance the utilization rate of H_2O_2 by 3.87 times while reducing H_2O_2 consumption by 80%. Additionally, H_2O_2-microwave pretreatment proves to be highly effective in pathogen reduction, rendering the pretreated sludge suitable for Class A biosolids for landfill application (Table 6.1).

Using only microwave pretreatment is a highly energy-intensive process for achieving a significant degree of sludge solubilization. However, incorporating alkalization into the microwave method can provide benefits in terms of reducing the energy consumption of microwave and the associated treatment costs required to achieve a comparable level of solubilization yield as microwave alone.

6.4 Economic Analysis of Sewage Sludge Pretreatments Methods

Comparing results of sludge cotreatments between different studies is not straightforward due to variations in the anaerobic digestion process parameters (continuous or batch, temperature, HRT) and type of sludge (sludge concentration, waste activated, primary and sludge age. Therefore, to compare pretreatment methods, various factors need to be considered such as whether the treatment method targets activated or primary sludges, treatment effectiveness in terms of increasing degradation rate or bioavailability, cost of treatment including energy cost and secondary costs due to nutrient release or byproduct generation, and chemical consumption, especially for chemical treatment or oxidative.

In one study by Kim et al. [79], different pretreatment methods (thermal, chemical, ultrasonic, and thermochemical) were compared prior to batch anaerobic digestion. The thermochemical and thermal pretreatments resulted in the highest methane production (3367 LCH_4m^3WAS and 3390 LCH_4m^3WAS, respectively), followed by

Table 6.1 Sludge solubilization via microwave-chemical pretreatment

Combined pretreatment conditions	Sludge solubilization	References
85 °C with a dose of 2 g/L of alkali (NaOH)	COD solubilization was increased from 2% (control) to 21.7%	[77]
120–170 °C, 0.2 g NaOH/g dry solids dosage, 5 min reaction time	The treatment resulted in the solubilization of VSS (volatile suspended solids) in the range of 50–70% and COD (chemical oxygen demand) solubilization of 80%	[125]
160 °C, NaOH (pH 12)	The pretreatment caused a significant increase in COD solubilization, rising from 0.5% in the control to 34%. Additionally, compared to the control, there was an improvement of 43.5% in biogas generation	[44]
210 °C, 0.2 g NaOH/g SS, 35 min reaction time	The treatment resulted in the solubilization of VSS (volatile suspended solids) by 85.1%. Moreover, compared to the control reactor, there was an increase of 14% in methane production	[29]
Subjected to microwave-alkali treatment for 10 min with a dose of 1.5 g/L of NaOH, and a pH of 12, at a power of 600 W for 2 min	After subjecting the WAS to pretreatment, there was a notable rise in COD (chemical oxygen demand) solubilization from 0.33% in the control to 45%	[70]
Microwave-alkali with a power of 900 W for 95 s and a pH of 12	There was a significant increase in COD (chemical oxygen demand) solubilization from 0.5% in the raw sample to 52.5% after treatment with microwave-NaOH	[150]
Microwave (20 °C/min) for 5 min and 71 ml/l H_2O_2 dosage (30% by wt)	The solution was able to dissolve more than 96% of TCOD	[90]
80 °C-5 min, 1 ml H_2O_2 (30%)/1%TS dosage, 2.9% TS	SCOD increased by 25%	[75]
Microwave/H_2O_2 (1%)-AOP, 80 °C-3 min	COD solubilization increased by 18%	[99]
120 °C-10 min, 1 g H_2O_2 (30%v/v)/g TS-6.4%	The solubility of COD increased from 3% in the control group to 24%	[46, 47]

chemical and ultrasonic pretreatments (2827 LCH_4m^3WAS and 3007 LCH_4m^3WAS, respectively), while the production from raw sludge was 2507 LCH_4m^3WAS.

Overall, all pretreatment methods need significant resources, and their performance is reflected in the intensity of treatment. Mechanical and sonication pretreatment are characterized by lower energy consumption and primarily enhance the hydrolysis rate. Conversely, oxidation and thermal hydrolysis, which are high-impact methods, yield substantial improvements but require significant energy (and potentially capital) investment.

6.4.1 Methane Yield

The gaseous combination of carbon dioxide (35–45%) and methane (55–65%) is the result of sludge digesters producing biogas, which may also contain small quantities of ammonia and hydrogen sulfide. Additionally, biogas might contain minor amounts of nitrogen, hydrogen, carbon monoxide, oxygen, and halogenated or saturated hydrocarbons. Typically, the biogas is fully saturated with water vapor and include particulate matter and some compounds containing silicon, such as siloxanes [6].

Measuring the biogas yield/production is crucial for evaluating the effectiveness of anaerobic digestion system. Composition and quantity of biogas generated can be influenced by factors such as pH, temperature, moisture, and the makeup of the substrate. Monitoring biogas composition on a regular basis is essential because abnormal microorganism activity and production and can be identified by the concentrations detected. The gases emitted can also reveal how much substrate (organics) remains in the anaerobic digestion system. In a batch digester, biogas production serve as a guide for determining when the entire digestion process will be completed and when new substrate need to be reload. In contrast, the production of biogas in a continuously digester shows the overall performance of anaerobic digestion system and enables consistent production, on condition that an adequate biodegradable organic is provided to support microorganisms' maximum digestion capacity. Biogas production also highlight issues such as temperature variations or pH fluctuations, which can impede the anaerobic digester [54].

In sewage sludge of traditional WWTP, there is a specific load of COD that can be converted anaerobically, which is about 15 kg COD per inhabitant per annum. This COD can be directly transformed into methane, with a conversion rate of 0.35 Nm^3 CH_4 per kg COD. Additionally, this methane can be further converted into thermal energy, with a conversion rate of 10 kWh per m^3CH_4. These values can be used to calculate the amount of electrical and thermal energy that can be generated from methane produced in the anaerobic sludge reactors. Taking into account a total efficiency of 90% and an electrical efficiency of 35% in the combined power and heat equipment, it is possible to generate 16–20 kWh of electrical energy and 25–31 kWh of thermal energy per person per year from the sewage sludge [118]. The calorific value or energy capacity of biogas is determined by its methane concentration. Calorific value of biogas with 60% methane per cubic meter can produce1.5–3 kWh of electrical energy. However, lower values ranging from 1.3 to 2.3 are often obtained in practice [15].

With proper conditioning, the methane component of biogas can replace natural gas. Roughly 1% of the biogas that is utilized is treated to the same standard as natural gas to be transmitted through the natural gas network for heating and energy purposes [108]. To meet the standard of natural gas, biogas need be enriched with methane, and its carbon dioxide, water, sediment, and foam content must be eliminated before compressing and injecting into the electric power grid's distribution network. Biomethane or "green gas" is the term used to describe methane that has been purified and enriched through processes that raise its methane concentration [61]. This approach is widely used by various firms and countries. Biogas can also be upgraded to liquefied natural gas (LNG) and compressed natural gas (CNG) or for use in engine that run on such fuels [143].

Anaerobic digestion has been successfully utilized as a technology to treat sewage sludge, eliminating the need for any pre-treatment. Recently, Xu et al. [164] conducted a review of the main gaps in knowledge that need to be addressed to enhance anaerobic sludge digestion. The author suggests that future research should focus on investigating the structure and properties of sludge in order to improve methane production and overall anaerobic sludge digestion. Various other reviews have also covered the progress made in anaerobic digestion research, including biogas production [102], energy recovery from sewage sludge [101], and organic matter from municipal solid waste [83]. Interested readers are encouraged to refer to these valuable contributions for further information.

During anaerobic digestion process, the characteristics of sewage sludge undergo changes that not only generate biogas but also offer multiple advantageous outcomes for managing the sludge. This process enhances sludge stabilization, minimizes the release of unpleasant odors and pathogens, and decreases the solid contents of the sludge, which leads to a significant reduction in the final volume of sludge.. Researchers widely recognize the advantages of anaerobic digestion in the treatment of sewage sludge, and the technology has been firmly established in multiple countries [38].

Several studies have shown that a high dry solids (DS) content in sludge can reduce the energy required for pretreatment and can also increase the net profits of the facility, with a linear increase in electricity yield (at a rate of 0.2 kWh per kg of total solids increase). Therefore, sludge concentration is a crucial factor in achieving energy self-sufficiency in a WWTP. In fact, a sludge thickening of just over 50 g/L is sufficient for energy self-sufficiency in a WWTP that uses thermal hydrolysis in combination with a CHP system. However, in the absence of a heat integrated system, a minimum sludge concentration of over 250 g/L must be achieved [21, 71].

Previous research has focused on the impact of HPH pretreatment on sewage sludge solubilization and its effect on anaerobic digestion. However, previous studies only applied a sludge disintegration degree (DD_{COD}) of 8–44%, leaving a substantial amount of organic matter in the sludge solid. The contribution of particulate and soluble organics to biogas production and the underlying mechanisms are unclear, and there have been few reports on this topic. Therefore, it is crucial to investigate the impact of sludge solid and liquid fractions on biogas production after HPH pretreatment. In a study, Nabi et al. [113] pretreated sewage sludge using HPH at different pressures, and then separated the sludge into solid and liquid fractions for anaerobic digestion to produce biogas. Additionally, the solid and liquid fractions were characterized to determine the responsible mechanisms for the improvement of anaerobic digestion after HPH pretreatment.

Following HPH pretreatment, a portion of organic matter was released from the sludge solid to the liquid phase, but a significant portion remained in the solid phase.

The biogas production of both solid and liquid fractions is presented in Figs. 6.9 and 6.10. Similar to the anaerobic digestion of HPH-pretreated sludge, the biogas production from the solid fraction after HPH pretreatment increased with higher homogenization pressure. The DD_{COD} of the sludge with HPH pretreatment is a critical factor in improving biogas production, as reported in previous studies [82, 188]. However, the cumulative biogas production from the solid fraction was only 18–25% lower than that from the sludge without solid–liquid separation. This suggests that the solid fraction in sewage sludge has considerable potential for biogas production. The HPH pretreatment disrupted the sludge by breaking down larger particles into smaller ones [142], and the solid fraction of the sewage sludge still contained a significant amount of organic matter that could produce biogas.

Figure 6.10 depicts the production of biogas using the liquid fraction of sewage sludge. Biogas generation occurred solely during the first four days, after which it ceased. One factor behind this outcome was that the readily available organic materials were primarily present in the liquid fraction and were quickly consumed by the microbial consortia. Conversely, the quantity of organics in the liquid fraction was limited, resulting in the cumulative biogas produced from the liquid fraction representing only 7–14% of that generated from the sewage sludge without solid–liquid separation [23, 113].

Evaluating the efficiency of anaerobic digestion requires consideration of the methane content present in biogas. When analyzing biogas from sludge without solid–liquid separation, an increase in homogenization pressure resulted in a corresponding increase in methane content. The maximum methane content of 64% was achieved at a pressure of 60 MPa, as shown in Fig. 6.11.

In a similar vein, an increase in homogenization pressure led to a higher methane content in biogas from the solid fraction. According to Wang and Li [157], the anaerobic digestion system saw a significant improvement in methanogen abundance when using pretreated sewage sludge instead of raw sludge, which could be

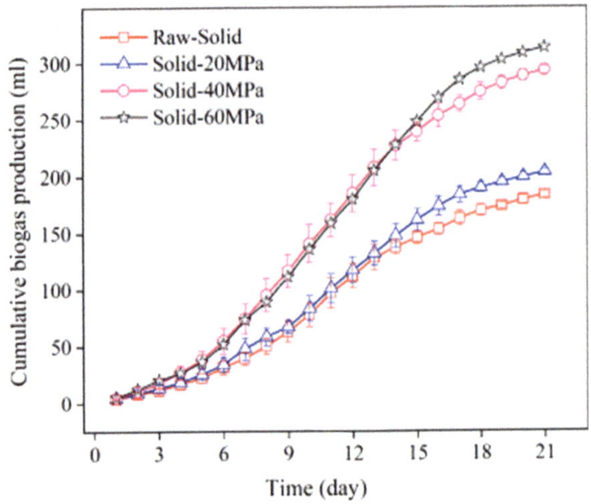

Fig. 6.9 Impact of homogenization pressure on biogas production from sludge solid fraction

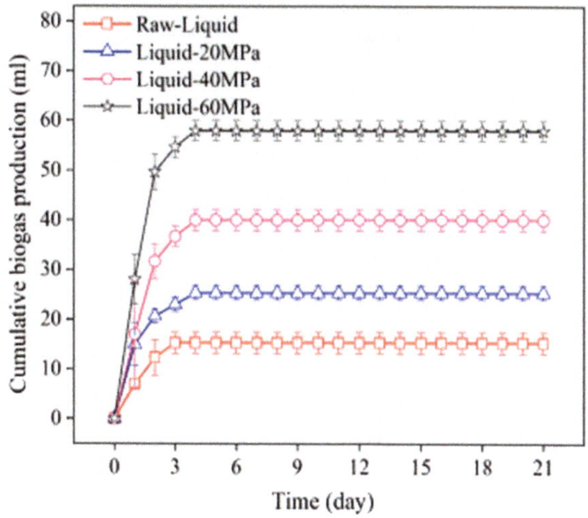

Fig. 6.10 Impact of homogenization pressure on biogas production from sludge liquid fraction

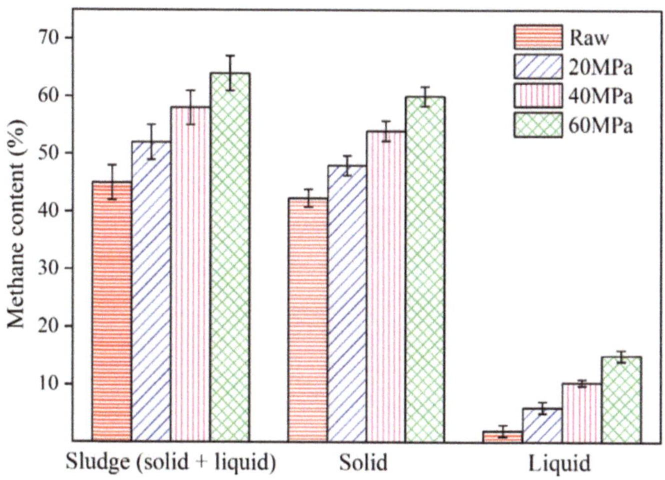

Fig. 6.11 Influence of homogenization pressure on methane content in biogas

a contributing factor to the rise in methane content. However, the biogas methane content from the liquid fraction was low, measuring below 15%. Microorganisms could have metabolized the small amount of soluble organic matter in the liquid phase, leaving insufficient organic matter for methane production [170]. As shown in Fig. 6.12, methane production increased alongside the homogenization pressure increase [113].

It was evident that the liquid fraction yielded very little methane due to low biogas production and methane content. The methane yield per unit VS showed a similar trend to methane production, but the methane yield from the liquid fraction

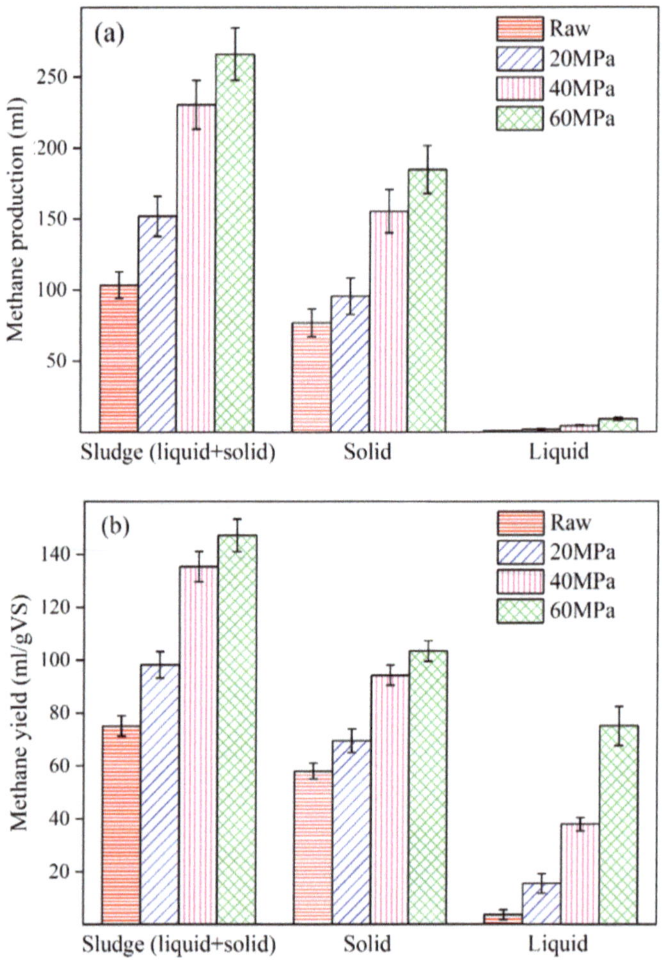

Fig. 6.12 Effect of homogenization pressure on **a** methane production and **b** methane yield

increased noticeably, indicating significant conversion of organics to methane. The liquid fraction was found to be more readily biodegradable than the solid fraction, regardless of the biogas production method used. The lower methane yield from both the solid and liquid fractions could be attributed to substrate conversion to carbon dioxide and hydrogen during acetogenesis and aceticlastic methanogenesis, along with less effective hydrogentrophic methanogenesis. In comparison to previous studies where a combination of HPH and NaOH pretreatment was used, the methane yield of sewage sludge in this study was lower. However, it was higher than the results of Li et al. [88], who employed ultrasonic pretreatment.

6.4.2 Energy Consumption During Pretreatment

6.4.2.1 Energy Consumption of Different Pretreatments

Table 6.2 provides an overview of different pretreatment methods for biogas production, comparing them based on various factors. The methods include Grinding, Pulsed fields, High pressure, Lysing centrifuge, Microwave, Sonication, γ and β radiations, Thermal, Chemical, Thermochemical, Combined, and Biological. The table evaluates each method's suitability in terms of energy demand, biogas production increase, volatile solids (VS) reduction, advantages, and disadvantages. It allows users to make informed decisions on the appropriate pretreatment method for their biogas production process.

(a) **HPH**

The homogenizer valve converts the applied energy into kinetic energy, which is then dispersed into the liquid. As a result, a significant amount of eddies are formed in the turbulent flow, causing microorganism cells in sewage sludge to experience shearing forces of varying scales and intensities. This dispersed energy creates turbulence in the liquid, leading to the formation of both large and small eddies. The small eddies play a crucial role in disrupting the cells [43].

The performance of HPH (high-pressure homogenization) is greatly influenced by the types of biomass being pretreated. Different microbial cells exhibit varying levels of resistance to HPH pretreatment for disintegration [141]. This resistance is attributed to the composition of the cell wall and cell membrane, such as the presence of peptidoglycan chains. For instance, gram-positive bacteria have cell walls consisting of 40 layers of peptidoglycan chains, making them more resistant to HPH pretreatment. On the other hand, gram-negative bacteria have cell walls with approximately 1–5 layers of peptidoglycan chains, which are more sensitive to HPH pretreatment [162]. In a study, a homogenization pressure of 250 MPa was applied to three different microbial strains: *E. coli*, *S. cerevisae*, and *L. delbrueckii*. The resulting inactivation of these strains achieved reductions of 6, 5, and 1 log-cycles, respectively [41]. Therefore, it is essential to consider the compositional characteristics of the biomass being treated. Additionally, when the biomass contains lignocellulose with a compact structure, it exhibits higher resistance to HPH disintegration, and the underlying mechanisms also differ. For instance, Chen et al. [28] pretreated sugarcane bagasse with HPH at 100 MPa and observed an increase in enzymatic digestibility from 29.5% to 59.4%.

The TS content present in the feeding sewage sludge has a significant impact on the effectiveness of HPH pretreatment. When the same homogenization pressure was used on the sewage sludge, the highest and lowest DDCOD (disintegration and dissolution chemical oxygen demand) values were observed at TS contents of 9.58 g/L and 24.60 g/L, respectively [179]. It was found that as the TS content of the feeding substrate increased, the DDCOD at the same homogenization pressure

Table 6.2 Summary of different pretreatment

Pretreatment	Full scale	Energy demand	Increase in biogas production	VS reduction	Advantages	Disadvantages
Grinding	Yes	Electrical	Yes	–	Simple	High capital expenditure (capex)
Pulsed fields	Yes	Electrical	Yes	9%		Significant energy demand
High pressure	Yes	Electrical	Yes	7–138%	Short retention time	Excessive energy requirements
Lysing centrifuge	Yes	Electrical	Yes			Substantial costs
Microwave	No	Electrical	Yes	23–48%	Minimal retention time	Intense heat demand
Sonication	Yes	Electrical	Yes	6–47%		Potential for the formation of recalcitrant compounds
γ and β radiations	No	Electrical	Yes	–	Affordable	Elevated capex
Thermal	Yes	Heat	Yes	7–105%		Digestate contamination by chemicals and the risk of inhibitor formation
Chemical	No	Chemical productions	Yes	5–133%	Low energy requirement	Concerns regarding chemical contamination of digestate and inhibitor formation
Thermo chemical	No	Chemicals productions + heat	Yes	11–72%		Considerable expenses
Combined	No	Electricity + chemical production	Yes	11–262%	Limited retention time	Risks associated with digestate contamination and inhibitor formation
Biological	Yes	Heat	Yes	11–58%		Lengthy exposure time

decreased. In other words, higher TS content led to lower DDCOD values under the same homogenization pressure conditions.

The performance of HPH (high-pressure homogenization) pretreatment is influenced by two key factors: homogenization pressure and cycles. As the homogenization pressure increases, there is an increase in the disintegration and dissolution chemical oxygen demand (DDCOD) of the substrate. Achieving cell disruption through a single pass of HPH requires very high pressure, which results in higher costs for energy, equipment, and maintenance. Alternatively, the effectiveness of HPH pretreatment can be enhanced by using multiple cycles [42, 103]. Increasing the number of cycles favors the disintegration of sludge; however, multiple cycles also require a significant amount of energy. The DDCOD of sewage sludge resulting from HPH pretreatment is influenced by two primary parameters: homogenization pressure (P) and cycle number (N), as indicated in Eq. (6.3) [179].

$$\mathrm{DDCOD} = \mathrm{kN^aP^b} \tag{6.3}$$

where k represents the rate constant governing the disintegration of sludge, a represents the exponent associated with the homogenization cycle, and b represents the exponent related to the homogenization pressure.

(b) **Microwave**

Microwaves, within frequency range of 300–300 GHz, are a segment of the electromagnetic spectrum [130]. By employing microwave heating, the rate of biodegradation and solubilization can be greatly enhanced, resulting in a notable reduction in the HRT. The significant reduction, such as transitioning from a 20-day HRT to just five days, allows for the treatment of sludge in smaller reactors. As a result, there are notable decreases in both operational costs and capital [148]. Despite the fact that microwave pretreatment of sewage sludge enhances biogas production, it is important to acknowledge that it requires a substantial amount of energy [10].

Various studies have examined the energy balance and consumption associated with microwave pretreatment. According to the analysis of energy balance, the sludge microwave pretreatment resulted in a negative net energy production of − 466.02 kWh/ton of sludge [13]. Tang et al. [145] reported that in order to rupture the cell membrane of sewage sludge, a minimum microwave energy of 1000 kJ/kg is necessary. Indeed, as highlighted in the study conducted by Chang et al. [27], prolonged pretreatment durations and high-intensity microwave irradiation have the potential to improve the release of intracellular material. Nevertheless, it is crucial to acknowledge that these conditions also result in increased energy consumption. According to the observations made by Climent et al. [34], it was found that an energy input of 13,000 kJ/kg of sewage sludge was necessary to achieve a substantial 311% increase in the FVS/TVS (fixed VS to total VS) ratio. On the other hand, an input of 7800 kJ/kg of sewage sludge resulted in a comparatively lower increase of 211% in the FVS/TVS ratio. The researchers identified a specific energy input of 13,000 kJ/kg of sewage sludge as the maximum limit to avoid sludge boiling. In the energy balance study conducted by Appels et al. [9], it was discovered that microwave pretreatment of 1 m^3 of sludge required an energy input of 336,000 kJ. This process led to an increase in biogas production by 2760 l, resulting in bioenergy output of 57,141 kJ. However, despite the energy output, the net energy production remained negative at – 278,859 kJ, indicating a lack of self-sufficiency.

Kavitha et al. [74] conducted an analysis of energy that compared the energy efficiency of hybrid ultrasonic-microwave pretreatment with that of single microwave pretreatment for sludge solubilization. For achieving a 20% sludge disintegration, the single microwave pretreatment required an energy input of 362.7 kWh. On the other hand, the combined ultrasonic and microwave pretreatment only needed 189 kWh of energy for the same level of disintegration. The total energy consumption of the ultrasonic-microwave system was 425 kWh, while the single microwave system consumed 598 kWh. Nevertheless, when considering both systems combined, the total energy produced amounted to 461 kWh. The study revealed energy ratios of 1.08 and 0.77 for the combined ultrasonic-microwave and single microwave systems, respectively. The researchers determined that the combined ultrasonic-microwave process exhibited a higher methane yield of 0.3 l/gCOD in comparison to the single microwave process i.e., 0.2 l/gCOD. Nevertheless, microwave pretreatment led to a sludge solubilization that was four to five times more significant than that achieved with ultrasound pretreatment, despite the former consuming 2620 kJ/kg TS while the latter only utilized 2370 kJ/kg TS [25]. One of the primary drawbacks of microwave pretreatment that deserves significant attention is its high energy consumption [48].

According to Bougrier et al. [19], several key factors, including substrate type, energy input, and ultrasonic frequency, have a notable influence on the performance of ultrasonic pretreatment. During the process of ultrasonic pretreatment, two energy conversions occur: the conversion of electrical energy into mechanical energy, and further conversion of mechanical energy into cavitation. These conversions result in significant energy losses. The energy consumption is determined by factors such as sonication time, ultrasonic power, sludge volume and TS concentration. When dealing with sludge containing 2–4%TS, ultrasonic pretreatment exhibits negative energy balance due to its high electricity demand rather than generating heat, as highlighted by Pérez-Elvira et al. [124].

Although electrokinetic pretreatment is considered an advanced method for sludge pretreatment, it has demonstrated the potential to significantly enhance production of biogas by up to 2.5 folds when compared to the control sample (unpretreated sludge). It has gained significant adoption in various industries. However, there are limitations associated with this process, such as insufficient research, electrode erosion, and a high energy requirement [30, 134, 149].

(c) **Thermal pretreatment**

Temperatures exceeding 100 °C lead to a more extensive rupture of cells, thereby increasing the intracellular content and enhancing biodegradability. Initially, high temperature pretreatment was employed to sterilize the sewage sludge and produce Class A biosolids. But, the majority of energy was consumed in water vaporization, making it unfeasible approach, as stated by Ariunbaatar et al. [11]. To achieve high-temperature pretreatment, steam injection or heat exchangers are utilized to elevate the sludge temperature. Although the demand for heat energy is substantial, it can be offset by the improved sludge degradation and biogas production enhancement, that can be harnessed for electricity and heat generation, as indicated by Ariunbaatar et al. [11]. The augmented energy demands can be balanced through utilizing the residual heat from the sludge to sustain the temperature of the digester, as highlighted by Haug et al. [63]. The implementation of a heat exchanger can effectively mitigate the expenses linked to energy demand of thermal pretreatment. Furthermore, employing higher pressures (600–800 kPa) and temperatures (160–170 °C) in pretreatment leads to greater energy gains due to increased biogas production, as reported by Appels et al. [9].

(d) **Acid alkali**

This particular group of pretreatments involves the introduction of acids or alkalis to facilitate the degradation process and enhance the bioavailability of intracellular content. However, these pretreatment methods are not extensively utilized due to the cost of reagents and the potential formation of toxic by-products, as indicated by reference [153]. This is evident from the limited number of studies that have employed chemical pretreatments. For instance, Ambrose et al. [4] employed a chemical pretreatment in conjunction with a prior microwave pretreatment, where the sample was subjected to 660 W microwave irradiation followed by the addition of 1% w/w hydrogen peroxide. Similarly, Gulsen Akbay et al. [56] investigated the impact of acid (H_2SO_4 0.1–1 M) and alkali (NaOH 0.1–1 M) by applying them to assess the degradation of organic matter.

(e) **Ozonation**

The strong oxidizing nature of ozone leads to the disintegration of the microbial cell wall, as highlighted by Yan et al. [165, 166]. When sludge is exposed to high oxidation condition, the cell wall ruptures, resulting in the release of SCOD. The reaction between ozone and polysaccharides, proteins and lipids produces smaller molecular weight compounds, as reported by Goel et al. [53]. Under high ozone dosages, mineralization of the released intercellular organics (materials) also takes place. Bougrier et al. [19] found out that with 0.10–16 g O_3/g TS of ozone dose, biogas yield of 300 ml/g COD could be produced in 15–18 days, whereas without ozonation, the process took approximately 24 days. In an energy balance study conducted by Boehler and Siegrist [16] on a Swiss WWTP, it was found that the energy consumed by the generation of liquid oxygen and ozonation was 0.5 kWh/Nm^3 O_2 and 12.5 kWh/kg O_3, respectively. 30% of sludge reduction could augmented the energy consumption of the plant by 20%. Hodaei et al. [67] studied the impact of ozonation pretreatment on WAS disintegration and methane generation, while also examining the energy analysis of ozonation process to assess its sustainability. Following equation was used to calculate energy consumed in ozonation:

$$E_{Ozonation} = W \times t \qquad (6.4)$$

In the equation, W represents the power of the ozone generator (watts), and t denotes the duration of pretreatment time (seconds).

Energy generated in anaerobic digestion system was determined by applying the ideal gas law. Initially, the methane concentration was computed through following equation:

$$\rho\text{act } CH_4 = \rho\text{std } CH_4 * \text{Pact}/\text{Pstd} * \text{Tact}/\text{Tstd} \tag{6.5}$$

In this equation, ρact CH_4 represents the methane density at STP (standard temperature and pressure), which is 0.72 kg/m^3. Tstd refers to standard temperature of 273 K and Pstd represents standard pressure of 101.325 kPa Tact corresponds to the gas temperature, measured at 303 K, while Pact signifies the total pressure, which is the sum of the air pressure and gauge pressure.

The calculation of the total energy generated from biogas was performed using the following equations:

$$\text{Hu.act} = VCH_4/\text{Vtotal} * \rho\text{act } CH_4 * \text{Hu, n} \tag{6.6}$$

$$E_{\text{biogas}} = \text{Hu.act} \times V_{\text{biogas}} \tag{6.7}$$

Hu,n represents the biogas calorific value (i.e., 50,000 kJ/kg). Hu,act represents the actual calorific value of the specific biogas sample (kJ/kg), and V_{biogas} represents the volume of biogas in liters. Finally, through following equation energy balance was determined.

$$\text{Energy balance} = E_{\text{ozonation}} - E_{\text{biogas}} \tag{6.8}$$

The authors' findings indicated that ozonating sludges with dosages of 0.1 gO_3/gTS and 0.05 gO_3/gTS resulted in only 29 and 38% of the energy being provided, respectively. The overall net energy balance was negative, indicating that the sludge ozonation requires high energy input. However, it leads to improved sludge dewaterability and increased organics solubilization. The study concluded that at high ozone dosages, biogas production decreased while energy demand increased. The higher dose of ozone promoted hydrolysis, leading to the VFAs accumulation and methanogens' performance inhibition under low pH conditions. In another investigation, conducted by Yukesh Kannah et al. [174], stated that dispersion-integrated ozonation yielded positive net energy balance i.e., 152.65 kWh/ton, as compared to single ozonation pretreatment (− 12.42 kWh/ton). Thermochemical sludge pretreatment requires an energy of 1450 kWh for 30% COD solubilization. However, combining thermochemical pretreatment with ozonation only consumed 607 kWh of energy. Therefore, combining ozonation with other pretreatments can significantly reduce the energy input required [73].

Fenton oxidation presents certain drawbacks, including safety risks related to the use of H_2O_2, the requirement to significantly lower the pH and subsequently neutralize it, and the corrosiveness of the system due to the low pH conditions. These factors can contribute to higher operating and maintenance expenses.

(f) **Biological**

The primary challenge for implementing industrial applications lies in sourcing enzymes and microbes, as pure biological additives are prohibitively expensive. In the context of biological pretreatment, the addition of both pronase E and cellulose improved VFA consumption in all experiments. However, individual enzymes had minimal impact on sludge disintegration, as observed by Zagrodnik et al. [175]. In a study by Yu et al. [172], the impact of EPS content on various enzymes was investigated, revealing that the highest concentration of extracellular enzymes (excluding α-amylase) was found in tightly bound EPS layers and pellet, as most of the organic matter transformation within the sludge occurred within microbial cells. Conversely, lower quantities of enzymes were detected in slime layers and loosely bound EPS.

The utilization of surfactants may enhance the solubility of hydrolases. However, it is crucial to promptly develop methods for removing excess surfactants, as they possess high toxicity and can inhibit methane production, ultimately increasing costs. Guan et al. [55] and Liu et al. [91] have highlighted the need for careful surfactant management in order to mitigate these issues.

6.4.2.2 Overview

One of the most significant inputs, environmentally, and financially is energy. While the cost of treatment may be disposal driven, in energy terms, energy utilized should hopefully match the energy produced by increases in biogas production. The energy input depends heavily on method, and may be a function of sludge composition, operating and ambient conditions, and equipment used, amongst others. Due to the additional energy input required for pretreatments that increase methane production, the assessment of the effectiveness of WAS pretreatment is influenced by the increase in methane alone. Therefore, it is crucial to determine if the energy recovered from the pretreated AD process can offset the energy consumed. Assessing the energy advantages of various pretreatment methods may be better achieved by evaluating the net energy production (NEP), which quantifies the disparity between energy input and output within the anaerobic digestion system [59].

Electricity is frequently employed as a synergistic agent in various pretreatment methods such as high pulse electric field, sonication, dispersion, ozonation, microwave, and more, to impact the process of sludge solubilization. In general, stirring and pumping are extensively utilized for phase-separated pre-treatment and anaerobic digestion processes, which are attributable to energy depletion. However, the generation of biomethane is considered a beneficial output energy obtained from the pretreated sewage sludge. Based on the fact that most of the studies have been conducted at bench scale, it appears that pretreatment technologies hold great potential for implementation at full scale, primarily due to the significant increase in bioenergy recovery from excess sludge (Fig. 6.13). However, a thorough energy feasibility study conducted by Cano et al. [21] found that most of the pretreatment methods explored in scientific literature were not viable for full-scale implementation due to their high energy demands for pretreatment, which usually involve electricity. Although individual pretreatment processes alone may not be able to achieve energy recovery from WAS, some researches have tried to convert negative net energy values into positive ones by improving the pretreatment strategies. For example, Rajesh Banu et al. [129] demonstrated that by combining microwave with biosurfactant and alkali, NEP value per energy produced by methane (NM) [156] became positive, as the alkali and biosurfactant contributed to increasing SCOD and breaking down EPS. Similarly, Kavitha et al. [74] found that the combination of ultrasound and microwave could also yield positive NM values, due to the reduced energy consumption of microwave. It is important to note that heat recovery can be an effective supplementary approach to reverse negative energy generation outcomes from thermally pretreated anaerobic digestion. For example, while increasing temperature of thermal pretreatment may lead to negative energy balance, Liu et al. [94] were able to achieve positive NM outcomes by 85% recovery of the heat generated during thermal pretreatment, indicating a promising approach for energy conservation. The anaerobic treatment of sludge produced from various sources is of paramount importance. Integrating pretreatment for this sludge within the sludge cycle process is essential to enhance bioenergy production from sludge (Fig. 6.13).

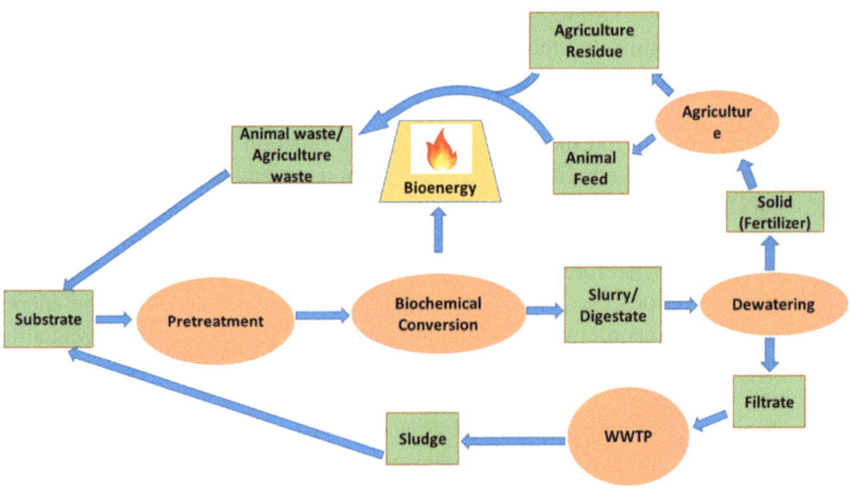

Fig. 6.13 Integration of pretreatment in sludge cycle process

6.5 Prospects

- To enhance enzymatic hydrolysis and minimize operational expenses, utilizing a combination of inexpensive mechanical and chemical pretreatments proves to be a favorable choice. Among these approaches, the synergistic effect of mechanical pretreatment combined with FNA (chemical pretreatment) demonstrates superior effectiveness compared to individual pretreatment methods. Based on several research findings, it is suggested that further investigations should focus on integrating HPH pretreatment with other chemical pretreatments to accomplish the desired objectives. It is important to highlight that comprehensive trials on a larger scale are necessary to thoroughly evaluate the potential of this FNA-based technology in the future. Additionally, it is recommended to combine HPH with other affordable and efficient chemicals for effective sludge pretreatment.
- Currently, the assessment of sludge anaerobic treatment relies solely on the biochemical methanogenic potential assessment, which has been utilized for almost three decades years. However, this method suffers from significant drawbacks including lengthy experimental periods, high experimental variability, and stringent experimental conditions. In recent years, alternative approaches such as the three-dimensional excitation-emission matrix fluorescence and UV–vis spectral method combined with fluorescence regional integration have been explored to evaluate the biochemical conversion of sludge organics. Nevertheless, these investigations have shortcomes in accurately reflecting anaerobic degradation. Hence, there is an urgent need to develop more efficient methods for characterizing sludge anaerobic degradability. One potential research direction is to investigate the combination of biochemical methanogenic potential testing and fluorescence methods.
- To protect the microbial consortia employed in anaerobic digestion of pretreated sludge, it becomes crucial to address the potential harm caused by the chemical reagents used during pretreatment. Hence, there is a need to detoxify these toxic compounds using methods that are cost-effective, technologically advanced, and highly efficient. Utilizing porous, activated carbon for adsorption-based detoxification could serve as an effective approach to promote the sustainable utilization of sludge through microbial fermentation.
- The effectiveness of different anaerobic bioprocesses is hindered by the limited bioavailability of sludge for bioenergy conversion. Hence, it is essential to acquire a thorough comprehension of the resistant characteristics of sludge and develop efficient pretreatment methods. This understanding serves as a crucial factor in enhancing the overall productivity, yield, and economic viability of bioenergy production through diverse anaerobic bioprocesses.

References

1. Ahn J-H, Shin SG, Hwang S (2009) Effect of microwave irradiation on the disintegration and acidogenesis of municipal secondary sludge. Chem Eng J 153(1):145–150. https://doi.org/10.1016/j.cej.2009.06.032
2. Alqaralleh RM, Kennedy K, Delatolla R (2019) Microwave vs. alkaline-microwave pretreatment for enhancing thickened waste activated sludge and fat, oil, and grease solubilization, degradation and biogas production. J Environ Manag 233:378–392. https://doi.org/10.1016/j.jenvman.2018.12.046
3. Ambrose HW, Chin CT-L, Hong E, Philip L, Suraishkumar GK, Sen TK, Khiadani M (2020) Effect of hybrid (microwave-H_2O_2) feed sludge pretreatment on single and two-stage anaerobic digestion efficiency of real mixed sewage sludge. Process Safety Environ Prot 136:194–202. https://doi.org/10.1016/j.psep.2020.01.032
4. Ambrose HW, Philip L, Suraishkumar GK, Karthikaichamy A, Sen TK (2020) Anaerobic co-digestion of activated sludge and fruit and vegetable waste: evaluation of mixing ratio and impact of hybrid (microwave and hydrogen peroxide) sludge pre-treatment on two-stage digester stability and biogas yield. J Water Process Eng 37:101498. https://doi.org/10.1016/j.jwpe.2020.101498
5. Anand H, Balasundaram B, Pandit A, Harrison S (2007) The effect of chemical pretreatment combined with mechanical disruption on the extent of disruption and release of intracellular protein from E. coli. Biochem Eng J 35(2):166–173. http://www.sciencedirect.com/science/article/pii/S1369703X07000162
6. Angelidaki I, Treu L, Tsapekos P, Luo G, Campanaro S, Wenzel H, Kougias PG (2018) Biogas upgrading and utilization: current status and perspectives. Biotechnol Adv 36(2):452–466. https://doi.org/10.1016/j.biotechadv.2018.01.011
7. Anjum M, Kumar R, Al-Talhi HA, Mohamed SA, Barakat MA (2018) Valorization of biogas production through disintegration of waste activated sludge using visible light ZnO-ZnS/Ag_2O-Ag_2S photocatalyst. Process Saf Environ Prot 119:330–339. https://doi.org/10.1016/j.psep.2018.08.022
8. Appels L, Baeyens J, Degrève J, Dewil R (2008) Principles and potential of the anaerobic digestion of waste-activated sludge. Progress Energy Combust Sci 34(6):755–781. https://doi.org/10.1016/j.pecs.2008.06.002
9. Appels L, Houtmeyers S, Degrève J, Van Impe J, Dewil R (2013) Influence of microwave pre-treatment on sludge solubilization and pilot scale semi-continuous anaerobic digestion. Bioresour Technol 128:598–603. https://doi.org/10.1016/j.biortech.2012.11.007
10. Ara E, Sartaj M, Kennedy K (2014) Effect of microwave pre-treatment of thickened waste activated sludge on biogas production from co-digestion of organic fraction of municipal solid waste, thickened waste activated sludge and municipal sludge. Waste Manag Res 32(12):1200–1209. https://doi.org/10.1177/0734242x14554641
11. Ariunbaatar J, Panico A, Esposito G, Pirozzi F, Lens PNL (2014) Pretreatment methods to enhance anaerobic digestion of organic solid waste. Appl Energy 123:143–156. https://doi.org/10.1016/j.apenergy.2014.02.035
12. Badawy MI, Ali MEM (2006) Fenton's peroxidation and coagulation processes for the treatment of combined industrial and domestic wastewater. J Hazard Mater 136(3):961–966. https://doi.org/10.1016/j.jhazmat.2006.01.042
13. Banu JR, Kavitha S (2017) Various sludge pretreatments: their impact on biogas generation. In: Singh L, Kalia VC (eds) Waste biomass management—a holistic approach. Springer International Publishing, pp 39–71. https://doi.org/10.1007/978-3-319-49595-8_3
14. Belmonte M, Hsieh CF, Figueroa CM, Campos JL, Vidal GJEJoB (2011) Effect of free ammonia nitrogen on the methanogenic activity of swine wastewater 14:2–2
15. Biasi CA, Mariani LF, Picinatto AG, Zank JCJFdIIB (2018) Energias renováveis na área rural da região sul do Brasil
16. Boehler M, Siegrist H (2006) Potential of activated sludge disintegration. Water Sci Technol 53(12):207–216. https://doi.org/10.2166/wst.2006.423

17. Bolzonella D, Pavan P, Zanette M, Cecchi F (2007) Two-phase anaerobic digestion of waste activated sludge: effect of an extreme thermophilic prefermentation. Ind Eng Chem Res 46. https://doi.org/10.1021/ie061627e
18. Borge ESM (2009) Effect of thermal treatment of anaerobic sludge on the bioavailability and biodegradability characteristics of the organic fraction. Braz J Chem Eng 26(03):11
19. Bougrier C, Albasi C, Delgenès JP, Carrère H (2006) Effect of ultrasonic, thermal and ozone pre-treatments on waste activated sludge solubilisation and anaerobic biodegradability. Chem Eng Process Process Intensification 45(8):711–718. https://doi.org/10.1016/j.cep.2006.02.005
20. Campo G, Riggio V, Cerutti A, Ruffino B, Panepinto D, Zanetti MJD, Treatment W (2017) Thermal, alkali and thermo-alkali pretreatments applied to monospecific microalgal biomass to improve anaerobic biogas production 127:165–170
21. Cano R, Pérez-Elvira SI, Fdz-Polanco F (2015) Energy feasibility study of sludge pretreatments: a review. Appl Energy 149:176–185. https://doi.org/10.1016/j.apenergy.2015.03.132
22. Carrère H, Bougrier C, Castets D, Delgenès JP (2008) Impact of initial biodegradability on sludge anaerobic digestion enhancement by thermal pretreatment. J Environ Sci Health Part A 43(13):1551–1555. https://doi.org/10.1080/10934520802293735
23. Carrère H, Dumas C, Battimelli A, Batstone D, Delgenès J, Steyer J, Ferrer I (2010) Pretreatment methods to improve sludge anaerobic degradability: a review. J Hazard Mater 183(1):1–15
24. Celebi EB, Aksoy A, Sanin FD (2021) Maximizing the energy potential of urban sludge treatment: an experimental study and a scenario-based energy analysis focusing on anaerobic digestion with ultrasound pretreatment and sludge combustion. Energy 221, Article 119876. https://doi.org/10.1016/j.energy.2021.119876
25. Cella MA, Akgul D, Eskicioglu C (2016) Assessment of microbial viability in municipal sludge following ultrasound and microwave pretreatments and resulting impacts on the efficiency of anaerobic sludge digestion. Appl Microbiol Biotechnol 100(6):2855–2868. https://doi.org/10.1007/s00253-015-7139-3
26. Chan WI, Wong WT, Liao PH, Lo KV (2007) Sewage sludge nutrient solubilization using a single-stage microwave treatment. J Environ Sci Health Part A 42(1):59–63. https://doi.org/10.1080/10934520601015735
27. Chang C-J, Tyagi VK, Lo S-L (2011) Effects of microwave and alkali induced pretreatment on sludge solubilization and subsequent aerobic digestion. Bioresour Technol 102(17):7633–7640. https://doi.org/10.1016/j.biortech.2011.05.031
28. Chen D, Guo Y, Huang R, Lu Q, Huang J (2010) Pretreatment by ultra-high pressure explosion with homogenizer facilitates cellulase digestion of sugarcane bagasses. Bioresour Technol 101(14):5592–5600. http://www.sciencedirect.com/science/article/pii/S0960852410002749
29. Chi Y, Li Y, Ji M, Qiang H, Deng HW, Wu YPJAMR (2010) Use of combined NaOH-microwave pretreatment for enhancing mesophilic anaerobic digestibility of thickened waste activated sludge 113–116:459–468
30. Chiavola A, Ridolfi A, D'Amato E, Bongirolami S, Cima E, Sirini P, Gavasci R (2015) Sludge reduction in a small wastewater treatment plant by electro-kinetic disintegration. Water Sci Technol 72:364–370. https://doi.org/10.2166/wst.2015.148
31. Chowdhury MMI, Nakhla G, Zhu J (2017) Ultrasonically enhanced anaerobic digestion of thickened waste activated sludge using fluidized bed reactors. Appl Energy 204:807–818. https://doi.org/10.1016/j.apenergy.2017.07.057
32. Chu CP, Lee DJ, Chang B-V, You CS, Tay JH (2002) "Weak" ultrasonic pre-treatment on anaerobic digestion of flocculated activated biosolids. Water Res 36(11):2681–2688. https://doi.org/10.1016/S0043-1354(01)00515-2
33. Chu L, Yan S, Xing X-H, Sun X, Jurcik B (2009) Progress and perspectives of sludge ozonation as a powerful pretreatment method for minimization of excess sludge production. Water Res 43(7):1811–1822. https://doi.org/10.1016/j.watres.2009.02.012
34. Climent M, Ferrer I, Baeza MdM, Artola A, Vázquez F, Font X (2007) Effects of thermal and mechanical pretreatments of secondary sludge on biogas production under thermophilic conditions. Chem Eng J 133(1):335–342. https://doi.org/10.1016/j.cej.2007.02.020

35. Coelho NMG, Droste RL, Kennedy KJ (2011) Evaluation of continuous mesophilic, thermophilic and temperature phased anaerobic digestion of microwaved activated sludge. Water Res 45(9):2822–2834. https://doi.org/10.1016/j.watres.2011.02.032
36. Das S, Dagnew M, Ray MB (2021) Anaerobic digestibility of resin acids in primary sludge: effect of ozone pretreatment. Chemosphere 262, Article 128316. https://doi.org/10.1016/j.chemosphere.2020.128316
37. De Vrieze J, Gildemyn S, Arends JBA, Vanwonterghem I, Verbeken K, Boon N, Verstraete W, Tyson GW, Hennebel T, Rabaey K (2014) Biomass retention on electrodes rather than electrical current enhances stability in anaerobic digestion. Water Res 54:211–221. https://doi.org/10.1016/j.watres.2014.01.044
38. Deena SR, Vickram AS, Manikandan S, Subbaiya R, Karmegam N, Ravindran B, Chang SW, Awasthi MK (2022) Enhanced biogas production from food waste and activated sludge using advanced techniques—a review. Bioresour Technol 355:127234. https://doi.org/10.1016/j.biortech.2022.127234
39. Deleris S (2003) Innovative strategies for the reduction of sludge production in activated sludge plant: BIOLYSIS® O and BIOLYSIS® E. N. p, Norway, p 7
40. Dewil R, Appels L, Baeyens J, Degrève J (2007) Peroxidation enhances the biogas production in the anaerobic digestion of biosolids. J Hazard Mater 146(3):577–581. https://doi.org/10.1016/j.jhazmat.2007.04.059
41. Donsì F, Ferrari G, Lenza E, Maresca P (2009) Main factors regulating microbial inactivation by high-pressure homogenization: operating parameters and scale of operation. Chem Eng Sci 64(3):520–532. http://www.sciencedirect.com/science/article/pii/S0009250908005198
42. dos Santos Aguilar JG, Cristianini M, Sato HH (2018) Modification of enzymes by use of high-pressure homogenization. Food Res Int 109:120–125. https://doi.org/10.1016/j.foodres.2018.04.011
43. Doulah M, Hammond T, Brookman J (1975) A hydrodynamic mechanism for the disintegration of *Saccharomyces cerevisiae* in an industrial homogenizer. Biotechnol Bioeng 17(6):845–858
44. Doğan I, Sanin FD (2009) Alkaline solubilization and microwave irradiation as a combined sludge disintegration and minimization method. Water Res 43(8):2139–2148. https://doi.org/10.1016/j.watres.2009.02.023
45. Erden G, Filibeli A (2010) Ultrasonic pre-treatment of biological sludge: Consequences for disintegration, anaerobic biodegradability, and filterability. J Chem Technol Biotechnol 85:145–150. https://doi.org/10.1002/jctb.2298
46. Eskicioglu C, Kennedy KJ, Droste RL (2008) Initial examination of microwave pretreatment on primary, secondary and mixed sludges before and after anaerobic digestion. Water Sci Technol 57(3):311–317. https://doi.org/10.2166/wst.2008.010
47. Eskicioglu C, Prorot A, Marin J, Droste RL, Kennedy KJ (2008) Synergetic pretreatment of sewage sludge by microwave irradiation in presence of H_2O_2 for enhanced anaerobic digestion. Water Res 42(18):4674–4682. https://doi.org/10.1016/j.watres.2008.08.010
48. Eswari AP, Kavitha S, Banu JR, Karthikeyan OP, Yeom I-T (2017) H_2O_2 induced cost effective microwave disintegration of dairy waste activated sludge in acidic environment for efficient biomethane generation. Bioresour Technol 244:688–697. https://doi.org/10.1016/j.biortech.2017.07.078
49. Fang W, Zhang P, Shang R, Ye J, Wu Y, Zhang H, Liu J, Gou X, Zeng G, Zhou S (2017) Effect of high pressure homogenization on anaerobic digestion of the sludge pretreated by combined alkaline and high pressure homogenization. Desalin Water Treat 62:168–174
50. Fang W, Zhang P, Zhang G, Jin S, Li D, Zhang M, Xu X (2014) Effect of alkaline addition on anaerobic sludge digestion with combined pretreatment of alkaline and high pressure homogenization. Bioresour Technol 168:167–172. http://www.sciencedirect.com/science/article/pii/S0960852414003551
51. Ge H, Jensen PD, Batstone DJ (2011) Increased temperature in the thermophilic stage in temperature phased anaerobic digestion (TPAD) improves degradability of waste activated sludge. J Hazard Mater 187(1):355–361. https://doi.org/10.1016/j.jhazmat.2011.01.032

52. Geng Y-K, Zhou Y (2022) Reduction of refractory Maillard reaction products by Fe^{3+} during thermal hydrolysis pretreatment and enhanced sludge biodegradability. J Hazard Materi 430:128400. https://doi.org/10.1016/j.jhazmat.2022.128400
53. Goel R, Yasui H, Shibayama C (2003) High-performance closed loop anaerobic digestion using pre/post sludge ozonation. Water Sci Technol 47(12):261–267. https://doi.org/10.2166/wst.2003.0655
54. Gould MC (2015) Chapter 18—bioenergy and anaerobic digestion. In: Dahiya A (ed) Bioenergy. Academic Press, pp 297–317. https://doi.org/10.1016/B978-0-12-407909-0.00018-3
55. Guan R, Yuan X, Wu Z, Wang H, Jiang L, Li Y, Zeng G (2017) Functionality of surfactants in waste-activated sludge treatment: a review. Sci Total Environ 609:1433–1442. https://doi.org/10.1016/j.scitotenv.2017.07.189
56. Gulsen Akbay HE, Dizge N, Kumbur H (2021) Enhancing biogas production of anaerobic co-digestion of industrial waste and municipal sewage sludge with mechanical, chemical, thermal, and hybrid pretreatment. Bioresour Technol 340:125688. https://doi.org/10.1016/j.biortech.2021.125688
57. Guo X, Liu J, Xiao B (2013) Bioelectrochemical enhancement of hydrogen and methane production from the anaerobic digestion of sewage sludge in single-chamber membrane-free microbial electrolysis cells. Int J Hydrogen Energy 38(3):1342–1347. https://doi.org/10.1016/j.ijhydene.2012.11.087
58. Guo Y, Guo Y, Gong H, Fang N, Tan Y, Zhou W, Huang J, Dai L, Dai X (2021) Variations of heavy metals, nutrients, POPs and particle size distribution during "sludge anaerobic digestion-solar drying-land utilization process": case study in China. Sci Total Environ 801:149609. https://doi.org/10.1016/j.scitotenv.2021.149609
59. Han D, Lee C-Y, Chang SW, Kim D-J (2017) Enhanced methane production and wastewater sludge stabilization of a continuous full scale thermal pretreatment and thermophilic anaerobic digestion. Bioresour Technol 245:1162–1167. https://doi.org/10.1016/j.biortech.2017.08.108
60. Han Y, Dague RR (1997) Laboratory studies on the temperature-phased anaerobic digestion of domestic primary sludge. Water Environ Res 69(6):1139–1143. http://www.jstor.org/stable/25044974
61. Hardi E, Latta P (2009) Biogas injection in the local natural gas grid–issues and solutions, monitoring and control in pilot case. 24th World gas conference (WGC 2009), Buenos Aires, Argentina
62. Harmsen P, Huijgen W, Bermudez L, Bakker R (2010) Literature review of physical and chemical pretreatment processes for lignocellulosic biomass
63. Haug RT, LeBrun TJ, Tortorici LD (1983) Thermal pretreatment of sludges: a field demonstration. J (Water Pollut Control Fed) 55(1):23–34. http://www.jstor.org/stable/25041794
64. Haug RT, Stuckey DC, Gossett JM, McCarty PL (1978) Effect of thermal pretreatment on digestibility and dewaterability of organic sludges. J (Water Pollut Control Fed) 50(1):73–85. http://www.jstor.org/stable/25039508
65. Hii K, Baroutian S, Parthasarathy R, Gapes DJ, Eshtiaghi N (2014) A review of wet air oxidation and Thermal Hydrolysis technologies in sludge treatment. Bioresour Technol 155:289–299. https://doi.org/10.1016/j.biortech.2013.12.066
66. Hoang SA, Bolan N, Madhubashani AMP, Vithanage M, Perera V, Wijesekara H, Wang H, Srivastava P, Kirkham MB, Mickan BS, Rinklebe J, Siddique KHM (2022) Treatment processes to eliminate potential environmental hazards and restore agronomic value of sewage sludge: a review. Environ Pollut 293:118564. https://doi.org/10.1016/j.envpol.2021.118564
67. Hodaei M, Ghasemi S, Khosravi A, Vossoughi M (2021) Effect of the ozonation pretreatment on biogas production from waste activated sludge of tehran wastewater treatment plant. Biomass Bioenergy 152:106198. https://doi.org/10.1016/j.biombioe.2021.106198
68. Hong S, Park J, Teeradej N, Lee Y, Cho YK, Park CH (2006) Pretreatment of sludge with microwaves for pathogen destruction and improved anaerobic digestion performance. Water Environ Res A Res Publ Water Environ Fed 78:76–83. https://doi.org/10.2175/106143005X84549

69. Hu M, Hu H, Ye Z, Tan S, Yin K, Chen Z, Guo D, Rong H, Wang J, Pan Z, Hu Z-T (2022) A review on turning sewage sludge to value-added energy and materials via thermochemical conversion towards carbon neutrality. J Clean Prod 379:134657. https://doi.org/10.1016/j.jclepro.2022.134657
70. Jang J-H, Ahn J-H (2013) Effect of microwave pretreatment in presence of NaOH on mesophilic anaerobic digestion of thickened waste activated sludge. Bioresour Technol 131:437–442. https://doi.org/10.1016/j.biortech.2012.09.057
71. Jenicek P, Kutil J, Benes O, Todt V, Zabranska J, Dohanyos M (2013) Energy self-sufficient sewage wastewater treatment plants: is optimized anaerobic sludge digestion the key? Water Sci Technol 68(8):1739–1744. https://doi.org/10.2166/wst.2013.423
72. Jiang Z, Fei B, Li Z (2016) Pretreatment of bamboo by ultra-high pressure explosion with a high-pressure homogenizer for enzymatic hydrolysis and ethanol fermentation. Bioresour Technol 214:876–880. http://www.sciencedirect.com/science/article/pii/S0960852416306654
73. Kannah RY, Kavitha S, Rajesh Banu J, Yeom IT, Johnson M (2017) Synergetic effect of combined pretreatment for energy efficient biogas generation. Bioresour Technol 232:235–246. https://doi.org/10.1016/j.biortech.2017.02.042
74. Kavitha S, Rajesh Banu J, Kumar G, Kaliappan S, Yeom IT (2018) Profitable ultrasonic assisted microwave disintegration of sludge biomass: modelling of biomethanation and energy parameter analysis. Bioresour Technol 254:203–213. https://doi.org/10.1016/j.biortech.2018.01.072
75. Kenge AA, Liao PH, Lo KV (2008) Role of mixing on microwave-enhanced advanced oxidation process in treating sewage sludge. J Environ Sci Health Part A 43(12):1431–1440. https://doi.org/10.1080/10934520802232170
76. Kennedy KJ, Thibault G, Droste RLJWS (2009) Microwave enhanced digestion of aerobic SBR sludge 33:261–270
77. Kennedy KJ, Thibault G, Droste RLJWS (2007) Microwave enhanced digestion of aerobic SBR sludge 33(2)
78. Khanh Nguyen V, Kumar Chaudhary D, Hari Dahal R, Hoang Trinh N, Kim J, Chang SW, Hong Y, Duc La D, Nguyen XC, Hao Ngo H, Chung WJ, Nguyen DD (2021) Review on pretreatment techniques to improve anaerobic digestion of sewage sludge. Fuel 285:119105. https://doi.org/10.1016/j.fuel.2020.119105
79. Kim J, Park C, Kim T-H, Lee M, Kim S, Kim S-W, Lee J (2003) Effects of various pretreatments for enhanced anaerobic digestion with waste activated sludge. J Biosci Bioeng 95(3):271–275. https://doi.org/10.1016/S1389-1723(03)80028-2
80. Kinnunen V, Ylä-Outinen A, Rintala J (2015) Mesophilic anaerobic digestion of pulp and paper industry biosludge–long-term reactor performance and effects of thermal pretreatment. Water Res 87:105–111. https://doi.org/10.1016/j.watres.2015.08.053
81. Kondaveeti S, Min B (2015) Bioelectrochemical reduction of volatile fatty acids in anaerobic digestion effluent for the production of biofuels. Water Res 87:137–144. https://doi.org/10.1016/j.watres.2015.09.011
82. Kuglarz M, Karakashev D, Angelidaki I (2013) Microwave and thermal pretreatment as methods for increasing the biogas potential of secondary sludge from municipal wastewater treatment plants. Bioresour Technol 134:290–297. https://doi.org/10.1016/j.biortech.2013.02.001
83. Kumar A, Samadder SR (2020) Performance evaluation of anaerobic digestion technology for energy recovery from organic fraction of municipal solid waste: a review. Energy 197:117253. https://doi.org/10.1016/j.energy.2020.117253
84. Lee I-S, Rittmann BE (2011) Effect of low solids retention time and focused pulsed pretreatment on anaerobic digestion of waste activated sludge. Bioresour Technol 102(3):2542–2548. https://doi.org/10.1016/j.biortech.2010.11.082
85. Li Y-Y, Noike T (1992) Upgrading of anaerobic digestion of waste activated sludge by thermal pretreatment. Water Sci Technol 26(3–4):857–866. https://doi.org/10.2166/wst.1992.0466

86. Li N, Xiao X, Li C, Sheng X, Zhang J, Ping Q (2022) Boosting VFAs production during the anaerobic acidification of lignocellulose waste pulp and paper mill excess sludge: ultrasonic pretreatment and inoculating rumen microorganisms. Ind Crops Prod 188:115613. https://doi.org/10.1016/j.indcrop.2022.115613
87. Li W, Yu N, Fang A, Liu B, Ren N, Xing D (2019) Co-treatment of potassium ferrate and ultrasonication enhances degradability and dewaterability of waste activated sludge. Chem Eng J 361:148–155. https://doi.org/10.1016/j.cej.2018.12.058
88. Li X, Guo S, Peng Y, He Y, Wang S, Li L, Zhao M (2018) Anaerobic digestion using ultrasound as pretreatment approach: changes in waste activated sludge, anaerobic digestion performances and digestive microbial populations. Biochem Eng J 139:139–145. https://doi.org/10.1016/j.bej.2017.11.009
89. Li X, Zhao J, Wang D, Yang Q, Xu Q, Deng Y, Yang W, Zeng G (2016) An efficient and green pretreatment to stimulate short-chain fatty acids production from waste activated sludge anaerobic fermentation using free nitrous acid. Chemosphere 144:160–167. https://doi.org/10.1016/j.chemosphere.2015.08.076
90. Liao PH, Lo KV, Chan WI, Wong WT (2007) Sludge reduction and volatile fatty acid recovery using microwave advanced oxidation process. J Environ Sci Health Part A 42(5):633–639. https://doi.org/10.1080/10934520701244417
91. Liu J-W, Wei K-H, Xu S-W, Cui J, Ma J, Xiao X-L, Xi B-D, He X-S (2021) Surfactant-enhanced remediation of oil-contaminated soil and groundwater: a review. Sci Total Environ 756:144142. https://doi.org/10.1016/j.scitotenv.2020.144142
92. Liu W, Cai W, Guo Z, Wang L, Yang C, Varrone C, Wang A (2016) Microbial electrolysis contribution to anaerobic digestion of waste activated sludge, leading to accelerated methane production. Renew Energy 91:334–339. https://doi.org/10.1016/j.renene.2016.01.082
93. Liu X, Huang X, Wu Y, Xu Q, Du M, Wang D, Yang Q, Liu Y, Ni B-J, Yang G, Yang F, Wang Q (2020) Activation of nitrite by freezing process for anaerobic digestion enhancement of waste activated sludge: performance and mechanisms. Chem Eng J 387:124147. https://doi.org/10.1016/j.cej.2020.124147
94. Liu X, Wang Q, Tang Y, Pavlostathis SG (2021) Hydrothermal pretreatment of sewage sludge for enhanced anaerobic digestion: resource transformation and energy balance. Chem Eng J 410:127430. https://doi.org/10.1016/j.cej.2020.127430
95. Liu X, Xu Q, Wang D, Wu Y, Fu Q, Li Y, Yang Q, Liu Y, Ni B-J, Wang Q, Yang G, Li H, Li X (2019) Microwave pretreatment of polyacrylamide flocculated waste activated sludge: effect on anaerobic digestion and polyacrylamide degradation. Bioresour Technol 290:121776. https://doi.org/10.1016/j.biortech.2019.121776
96. Liu X, Xu Q, Wang D, Zhao J, Wu Y, Liu Y, Ni B-J, Wang Q, Zeng G, Li X, Yang Q (2018) Improved methane production from waste activated sludge by combining free ammonia with heat pretreatment: performance, mechanisms and applications. Bioresour Technol 268:230–236. https://doi.org/10.1016/j.biortech.2018.07.109
97. Lizama AC, Figueiras CC, Herrera RR, Pedreguera AZ, Ruiz Espinoza JE (2017) Effects of ultrasonic pretreatment on the solubilization and kinetic study of biogas production from anaerobic digestion of waste activated sludge. Int Biodeterior Biodegradation 123:1–9. https://doi.org/10.1016/j.ibiod.2017.05.020
98. Lizama AC, Figueiras CC, Pedreguera AZ, Ruiz Espinoza JE (2018) Effect of ultrasonic pretreatment on the semicontinuous anaerobic digestion of waste activated sludge with increasing loading rates. Int Biodeterior Biodegradation 130:32–39. https://doi.org/10.1016/j.ibiod.2018.03.013
99. Lo KV, Liao PH, Yin GQ (2008) Sewage sludge treatment using microwave-enhanced advanced oxidation processes with and without ferrous sulfate addition 83(10):1370–1374. https://doi.org/10.1002/jctb.1965
100. Lv W, Schanbacher F, Yu Z (2010) Putting microbes to work in sequence: recent advances in temperature-phased anaerobic digestion processes. Biores Technol 101:9409–9414. https://doi.org/10.1016/j.biortech.2010.07.100

101. Ma H, Guo Y, Qin Y, Li Y-Y (2018) Nutrient recovery technologies integrated with energy recovery by waste biomass anaerobic digestion. Bioresour Technol 269:520–531. https://doi.org/10.1016/j.biortech.2018.08.114
102. Mao C, Feng Y, Wang X, Ren G (2015) Review on research achievements of biogas from anaerobic digestion. Renew Sustain Energy Rev 45:540–555. https://doi.org/10.1016/j.rser.2015.02.032
103. Maresca P, Donsì F, Ferrari G (2011) Application of a multi-pass high-pressure homogenization treatment for the pasteurization of fruit juices. J Food Eng 104(3):364–372. http://www.sciencedirect.com/science/article/pii/S0260877410006308
104. Mawson AJ, Earle RL, Larsen VF (1991) Degradation of acetic and propionic acids in the methane fermentation. Water Res 25(12):1549–1554. https://doi.org/10.1016/0043-1354(91)90187-U
105. Mehdizadeh SN, Eskicioglu C, Bobowski J, Johnson T (2013) Conductive heating and microwave hydrolysis under identical heating profiles for advanced anaerobic digestion of municipal sludge. Water Res 47(14):5040–5051. https://doi.org/10.1016/j.watres.2013.05.055
106. Morello R, Di Capua F, Esposito G, Pirozzi F, Fratino U, Spasiano D (2022) Sludge minimization in mainstream wastewater treatment: Mechanisms, strategies, technologies, and current development. J Environ Manag 319:115756. https://doi.org/10.1016/j.jenvman.2022.115756
107. Morello R, Di Capua F, Pontoni L, Papirio S, Spasiano D, Fratino U, Pirozzi F, Esposito G (2021) Microaerobic digestion of low-biodegradable sewage sludge: effect of air dosing in batch reactors 13(17):9869. https://www.mdpi.com/2071-1050/13/17/9869
108. Moss LH, Donovan JF, Carr S, Stone L, Polo C, Khunjar W, Latimer R, Jeyanayagam S, Beecher N, McFadden L (2013) Enabling the future: advancing resource recovery from biosolids. Water Environ Res Found
109. Nabi M, Zhang G, Li F, Zhang P, Wu Y, Tao X, Bao S, Wang S, Chen N, Ye JJDWT (2020) Enhancement of high pressure homogenization pretreatment on biogas production from sewage sludge: a review 175:341–351
110. Nabi M, Gao D, Liang J, Cai Y, Zhang P (2022) Combining high pressure homogenization with free nitrous acid pretreatment to improve anaerobic digestion of sewage sludge. J Environ Manag 318:115635. https://doi.org/10.1016/j.jenvman.2022.115635
111. Nabi M, Liang J, Zhang P, Wu Y, Fu C, Wang S, Ye J, Gao D, Shah FA, Dai J (2021) Anaerobic digestion of sewage sludge pretreated by high pressure homogenization using expanded granular sludge blanket reactor: feasibility, operation optimization and microbial community. J Environ Chem Eng 9(1):104720. https://doi.org/10.1016/j.jece.2020.104720
112. Nabi M, Liang J, Zhang P, Wu Y, Fu C, Wang S, Ye J, Gao D, Shah FA, Dai JJJoece (2020) Anaerobic digestion of sewage sludge pretreated by high pressure homogenization using expanded granular sludge blanket reactor: feasibility, operation optimization and microbial community 104720
113. Nabi M, Zhang G, Zhang P, Tao X, Wang S, Ye J, Zhang Q, Zubair M, Bao S, Wu Y (2019) Contribution of solid and liquid fractions of sewage sludge pretreated by high pressure homogenization to biogas production. Bioresour Technol 121378
114. Neumann P, Pesante S, Venegas M, Vidal G (2016) Developments in pre-treatment methods to improve anaerobic digestion of sewage sludge. Rev Environ Sci Bio/Technol 15(2):173–211. https://doi.org/10.1007/s11157-016-9396-8
115. Neyens E, Baeyens J, Weemaes M, Heyder B (2002) Advanced biosolids treatment using H_2O_2 oxidation. Environ Eng Sci 19:27–35. https://doi.org/10.1089/109287502753590214
116. Neyens E, Baeyens J, Weemaes M, De heyder B (2003) Hot acid hydrolysis as a potential treatment of thickened sewage sludge. J Hazard Mater 98(1):275–293. https://doi.org/10.1016/S0304-3894(03)00002-5
117. Nielsen H, Thygesen A, Bjerre A, Schmidt J (2011) Anaerobic digestion of waste activated sludge—comparison of thermal pretreatments with thermal inter-stage treatments. J Chem Technol Biotechnol 86:238–245. https://doi.org/10.1002/jctb.2509

118. Nowak O, Enderle P, Varbanov P (2015) Ways to optimize the energy balance of municipal wastewater systems: lessons learned from Austrian applications. J Cleaner Prod 88:125–131. https://doi.org/10.1016/j.jclepro.2014.08.068
119. Onyeche T (2007) Economic benefits of low pressure sludge homogenization for wastewater treatment plants
120. Pant D, Singh A, Van Bogaert G, Irving Olsen S, Singh Nigam P, Diels L, Vanbroekhoven K (2012) Bioelectrochemical systems (BES) for sustainable energy production and product recovery from organic wastes and industrial wastewaters. RSC Adv 2(4):1248–1263. https://doi.org/10.1039/C1RA00839K
121. Parkin GF, Owen WF (1986) Fundamentals of anaerobic digestion of wastewater sludges 112(5):867–920. https://doi.org/10.1061/(ASCE)0733-9372(1986)112:5(867)
122. Pilli S, More TT, Yan S, Tyagi RD, Surampalli RY (2016) Fenton pre-treatment of secondary sludge to enhance anaerobic digestion: energy balance and greenhouse gas emissions. Chem Eng J 283:285–292. https://doi.org/10.1016/j.cej.2015.07.056
123. Poirier S, Chapleur O (2018) Inhibition of anaerobic digestion by phenol and ammonia: effect on degradation performances and microbial dynamics. Data Brief 19:2235–2239. https://doi.org/10.1016/j.dib.2018.06.119
124. Pérez-Elvira SI, Fernández-Polanco F, Fernández-Polanco M, Rodríguez P, Rouge P (2008) Hydrothermal multivariable approach: full-scale feasibility study. Electron J Biotechnol 11:7–8. http://www.scielo.cl/scielo.php?script=sci_arttext&pid=S0717-34582008000400007&nrm=iso
125. Qiao W, Wang W, Xun R, Lu W, Yin K (2008) Sewage sludge hydrothermal treatment by microwave irradiation combined with alkali addition. J Mater Sci 43(7):2431–2436. https://doi.org/10.1007/s10853-007-1957-3
126. Rabinowitz B, Stephenson R (2006) Full-scale demonstration of waste activated sludge homogenization at the Los Angeles county joint water pollution control plant. Proc Water Environ Fed 2006:445–459. https://doi.org/10.2175/193864706783797429
127. Ragi KB, Ekka B, Mezule L (2022) Zero pollution protocol for the recovery of cellulose from municipal sewage sludge. Bioresour Technol Rep 20:101222. https://doi.org/10.1016/j.biteb.2022.101222
128. Rai CL, Rao PG (2009) Influence of sludge disintegration by high pressure homogenizer on microbial growth in sewage sludge: an approach for excess sludge reduction. Clean Technol Environ Policy 11(4):437–446. https://doi.org/10.1007/s10098-009-0202-y
129. Rajesh Banu J, Kannah RY, Kavitha S, Gunasekaran M, Kumar G (2018) Novel insights into scalability of biosurfactant combined microwave disintegration of sludge at alkali pH for achieving profitable bioenergy recovery and net profit. Bioresour Technol 267:281–290. https://doi.org/10.1016/j.biortech.2018.07.046
130. Remya N, Lin J-G (2011) Current status of microwave application in wastewater treatment—a review. Chem Eng J 166(3):797–813. https://doi.org/10.1016/j.cej.2010.11.100
131. Riau V, De la Rubia M, Perez M (2015) Upgrading the temperature-phased anaerobic digestion of waste activated sludge by ultrasonic pretreatment. Chem Eng J 259:672–681. https://doi.org/10.1016/j.cej.2014.08.032
132. Riau V, De la Rubia M, Perez M, Martín A, Borja R (2012) Modelling of the temperature-phased batch anaerobic digestion of raw sludge from an urban wastewater treatment plant. J Environ Sci Health Part A Toxic/Hazard Subst Environ Eng 47:221–227. https://doi.org/10.1080/10934529.2012.640584
133. Riau V, De la Rubia MÁ, Pérez M (2010) Temperature-phased anaerobic digestion (TPAD) to obtain class a biosolids: a semi-continuous study. Bioresour Technol 101(8):2706–2712. https://doi.org/10.1016/j.biortech.2009.11.101
134. Rittmann B, Lee H-S, Zhang H, Alder J, Banaszak J, Lopez R (2008) Full-scale application of focused-pulse pre-treatment for improving biosolids digestion and conversion to methane. Water Sci Technol J Int Assoc Water Pollut Res 58:1895–1901. https://doi.org/10.2166/wst.2008.547

135. Şahinkaya S, Kalıpcı E, Aras S (2015) Disintegration of waste activated sludge by different applications of Fenton process. Process Saf Environ Prot 93:274–281. https://doi.org/10.1016/j.psep.2014.05.010
136. Sakai Y, Fukase T, Yasui H, Shibata M (1997) An activated sludge process without excess sludge production. Water Sci Technol 36(11):163–170. https://doi.org/10.1016/S0273-1223(97)00704-X
137. Sasaki K, Morita M, Hirano S-I, Matsumoto N, Watanabe A, Ohmura N, Igarashi Y (2011) A bioelectrochemical reactor containing carbon fiber textiles enables efficient methane fermentation from garbage slurry. Biores Technol 102:6837–6842. https://doi.org/10.1016/j.biortech.2011.04.022
138. Sasaki D, Sasaki K, Watanabe A, Morita M, Matsumoto N, Igarashi Y, Ohmura N (2013) Operation of a cylindrical bioelectrochemical reactor containing carbon fiber fabric for efficient methane fermentation from thickened sewage sludge. Bioresour Technol 129:366–373. https://doi.org/10.1016/j.biortech.2012.11.048
139. Shi Y, Yang J, Yu W, Zhang S, Liang S, Song J, Xu Q, Ye N, He S, Yang C, Hu J (2015) Synergetic conditioning of sewage sludge via Fe^{2+}/persulfate and skeleton builder: effect on sludge characteristics and dewaterability. Chem Eng J 270:572–581. https://doi.org/10.1016/j.cej.2015.01.122
140. Skiadas I, Gavala H, Lu J, Ahring B (2005) Thermal pre-treatment of primary and secondary sludge at 701 °C prior to anaerobic digestion. Water Sci Technol J Int Assoc Water Pollut Res 52:161–166. https://doi.org/10.2166/wst.2005.0512
141. Spiden EM, Yap BHJ, Hill DRA, Kentish SE, Scales PJ, Martin GJO (2013) Quantitative evaluation of the ease of rupture of industrially promising microalgae by high pressure homogenization. Bioresour Technol 140:165–171. https://doi.org/10.1016/j.biortech.2013.04.074
142. Stephenson R, Rabinowitz B, Laliberte S, Elson P (2005) Teaching an old digester new tricks: full-scale demonstration of the MicroSludge™ process to liquefy municipal waste activated sludge. Proc Water Environ Fed 2005(2):386–411
143. Svensson M (2013) 18 - Biomethane for transport applications. In: Wellinger A, Murphy J, Baxter D (eds) The biogas handbook. Woodhead Publishing, pp 428–443. https://doi.org/10.1533/9780857097415.3.428
144. Takashima M, Tanaka Y (2014) Acidic thermal post-treatment for enhancing anaerobic digestion of sewage sludge. J Environ Chem Eng 2(2):773–779. https://doi.org/10.1016/j.jece.2014.02.018
145. Tang B, Yu L, Huang S, Luo J, Zhuo Y (2010) Energy efficiency of pre-treating excess sewage sludge with microwave irradiation. Bioresour Technol 101(14):5092–5097. https://doi.org/10.1016/j.biortech.2010.01.132
146. Tian X, Ng WJ, Trzcinski AP (2018) Optimizing the synergistic effect of sodium hydroxide/ultrasound pre-treatment of sludge. Ultrason Sonochem 48:432–440. https://doi.org/10.1016/j.ultsonch.2018.07.005
147. Tony MA, Zhao YQ, Fu JF, Tayeb AM (2008) Conditioning of aluminium-based water treatment sludge with Fenton's reagent: effectiveness and optimising study to improve dewaterability. Chemosphere 72(4):673–677. https://doi.org/10.1016/j.chemosphere.2008.03.032
148. Toreci I, Kennedy KJ, Droste RL (2009) Evaluation of continuous mesophilic anaerobic sludge digestion after high temperature microwave pretreatment. Water Res 43(5):1273–1284. https://doi.org/10.1016/j.watres.2008.12.022
149. Tyagi VK, Lo S-L (2011) Application of physico-chemical pretreatment methods to enhance the sludge disintegration and subsequent anaerobic digestion: an up to date review. Rev Environ Sci Bio/Technol 10(3):215–242. https://doi.org/10.1007/s11157-011-9244-9
150. Tyagi VK, Lo S-L (2012) Enhancement in mesophilic aerobic digestion of waste activated sludge by chemically assisted thermal pretreatment method. Bioresour Technol 119:105–113. https://doi.org/10.1016/j.biortech.2012.05.134

151. Ukwatta A, Mohajerani A, Setunge S, Eshtiaghi N (2015) Possible use of biosolids in fired-clay bricks. Constr Build Mater 91:86–93. https://doi.org/10.1016/j.conbuildmat.2015.05.033
152. Ushani U, Rajesh Banu J, Tamilarasan K, Kavitha S, Tae Yeom I (2017) Surfactant coupled sonic pretreatment of waste activated sludge for energetically positive biogas generation. Bioresour Technol 241:710–719. https://doi.org/10.1016/j.biortech.2017.05.201
153. Volschan Junior I, de Almeida R, Cammarota MC (2021) A review of sludge pretreatment methods and co-digestion to boost biogas production and energy self-sufficiency in wastewater treatment plants. J Water Process Eng 40:101857. https://doi.org/10.1016/j.jwpe.2020.101857
154. Wahidunnabi AK, Eskicioglu C (2014) High pressure homogenization and two-phased anaerobic digestion for enhanced biogas conversion from municipal waste sludge. Water Res 66:430–446. https://doi.org/10.1016/j.watres.2014.08.045
155. Wang DB, Duan YY, Yang Q, Liu YW, Ni BJ, Wang QL, Zeng GM, Li XM, Yuan ZG (2018) Free ammonia enhances dark fermentative hydrogen production from waste activated sludge. Water Res 133:272–281. https://doi.org/10.1016/j.watres.2018.01.051
156. Wang X, Jiang C, Wang H, Xu S, Zhuang X (2023) Strategies for energy conversion from sludge to methane through pretreatment coupled anaerobic digestion: potential energy loss or gain. J Environ Manage 330:117033. https://doi.org/10.1016/j.jenvman.2022.117033
157. Wang J, Li Y (2016) Synergistic pretreatment of waste activated sludge using CaO_2 in combination with microwave irradiation to enhance methane production during anaerobic digestion. Appl Energy 183:1123–1132
158. Wang Q, Ye L, Jiang G, Jensen PD, Batstone DJ, Yuan Z (2013) Free nitrous acid (FNA)-based pretreatment enhances methane production from waste activated sludge. Environ Sci Technol 47(20):11897–11904
159. Wang JF, Zhou HZ, Tang GH, Huang JW, Liu H, Cai ZX, He ZW, Zhu H, Song XS (2023) Reducing the inhibitive effect of fluorine and heavy metals on nitrate reduction by hydroxyapatite substrate in constructed wetlands. J Hazard Mater 446:130692. https://doi.org/10.1016/j.jhazmat.2022.130692
160. Wei L, Zhu F, Li Q, Xue C, Xia X, Yu H, Zhao Q, Jiang J, Bai S (2020) Development, current state and future trends of sludge management in China: based on exploratory data and CO(2)-equivaient emissions analysis. Environ Int 144:106093. https://doi.org/10.1016/j.envint.2020.106093
161. Wei Y, Van Houten RT, Borger AR, Eikelboom DH, Fan Y (2003) Minimization of excess sludge production for biological wastewater treatment. Water Res 37(18):4453–4467. https://doi.org/10.1016/S0043-1354(03)00441-X
162. Wuytack EY, Diels AM, Michiels CW (2002) Bacterial inactivation by high-pressure homogenisation and high hydrostatic pressure. Int J Food Microbiol 77(3):205–212. http://www.sciencedirect.com/science/article/pii/S0168160502000545
163. Xu XJ, Wang WQ, Chen C, Xie P, Liu WZ, Zhou X, Wang XT, Yuan Y, Wang AJ, Lee DJ, Yuan YX, Ren NQ (2020) Bioelectrochemical system for the enhancement of methane production by anaerobic digestion of alkaline pretreated sludge. Bioresour Technol 304, Article 123000. https://doi.org/10.1016/j.biortech.2020.123000
164. Xu Y, Lu Y, Zheng L, Wang Z, Dai X (2020) Perspective on enhancing the anaerobic digestion of waste activated sludge. J Hazard Mater 389:121847. https://doi.org/10.1016/j.jhazmat.2019.121847
165. Yan S-T, Chu L-B, Xing X-H, Yu A-F, Sun X-L, Jurcik B (2009a) Analysis of the mechanism of sludge ozonation by a combination of biological and chemical approaches. Water Res 43(1):195–203. https://doi.org/10.1016/j.watres.2008.09.039
166. Yan S-T, Zheng H, Li A, Zhang X, Xing X-H, Chu L-B, Ding G, Sun X-L, Jurcik B (2009b) Systematic analysis of biochemical performance and the microbial community of an activated sludge process using ozone-treated sludge for sludge reduction. Bioresour Technol 100(21):5002–5009. https://doi.org/10.1016/j.biortech.2009.05.029
167. Yang YF, Zhang YB, Li ZY, Zhao ZQ, Quan X, Zhao ZS (2017) Adding granular activated carbon into anaerobic sludge digestion to promote methane production and sludge decomposition. J Clean Prod 149:1101–1108. https://doi.org/10.1016/j.jclepro.2017.02.156

168. Yang G, Zhang G, Wang H (2015) Current state of sludge production, management, treatment and disposal in China. Water Res 78:60–73. https://doi.org/10.1016/j.watres.2015.04.002
169. Yang XJ, Nie JM, Wang D, Zhao ZW, Kobayashi M, Adachi Y, Shimizu K, Lei ZF, Zhang ZY (2019) Enhanced hydrolysis of waste activated sludge for methane production via anaerobic digestion under N-2-nanobubble water addition. Sci Total Environ 693, Article 133524. https://doi.org/10.1016/j.scitotenv.2019.07.330
170. Ye J, Hu A, Ren G, Chen M, Tang J, Zhang P, Zhou S, He Z (2018) Enhancing sludge methanogenesis with improved redox activity of extracellular polymeric substances by hematite in red mud. Water Res 134:54–62. https://doi.org/10.1016/j.watres.2018.01.062
171. Yellezuome D, Zhu X, Wang Z, Liu R (2022) Mitigation of ammonia inhibition in anaerobic digestion of nitrogen-rich substrates for biogas production by ammonia stripping: a review. Renew Sustain Energy Rev 157:112043. https://doi.org/10.1016/j.rser.2021.112043
172. Yu G-H, He P-J, Shao L-M, Zhu Y-S (2008) Extracellular proteins, polysaccharides and enzymes impact on sludge aerobic digestion after ultrasonic pretreatment. Water Res 42(8):1925–1934. https://doi.org/10.1016/j.watres.2007.11.022
173. Yuan H, Guan R, Chufo Wachemo A, Zhu C, Zou D, Li Y, Liu Y, Zuo X, Li X (2018) Enhancing methane production of excess sludge and dewatered sludge with combined low frequency CaO-ultrasonic pretreatment. Bioresour Technol. https://doi.org/10.1016/j.biortech.2018.10.040
174. Yukesh Kannah R, Kavitha S, Rajesh Banu J, Parthiba Karthikeyan O, Sivashanmugham P (2017) Dispersion induced ozone pretreatment of waste activated biosolids: arriving biomethanation modelling parameters, energetic and cost assessment. Bioresour Technol 244:679–687. https://doi.org/10.1016/j.biortech.2017.08.001
175. Zagrodnik R, Duber A, Seifert K (2021) Hydrogen production during direct cellulose fermentation by mixed bacterial culture: the relationship between the key process parameters using response surface methodology. J Cleaner Prod 314:127971. https://doi.org/10.1016/j.jclepro.2021.127971
176. Zahedi S, Icaran P, Yuan Z, Pijuan M (2017) Effect of free nitrous acid pre-treatment on primary sludge at low exposure times. Bioresour Technol 228:272–278. https://doi.org/10.1016/j.biortech.2016.12.117
177. Zamanzadeh M, Parker WJ, Verastegui Y, Neufeld JD (2013) Biokinetics and bacterial communities of propionate oxidizing bacteria in phased anaerobic sludge digestion systems. Water Res 47(4):1558–1569. https://doi.org/10.1016/j.watres.2012.12.015
178. Zhang LG, Duan HR, Ye L, Liu L, Batstone DJ, Yuan ZG (2019) Increasing capacity of an anaerobic sludge digester through FNA pre-treatment of thickened waste activated sludge. Water Res 149:406–413. https://doi.org/10.1016/j.watres.2018.11.008
179. Zhang Y, Zhang P, Ma B, Wu H, Zhang S, Xu X (2012) Sewage sludge disintegration by high-pressure homogenization: a sludge disintegration model. J Environ Sci 24(5):814–820
180. Zhang Y, Zhang P, Zhang G, Ma W, Wu H, Ma B (2012) Sewage sludge disintegration by combined treatment of alkaline+high pressure homogenization. Bioresour Technol 123:514–519. https://doi.org/10.1016/j.biortech.2012.07.078
181. Zhao Q, Li M, Zhang K, Wang N, Wang K, Wang H, Meng S, Mu R (2020) Effect of ultrasound irradiation combined with ozone pretreatment on the anaerobic digestion for the biosludge exposed to trace-level levofloxacin: Degradation, microbial community and ARGs analysis. J Environ Manag 262:110356. https://doi.org/10.1016/j.jenvman.2020.110356
182. Zhen G, Lu X, Kato H, Zhao Y, Li Y-Y (2017) Overview of pretreatment strategies for enhancing sewage sludge disintegration and subsequent anaerobic digestion: current advances, full-scale application and future perspectives. Renew Sustain Energy Rev 69:559–577. https://doi.org/10.1016/j.rser.2016.11.187
183. Zhen G-Y, Lu X-Q, Li Y-Y, Zhao Y-C (2013) Innovative combination of electrolysis and Fe(II)-activated persulfate oxidation for improving the dewaterability of waste activated sludge. Bioresour Technol 136:654–663. https://doi.org/10.1016/j.biortech.2013.03.007
184. Zhen G, Lu X, Li Y-Y, Zhao Y (2014) Combined electrical-alkali pretreatment to increase the anaerobic hydrolysis rate of waste activated sludge during anaerobic digestion. Appl Energy 128:93–102. https://doi.org/10.1016/j.apenergy.2014.04.062

185. Zhen G, Lu X, Li Y, Zhao Y, Wang B, Song Y, Chai X, Niu D, Cao X (2012) Novel insights into enhanced dewaterability of waste activated sludge by Fe(II)-activated persulfate oxidation. Bioresour Technol 119:7–14. https://doi.org/10.1016/j.biortech.2012.05.115
186. Zhou X, Wang Q, Jiang G (2015) Enhancing methane production from waste activated sludge using a novel indigenous iron activated peroxidation pre-treatment process. Bioresour Technol 182:267–271. https://doi.org/10.1016/j.biortech.2015.01.132
187. Zhou X, Wang Q, Jiang G, Liu P, Yuan Z (2015) A novel conditioning process for enhancing dewaterability of waste activated sludge by combination of zero-valent iron and persulfate. Bioresour Technol 185:416–420. https://doi.org/10.1016/j.biortech.2015.02.088
188. Zubrowska-Sudol M, Podedworna J, Sytek-Szmeichel K, Bisak A, Krawczyk P, Garlicka A (2018) The effects of mechanical sludge disintegration to enhance full-scale anaerobic digestion of municipal sludge. Thermal Sci Eng Prog 5:289–295. https://doi.org/10.1016/j.tsep.2017.11.009

MIX
Papier aus verantwortungsvollen Quellen
Paper from responsible sources
FSC® C105338

If you have any concerns about our products,
you can contact us on
ProductSafety@springernature.com

In case Publisher is established outside the EU,
the EU authorized representative is:
Springer Nature Customer Service Center GmbH
Europaplatz 3, 69115 Heidelberg, Germany

Printed by Libri Plureos GmbH
in Hamburg, Germany